SEFTON'S DYNAMIC COAST

Proceedings of the conference on coastal geomorphology,
biogeography and management 2008.

Edited by A.T. Worsley, G. Lymbery, V. J.C. Holden and M. Newton

2010 published by:
Coastal Defence: Sefton MBC Technical Services Department
Ainsdale Discovery Centre, The Promenade, Shore Road, Ainsdale-on-Sea, Southport PR8 2QB

Copyright © 2010 individual authors and Sefton MBC

Cover illustration courtesy of
Dave McAleavy

The Ordnance Survey mapping included within this publication is
provided by Sefton Council under licence from the Ordnance Survey
in order to fulfil its public function to support coastal monitoring.
Persons viewing this mapping should contact Ordnance Survey
copyright for advice where they wish to licence Ordnance Survey
mapping for their own use.
© Crown Copyright.
All rights reserved Sefton Council Licence no 100018192 2010

Designed and printed
by
Mitchell & Wright (Printers) Limited, Southport, PR8 5AL

SEFTON'S DYNAMIC COAST

Proceedings of the conference on coastal geomorphology,
biogeography and management 2008.

Edited by A.T. Worsley, G. Lymbery, V. J.C. Holden and M. Newton

2010 published by:
Coastal Defence: Sefton MBC Technical Services Department
Ainsdale Discovery Centre, The Promenade, Shore Road, Ainsdale-on-Sea, Southport PR8 2QB

ISBN 978-0-9566350-0-6

Preface and Acknowledgements

Like many ideas the idea to hold a conference to celebrate thirty years of working in partnership on the Sefton Coast seemed like a good one at the time; it has however required a significant amount of work, patience and occasional frustration to organise the conference and to produce these proceedings. I have no doubt that all this effort will be worthwhile as this publication brings our knowledge and understanding of this coast up to date. The value of these proceedings should not be underestimated as they will provide a key reference for further research and for the development of management strategies.

At the time of inviting papers for the conference there were both more papers submitted than could be presented and also papers submitted from individuals whose preference was not to present them. Given this the decision was taken to include all the papers submitted rather than lose the opportunity to capture this knowledge. Papers were also solicited from non-academics in order to capture a wider range of experience. All the papers submitted have been peer reviewed and are presented in the proceedings within three sections; geomorphology; biogeography; and management. The keynote talk from Phil Dyke of the National Trust is kept in the format of a talk and presents a national context for some of the issues being faced within coastal management at the time. Jen Lewis was invited to write an introductory chapter to provide a historical context for the proceedings.

A key coastal management issue at this time is adaptation to coastal change, particularly with reference to the potential for accelerated change due to climate change. The Council started to examine this particular issue as a Partner in a European project (IMCORE which stands for Innovative Management of Europes changing COastal REsource) funded under the Interreg IVB programme and this has subsequently led to additional funding from the Environment Agency and the Department of Environment, Food and Rural Affairs (DEFRA). These proceedings are important for this issue in that they provide evidence to support both our current understanding of our dynamic coast and our understanding of the implications of any actions we propose. They also provide the technical resource from which we can develop non-technical communications to help other coastal stakeholders to understand how the coast works and why it is so important.

I hope you enjoy reading these proceedings and would ask you to spare a thought of gratitude for the authors of individual papers, editors, designers, printers, conference organisers and speakers who have contributed to them and to the IMCORE project and Partners on the Sefton Coast for funding them.

Regards
Graham Lymbery
Project Leader Coastal Defence
Sefton MBC

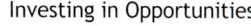

Contents

		Page
Keynote		
P. Dyke	Shifting Shores – living with a changing coastline	07
Introduction		
J.M. Lewis	Archaeology and history of a changing coastline	11
Section A: Geomorphology		**27**
A.J. Plater and J. Grenville	Liverpool Bay: Linking the eastern Irish Sea to the Sefton Coast	28
V.J.C. Holden	The historic development of the north Sefton Coast	55
A.J. Plater, D. Hodgson, M. Newton and G. Lymbery	Sefton South Shore: Understanding Coastal Evolution from past changes and present dynamics	83
D. Clarke, R.K. Abbott and S. Sanitwong Na Ayutthaya	Effects of climate change on wet slacks in the dune system at Ainsdale	107
J.A. Millington, C.A. Booth, M.A. Fullen, I.C. Trueman and A.T. Worsley	Distinguishing dune environments based on topsoil characteristics: a case study on the Sefton Coast	116
K. Pye and S.J. Blott	Geomorphology of the Sefton Coast sand dunes	131
Section B: Biogeography		**161**
S. White	The birds of the Sefton Coast: A review	162
P.H. Smith	Dragonflies (Odonata)	175
S.E. Edmondson	Dune slacks on the Sefton Coast	178
P.H. Smith	Grasshoppers and crickets (Orthoptera)	188
G. Russell and C. Felton	Driftweed of the Sefton Coast: Its composition and origins	189

Contents

		Page
P.H. Smith	Birkdale Green Beach	194
S.E. Edmondson	Non-native plants on the Sefton Coast sand-dunes	201
D. Hardaker	The status of the Sand Lizard *Lacerta agilis* on the Sefton Coast: A recent study	214
P.H. Smith	Sefton Coast rare plants	225
R. Burkmar	A potential role for habitat suitability mapping in monitoring rare animals on the Sefton Coast	239
P.H. Smith	An inventory of vascular plants for the Sefton Coast	248
D. Hardaker	Common Lizard *Zootoca vivipara*	251
D.T. Holyoak	Mosses and liverworts of the Sefton Coast	253
P.H. Smith	The Hall Road - Hightown shingle beach	258
S. Judd	Northern Dune Tiger Beetle *Cicindela hybrida* L. a 'flagship species' of high conservation value on the Sefton Coast sand dunes	263

Section C: Management — **278**

G. Lymbery	Marine Drive, Southport – sea defence improvements	279
J. Houston	The development of Integrated Coastal Zone Management (ICZM) in the UK: the experience of the Sefton Coast	289
D. Bill	The Pinewoods Project – an innovative approach to improving consultation with younger members of the community	306
P. Sandman	Marketing Sefton's natural coast – evaluating the impact	311
D. McAleavy	Sefton beach management – twenty years of progress	318
G. Lymbery, M. Newton and P. Wisse	Sefton's Dynamic Coast – our response	327

Sefton Metropolitan Borough Council — Sefton's Dynamic Coast

Keynote
Shifting Shores – living with a changing coastline

Phil Dyke

The National Trust's Coast and Marine Advisor outlines the key challenges that face the Trust and wider society in managing climate change related impacts on our coastline.
phil.dyke@nationaltrust.org.uk

INTRODUCTION

In the UK nobody lives further than 75 miles from the coast, but distance does not seem to lessen the pull that most of us feel for the beaches and cliffs which mark the end of our land and the start of the sea.

For over 110 years The National Trust has been protecting special places for everyone. This does not stop at the walls of a mansion house estate as many people may think. These special places span our countryside, all the way to the coast. In fact, the very first piece of land given to the Trust in 1895 was five acres of gorse-covered clifftop at Dinas Oleu overlooking Cardigan Bay in Wales. Today the Trust cares for almost a tenth of the coastline in England, Wales and Northern Ireland - an impressive 707 miles (1130km).

This growth in our coastal holding is thanks, in part, to the Neptune Coastline Campaign – an enduring and inspiring collaboration between the Trust and our supporters – which has enabled us to acquire and manage coastal land.

With this growth has come challenges. Our work does not end with the border of our landholding and we increasingly need to work in partnership across boundaries. In my time with the Trust I have worked with a myriad of partners and coastal communities, helping to promote and deliver more joined-up thinking in coastal management.

Another challenge is our changing coastline. The tides and the wind are constantly altering the shape of our island, making our coast perpetually dynamic and a 'canary' for climate change. This provides us with a juxtaposition to the images of melting glaciers and storms from far off lands and reminds us that we are already seeing the impacts of our changing weather patterns here in the UK today.

We understand that with our changing climate, this kind of transformation to our land is inevitable. At the National Trust we are working to foster a better understanding of the realities of climate change. On the coast this means facing up to changes associated with sea-level rise, in particular, an increase in coastal erosion and flooding.

Understanding coastal change

Future forecasts say climate change will lead to sea-level rise and increased storminess which will, in turn, accelerate the scale and pace of coastal change.

To help plan for this uncertain future the Trust has commissioned research looking at how our coastline is likely to change over the next 100 years. The results paint an interesting picture and, perhaps, an unexpected picture of the future.

Many of the Trust's important and well loved sites are at risk from coastal erosion and flooding, including iconic sites such as the White Cliffs of Dover, St Michael's Mount in Cornwall, Formby sands in Lancashire, Rhossili Bay in Wales and The Giant's Causeway in Northern Ireland, along with nearly 200 other less familiar, but equally valued, sites.

This research led to Shifting Shores 2005 – an unprecedented policy document that set out the organisation's position on working with a changing coast in England. In 2007 we published a companion version looking at coastal change in Wales and in January 2008 a similar document was launched covering a number of key Trust sites in Northern Ireland.

Hard choices and coastal squeeze

The key message in Shifting Shores is that it is unrealistic to think that we can continue to build our way out of trouble on the coast as we have for the past 150 years.

Ever since the civil engineering advances of the 19th century made it possible, we have reacted to coastal change by building hard coastal defences. Indeed we have gone further by pushing out into the sea - reclaiming land for agriculture, ports and urban settlements. Our pre-Victorian forebears would have cautioned against such an audacious approach, counselling us to work with, not against nature and ensure that coastal infrastructure was placed beyond the grasp of the sea.

In fact, with sea level documented as having risen by 1-2mm a year throughout the 20th century and accelerating to 3mm per year over the past 15 years we have arguably been building our way into trouble on the coast. The Trust is no exception to this and our entire coastline, not just the soft low-lying parts, will see change.

We are actively seeking to understand how we can work with natural coastal processes because it is these very processes that could save our coastline. Newly eroded material can be retained at the shoreline to provide a natural sea defence in the form of dunes and salt marsh. However, in some areas, this valuable sediment is being washed out to sea in a perverse side-effect of engineered coastal defences – an example of a phenomenon known as 'coastal squeeze'.

Where a shoreline feature, such as a sand dune, is free to move inland it will do so automatically as sea levels rise – maintaining a natural and dynamic form of sea defence. If constrained by a man made sea wall on one side and a rising sea level on the other, the beach material is squeezed into suspension in the sea water, transported offshore and often lost. What is left behind is a steepened beach profile and an undermined sea wall vulnerable to collapse.

At the Trust we have taken a number of decisions, following careful consideration, to remove failing or counter productive sea defences and to allow natural coastal processes to take over.

The Trust is not alone in highlighting this concern. In 2008 the Marine Climate Change Impacts Partnership (MCCIP – is a partnership between scientists, government, its agencies and non-government organisations) published their 2007-8 annual report card (www.mccip.org.uk). They highlighted that almost two thirds of the beach profiles in England and Wales have steepened over the past 100 years. They went on to note that the steepening of beaches is particularly prevalent on coasts protected by hard engineering structures.

In the future there may be a place for appropriate sea-defences but these will be structures which enable us to buy time so we can develop long-term and sustainable approaches to managing our future coast.

Adopting a preference to work with natural processes in the management of coastal change inevitably means that some of the decisions we face will be difficult and on occasion controversial.

Making time and space for change

On top of this we need to start thinking in the long-term to deal effectively with a changing coast. Thinking in twenty, fifty and one hundred year time frames needs to become the norm, to replace our current practice of thinking in short five year planning and political cycles.

Since 2005 staff at the Trust's coastal properties have been putting this thinking into practice – challenging ourselves to think beyond the short-term. Currently we are working on more detailed assessments of the impacts of sea-level rise at every one of our coastal locations. These risk assessments will help us to gain a clearer idea of the vulnerability of wildlife habitats, historic building and archaeological features in our care and threats to the associated visitor infrastructure.

We are also able to estimate the likely timescales in which changes will occur. This valuable information will help us to plan ahead. Then we can maximise the useful life of a building before it is lost or create wildlife habitat further inland of today's coast to secure a future for coastal habitats that are coming under increasing pressure.

The Trust believes that by embracing an adaptive approach, harnessing creativity and innovation, we stand the best chance of making the right decisions about the future management of our coast.

Below are a number of examples of how the Trust is putting these words into action through novel approaches to the conservation of buildings and management of visitor infrastructure, creating opportunities for wildlife and ensuring we understand the significance of archaeological features as they are exposed.

Buildings and visitors

The Trust's activity centre at **Brancaster** on the North Norfolk coast is periodically flooded by high tides. The centre has led the way in showing how a building can be adapted to accommodate flooding. Power cables are routed to sockets located a metre off the ground instead of a few inches and floors are covered with washable materials. This 'wash and go' approach is being incorporated into many buildings across the Trust as re-servicing projects take place – with the result that we can extend the usable life of a building beyond what it otherwise would have been.

At **Portstewart Strand**, County Londonderry, a new wooden visitor reception building has been designed to be 'demountable' so that ultimately it can be removed and relocated with minimal impact.

Wildlife

At **Porlock**, Somerset, the sea has been allowed to breach the shingle ridge and flood inland creating new saltmarsh and attracting waders, ducks and plants that had previously been rare visitors or completely absent.

As well as winners there will also be losers as wildlife and habitats react to a changing coast. The loss of freshwater marsh adjacent to the coast at **Blakeney** in North Norfolk urgently needs to be compensated for by finding space to create new freshwater habitat inland.

At **Strangford Lough**, Northern Ireland, summer breeding seabirds, especially Artic, Common and Sandwich Terns will be affected as the availability of nest sites on shingle or rocky shores of the islands on the Lough diminishes.

At **Newtown Harbour**, on the Isle of Wight, the Trust is developing a long-term plan for retreat in the face of increased coastal flooding. We have purchased two areas of farmland beside the estuary and these will be allowed to flood as sea levels rise, creating space for the saltmarsh and its associated wildlife.

Archaeology

While some archaeology sites may be physically lost there are opportunities to learn more about them as they erode. A photographic record of the medieval village at **Rhossili**, Gower, is the only sure way to ensure future generations understand it's significance.

In the historic environment change is not all about loss as is shown by the recent revealing of a shipwreck at **Llangennith**, Gower, and the find of a beautiful Bronze-age bracelet at **Pistyll**, Llyn, revealed by a coastal landslip.

Archaeology helps us understand the constant nature of coastal change through time. The medieval village at **Dunwich, Suffolk,** now lies under the North Sea and is a testament to how society has and can continue to adapt to a changing coast.

The wider challenges

Climate change driven coastal change is happening now and there is a clear need for wider society to consider how it can help vulnerable communities and individuals to adapt to a changing coast in a cost-effective and equitable manner. In partnership with communities and government the Trust can play a part in developing innovative and sustainable solutions to coastal change.

There is a need for national governments to invest further in coastal change monitoring. Some excellent examples exist, such as the Channel Coast Observatory (www.channelcoast.org.uk), but around many parts of the coast of England, Wales and Northern Ireland there are large gaps in coastal change data and monitoring.

This lack of basic information is a poor basis for decision makers engaged in, for example, the development of planning policy and local development plans. We need good data on predicted flood and erosion risk so that coastal change can be properly taken into account.

Finally, and in a rather unexpected way, I find myself returning to where I started this piece on the subject of the Neptune Coastline Campaign. Coastal change gives a new opportunity for the Trust and its supporters to apply the successful formula of the Neptune Coastal Campaign in a new context - to help us manage and continue to enjoy the wonders of our coast as it continues to change.

Introduction
Archaeology and history of a changing coastline

Dr Jennifer M. Lewis

Honorary Research Fellow, School of Archaeology, Classics and Egyptology, The University of Liverpool
Fellow, Centre for Lifelong Learning, The University of Liverpool
Turkey Lodge, Knowsley Park, Prescot, L34 4AQ jennifer.lewis@liv.ac.uk

ABSTRACT

Since the 1980s a combination of historical research, archaeological recording and environmental sampling has produced evidence for 10,000 years of coastal change. On the shore, the footprints of prehistoric people, mammals and birds are imprinted on ancient sediments washed clean by the tide. From time to time historic documents tell of damage to settlement and loss of livelihood due to natural forces. A combination of wind and wave action on the dune face exposes a sequence of buried land surfaces dating from prehistoric times to the present day. Archaeologically-sensitive deposits can appear or disappear within a few hours in response to the dynamics of erosion and accretion. Strategies for consistent archaeological monitoring and recording are needed to mitigate the demands of coastal management.

Archaeological and historical evidence – Ice Age to modern age

In 1953, when the British Association for the Advancement of Science met in Liverpool, it was claimed that 'to [prehistoric] people whose material equipment included no means of clearing dense and strong-growing vegetation, great areas of lowland Lancashire and Cheshire offered no means of livelihood …' (Powell 1953, 210). Forty years later, when the proceedings of the Sefton Coast Research Seminar in 1991 appeared in print, there was little new evidence to disprove this notion. As there was almost no published information about the coast's archaeology, emphasis was placed on the historical evidence of the last 1,000 years, rather than that of earlier centuries (Jones *et al.* 1993, 3-20).

Until the late 1970s, when archaeological research on our coast was formalised through Merseyside's Sites and Monuments Record (now Merseyside's Heritage Environment Record - HER), there had been only the occasional reference to archaeological discoveries – a prehistoric stone axe or a stray Roman coin. Since then targeted programmes to identify and record the archaeology and palaeo-environment of the coast and its hinterland have produced impressive evidence for post-glacial landscapes associated with human activity. A coastal environment seemingly of little human interest until the centuries prior to Domesday has now been revealed as one through which people have moved – and settled – for 10,000 years.

Isolated coast

Settlement on the coast is topographically constrained to a narrow strip of land, bounded on the west by the shore, sand dunes or tidal marshes. To the east peat mosslands, punctuated occasionally by low, sandy islands run almost uninterrupted for up to 8km until rising ground is reached at about 7m OD. Natural drainage in the south comes from the Rimrose Brook, river Alt and Downholland Brook and in the north by brooks running off Martin Mere. Historical records tell of extensive areas of permanent or semi-permanent standing water and the need for the creation and careful maintenance of numerous artificial ditches and their protective banks (Figure 1). Even after the Alt Drainage Act was passed in 1779 problems persisted for many generations on the 'damaged lands' of the river Alt and Downholland Brook (Lewis 2002, 5-8). Today, in times of heavy rainfall coupled with high tides, these lands remain vulnerable to flooding (Plate 1).

Effectively, therefore, for much of the year the coastal strip might be isolated and over the millennia people

Figure 1. Part of John Speed's map of Lancashire.

used the shore as their preferred route through the landscape. Certainly in the 17th century the largest inland water, Martin Mere, was said to have "parted many a man and his mare" and be "very hazardous for Strangers to pass by it …". In the mid-19th century the "perfectly smooth and hard" beach at Southport was described as extending for many miles and to be "perfectly free from those disagreeable and sometimes dangerous accompaniments, - quicksands, stones, and pools of water". It was, therefore, "well adapted for riding, and has been compared to an immense natural

Plate 1. Flooded landscape looking east across the main drainage ditch to the mosslands. October 2000. (Photograph by J.M. Lewis).

road". All the same, the journey along the beach between Liverpool and Little Crosby, if not elsewhere, was not without its hazards as travellers negotiated the anchor lines of boats moored off shore (Gonzalez *et al.* 2002, 578-579; Lewis 2002, 5, 8; Robinson 1848, 10; Tyrer 1968, 93).

Dynamic coast

Under the natural influences of wave patterns, sediment drift and storms Sefton's soft coastline of marshes and sand dunes is subject to constant change. In the last 40 years or so, the loss of at least 200m of the dune front around Formby Point has been compensated by the creation of Birkdale's new 'green beach' (see Smith, this volume). Human intervention has also played its part in changing the shape of the coast through the construction of riverine defences, docklands and dredging and reclamation of estuarine marshes on the Ribble and Alt. Historic documents and maps, coastal navigation charts and more recent aerial photographs, provide further evidence for the form, location – and rate – of change.

For example, in the 17th and 18th centuries Fairclough's Lake, an inlet at North Meols, was favoured by smugglers as an offshore anchorage. The lake, a linear 'lagoon' between sand banks some of which were dry at low tide, was two miles long and half a mile wide. It was seven fathoms deep and ships could be anchored within half a mile of the shore when they plied their trade around the Irish Sea coast (Figure 2; Aughton 1988, 48; Lewis 2002, 15; 2008a, 9-10). The lake seems to have been lost to coastal accretion by the mid-19th century when the first Ordnance Survey maps were produced and today it is probably buried in the marshy landscape at Southport's Marshside Road. A chronological sequence of maps and navigation charts also demonstrates the sequence of change at the outfall of the river Alt. Here, a substantial sandstone revetment recently recorded archaeologically near the Alt outfall in Hightown, may have its origins in the late 16th century when a Duchy of Lancaster Commission recommended construction of a 10 foot high stone wall to strengthen the river banks (Lewis, in preparation; Maddock 1999, 71-73) (Plate 2). In 1860,

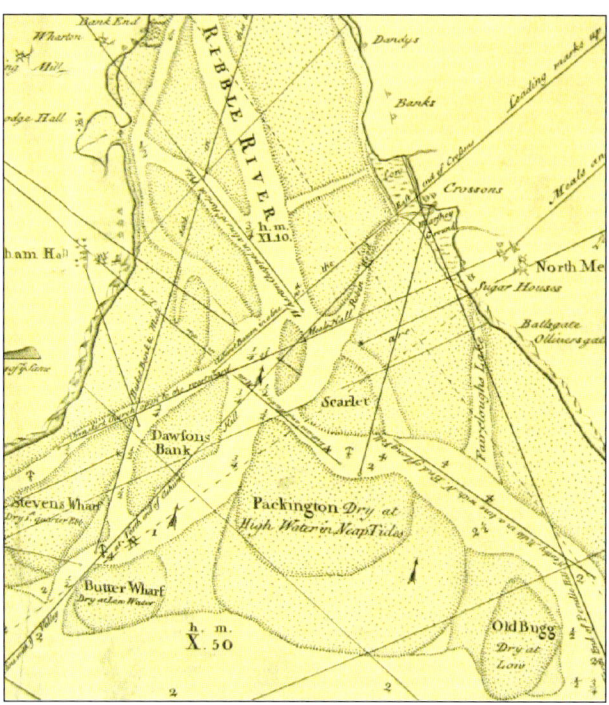

Figure 2. Part of Fearon and Eyes' chart of the Sea Coast from Formby Point to the Harbour of the Wyer, 1738. (Courtesy of National Museums, Liverpool).

reclaimed land known as Ballings Wharf on the west side of the estuary was taken over for military use.

These are but two examples of recent change. When archaeological evidence is combined with that of the palæo-environment, the story of coastal dynamics takes us back many thousands of years, to the end of the last Ice Age (Figure 3). The dune landscape so characteristic of our coast had yet to appear. Ten thousand years ago, as the ice retreated and temperatures started to rise, sea level was 20m lower than at present; a landscape

Plate 2. Sandstone revetment at Alt Grange, Hightown, revealed in July 2007. (Photograph by J.M. Lewis).

characterised by marshes and lagoons, perhaps comparable with those of Birkdale's 'green beach' and Marshside today, extended many kilometres west of the present shoreline. Over the following centuries climatic change and rising sea levels resulted in at least three significant marine transgressions, the earliest of which dates to about 7000-5600 years BP[1]. These inundations, each of which may have endured for several hundred years, extended several miles inland leaving deposits of marine silts and muds. Today, when the mossland ditches are scoured and cleaned of vegetation, the evidence can be seen in alternating horizons of soils, peats and blown sands that developed as the sea retreated and a freshwater environment emerged (Plate 3). A fourth transgression some 4550 years BP deposited silts in embayments at Formby and around the present estuary of the Alt. Layers of organic sand containing alder roots (dated to 3230 years BP) and lying above a thin deposit of blue silt are considered to relate to a final incursion in the Iron Age (Table 1; Cowell 2008, 20; Gonzalez *et al.* 2002, 574, 576).

Radio-carbon dating of peat or wood found in interleaved organic dune-slack layers has indicated that the line of the present dune system became roughly fixed some 4500 - 2500 years BP ago since when it has been breached intermittently and locally (Plater *et al.* 1993, 29; Cowell 2008, 20-21). It seems also, however, that over a fairly short distance between Waterloo and Formby the dunes did not all stabilise at the same time (Table 2; Innes 2002, 355-356).

Evidence for an 'ancient' environment has been observed from at least as early as the 17th century. The fossilised remains of trees such as fir, birch, oak and ash were recognised lying randomly 'not keeping any Order at all' and the head of an elk was recovered from 'four Yards within Marle' under the moss at Meales (presumed to be North Meols) (Leigh 1700, Book 1, 61-63). Attributed by Leigh and his contemporaries to destruction by Noah's biblical deluge, these antiquarians were perhaps not far off the mark! Today the remains of trees and plants such as Royal Fern (*Osmunda regalis*)[2] can be seen at the outfall of the river Alt at Hightown where an extensive peat bed has been dated to about 4000 BP so providing an indication

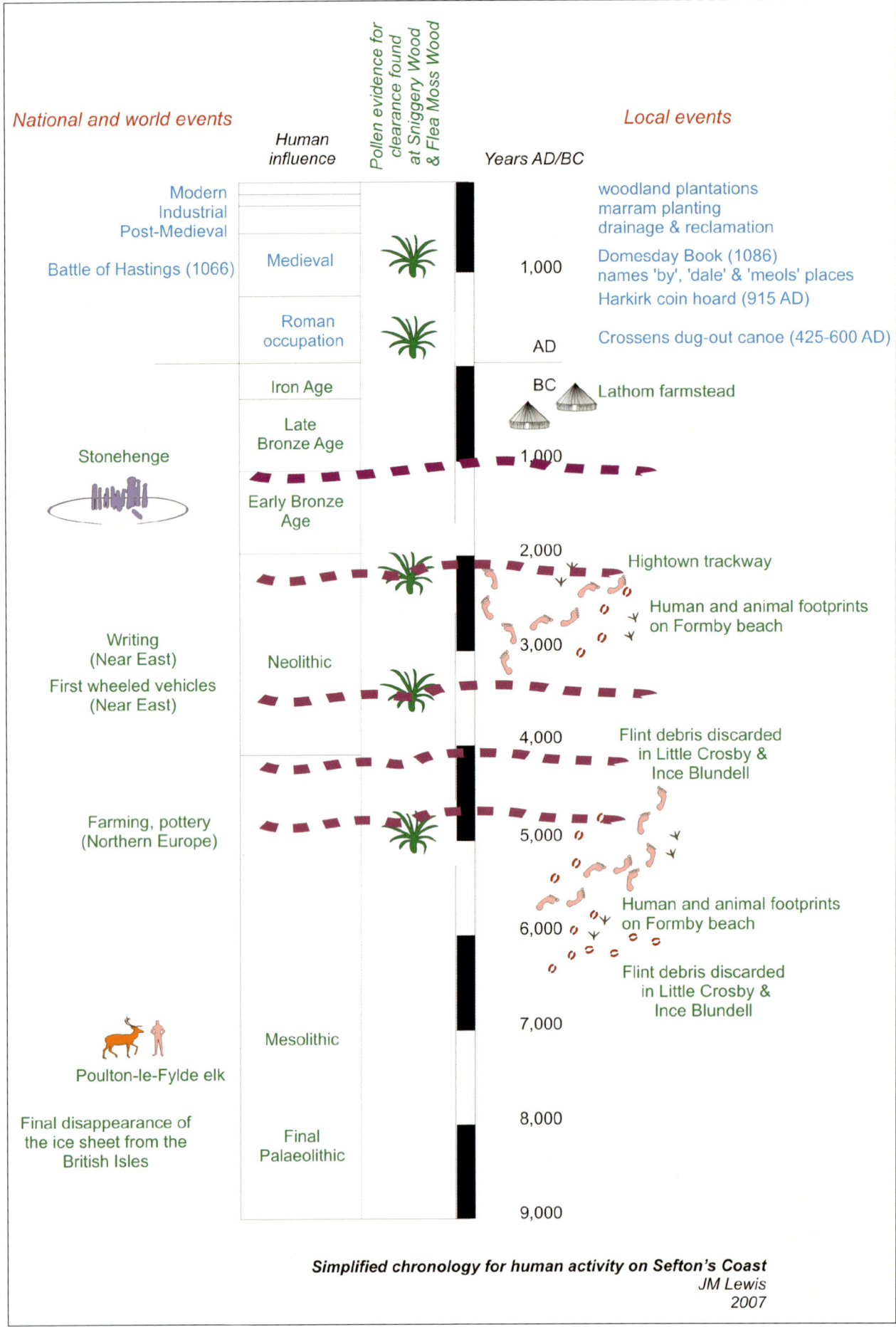

Figure 3. Chronological sequence of natural and human 'events' on the Sefton Coast.

Plate 3. Sequence of silt and peat horizons seen in the edge of a ditch on Downholland Moss, 1984. (Photograph by J.M. Lewis).

of the environment in Neolithic times. Somewhat surprisingly, comparison with an illustration published in 1877 suggests a remarkable survival of these prehistoric deposits (Figure 4 and Plate 4).

From time to time the chronological sequence of coastal change is visible on the dune face at Formby as wind and tidal action expose a series of buried land surfaces. These appear as horizons of dark sandy peat or firm silty sands alternating with those of wind-blown sand (Plate 5). On the shore itself, the sequence is taken back even earlier as sand is washed from the surface of thinly laminated beds of fine silts (Plate 6). Pollen grains trapped in these silts and in peat deposits recovered from Little Crosby and Waterloo demonstrate a complex and fluctuating prehistoric environment of lagoons, dunes, freshwater slacks and marshes, scrubland, heath and woodland as the landscape responded to the effects of marine transgression and retreat over many thousands of years (Cowell 2008, 26; Gonzalez *et al.* 2002, 569-588; Jones *et al.* 1993, 4, figure 1.1; Roberts and Worsley 2008, 29, 38-43, figures 4a-4c).

Table 1. Simplified summary of the evidence for marine transgressions. (Based on Plater et al. 1993, 25-26; Gonzalez et al. 2002, 574, 576)

Transgression	Sample site(s)	Sample type	Years BP	Conjectured duration of transgression (in years)
V (inferred)	Formby foreshore (Lifeboat Road)	alder roots in organic sand overlying blue silt	prior to 3649-3230	
IV	Altmouth, Formby and Birkdale	sands, silts and clays	4800 to 4545	c. 255
DM III	Downholland Moss and Crossens	silty clay	5900 to 5615 +/- 45	c. 285
DM II	Downholland Moss	sediments	6500 to 6050 +/- 65	c. 450
DM I	Downholland Moss	blue clay	6980 to 6760	c. 220

Table 2. Phases of dune development suggested by environmental sampling. (Based on Innes 2002, 355-356)

Sniggery Wood, Little Crosby	4510 +/- 50 years BP	late Neolithic
Murat Street, Waterloo	2680 +/- 50 years BP	Iron Age
Lifeboat Road, Formby	2335 +/- 120 years BP	Iron Age

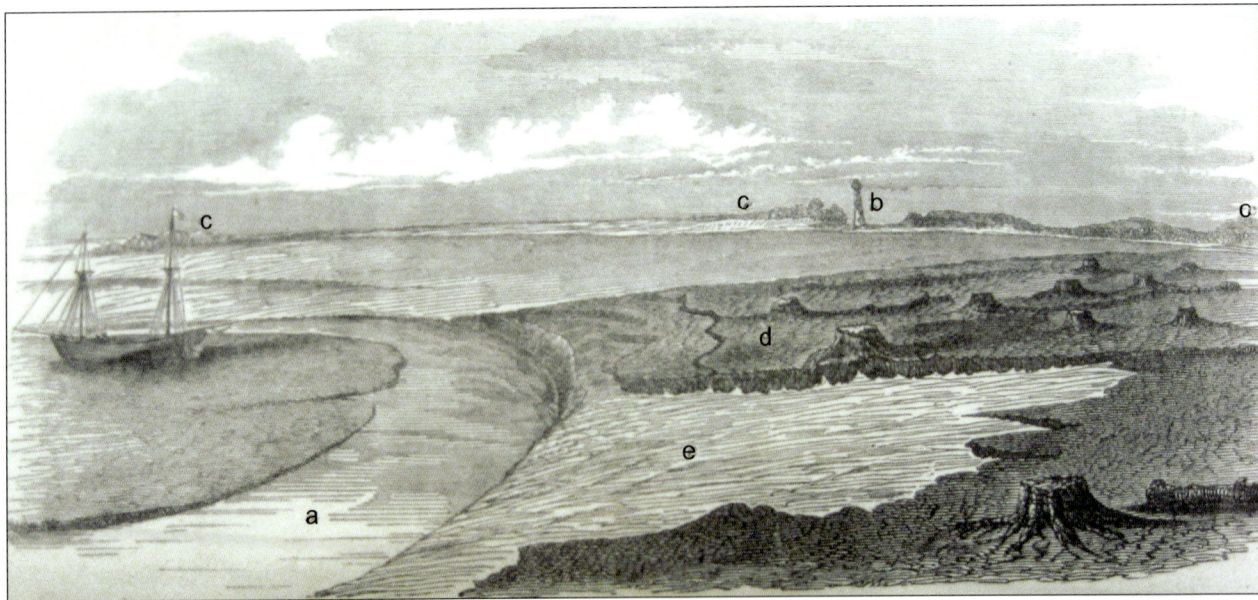

Figure 4. Remains of the prehistoric forest peat bed at Hightown c. 1877 (after de Rance, republished in Gonzalez et al. 2002, 583).
Key: a- river Alt; b – Formby Old Lighthouse; c – sandhills; d – peat beds; e - silts

Plate 4. Remains of the prehistoric forest bed at Hightown, June 2008. (Photograph by J.M. Lewis).

Prehistoric people on the coast

Prehistoric people were attracted to this watery landscape, with its rich resource of fish, mammals and waterfowl. Since 1989 no fewer than 219 trails of the footprints of prehistoric men, women and children have been recorded over a distance of 4km around Formby Point (Plate 6). Intermingled with tracks of the extinct aurochs (wild cattle), red and roe deer, wild boar, ponies and dog-like mammals together with those of wading birds, the prints are considered to represent evidence for hunting and fishing in a fluctuating landscape of tidal marshes and freshwater lagoons (Roberts and Worsley 2008, 31-36). The earliest print-bearing sediments may have been laid down in Mesolithic times some 7000–6000 years ago when the intertidal mud flats were protected by a sand barrier some distance further west and now below the sea. Later print-bearing sediments appeared some 1,000 years later, in about 5150 years BP. In c.4500 BP, a trackway was laid across the marshy landscape at Hightown on the shore south of

Plate 5. Chronological sequence of horizons exposed at Lifeboat Road, Formby, April 2008. (Photograph by J.M. Lewis, 2008).

Plate 6. Deposits of prehistoric laminated silts on the shore at Formby; modern tyre tracks run towards a trail of prehistoric human footprints in the foreground. (Photograph by J.M. Lewis, 2008).

the Alt outfall. It was 1.4m wide and ran in a south-westerly direction from the shoreline surviving for about 60 metres. Constructed of small branches of alder, hazel and oak with a lesser amount of ash, elm, lime, 'hawthorn type', alder buckthorn and birch, such evidence suggests the nature of the prevailing woodland environment in Neolithic times as people moved across the estuarine marshes - perhaps to reach their coracles or fishtraps. It has recently been suggested that intercalated silts, a few hundred metres north of the trackway, also contain the imprinted trails and tracks of prehistoric animals and people (J. Taylor, pers. comm.). However, despite the abundance of human prints almost no artefacts have been recovered from the shore and the sediments have produced only a few animal remains such as a set of Neolithic red deer antlers (4450+/- 45 BP) and a Bronze Age jawbone of a dog (3649 +/- 109 BP) (Cowell 2008, 23, 25; Roberts and Worsley 2008, 36-38).

Despite all this activity, no archaeological deposits have yet been exposed to show us where these prehistoric people lived. Though many thousands of flint artefacts have been recovered from sandy islands on the mosslands of Little Crosby and Ince Blundell, the suspicion remains that these represent discarded waste as people, perhaps on a seasonal basis, stopped to create flint tools as they travelled to hunt and fish on the coast. On the other hand, fragments of charcoal found in beach sediments at Formby and Hightown coupled with pollen grains from peat deposits in Little Crosby provide a tantalising glimpse of people who may have been involved in small scale woodland clearance and the first attempts at cultivation on our coast (Cowell and Innes 1994, 98-114; Cowell 2008, 23; Gonzalez *et al.* 2002, 588; Roberts and Worsley 2008, 40-42).

The later coastal landscape

Evidence for people living on the coast in the Roman and pre-Conquest periods is even more elusive and widely dispersed. Occasionally coins have been recovered by chance at various locations along the shoreline[3]. From the fringes of Martin Mere no fewer than 16 dugout canoes have been recovered over the years as land was reclaimed and brought into cultivation. One of these boats, discovered in 1899, is now in the Botanic Gardens Museum in Southport and has been radiocarbon dated to the middle of the 6th-century AD[4]. The discovery in Little Crosby of a hoard of early 10th century Viking silver coins and ingots dated to AD 915 suggests their concealment perhaps at a time when communities on the coast were under stress. Many of our coastal settlements, first recorded in the Domesday Survey of 1086, bear place-names of Scandinavian origin indicative of topographical features in the landscape (*dale* = valley, *meols* = sand bank or sand hill) and of farmsteads (*by* = farm or village) together with a hint of religious foci (*kross*). Rather than arriving directly from Scandinavian countries it is thought that those who brought the place-name language to our shores came from the Isle of Man or Ireland, perhaps in the opening years of the 10th century, and to have settled in a sparsely populated landscape, (Lewis 2000, 15; 2002, 12; 2008, 8, 13-14; Philpott 2008, 49-52, 54). But, as with the preceding thousands of years, convincing evidence for the homesteads of those who lived on the coast has yet to be found.

The later medieval economy – and perhaps that which survived to the opening years of the 20th century – was based on open-field arable cultivation of farming in strips together with breeding of sheep, cattle, horses and pigs, rabbits and fishing (Lewis 2002, 42; Philpott 2008,

57-59). But by the time the first detailed Ordnance Survey maps were drawn in the mid-19th century there was little visible evidence for the former common fields (First Edition 6" Sheets numbers 75, 82-83, 90-91, 98-99). Except at Formby, where a highly-organised open field system survived until the latter years of the 19th century (Lancs. RO DRL 1/27), the coastal landscape had become one of predominantly rectangular enclosures that encroached on the rabbit warrens and rough pasture that separated the agricultural land from the dunes. No doubt a complex series of factors influenced the change from an apparently sophisticated and highly-organised system of medieval open field cultivation where many tenants held strips of land in a single field to one of separate enclosures held in individual tenancies. Changing social and economic trends would have played their part but archaeological, historical and palaeo-environmental evidence has shown that both wind and sea wrought havoc on the landscape (Table 3).

Over one thousand years since the Norman Conquest centuries, there are hints of damage and loss as the livelihood of those living on the coast was threatened by sand and sea. The farming community at Ainsdale had probably been badly affected by 1555-6 when it was claimed "…there was a certain town in times past called Aynesdale …[and that] … the town time out of mind had been and still was 'overflowen' with the sea so that there remains no remembrance thereof now". In the 14th and 16th centuries there was probably also some loss as communities at Argarmeols (Birkdale), Ravenmeols and Formby claimed that they had been destroyed by "sand and sea". More realistically, such communities may have been forced to move to less vulnerable ground. For example, in 1428 and 1442 reference was made to the highway between Altcar and "Olde Forneby"; in 1499 a lease was made on property in "Holde Forneby" and in 1532 part of Formby was referred to as the "old town" (perhaps even a survival from the Domesday place-name element forn with a meaning of 'old'). Two centuries later, between 17-20 December 1720 the diarist Nicholas Blundell of Little Crosby wrote that it was "very wet and extreamly Windy the like scarce ever know and never so high a Tide known as was these four dayes … chiefly at the Meales, Alker, Alt-Grange and towards Lancaster". Around the Alt estuary and at North Meols (now Churchtown in Southport) many houses were destroyed or badly damaged and over 6,000[5] customary acres were flooded. Within a couple of decades the medieval chapel at Ravenmeols had been buried by sand leaving only its graveyard as evidence for its location (Fellows-Jensen 2007, 22; Lewis 2002, 14-15, 41-44; Tyrer 1972, 31-32).

At North Meols the landscape north and west of the church had probably always been under pressure squeezed as it is between Martin Mere and Marshside. From at least the 13th century, local people were taking steps to protect their environment by creating a series of earthen embankments or 'cops'. Arguably, the oldest of these survives close to St Cuthbert's Church, just inside the wall of the Botanic Gardens where, perhaps, the embankment acted as a barrier against flooding from Martin Mere rather than the sea. Seawards, either side of Marshside Road, a series of turf-built cops that run parallel to the shoreline marks a series of encroachments as the marshes were enclosed. Such encroachments may have advanced, perhaps, as tidal change altered the pattern of sand banks and lagoons of which Fairclough's Lake had once been a feature.

Managing the coast – protecting its archaeology and history?

The coastal environment is – and always has been – vulnerable; over many millennia those living here have survived alongside tidal and riverine inundation and shifting sand. But such natural events do not necessarily destroy archaeological remains. Timber artefacts, such as the Neolithic trackway at Hightown and environmental evidence preserved within buried organic deposits, are likely to survive well in the intertidal zone. Livelihoods may have been threatened, perhaps entirely destroyed, as blowing sand laid agricultural land to waste and damaged buildings and settlements, but evidence of past landscapes can be recovered from buried soils (Davidson 2002, 3).

Dune management to assist stabilisation and reduce the

Table 3. Summary of historical and archaeological evidence for the coastal environment.

*c.*535	a radiocarbon date of one of 16 dug out canoes recovered from the Martin Mere area indicates use of the mere for fishing
*c.*915	hoard of Norse Viking silver coins and ingots concealed at Little Crosby; the hoard was discovered in 1611 as a new burial ground was created at the Harkirk
1002-1004	annual supply of 3,000 shad[6] required from Wulfric Spot's estate between the Ribble and Mersey
1086	Domesday Book records settlements at North Meols, Argarmeols (Birkdale), Ainsdale, Formby, Ince Blundell, Crosby and Bootle; Altcar was described as 'waste' (as the land here had no agricultural value, was Altcar damaged by flooding in the mid 11th century?)
early 13th C	mossland at Ainsdale is cleared of sand and a well-organised system of common field cultivation is established; sheep, cattle and horses kept at Altcar, Ravenmeols and Ainsdale
1204-11	reference to salt making in North Meols
*c.*1212	by this time there was a watermill on the river Alt, probably near Alt Grange; eel fisheries also recorded on the Alt and at Otterpool in North Meols
1238	cattle taken to graze on the pasture at Alt Marsh
pre 1242	Alt Marsh had been drained by dykes; the marsh between Ince and Scholes (probably at or near Alt Grange) was to remain "untilled for ever, as common pasture"
c. 1279	first documentary reference to Little Crosby's Harkirk chapel
1289	half of Ravenmeols lost by this time
1346	loss of Argarmeols (Birkdale) to the sea by this time
1442	possible loss of settlement at "Oldforneby" by this time
1451	fisheries – or foreshore rights to fishing - recorded at Ainsdale and Formby
1550s	loss of revenue from Ainsdale "overflowen with the sea" suggesting that the common-field farming community had been seriously damaged; claims that fishing stalls and grazing have been ruined by sea and sand at Formby and Ravenmeols
mid 16th C	historic maps suggest that the river Alt flowed directly westwards
1560	covenants put in place regarding preservation of star grass (marram) at Ravenmeols
1589	a Duchy of Lancaster Commission enquires into the danger from "the breakinge oute of the sea banckes called Altemouthe"; recommends that a 10' high stone wall should be built to strengthen the banks
1596	fish are bought at Formby's fish market; this, presumably, has recovered after the damage earlier in the century
1500s onwards	records of 300 ships wrecked on the Southport sands
early 17th C	planting is used to trap sand and prevent drainage ditches from clogging up in Little Crosby
1611	hoard of 10th-century Norse Viking silver discovered at the Harkirk, Little Crosby
1626	first recorded theft of rabbits
1626, 1633	record of ships called The Bennett, The Peeter, The Angell, The John, The Henrie, The Ellen (Elline), The Merry Johnes (Marrie John), The Unitie, The Patericke – all of Formby; record of The Bartholomew of Altcar; The Trinitie of Altcar in 1626 but of Formby in 1633
1665	shipping records cite the Elizabeth of Formby owned by Robert T[h]ompson

1666-67	division of the warrens agreed by Formby's two landowners
1686	Fairclough's Lake at North Meols is used for anchorage and unloading wine and brandy from France
1703-38	issue of licences for fishing stalls
early 18th C	between Little Crosby and Liverpool boats are anchored from the shore and floated off at high tide
1710	Peter White's ship called the Betty anchored at Fairclough's Lake, North Meols; Peter White lands near the Grange at Altmouth and walks to Crosby Hall
1714	brigantine Suckcess shipwrecked, seemingly near Hightown
1716	a drawing of Crosby shows the formal arrangement of settlement and land use between the sea and village
1717	shipwreck with cargo of butter; probably between Liverpool and Great Crosby
mid Dec 1720	high winds accompanied by high tides cause flooding over 6,000 customary acres of land, loss of 157 houses and a further 200 houses badly damaged around the Alt and Ribble estuaries
1736	encroachments on Churchtown marshes shown on an estate map
1736-7	navigation chart of the coast by Fearon and Eyes' (published 1738) shows Fairclough's Lake as a long narrow inlet running shorewards to Sugar Houses at North Meols; sight lines drawn on the church at North Meols to Lathom 'Hall' and from North Meols windmill to Ormskirk church suggest that the inlet ran shorewards on the north side of Marshside Road; sand accumulation on the seaward side of the Alt outfall
1739	the medieval chapel at Ravenmeols is finally abandoned due to sand inundation
1755-1908	Lawson Booth list of 91 ships wrecked or damaged off southern part of the Sefton coast (Altcar – 1; Crosby – 56; Formby – 34)
1774	the Liverpool to Gathurst (near Wigan) section of the Leeds and Liverpool Canal opens providing city dwellers with easier access to the coast
1769	a group of 'salt fields' named on an Altcar estate map close to the confluence of the Downholland Brook and the Alt
1779	the Alt Drainage Act is passed to ameliorate the effects of flooding between Aintree and Altmouth; problems with flooding persist until the early 20th century
1780	Bootle Marsh is 'improved'
1786	William Yates' map shows the outfall of the river Alt developing a southerly trend; Fairclough's Lake not shown on Yates' map
1792	the first bathing house is established at South Hawes (Southport) to provide some form of shelter and changing facilities for bathers
1816	enclosure of the Great Marsh at Crosby
1821	400 yards of coastline lost in the vicinity of Formby's lifeboat house
1833	high tide 7.18m above OD at Hesketh Arms on 31st December
1839	possible coastline recovery at Formby's lifeboat house
1845-48	Navigation charts and Ordnance Survey map evidence indicate dune formation and southerly trend of coastal accretion at Ballings Wharf on the Alt outfall

1852	'night soil' transported by train from Liverpool to Freshfield to improve Formby's warrens and bring them into cultivation
1855	reclamation of land for grazing at Ballings Wharf on the north bank of the river Alt turns the outfall southwards
1870-1900	reclamation of Massams Slack, Formby
1883-1950	ships wrecked off the Southport shore - Star of Hope (1883), Mexico (1886), The Zelandra (1917), The Chrysopolis (1918), The Endymion (1933), The Charles Livingston and The Ionic Star (1939), Happy Harry (1950)
1893	the beginnings of small-scale 'enclosure' of the rabbit warrens at Formby are indicated on Ordnance survey maps
early 20th C	extensive sand extraction at Hightown, Formby and Ainsdale leads to the removal of the largest sand dunes on the coast
1907-1948	houses between Hall Road and Blundellsands are under severe threat due to migration of the river Alt
1927	serious damage is caused by a storm surge at 6.1m above OD on 28-29 November
1939	River Alt diverted by training wall north of Hall Road
1942	rubble from Liverpool's bomb-damaged buildings brought to the Hightown/Crosby coast to stabilise the landward side of the Alt estuary
1952-1963	816,467 tons of sand extracted from dunes north and south of Shore Road, Ainsdale
1956-1974	tobacco waste tipped on the dune landscape south of Victoria Road, Formby
1959	relocation of Pinetrees caravan site at Victoria Road, Formby due to marine erosion; Formby Urban District Council enacts a Coast Protection Order to limit sand extraction at Formby
1961	the Pine Tree Café at Victoria Road, Formby falls into the sea
1966-1973	570,135 tons of sand removed from the foreshore north of Shore Road, Ainsdale
1967-1968	numerous storm surges increase the rate of erosion on the Formby coastline
1972	start of sand winning from the Horse Bank
1973	removal of Little Balls Hill sand dune at Ainsdale and dunes stabilised
1977	on 11 November there is a storm surge 6.11m above OD
1981	caravan site at Victoria Road, Formby relocated further inland
1983	a storm surge on 31 January reached 6.0m above OD
1990	a storm surge on February 26th-28th reaches 6.3m above OD
2000	in October excessive rainfall leads to flooding over large areas of the mosslands making the causewayed roads almost impassable

effects of wind blow can help to protect underlying archaeological deposits. Similarly, material such as industrial waste brought to the shore at Blundellsands from a tin-smelting works in Bootle in the 1930s and, from 1942 onwards, rubble from bomb-damaged buildings deposited on the shore south of the Alt, also help to protect the ancient landscape (Gresswell 1953, 94-97; Jones et al. 1993, 19; Smith 1999, 88). Here, Hightown's prehistoric trackway may survive for some distance buried beneath this shingle beach. However, the effect on the archaeological remains of the new deep rooting plants that now colonise the shingle is yet to be evaluated. Elsewhere, thousands of tons of tobacco waste deposited in the dunes at Victoria Road in

Formby and surfaces for hard standing at various places along the coast help to protect earlier cultivated land surfaces. Here, it is wave and wind action that causes the damage (Plate 7).

The most extensive destruction to our buried archaeological deposits has probably come from extractive processes in many parts of the dune backlands. These probably began in the late 19th century as small-scale and localised clearances to improve the warrens and bring areas into cultivation. Greater change has been brought about by recognition of the commercial value of the sands for construction and industrial purposes. During the first half of the 20th century many millions of tons of sand were removed for use in the manufacture of glass and bricks, to make moulds for metal casting and, in WWII, for sandbags. It was only with recognition of the severe threat of extraction on the frontal dunes at Formby, and the possible consequence of marine flooding, that the Formby Coast Protection Order of 1959 was finally put into effect (Bulman 2003, 3-8; Jones *et al.* 1993, 13-14; Lewis 2008b). But in many areas sand had already been removed down to the water table and only the most deeply buried archaeological deposits and those below the surviving dunes may still remain undisturbed.

Finding out more about the coastal landscape

Over the last twenty years or so, spectacular new evidence has emerged as archaeologists, palaeo-environmentalists and others have recovered evidence that is starting to improve our understanding of coastal change and how it has affected those living on the coast. The surviving dune belt and the undisturbed dune backlands have an important role in protecting earlier landscapes and the high archaeological significance of these buried land surfaces has been recognised (HLC 2001, 35-37; Houston 1993, ix; Smith 1999, 26-27). We now know that these early deposits contain important evidence for past environments in the form of pollen, micro- and macro-fossil remains. We know that they may also contain material suitable for dating such remains.

Plate 7. Rubble remains of hard-standing visible high in the dune face at Victoria Road, Formby above plough turves and earlier horizons of developing soils. (Photograph by J.M. Lewis, 2008).

In areas of the dune backlands undisturbed by ploughing or sand extraction, ancient land surfaces - possibly those representing settlements lost to coastal change over the centuries – may still survive. On the shore we can be confident that new exposures of prehistoric deposits will continue to appear as the wind and tide redistribute beach sand exposing sediment beds bearing the 'fossilised' imprints of ancient people, animals and birds. At Hightown, comparison of an image published in 1877 with a photograph taken in early June 2008 gives hope that though tidal and riverine action will continue to destroy the exposed face of the peat beds these are, perhaps, more durable than might at first be imagined (see Plate 4 and Figure 4). More fragile is the dune face where wind and wave action expose a sequence of buried silts, organic deposits and dune sand.

But our archaeological and palaeo-environmental interpretation is constrained to evidence from 'key hole' investigations at specific locations such as Little Crosby, Hightown and Formby. On the other hand, nothing is known of the manner of coastal change at the Alt outfall prior to the reclamation of Ballings Wharf in the 19th century or of its affect on prehistoric and medieval communities living beside the Crossens and Marshside marshes.

The archaeological community in north-west England has expressed a need for greater understanding of the coastal zone for all periods; particular emphasis has been placed on the integration of archaeology into Shoreline Management Plans (Brennand 2007, 187-189). The same need for understanding applies to all - land managers, members of the public, professional archaeologists and historians, palaeo-environmentalists, geomorphologists and sedimentologists - concerned with the effects of climate change and programmes such as that established by ShoreWatch in Scotland perhaps provide a model that we could apply (Fraser *et al.* 2003; Philpott 2008, 62-63). The dynamics that have created our landscape present us all with a challenge to recognise, record, interpret, understand and communicate our interpretation of how the coast has changed over many thousands of years – and, using such information, to suggest how it will continue to change.

ACKNOWLEDGEMENTS

My grateful thanks to Antonio da Cruz, Map Curator for the Department of Geography at The University of Liverpool, for assistance in locating the Fearon and Eyes Navigation Chart of 1738 and to John Moore of National Museums, Liverpool for access to the original chart.

REFERENCES AND FURTHER READING

Act (1779) An Act for draining, improving, and preserving, the Low Lands in the Parishes of Altcar, Sefton, Halsall, and Walton upon the Hill, in the County Palatine of Lancaster.

Adams, M. (2007) Sefton Coast Archaeological Assessment (unpublished).

Atkinson, D. and **Houston, J.** (eds) (1993) *The Sand Dunes of the Sefton Coast*, National Museums and Galleries on Merseyside/Sefton Metropolitan Borough Council.

Aughton, P. (1988) *North Meols and Southport: a history*, Carnegie Publishing, Lancaster.

Booth, J.H.L. (1947) *Sea casualties on the Southport coast 1745-1946*, Botanic Gardens Museum, Southport.

Booth, J.H.L. (1949) *History of the Southport lifeboats,* Southport Museum, Southport.

Brennand, M. (ed.) (2007) *Research and Archaeology in North West England: An Archaeological Research Framework for North West England – Volume 2 Research Agenda and Strategy*, Archaeology North West Volume 9 (Issue 19), The Association for Local Government Archaeological Officers North West and English Heritage with The Council for British Archaeology North West.

Bulman, J. (2003) *My Hightown 1897-1969, being personal reminiscences,* 3rd (revised) edition, Sefton Libraries.

Cowell, R. and **Innes, J.B.** (1994) *The Wetlands of Merseyside,* North West Wetlands Survey 1, Lancaster Imprints 2, Lancaster University Archaeological Unit.

Cowell, R. (2008) 'Coastal Sefton in the Prehistoric period' in J. Lewis and J. Stanistreet (eds) Sand and Sea – *Sefton's Coastal Heritage: archaeology, history and environment of a landscape in north-west England*, Sefton Council, Leisure Services Department (Libraries), 20-27.

Davidson, A. (2002) 'Introduction' in A. Davidson, ed. *The coastal archaeology of Wales,* CBA Research Report 131, Council for British Archaeology, 1-6.

Fearon, S. and **Eyes, J.** (1738) *Chart of the Sea Coast from Formby Point to the Harbour of the Wyer ... surveyed in 1736*, 1737 (an original copy held in the Maritime Archives Collection, National Musuems, Liverpool).

Fellows-Jensen, G. (2007) 'Another look at the Danes in Lancashire' in Z. Lawson (ed.) *Aspects of Lancashire History: Essays in Memory of Mary Higham*, Lancashire Local Historian 20 (2007-2008), 18-27.

Fraser, S.M., Gilmour, S. and **Dawson, T.** (2003) 'Shorewatch: monitoring Scotland's coastal archaeology' in T. Dawson (ed) *Coastal Archaeology and Erosion in Scotland*, Historic Scotland, Edinburgh, 197-202.

Gonzalez, S. and **Huddart, D.** (2002) 'Formby Point' and 'Hightown' in D. Huddart and N.F. Glasser *Quaternary of Northern England* Joint Nature Conservation Committee, 569-588.

Gresswell, R.K. (1953) *Sandy Shores in South Lancashire*, Liverpool University Press.

Griffiths, D., Philpott, R.A. and **Egan, G.** (2007) *Meols: The Archaeology of the North Wirral Coast: Discoveries and Observations in the 19th and 20th Centuries, With a Catalogue of Collections.* Oxford University School of Archaeology Monograph 68, Oxford University School of Archaeology.

HLC (2001) Lancashire Historic Landscape Characterisation Programme: Sefton Council Area – A study carried out by Lancashire County Council on behalf of Sefton Council and English Heritage (second draft, unpublished).

Houston, J. (1993) 'Introduction to the Coast' in D. Atkinson and J. Houston (eds) T*he Sand Dunes of the Sefton Coast*, National Museums and Galleries on Merseyside/Sefton Metropolitan Borough Council, ix-xiv.

Innes, J.B. (2002) 'Introduction' in D. Huddart and N.F. Glasser 'The Holocene (Flandrian) history and record of northern England' *Quaternary of Northern England* Geological Conservation Review Series, Joint Nature Conservation Committee, 351-364.

Jones, C.R., Houston, J.A. and **Bateman, D.** (1993) 'A History of Human Influence on the Coastal Landscape' in D. Atkinson and J. Houston (eds) *The Sand Dunes of the Sefton Coast*, National Museums and Galleries on Merseyside/Sefton Metropolitan Borough Council, 3-20.

Leigh, C. (1700) *The Natural History of Lancashire, Cheshire and the Peak in Derbyshire*. 1st Edition Oxford.

Lewis, J.M. (2000) *Medieval Earthworks of the Hundred of West Derby* British Archaeological Reports 310.

Lewis, J.M. (2002) 'Sefton Rural Fringes' in *The Archaeology of a Changing Landscape: The Last Thousand Years in Merseyside* J. Merseyside Archaeological Society 11, 5-88.

Lewis, J.M. (2008a) 'Archaeology and history on a vulnerable coast' in J. Lewis and J. Stanistreet (eds) *Sand and Sea – Sefton's Coastal Heritage: archaeology, history and environment of a landscape in north-west England.* Sefton Council Leisure Services Department (Libraries), 3-19.

Lewis, J.M. (2008b) Archaeological Evaluation at Formby, 2007, Aggregate Levy Sustainability Fund (unpublished).

Lewis, J.M. (in preparation) Archaeological and documentary evidence for coastal change at the Alt estuary, Lancashire.

Maddock, A. (1999) 'Watercourse management and flood prevention in the Alt Level, Lancashire, 1589-1779'. *Transactions of the Historic Society of Lancashire and Cheshire,* 148, 59-94.

Philpott, R. (2008) 'Searching for Lost Settlements – the example of Meols' in J. Lewis and J. Stanistreet (eds) *Sand and Sea – Sefton's Coastal Heritage: archaeology, history and environment of a landscape in north-west England*. Sefton Council Leisure Services Department (Libraries), 44-63.

Plater, A.J., Huddart, D., Innes, J.B., Pye, K., Smith, A.J., and **Tooley, M.J.** (1993) 'Coastal and Sea-level Changes' in D. Atkinson and J. Houston (eds) *The Sand Dunes of the Sefton Coast*, National Museums and Galleries on Merseyside/Sefton Metropolitan Borough Council, 23-34.

Powell, T.G.E. (1953) 'Prehistoric Archaeology' in W. Smith (ed.) *A Scientific Survey of Merseyside*, British Association for the Advancement of Science, University Press of Liverpool, 210-213.

Rance, C.E. de (1877) *The Superficial Geology of the country adjoining the coast of south-west Lancashire*, Memoirs of the Geological Survey of the UK, HMSO.

Roberts, G. and **Worsley, A.** (2008) 'Evidence of human activity in mid-Holocene coastal palaeoenvironments of Formby, north west England' in J. Lewis and J. Stanistreet (eds) *Sand and Sea – Sefton's Coastal Heritage: archaeology, history and environment of a landscape in north-west England*. Sefton Council Leisure Services Department (Libraries), 28-43.

Robinson, F.W. (1848) *A Descriptive History of Southport*, Arthur Hall and Co. (reprinted to mark the bi-centenary of Southport 1792-1992, Sefton Libraries).

Smith, E.H. (1959) 'Lancashire Long Measure' *Transactions of the Historic Society of Lancashire and Cheshire*, 110, 1-14.

Smith, P.H. (1999) *The Sands of Time: an introduction to the sand dunes of the Sefton coast*, National Museums and Galleries on Merseyside/Sefton Metropolitan Borough Council.

Tyrer, F. (1968, 1970, 1972) *The Great Diurnal of Nicholas Blundell of Little Crosby, Lancashire, 1702-1728*, 3 vols, Record Society of Lancashire and Cheshire.

Yates, W. (1786) *Map of Lancashire.*

ABBREVIATIONS
BP   Before Present

HLC   Historic Landscape Characterisation, Sefton 2001

Lancs RO   Lancashire Record Office, Preston

END NOTES
1. Before Present (in radiocarbon citations, taken as the year AD 1950).

2. Species identified include *Pinus sylvestris, Pinus sp., Myrica gale, Quercus sp., Betula sp, Alnus Glutinosa, Corylus avellana, Tilia europaea, Salix cinerea, Salix aurita, Salix sp.* and *Ilex aquifolium* (Gonzalez *et al.* 2002, 587). Of these, the majority are mainly oak and birch (P. Smith, pers.comm).

3. The discovery and reporting of artefacts such as coins to the Portable Antiquities Scheme at Liverpool Museum, is helping to build up a picture of how the coast was used in Roman and pre-Conquest times.

4. Radiocarbon date calibrated to 535 AD (Lewis 2002, 12).

5. A customary acre was measured according to the length of the rod (pole) used. This could differ from one estate to another. In Lancashire in 1697 the rod varied in length between 18' – 24' (6 – 8 yards), the most favoured being 24' on land south of the Ribble though six other different lengths are known from this area (Smith 1959, 5-6). Whatever the measurement used, the result was always considerably more than that of a statute acre of 4840 sq. yards (4046.9m^2).

6. Middle English *schad*, from Old English *sceadd*. Any of several saltwater food fishes of the herring family (*Clupeidae*); related to the herrings but atypical in swimming up streams from marine waters to spawn. The Allis (or Allice) shad (*A. alosa*) of Europe is about 30 in. (75 cm) long and weighs about 8 lbs (3.6 kg) (*Britannica Concise Encyclopedia*, 2007).

SECTION A: GEOMORPHOLOGY

Liverpool Bay: Linking the eastern Irish Sea to the Sefton Coast

Andrew J. Plater and John Grenville

Department of Geography, University of Liverpool, Roxby Building, Chatham Street, Liverpool, Merseyside, L69 7ZT, UK gg07@liverpool.ac.uk

ABSTRACT

The present-day characteristics of the Sefton Coast have to be considered in the context of processes and factors that operate in the eastern Irish Sea and Liverpool Bay over a variety of scales in time and space. The character and distribution of the seabed and coastal sediments are largely inherited from the effects of post-glacial sea-level rise. In attaining present sea level about 5000 years ago, a dune barrier coast evolved, fronting a series of estuary and lagoonal wetlands. Today, tidal currents in the well-mixed waters of the eastern Irish Sea cause a net onshore movement of sediment from Liverpool Bay. These tidal currents weaken towards the shore where shallow-water waves, driven by the dominant westerly and north-westerly winds, then influence sediment transport over the tidal flats, low-amplitude foreshore ridges and beaches. The combination of tides, waves and coastal morphology lead to erosion at Formby Point and redistribution of this sediment alongshore and towards the estuaries of the Mersey and the Ribble, which have acted as sediment sinks for much of the last 10,000 years. Although net offshore sediment transport takes place during storm events, this erosion is generally made good by net onshore transport during the inter-storm period. Historical evidence and observational data show that the coastal sediment budget has been affected by human actions during the last few centuries to maintain navigation and to protect the coast, thus leading to considerable changes in coastal morphology and sediment transport pathways. The present situation of human intervention to manage what is, in essence, the recycling of a finite pool of sediment leaves the Sefton Coast vulnerable to increases in the frequency of extreme events.

INTRODUCTION

This chapter recognises the coupled nature of shoreline evolution with the dynamics of the adjacent sea in which coastal morphology, at any given time, is a response to material and energy interaction that results from a variety of processes operating over a range of temporal and spatial scales (Carter and Woodroffe, 1997). Whilst a snap-shot survey of coastal processes will generally reveal the way in which sediment is eroded, transported and deposited in response to prevailing tidal and wave dynamics, it is unlikely to capture cyclical changes in these dynamics, such as spring and neap tidal cycles or seasonal changes in the prevailing wind and wave climate, thermal characteristics of the water column or properties of freshwater inflow. Similarly, sudden or episodic shifts in the system are not easily detectable over short timescales of observation, thus not representing the effects of storms on tidal elevation or coupled tide-wave currents and hence net sediment transport and coastline accretion. Extending the time period of observation enables recognition of long-term processes which may have a bearing on the coast through inheritance or significant temporal changes in prevailing material and energy inputs, i.e. sea-level rise onto a formerly glaciated seabed.

In this chapter we recognise that the characteristics of the Sefton Coast and Liverpool Bay are heavily influenced by the dynamics of the eastern Irish Sea over timescales of up to several thousands of years and spatial scales of up to several hundreds of kilometres (Dobson, 1977). It is shown that the characteristics of the seabed are a product of present-day dynamics superimposed on the consequences of long-term climate and sea-level change. Similarly, the present-day dynamics of the Irish Sea are shown to be the result of forcing by tides, waves, storms, temperature, density, sediment grain size and distribution, coastal morphology and bathymetry, and, of course, human intervention (e.g. Bowden and Sharaf El Din, 1966; Heaps and Jones, 1969; Kershaw et al., 1988; Jones and Davies, 1997; Holt and Proctor, 2003; Blott et al., 2006). Predicting the response of the coast to any given event therefore requires a broad knowledge of the Irish Sea and Liverpool Bay over temporal and spatial scales greater than that of the event in question.

MORPHOLOGY AND CHARACTER OF THE IRISH SEA AND LIVERPOOL BAY

The Irish Sea is a semi-enclosed shelf sea occupying an area of approximately 4000 km^2 and extending more than 300 km from the southern margins of St. George's Channel to the North Channel between Northern Ireland and southwest Scotland (Figure 1). Although one continuous body of water, the Irish Sea can be divided into three distinct physiographic regions: (i) a deep (up to 250 m) central trough connecting St. George's Channel with the North Channel; (ii) a series of shallow platforms; and (iii) the coastal waters (Eyles and McCabe, 1989). Indeed, a large part of the Irish Sea seabed is made up of extensive platforms that lie no more than 50-60 m below the sea surface, although there are exceptions such as the "Lune Deep" which is a relict incision feature formed at the end of the last glaciation (McLaren, 1989). The relief of the Irish Sea floor is subdued; Pantin (1977) describes low amplitude features with lengths of 10-15 km, a width of 5 km and heights of no more than 10 m. These features are thought to result from differential compaction of the underlying late Quaternary sediment formations.

The coast of the eastern Irish Sea extends from just south of the Solway Firth in the north to Liverpool in the south, including the coastlines of Cumbria, Morecambe Bay, the Fylde, Sefton and Liverpool Bay, and continues westward along the North Wales shore as far as Anglesey (Figure 1). The present coast is dominated by estuaries and coastal embayments created by post-glacial sea-level rise. The seabed and the shore are characterised by a wide array of highly mobile sediments varying from muddy sands to gravels (Pantin, 1978), the distribution of which has also been largely determined by post-glacial relative sea-level trends and, more recently, sediment transport patterns determined by tide-, wave-, storm- and density-driven dynamics. Further influence on the eastern Irish Sea coast comes from the main rivers of the Dee, Mersey and Ribble, which have an average total input of freshwater to Liverpool Bay of approximately 220 m^3/s (Krivstov et al., 2008). Whilst the fluvial input shows seasonal variation in terms of volume, temperature and sediment load, the biology of the Irish Sea lies within the Boreal biogeographic province in which species associations of colder waters are favoured and the migration of warmer water species limited (Dinter, 2001).

QUATERNARY HISTORY: LONG-TERM DYNAMICS OF THE IRISH SEA BASIN

The physiography and seabed sediments of the Irish Sea are very much linked to the influxes of ice during the Pleistocene glaciations (Charlesworth, 1957). The glacial deposits left behind by the ice sheets now blanket most of the sea floor and are concealed by either reworked glacial or younger Holocene sediments. Quaternary sediments therefore cover the underlying rock across much of the seabed. These underlying strata host an important hydrocarbon reserve which has been exploited since 1974 following the discovery of the South Morecambe Field (Colter, 1997). In the western part of the eastern Irish Sea the geology is Palaeozoic and in the east it is Mesozoic (Dobson, 1977). The late Devonian closure of the ancient Iapetus Ocean resulted in a sub-crop of Lower Palaeozoic rocks in the central Irish Sea Basin, on top of which lies a thick sequence of Carboniferous marine limestone and mudstones. This region also has one of the thickest Upper Triassic sequences in North West Europe; the Mercia mudstone extending down as deep as 3 km. There are intermittent, isolated Jurassic sub-crops in the centre and north of the eastern area as well as evidence of Tertiary igneous activity similar to that found in the more extensive outcrops of Northern Ireland. The eastern Irish Sea Basin, occupied by Carboniferous rocks extending nearly 10 km in depth and covered by up to 6 km of Permian to Lower Jurassic lithologies, was once the largest and deepest basin in western Britain.

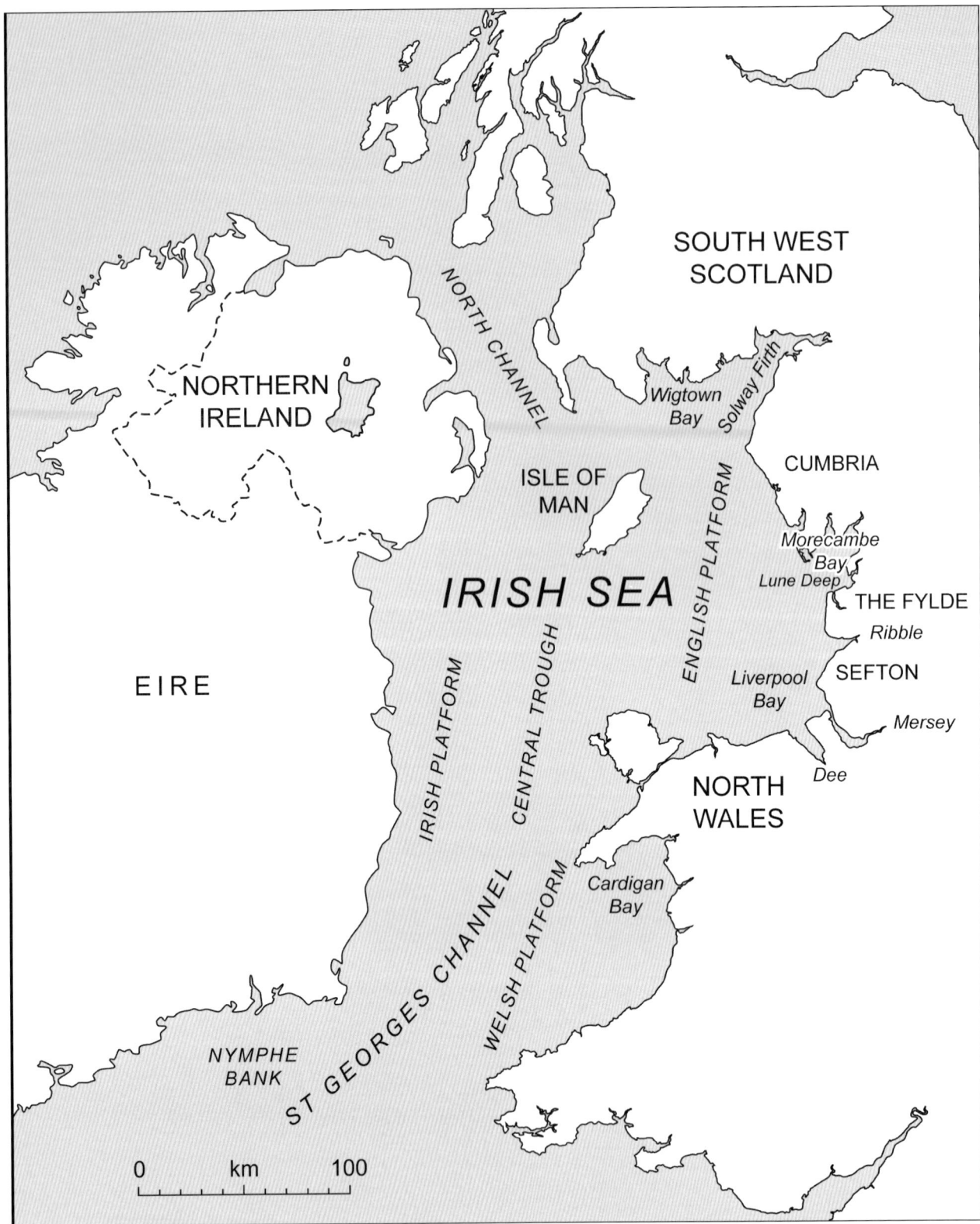

Figure 1. The Irish Sea, showing the three physiographic regions: (i) a deep central trough connecting the southern St. George's Channel with the North Channel; (ii) a series of shallow platforms; and (iii) the coastal waters. (After Eyles and McCabe, 1989).

The oldest Quaternary sediments in the region date back to the Middle Pleistocene, but the sediments from the last glacial and post-glacial sediments are most prevalent. At the maximum extent of the last major glaciation (approx. 18,000 years ago), global sea level was about 120 m lower than today (Fairbanks, 1989). Whilst this fall in sea level was offset to some degree by crustal subsidence due to ice loading (Eyles and Eyles, 1984; Thomas, 1985; Thomas and Dackombe, 1985; Eyles and McCabe, 1989; Austin and McCarroll, 1992;

McCarroll and Harris, 1992; McCabe, 1997; Roberts et al., 2006; McCabe et al., 2007; Brooks et al., 2008), the eastern shelf of the Irish Sea was occupied by ice lobes which deposited glacial and outwash sediments as they retreated during the subsequent period of climatic warming and sea-level rise. Because of the interaction between sea- and land-level movements, the extent and location of the ice limits at the last glacial maximum and during later re-advance have been much debated amongst Quaternary scientists (e.g. Mitchell, 1972; Huddart et al., 1977; Eyles and McCabe, 1989; McCarroll, 2001; Scourse and Furze, 2001; Thomas et al., 2004; Roberts et al., 2006; McCabe, 2008).

A comprehensive review of Quaternary sediments in the Irish Sea is that of Pantin (1977, 1978), who focuses on the northern part of the Irish Sea basin in its entirety and gives a detailed review of the bathymetry and varying surface sediment types found along the west coast of Britain and within the adjoining coastal estuaries of the eastern Irish Sea. The data are derived from a multi-faceted investigation utilising grab samples, cores, boreholes and acoustic (pinger) traverses. Quaternary sediments found in the central Irish Sea Basin can, in places, exceed more than 300 m. However, apart from infilled glacigenic incisions, e.g. the relict kettlehole of the Lune Deep, the Quaternary sediments tend not to extend deeper than 50 m of the marine platforms (Lamplugh, 1903; Blundell et al., 1969; Bott and Young, 1971). In addition, the Pleistocene and Holocene drift between Liverpool and Southport only reaches thicknesses of up to 30 m. The overall stratigraphy in the south-eastern sector of the eastern Irish Sea shows a sedimentary record linked to marine transgression of a formerly glaciated basin (Table 1, Figure 2) that led to considerable changes in tidal dynamics (Uehara et al., 2006) and thus influenced the pattern of sediment transport (van Landeghem et al., 2009). Roberts et al. (2006) suggest that shallow marine conditions may have prevailed in the central Irish Sea Basin between c.21,000-16,000 BP (years ago). The rising surf zone reworked the glacial tills and fluvioglacial outwash gravels, leaving a coarse gravel armour where the finer grains were winnowed away (Wright et al., 1971). The underlying till units are approximately 10-20 m in thickness, and are overlain by 1-3 m of gravelly sands, sandy gravels and well-bedded sandy muds, reaching as much as 10 m in some locations. The deposits are variably proglacial, fluvial, lagoonal and marine in their character, the latter increasing upward in their abundance. The grain size of these marine units also shows a general decrease upward as the gravel-dominated units give way to 2-3 m, and as much as 10 m, of gravelly sands, sands and sandy muds laid down in a shallow sea environment (Pantin, 1978). This upper marine facies also shows a marked degree of spatial variation in character, from tidal flats over much of Liverpool Bay becoming shoreface and beach sands towards the present-day coastline (BGS, 1992).

The pattern of Late Quaternary relative sea-level change in the Irish Sea region is best considered with reference to the output from geodynamic models of glacial isostatic adjustment (GIA) processes (Lambeck, 1996; Peltier et al., 2002; Shennan et al., 2006; Brooks et al., 2008). These are driven by crustal rebound in response to the unloading of ice sheets and the rise in sea-level from the melting Late Pleistocene ice sheets, as well as the response of the crust to loading by water from flooding (hydro-isostasy) (e.g. Lambeck, 1991; Lambeck and Purcell, 2001). A broad pattern can be identified for the eastern Irish Sea coast in which an early period of post-glacial relative sea-level fall results from the predominant effect of glacio-isostatic rebound in the form of crustal uplift (Figure 3). This gives way to a later period, after about 11,000 years ago, which is characterised by dominant eustatic sea-level rise. Hence, there is the potential for much of the eastern Irish Sea basin to have passed twice through shallow marine conditions. About 5,000 years ago, relative sea level reached the approximate altitude of that at the present day.

Lambeck (1996) noted particular limitations of knowledge with regard to the Irish and Irish Sea ice sheet, including: the location of ice limits at the time of maximum ice extent, the rates of ice retreat over southern Ireland and the Irish Sea, and the thickness of the ice at different times since the last glacial maximum.

Table 1. Late Quaternary stratigraphic sequence and palaeo-environments in the south-eastern Irish Sea (after Pantin, 1978; BGS, 1992).

Liverpool Bay Muddy sand (Gravelly) *Tidal flats*	**Coastal Waters** Sand (generally 2-3 m, as much as 10 m) *Beach sands*
Gravelly sand and sandy gravel with well-bedded sands and muds (generally 1-3 m, as much as 10 m) *Sub-tidal marine deposits overlying proglacial/waterlain/lagoonal deposits in outer Bay*	
Devensian till (10-20 m)	
Permo-Triassic basement	

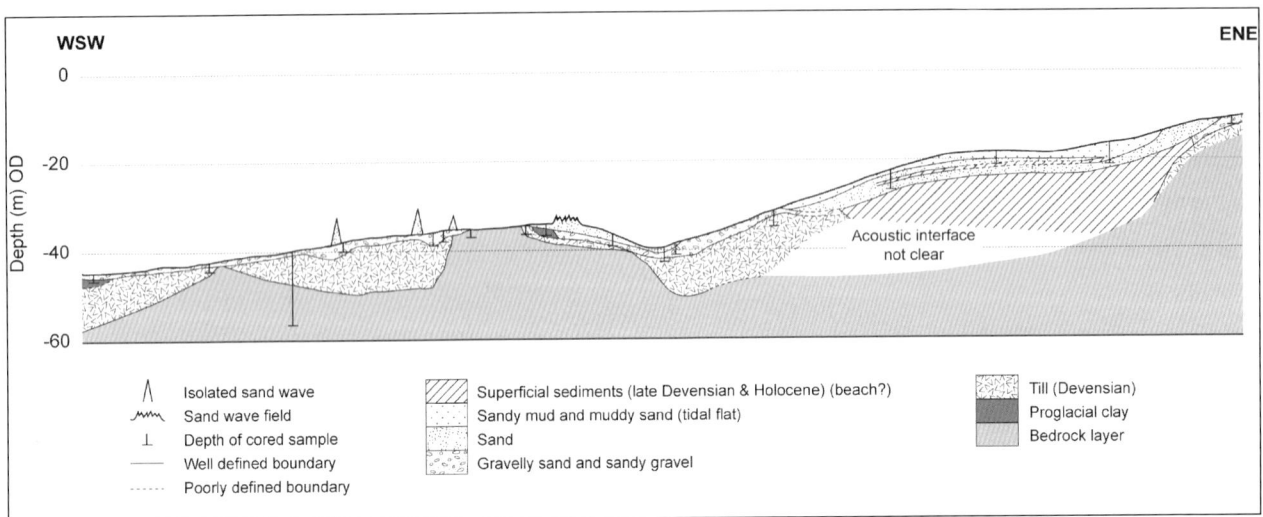

Figure 2. Cross-section of the Irish Sea floor. (After BGS, 1992).

This sparked debate and activity concerning relative sea-level trends for the period of c.21,000-13,000 BP. For the Irish Sea region, Shennan and Horton (2002) noted a significant discrepancy between model output and observations for Liverpool Bay and North Wales where there are no observations to support a mid-Holocene highstand (after Peltier et al., 2002). This, therefore, required some modification of the ice model that better conforms to the chronology of ice retreat in the Irish Sea (Shennan et al., 2002; 2006; Brooks et al., 2008).

Field evidence in north-west England for late-glacial relative sea-levels above the present is sparse. Wright (1914) considered a post-glacial raised shoreline (the 25-foot raised beach) extending from the Lleyn Peninsula, across the Wirral coast and into south-west Lancashire, evidence for which is found from Morecambe Bay, through Cumbria to the Solway Firth (Plater, 2004). Gresswell (1957) documented this feature over much of the Lancashire coast, which he identified as a mid-Holocene (6,000-5,000 BP) shoreline, the 'Hillhouse Coast', fronted by ancient beach sands (Shirdley Hill Sand) and tidal silts (Downholland Silt). Tooley (1978a) challenged this interpretation on the basis of the shoreline altitude and the palaeoenvironment of the Shirdley Hill Sand, which is late-glacial in age (Godwin, 1959; Tooley and Kear, 1977; Wilson et al., 1981; Innes et al., 1989). Plater et al. (1999) compared the altitude of the Hillhouse Coast with relative sea-level output from geodynamic models for the region (e.g. Lambeck, 1991, 1993; Wingfield 1992, 1995), noting the potential for a shoreline feature of this altitude to date from approximately 16,000 BP.

The sedimentary record of Holocene relative sea-level change on the eastern shores of Liverpool Bay has been studied since the mid-19th century, with the complex

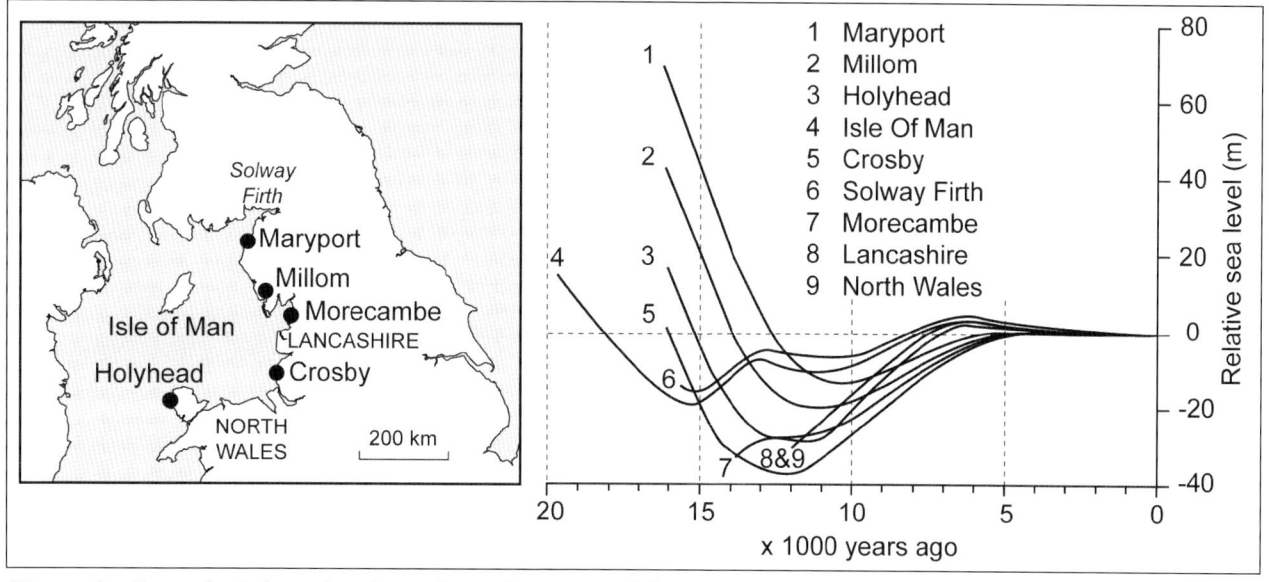

*Figure 3. Post-glacial sea-level trends on the coast of the eastern Irish Sea.
(After Lambeck, 1991; Shennan et al., 2006; Roberts et al., 2006; Brooks et al., 2008).*

intercalation of minerogenic and biogenic units being interpreted in the context of coastal and land-level change (Binney and Talbot, 1843; Picton, 1849; Reade, 1871; Travis, 1926; Erdtman, 1928; Blackburn (in Cope, 1939); Gresswell, 1937, 1953, 1964; Tooley, 1970, 1974, 1976, 1978a, 1978b, 1980, 1982, 1985, 1990, 1992; Huddart, 1992; Neal, 1993; Pye and Neal, 1993a, 1993b; Huddart et al., 1999a, 1999b). The relative sea-level curve is characterised by a rapid rise from approximately –20 to –7 m OD between 9,000 and 7,000 BP, followed by a slight subsequent increase to the present (e.g. Tooley 1978a). During the mid- to late Holocene, the importance of relative sea-level change in driving coastal response gives way to more local site-specific influences. The complex layering of sandflat, mudflat and fringing wetland environments during the mid- to late Holocene is thus the consequence of changing coastal morphology. Pye and Neal (1993a) suggest that periodic breaching of an offshore barrier enabled a sequence of marked changes in the connection with the sea and, thus, may have provided an important local control on sedimentation. Further detail on the Holocene evolution of the coastal barrier complex in the region of the Sefton Dunes is given in Pye and Blott (This volume) and Plater et al. (This volume).

PRESENT-DAY DYNAMICS

Tides, Stratification, Waves and Storm Surges

A considerable body of research has been undertaken on the dynamics of the Irish Sea, mainly because it is affected by all the important physical processes active in mid-latitude shelf seas (Holt and Proctor, 2003). These processes include tides, wind- and density-driven currents, thermal stratification and fronts, and river inputs (Heaps and Jones, 1969; Proctor, 1981; Aldridge and Davies, 1993; Jones and Davies, 1996; Lee and Davies, 2001; Wolf et al., 2002; Horsburgh and Hill, 2003).

The coasts of the eastern Irish Sea experience a semi-diurnal macrotidal regime, with mean spring tidal ranges of 8.22 and 8.45m at Liverpool and Heysham, respectively (Admiralty, 2000). Strong tidal currents reach up to 1 m/s on spring tides (Jones, 1980; Jones and Davies, 1996; Admiralty, 2000; Krivstov et al., 2008), which can exceed 1.5 m/s during storms at some locations (Jones and Davies, 2003a). The eastern Irish Sea has shallower waters than those to the west and, hence, experiences higher tidal currents (>1 m/s compared to <0.3 m/s) (Horsburgh et al., 2000). Non-tidal currents in the Irish Sea with a weak northward residual flow (c.1 cm/s) have been inferred from salinity distributions (Bowden, 1960), although Bowden and

Sharaf El Din (1966) also identified a residual current of the order of 5 cm/s flowing southwards along the coast of northwest England, turning in Liverpool Bay and continuing westwards along the North Wales coast. However, the northward residual drift at basin-scale has been confirmed in subsequent tracer observations and model output (e.g. Ramster and Hill, 1969; Prandle, 1984). At the coast, flood tidal currents diverge off Formby Point (Figure 4), leading to residual currents that transport sediment north and south into the estuaries of the Mersey and Ribble (Sly, 1966; Halliwell, 1973).

Whilst the Irish Sea is generally well-mixed by the strong tides, thermal stratification may develop in the western Irish Sea and parts of the eastern Irish Sea (between the Isle of Man and the Cumbrian coast) (Xing and Davies, 2001). In these areas, water depth is relatively deep (>100 m) and the semidiurnal tidal currents are less than 0.3 m/s, thus giving insufficient tidal mixing to overcome the input of surface buoyancy resulting from solar heating (Simpson, 1971). A tidal mixing front therefore develops, mainly between the southern point of the Isle of Man and the coastal waters off Dublin, where the stratified waters of the western Irish Sea mix with the shallow, fast flowing, well-mixed waters of the eastern Irish Sea (Simpson and Hunter, 1974). Holt and Proctor (2003) modelled thermal stratification and advection, concluding that horizontal and vertical temperature structures are due to a complex interaction of advection, diffusion and dynamical processes, with advective heating dominating in strongly stratified regions, e.g. western Irish Sea. The thermal structure of the Irish Sea can then be disturbed by tidal straining in Liverpool Bay where vertical changes in the tidal currents periodically destabilize the water column (Sharples and Simpson, 1995). Here, Simpson et al. (1990) identified that stratification is influenced by: (i) density-driven circulation due to the input of freshwater runoff, (ii) tidal straining by currents causing vertical shearing, and (iii) spring-neap tidal cyclicity favouring stratification during the neaps and mixing on the higher, faster spring tides. Seasonal cycles in heating and cooling and in freshwater runoff have also been recognised as causal factors (Sharples, 1992).

Deep-water wave height is determined by wind speed, directional variability and fetch of the wind blowing over the sea surface. At the coast, wave height is affected by wave attenuation and refraction across the continental shelf and especially in the nearshore zone (Short and Hesp, 1982). In general, the greater the refraction, the longer the length of the shoreline over which the wave energy is distributed and the lower the breaker wave energy per unit shoreline; the highest waves are associated with areas of least refraction. Hence, breaker wave energy is the deep-water wave energy less the amount lost to wave attenuation through shoaling and that redistributed by wave refraction and diffraction (e.g. Osuna et al., 2007).

The dominant winds and waves in the eastern Irish Sea approach from the west and west-north-west, with the greatest fetch (c.200 km) lying to west-north-west (Gresswell, 1953; Sly, 1966; HRS, 1969a; Bell et al., 1975). At Southport and Formby Point, Jay (1998) recognised a dominance of winds from the north-west and west-north-west. The resulting incident wave spectra have been the subject of previous studies (e.g. Darbyshire, 1958; Murthy and Cook, 1962; Draper and Blakey, 1969; HRS, 1969a, 1969b, 1977), whilst recent model output for the period June 1991-April 2002 illustrates that waves originate mainly from the west, followed by north-west and south-west directions (Associated British Ports, 2002). Parker (1975) noted that the most common wave conditions are those of moderate energy possessing a significant wave height of between 0.6 and 1.0 m with a period of between 4.0 and 4.5 s (c.6% of the time). Output from modelling using HINDWAVE (HR Wallingford, 2002) calibrated against wave buoy data off the North Wales coast (HR, 1986) give a 1-in-1 year Hs (significant wave height) of 4.7 m with a period of 6.8 s. These data are the highest for the Great Orme to Formby Point Shoreline Management Plan area (HR Wallingford, 2006a).

Combined tide and wave modelling (HR Wallingford, 2006b) for the South Sefton shoreline suggests that there is greater wave penetration into the upper reaches of Crosby Channel and onto the foreshore during peak flood conditions, whilst peak ebb flows lead to relatively low wave heights in Crosby Channel but with an increase in wave heights at the channel entrance and on the southern side of the Queen's Channel North Training Bank (Figure 4). Formby Bank is generally calm on peak ebb, and Taylor's Bank and Mad Wharf Sands are also calmer than under peak flood conditions. Spatial variation in wave height is also reflected in sediment transport modelling along the coast (HR Wallingford, 2006a), with modal wave heights of 0.8-2.0 m across the Formby foreshore, 0.0-0.4 m between Taylor's Bank and the Alt Estuary, and 0.4-0.8 m from Hightown to Crosby Marina.

Waves in the eastern Irish Sea are generally higher during the autumn and winter months, with the highest measured waves being recorded at 9.4 m with a period of 8.7 s (Tucker, 1963; Draper, 1966). These conditions agree well with the results from HRS (1969a), although Pye (1990) suggests that severe storm waves in the eastern Irish Sea only reach a height of 5.7 m with a period of 8.7 s. Estimates of maximum wave height (Hmax) for the Irish Sea range from 8 to 15 m (HRS, 1977; Devoy, 2000). However, the limited fetch restricts wave development while topographic and bathymetric features limit the incursion of swell from the Atlantic. Indeed, Osuna et al. (2007) note that the only possibility for waves to re-entrain sediment in Liverpool Bay is through the occurrence of northwesterly winds. Based on interpolated data for the Irish Sea, Hs50 (significant wave height with a recurrence period of 50 years) for Liverpool is c.8 m (Draper and Carter, 1982; Dept. of Energy, 1990; Draper, 1991).

Because of the shallow nature of the eastern Irish Sea, the incident waves are refracted as they pass over the nearshore sand banks. Those from the west and west-north-west are refracted as they approach the shoreline, converging on Formby Point and increasing the breaker height. There is also some wave reflection off Taylor's Bank towards the Formby shore (Figure 4). The present bathymetry and wind regime give rise to wave focussing at Formby Point and a net littoral drift divide in the region of Victoria Road (HRS, 1969a, b; Parker 1971, 1975; Pye and Neal, 1994), thus favouring a negative sediment budget at Formby Point and the longshore transport of sand (Pye and Blott, 2008). Consequently, wave crests are direct to the shore at Formby, but divergence leads to longshore drift to the north and south (Gresswell, 1953; HRS, 1969a; Pye 1990).

Much research has been completed on the genesis of storm surges along the west coast of Britain (Heaps and Jones, 1975, 1979; Heaps, 1983; Flather and Smith, 1993; Jones and Davies, 2003a). Several storm surge models have been established and tested in the Irish Sea achieving progressively improved resolution and agreement with observational data (Heaps, 1973; Heaps and Jones, 1979; Howarth and Jones, 1981; Davies and Jones, 1992; Jones and Davies, 1998, 2001, 2003b, 2004a, b). Jones and Davies (2004b) noted that strong uniform winds over the Irish Sea region have the potential to produce positive storm surges, whilst the generation of negative surges by local winds may be offset by 'far-field' inflow of water to the west coast region. Indeed, the nature of Irish Sea response to wind forcing depends very much on the pathway taken by depressions (Lennon, 1963). Computed surge elevations and currents in the eastern Irish Sea are a balance of flows to the north and, in particular, south of the Isle of Man (Jones and Davies, 2003b). Here, storm surge elevations in shallow water regions such as Liverpool Bay appear to be affected by the position of this inflow south of the Isle of Man. Furthermore, Jones and Davies (2003a) conclude that outflows following major storms in the eastern Irish Sea are a function of sea surface elevations and topography.

Storm surges are frequent and typically produce a positive elevation of approximately 1 m (Jones and Davies, 1997, 1998). Data from tide gauge sites (http://www.pol.ac.uk/ntslf/networks.html) show maximum positive storm surges of 2.60 m at Heysham on 26th February 1990 and 2.26 m at Liverpool on 27th October 2002. Evidence for any recent increase in storm frequency due to climate change is equivocal, although tidal data acquired since 1990 (Pye and Blott, 2008) show an increased frequency of extreme high tides when compared with that analysed previously for

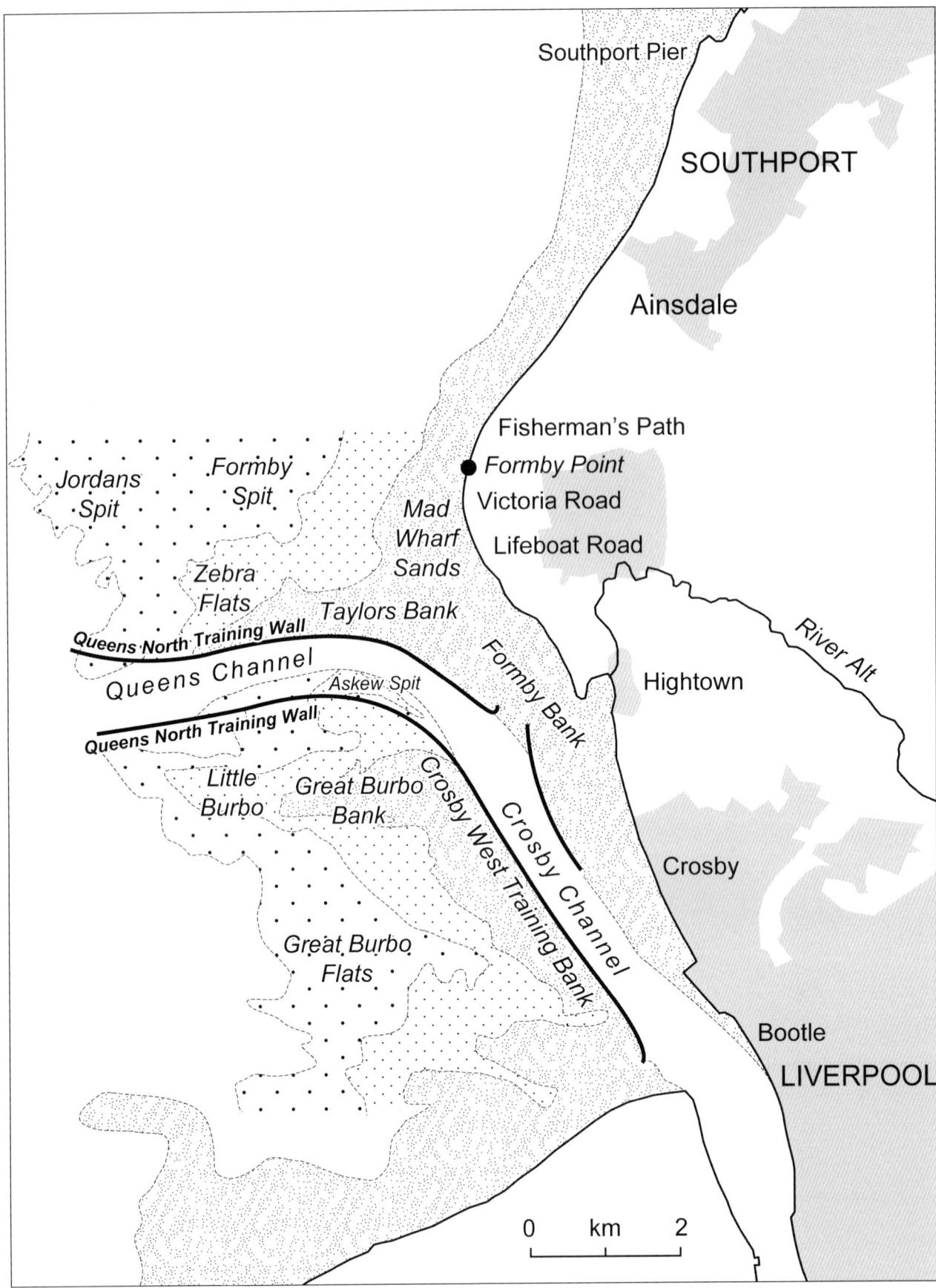

Figure 4. The Sefton Coast, showing locations at the coast and main morphological features in the intertidal zone.

the period 1963-1990 (Woodworth and Blackman, 2002). An extreme water level assessment based on data from POL (1995, 1997) and Coles and Tawn (1990) undertaken for the Liverpool Bay Shoreline Management Plan (Liverpool Bay Coastal Group, 1999) gave return periods for local extreme levels at Formby of 1 in 50 years for 6.2 m OD, 1 in 100 years for 6.3 m OD, and 1 in 200 years for 6.6 m OD. The joint probability of large waves and extreme water levels at Liverpool has also been calculated by HR Wallingford (1990, 2000).

Seabed Sediments and Sediment Dynamics

In terms of the present-day sediment distribution, the bed of the eastern Irish Sea is covered with fine or medium sand with varying amounts of mud and some areas of gravel (Sly, 1966; Cronan, 1969; Wright et al., 1971; Pantin, 1977; Jackson et al., 1995; Pye et al., 2006) (Figure 5). The main source of this sand is reworking of the fluvio-glacial sediments that immediately underlie the sea floor (Wright et al., 1971). Large-scale mapping of seabed sediments (e.g., ICES, 1988) shows a marked difference between the sediments of the western and eastern Irish Sea. The western Irish Sea is characterised by extensive areas of soft mud with a very high silt-clay content of up to 94% (Hensley, 1996). However, the eastern Irish Sea sediments tend to be much coarser, with exceptions such as the soft muds off the coast of Cumbria (Kershaw et al., 1988). The difference in particle size and sediment type across the Irish Sea floor may well be a reflection of the differing water depths and varying magnitude of tidal currents. Here, the east-west gradient from sand and gravel to soft mud-dominated sediment is most pronounced to the southwest of the Isle of Man in the area of the western Irish Sea tidal mixing front. In this respect, Gowen et al. (2000) note that biological productivity in Liverpool Bay makes only a small contribution to the seabed sediment budget.

Bedload transport in the eastern Irish Sea is dominated by an asymmetry in the current ellipse (Belderson and Stride 1969; Aldridge, 1997). Here, the maximum tidal velocity achieved in one direction is exceeded by the opposing tidal velocity, thus favouring the movement of sediment in one direction. In Liverpool Bay there is a net shoreward drift of sediment due to asymmetry between the flood and ebb tides. This tidal asymmetry has been used in conjuction with the asymmetry of sand waves to depict the bed load transport paths in the Irish Sea.

The offshore zone in the eastern Irish Sea is characterised by a series of sandbanks and channels superimposed on a seafloor that slopes gently westwards. Bedform distribution patterns of the seabed and the results of drifter investigations confirm that net sediment movement is in an easterly direction with a convergence of bedload transport on Liverpool Bay (Sly, 1966; Belderson and Stride, 1969) (Figure 6). This is evidenced by the model results of Price and Kendrick (1963) and field data (Best, 1972; Best et al., 1973; Halliwell, 1973; Pye, 1990). Work by BGS (1992) also confirms this pattern, showing extensive sand wave fields in the southern Irish Sea and a line of high amplitude forms trending northward from Great Orme.

Water movements in the region show a pattern of near-bed residual water flow over a wide area to the south and southeast into Liverpool Bay (Bowden, 1960; Sharaf El Din, 1970; Best et al., 1973; Halliwell, 1973). Indeed, Bowden and Sharaf El Din (1966) identified estuarine-type circulation extending to a distance of approximately 12 km from the mouth of the Mersey in which a net seaward flow in the upper waters and a shoreward flow near the bottom is seen in current and salinity profiles. Parker (1975) noted that in the offshore areas west of Jordan's Spit, the major tidal flood streams diverge (northeast to the Ribble and southeast to the Mersey) and the ebb streams converge. This area is linked to a corresponding divergence of bedload transport (Sly, 1966) and a divergence of near-bed residual water drift (Ramster and Hill, 1969). Residual currents, i.e. long-term averaged water flow, are able to transport muds in suspension and in a landward direction. These transport vectors were studied by Ramster and Hill (1969), showing a predominantly northward drift of surface water through Liverpool Bay towards the Solway Firth, with a predominant eastward drift near the bed towards the

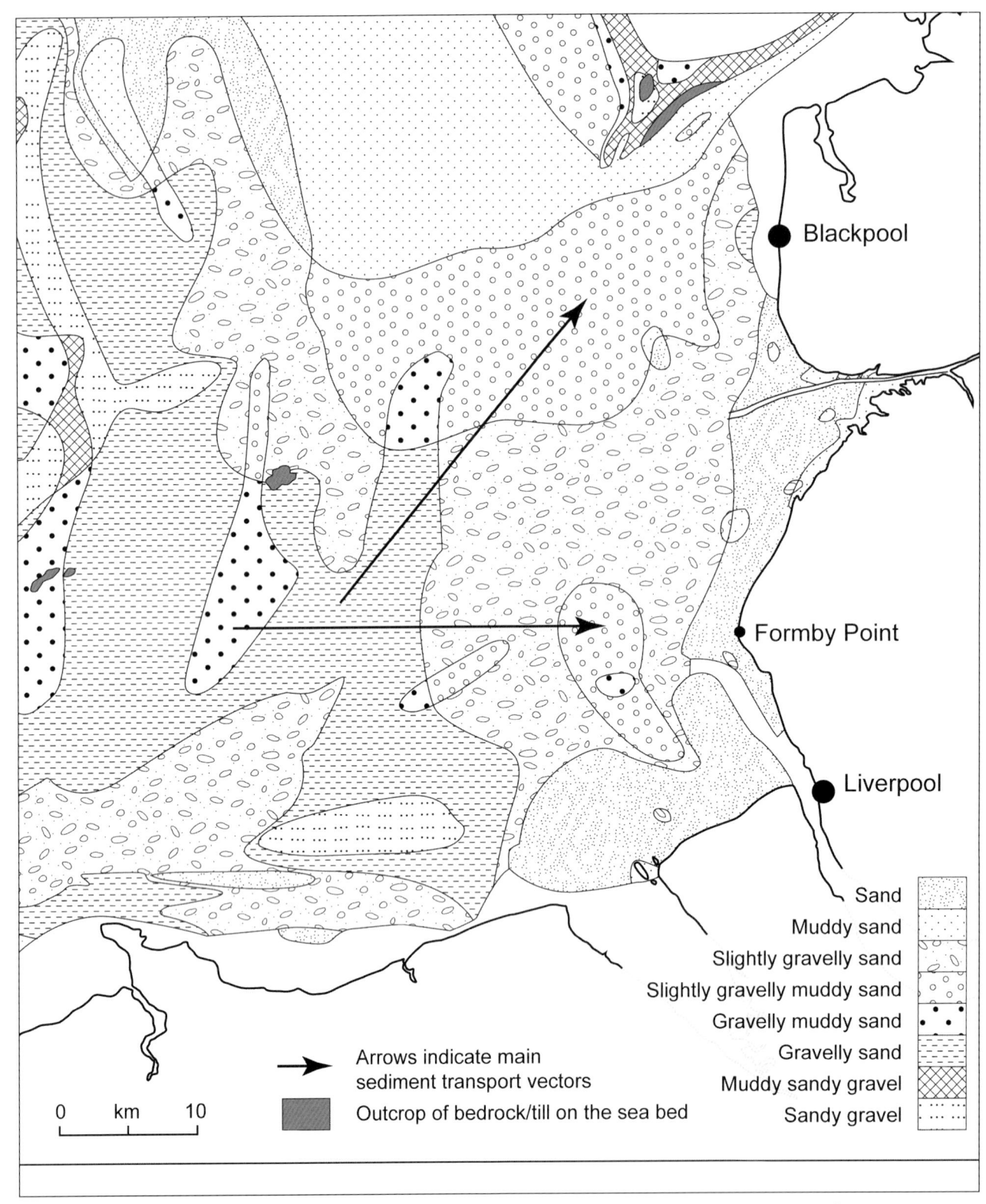

Figure 5. Seabed sediment distribution in the eastern Irish Sea (after BGS, 1992). Arrows indicate main sediment transport vectors.

North Wales and Lancashire coasts.

Kershaw et al. (1988) acknowledge that the sediment distribution pattern of the northern Irish Sea during the Holocene has been the subject of great discussion and note that this is particularly true of the fate of the finer grained sediments (Belderson, 1964; Belderson and Stride, 1969; Cronan, 1969; Pantin, 1977, 1978; Mauchline, 1980; Williams et al, 1981; Stride, 1982; Johnson, 1983; Kirby et al, 1983; Kershaw, 1986). Muddy sediments are found in areas characterised by areas of weak tidal currents (Belderson, 1964; Howarth,

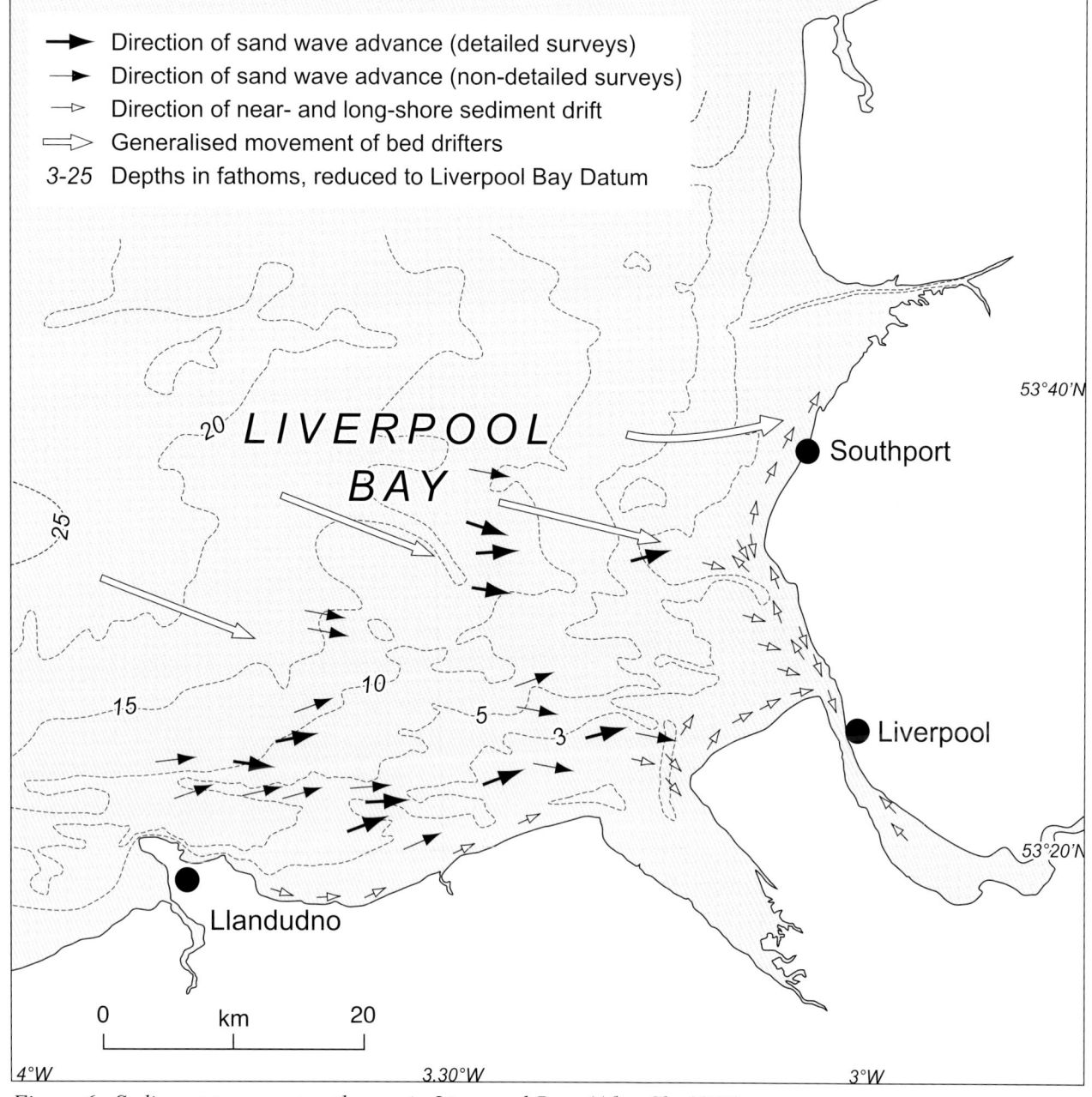

Figure 6. Sediment transport pathways in Liverpool Bay. (After Sly 1966).

1984) whilst stronger tides result in coarser sands and gravel. Eden et al. (1973) suggest that the finer particles from the areas of high tidal energy have been winnowed away. Observations on the bedforms suggest the winnowed fine material moves easterly to add to the mud belt that runs along the Cumbrian coast (Eden et al, 1973; Belderson and Stride, 1969; Stride, 1982). Pantin (1978) postulated that the source of this fine mud is the rivers of the Irish Sea Basin, coastal erosion and sub-tidal erosion, and that the compact muddy marine sediments off the Wigtown Bay-Solway Firth area are a potential source of sediment for the Cumbrian mud bank. Kirby (1987) recognised that the estuaries of the eastern Irish Sea are sinks for, rather than sources of, fine sediment. This is true for much of the Holocene period (van der Wal et al., 2002; Thomas et al., 2002; Blott et al., 2006). From their analysis of sedimentation rate, Kershaw et al. (1988) conclude that net sediment accumulation has occurred in at least three sites in the eastern Irish Sea mud belt; and at one study site deposition of the upper 20 cm appears to have taken place over the last 50 years. Indeed, Hetherington (1976) found the muds off the Cumbrian coast to contain freshly irradiated material that, once discharged offshore from the Sellafield nuclear waste reprocessing centre, is incorporated into the fine bed sediment and potentially transported shoreward.

In terms of sediment flux into Liverpool Bay, the long-term net transport of sediment from the eastern Irish Sea has reduced average depth (Pye, 1977; Neal, 1993) and has led to a net drift of sediment into the Inner Estuary (O'Connor, 1987). Considerable nearshore accretion has taken place in Liverpool Bay during the 19th and 20th centuries (HRS, 1958; Thomas et al., 2002; Blott et al., 2006), possibly due to dredge spoil reworking or to training wall construction (Cashin, 1949; HRS, 1958; McDowell and O'Connor, 1977; 1987; Smith, 1982). Price and Kendrick (1963) concluded that changes in tidal propagation due to human-induced changes in circulation patterns in the Outer Mersey Estuary, and by association the reclamation and construction of the main dock developments close to Liverpool, were not the trigger for estuarine accretion in the early 19th Century. Similarly, O'Connor (1987) concluded that the main dock reclamations at Seaforth and around Ince Banks (during the construction of the Manchester Ship Canal in the inner estuary) have had only a small affect on the tidal prism in the Mersey. Thomas et al. (2002) conclude from verified hydrodynamic modelling that the observed bathymetric changes in sediment transport and accretion resulted from the combined influences of training wall construction and dredging coupled with the effects of salinity-related circulation within the Mersey estuary/Liverpool Bay system. Further discussion of the impact of human activity on sediment accretion and coastal morphology along the South Sefton shore is described in Plater et al. (This volume).

SEDIMENT TRANSFER AT THE SHORELINE

The sediment transport dynamics of the Sefton shore have been studied by several workers (e.g. Gresswell, 1937, 1957; Parker, 1971, 1974, 1975; Wright, 1976). The nearshore in the region of Formby is characterised by a series of symmetrical sand ridges which are between 0.5 and 1.0 m high with a wavelength between 150 and 500 m, which are aligned sub-parallel to the intertidal zone (Plate 1). The lower foreshore is a muddy intertidal sandflat which attenuates in width to the north, whilst the upper foreshore comprises a series of intertidal ridges and runnels which lie obliquely to the curve of the high water mark such that they trend away from the shore along the South Sefton shore (Parker, 1975). Here, the runnels form unbroken channels. This ridge and runnel topography is separated from the dune complex by a planar slope.

King and Williams (1949) noted that sand ridges orientated parallel to the coast are common on many sandy beaches under a variety of dynamic states. Those found on beaches with a considerable tidal range were termed ridge and runnel beaches. Ridges move considerably with changing weather: the beach is generally smoother during rough weather, whilst ridge growth occurs during calmer weather. Onshore migration rates are generally 1-10 m per month, and only under optimal wave/tide conditions do they move onshore more than 1 m during a single tidal cycle (Kroon and Masselink, 2002). Masselink et al. (2006) recognise that intertidal ridge systems (a ridge and its associated trough located shorewards) are common on wave-dominated coasts with a tidal range in excess of 1 m, and that their morphology is dependent upon wave conditions, tidal range and, to a lesser degree, nearshore gradient. These features have been variably described as ridge and runnel, sand wave and swash bar. The term ridge and runnel is now mainly reserved for small-scale features found on intertidal mudflats and composed of cohesive sediments. The large-scale morphological features of the multi-barred foreshore at Formby (cf. Parker, 1975) are perhaps more accurately considered as multiple swash bars, and in particular low-amplitude ridges which are generally less than 1 m in height with a spacing of c.100 m and dissected in places by shore-normal drainage channels that occur across the intertidal profile throughout the year and are relatively static in their location.

Sediment transport and morphological response are controlled by tidal level on the beach which determines the type, intensity and duration of the shallow water wave processes operating on the cross-shore profile (King, 1972; King and Williams, 1949; Masselink, 1993; Masselink and Turner, 1999; Masselink and Anthony, 2001; Kroon and Masselink,

Plate 1. Low-amplitude intertidal ridges on the Sefton foreshore. The figure shows the oblique orientation of ridge crests in relation to the high-tide shoreline, encouraging the net southward and northward drift of sediment alongshore from Formby Point. (Source: Sefton Council).

2002). Cross-shore sediment transport under breaking waves, as found on the seaward slope and crest of an intertidal ridge, is the product of onshore-directed transport due to wave skewness and an offshore-directed component resulting from the bed return flow. Net transport directions over the longer term are largely determined by the balance between net onshore transport by waves during calm weather (Sunamura and Takeda, 1984) and net offshore transport during storms (Russell and Huntley, 1999). Shoreward of the breaker zone, onshore transport takes the form of turbulent bores in the surf zone which collapse on the beach in the form of swash, thus promoting net onshore sediment transport. Hence, there is considerable spatial (and temporal) variability in the net sediment transport direction under the influence of waves. Masselink (2004) describes a shape function that expresses sediment transport as a function of cross-shore location (Figure 7).

Tides play an important morphodynamic role by shifting the operational zone of waves up and down the beach profile, and thus determining the position and duration of different wave processes (Masselink and Turner, 1999). In essence, the rise and fall of the tides during the tidal cycle causes temporal changes in the cross-shore transport direction at any location on the intertidal profile, which becomes more pronounced as tidal range increases. Furthermore, spring tides induce a large spatial variation in water levels and reduce residence times for distinct processes, whilst neap tides narrow the intertidal area and increase the time for processes to operate at one location (Kroon and Masselink, 2002). As a general rule, the larger the relative tidal range (the ratio between tidal range and wave height), the shorter the residence times for swash and surf zone processes, the more important shoaling wave processes, and the more likely the occurrence of changes in the cross-shore sediment transport direction during the tidal cycle.

As observed by Parker (1975) for the low-amplitude ridges at Formby, longshore currents in the troughs between bars are driven by a combination of wave and tidal processes. When wave breaking occurs on a seaward ridge, the longshore current in the trough is a product of the height and incident angle of the waves. As tidal height increases and the wave height over the ridge crest is high (relative height of wave height, H, to water depth, h, exceeds 0.5), the longshore current is fed by propagating bores and swashes (Kroon and de Boer, 2001). Here, trough currents may be sufficient to prevent the movement of sand across the trough, causing the trough to act as an effective barrier to sediment transport onshore. This longshore sediment transport may be significant, and may lead to alongshore migration of the intertidal ridge configuration (King and Barnes, 1964).

The Sefton shore is predominantly medium and fine sands, with an increased silt component in the region of the Alt at Hightown and on saltmarsh deposits north of Southport pier (Pye et al., 2006). The sandy foreshore shows limited grain size variation alongshore, perhaps with a subtle fining trend to the north and south of Formby Point and Ainsdale, but a significant fining trend offshore. Indeed, the coarsest sediments are immediately in front of the dune toe after erosion. Sefton Council have maintained a record of accretion and erosion along its sand dune coastline and adjacent foreshore since the mid-1950s, and possesses records of beach cross-sections and shore surveys dating back to the 1880s as well as aerial photographic data. Much of this information is summarised by Smith (1982), Pye and Smith (1988), Neal (1993), Pye and Neal (1994), Newton et al., (2008), Pye and Blott (2008, This volume) and Sefton MBC (2010). Beach profiles show a high degree of variability over a range of timescales from successive tides, following storms, and on a seasonal basis. This is primarily a response to changing wave energy, which exhibits marked seasonality. Consequently, the beach tends to build up during summer, with a marked berm above the high tide mark, and flattens during winter when there is a net offshore sediment transport (HR Wallingford, 2006a). In addition, the low-amplitude ridges on the foreshore increase in amplitude during the summer months under conditions of moderate waves, but are less prominent during more energetic winter conditions.

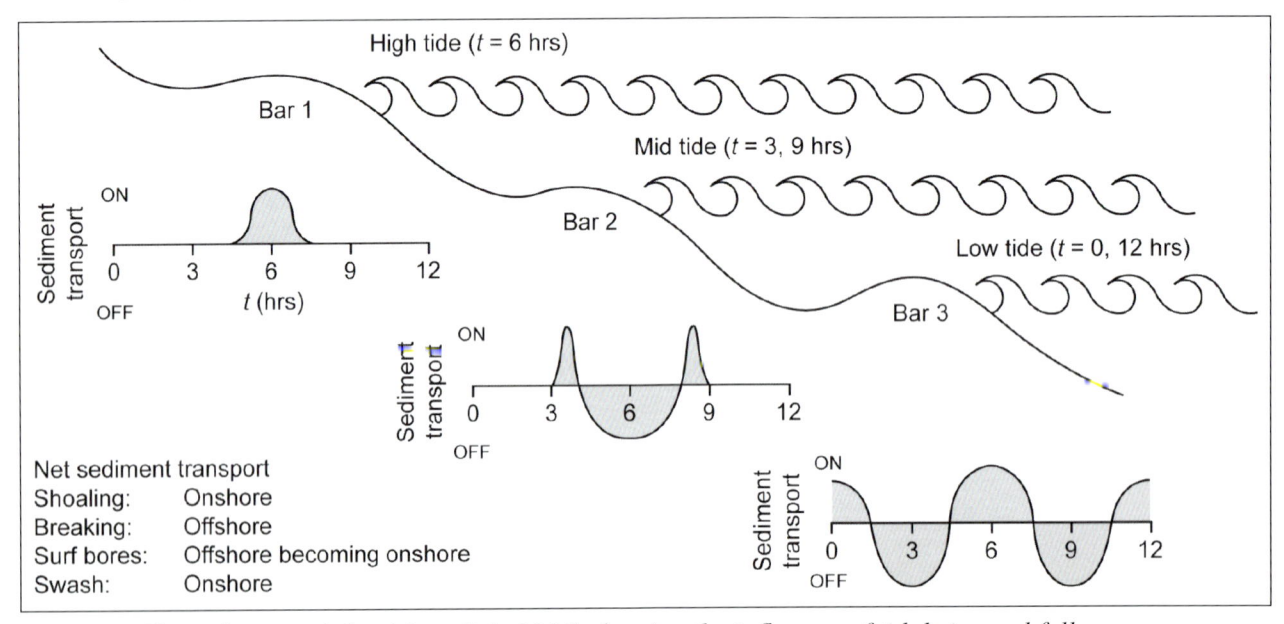

Figure 7. Shape function (after Masselink, 2004) showing the influence of tidal rise and fall on wave-induced onshore (ON) and offshore (OFF) sediment transport vectors across low-amplitude intertidal ridges at low-, mid- and high-tide locations on the foreshore.

In terms of offshore-nearshore-foreshore connectivity, combined wave, sediment flux and morphological modelling (HR Wallingford, 2006a) confirms a dynamic interaction between the offshore and the beach. Sediment transport in the offshore zone follows the predominant wave climate, which moves eastward into Liverpool Bay where it then is diverted at the coast and moves alongshore. Wave action is considerably reduced at the mouth of the Alt where the beach is protected to some degree by offshore training walls and Taylor's Bank, both of which act to dissipate incident wave energy.

CONCLUSIONS

The eastern Irish Sea and Liverpool Bay are areas of complex interaction between a series of dynamical processes. Predominant amongst these processes is perhaps the operation of tidal currents that exhibit some degree of asymmetry (flood>ebb). Hence, there is a net onshore movement of sediment to the coast from the seabed. This is greatest in mid-Bay, but tidal currents weaken towards the shore where wave-induced currents influence sediment transport, particularly over the nearshore banks and the foreshore. However, even on the foreshore, the location, duration and direction of wave-induced sediment transport are influenced by the rise and fall of the tide.

Humans have had a significant impact on coastal bathymetry and sediment budget over the scale of decades to centuries, but this effect is negligible when compared to long-term climate-induced sea-level rise, probably in two phases, onto an area once characterised by ice sheets and glacial meltwater streams, as evidenced by the late Quaternary stratigraphy of the eastern Irish Sea. Hence, the distribution of sediments in the nearshore zone owes as much to post-glacial sea-level rise and sediment reworking in the advancing surf zone as to the present dynamics of the eastern Irish Sea. In this respect, a great deal of what we see today has been inherited from a coastal setting that is no longer in operation. Indeed, it may be argued that the present dune barrier complex along the Sefton Coast is something of a relict coastline that is yet to achieve some level of equilibrium with the prevailing coastal dynamics and sediment flux; hence the eroding dunes, the tidal flat and saltmarsh accretion, and, thus, the need for human intervention in the form of dune management and engineering works to maintain the status quo. Given that the available sediment budget at the coast today is largely inherited from post-glacial sea-level rise and later supplemented by human actions, it is easy to over-estimate the availability of sediment for coastal accretion in response to dynamical forcing, particularly during extreme events. Indeed, the present erosion at Formby Point and the observed accretion along adjacent coasts to the north and south illustrate that reworking and redistribution of a finite sediment pool is a more accurate assessment of the present-day coast. However, the available evidence suggests that the onshore transport of sediment at least still operates in the form of bedform migration and accumulations of mud.

The dynamics of the Irish Sea are well-known and have been the subject of intensive study and computer modelling, illustrating the significance of a range of forcing factors from tide- and wave-induced circulation, thermal stratification, and characteristics linked to regions of freshwater influence. In the nearshore-foreshore zone our knowledge of coastal dynamics and understanding of change are further supplemented by a wealth of historical observations and monitoring work. The foreshore, beach and dune system are characterised by onshore and alongshore sediment transport during fair weather conditions. Therefore, offshore sediment transport arising from erosion during storm events is expected to be made good during the inter-storm period. This may well be the case for present storm magnitude/frequency and sediment budget, but any increase in the former or decrease in the latter (or both) may have dramatic consequences for the Sefton Coast. Aside from changes in sediment supply, an increase in the frequency of extreme levels will shorten the recovery period for the foreshore and beach between storms. Quantifying the available nearshore sediment supply and understanding past trends in storm frequency and tide-wave interaction are, therefore, foremost research questions for informing sustainable coastal zone management.

ACKNOWLEDGEMENTS

The authors should like to thank Jason Kirby for his comments on the initial draft, Michelle Newton and Graham Lymbery for the provision of data and information, and to a wide variety of colleagues who have undertaken research on the Irish Sea and Liverpool Bay.

REFERENCES

Aldridge, J.N. (1997) Hydrodynamic model predictions of tidal asymmetry and observed sediment transport paths in Morecambe Bay. *Estuarine, Coastal and Shelf Science 44*, 39-56.

Aldridge, J.N. and **Davies, A.M.** (1993) A high-resolution three-dimensional hydrodynamic tidal model of the Eastern Irish Sea. *Journal of Physical Oceanography 23, 207-224.*

Associated British Ports (2002) *Burbo Offshore Wind Farm Coastal Process Study*. Report No. R.962. ABP Marine Environmental Research Ltd., Southampton.

Admiralty (2000) *Admiralty Tide Tables*, Volume 1, (2000) European Waters including Mediterranean Sea. Hydrographer of the Navy, London.

Austin, W.E.N. and **McCarroll, D.** (1992) Foraminifera from the Irish Sea glacigenic deposits and Aderdaron, Western Lleyn, North Wales: palaeoenvironmental implications. *Journal of Quaternary Science 7*, 311-317.

Belderson, R.H. (1964) Holcene sedimentation in the western half of the Irish Sea. *Marine Geology 2*, 147-163.

Belderson, R.H. and **Stride, A.H.** (1969) Tidal currents and sand wave profiles in the north-eastern Irish Sea. *Nature* 222, 74-75.

Bell, M.N., Barber, P.C. and **Smith, D.G.E.** (1975) The Wallasey embankment. *Proceedings of the Institution of Civil Engineers* 58, 569-590.

Best, R. (ed.) (1972) *Out of Sight, Out of Mind*. Report of a working party on sludge disposal in Liverpool Bay, H.M.S.O., London, two volumes.

Best, R., Ainsworth, G., Wood, P.C. and **James, J.E.** (1973) Effects of sewage sludge on the marine environment: a case study in Liverpool Bay. *Proceedings of the Institution of Civil Engineers* 55, 43-66, with Discussion on pp.755-765.

BGS (British Geological Survey) (1992) *Marine Aggregate Survey Phase 4: Irish Sea*. BGS Marine Report WB/92/10.

Binney, E.W. and **Talbot, J.H.** (1843) On the petroleum found in the Downholland Moss, near Ormskirk. Paper read at the Fifth Annual General Meeting of the Manchester Geological Society, 6th October 1843. *Transactions of the Manchester Geological Society* 7, 41-48.

Blott, S.J., Pye, K., van der Wal, D. and **Neal, A.** (2006) Long-term morphological change and its causes in the Mersey Estuary, NW England. *Geomorphology* 81, 185-206.

Blundell, D.J., Griffiths, D.H. and **King, R.F.** (1969) Geophysical investigations of buried river valleys around Cardigan Bay. *Geological Journal* 6, 161-180.

Bott, M.P.H. and **Young, D.G.G.** (1971) Gravity measurements in the north Irish Sea. *Quarterly Journal of the Geological Society of London* 126, 413-434.

Bowden, K.F. (1960) Circulation and mixing in the Mersey Estuary. *International Association of Scientific Hydrology and Surface Water*, Publication 51, 352-360.

Bowden, K.F. and **Sharaf El Din, S.H.** (1966) Circulation and mixing processes in the Liverpool Area of the Irish Sea. *Geophysical Journal of the Royal Astronomical Society* 11, 279-292.

Brooks, A.J., Bradley, S.L., Edwards, R.J., Milne, G.A., Horton, B. and **Shennan, I.** (2008) Postglacial relative sea-level observations from Ireland and their role in glacial rebound modelling. *Journal of Quaternary Science* 23(2), 175-192.

Carter, R.W.G. and **Woodroffe, C.D.** (1997) *Coastal Evolution - Late Quaternary Shoreline Morphodynamics*, Cambridge University Press, Cambridge, 539p.

Cashin, J.A. (1949) Engineering works for the improvement of the estuary of the Mersey. *Journal of the Institution of Civil Engineers* 32, 355-396.

Charlesworth, J.K. (1957) *The Quaternary Era, with Special Reference to its Glaciation*. Edward Arnold, London, 700.

Coles, S.G. and **Tawn, J.A.** (1990) Statistics of coastal flood prevention. *Philosophical Transactions of the Royal Society* A 332, 457-476.

Colter, V.S. (1997) The East Irish Sea Basin- From caterpillar to butterfly, a thirty-year metamorphosis. In: Meadows, N.S., Trueblood, S.R., Hardman, N. and Cowan, G. (eds.), *Petroleum Geology of the Irish Sea and Adjacent Areas*, Geological Society of London, Special Publication 124, pp. 1-9.

Cope, F.W. (1939) Oil occurrences in South-West Lancashire. *Bulletin of the Geological Survey of Great Britain* 2, 18.

Cronan, D.S. (1969) *Recent sedimentation in the central north-eastern Irish Sea*. Report No. 69/8, Institute of Geological Sciences, H.M.S.O., London, 10p.

Darbyshire, M. (1958) Waves in the Irish Sea. *Dock and Harbour Authority* 39, 245-248.

Davies, A.M. and **Jones, J.E.** (1992) A three dimensional wind driven circulation model of the Celtic and Irish Seas. *Continental Shelf Research* 12, 159-188.

Department of Energy (1990) *Offshore Installations: Guidance on Design, Construction and Certification*. Section 11: Environmental Considerations (4th Edition). HMSO, London.

Devoy, R.J.N. (2000) Climate warming and the links to the coast. *ICE (Joint Ireland-Wales EU Interreg Project) Newsletter*.

Dinter, W.P. (2001) *Biogeography of the OSPAR Maritime Area*. Federal Agency for Nature Conservation in Germany, Bonn, 168p.

Dobson, M.R. (1977) The history of the Irish Sea Basins. In: Kidson, C. and Tooley, M.J. (eds.), *The Quaternary History of the Irish Sea*, Seal House Press, Liverpool, pp. 93-98.

Draper, L. (1966) The analysis and presentation of wave data – a plea for uniformity. *Proceedings of the 10th Conference on Coastal Engineering*, Tokyo, volume 1, pp. 1-11.

Draper, L. (1991) *Wave Climate Atlas of the British Isles*. Offshore Technology Report, OTH 89, HMSO for Department of Energy, London.

Draper, L. and **Blakey, A.** (1969) *Waves at the Mersey Bar light vessel*, National Institute of Oceanography, Internal Report No. A.37, Wormley, March 1969.

Draper, L. and **Carter, D.J.T.** (1982) *Waves at Morecambe Bay light vessel in 1957*, summary analysis and interpretation report: IOS Report No. 113.

Eden, R.A., Deegan, C.E., Rhys, G.H., Wright, J.E. and **Dobson, M.R.** (1973) *Geological investigations with a manned submersible in the Irish Sea and off western Scotland, 1971*. Report of the Institute of Geological Sciences No.73/2, H.M.S.O., London, 29p.

Erdtman, G. (1928) Studies of the Post-Arctic History of the Forests of North Western Europe I. Investigations in the British Isles. *Geologiska Föreningens i Stockholm Förhandlingar* March-April, 133.

Eyles, C.H. and **Eyles, N.** (1984) Glaciomarine sediments of the Isle of Man as a key to late Pleistocene stratigraphic investigations in the Irish Sea Basin. *Geology* 12, 359-364.

Eyles, N. and **McCabe, A.M.** (1989) The Late Devensian (less than 22,000 BP) Irish Sea Basin – The sedimentary record of a collapsed ice-sheet margin. *Quaternary Science Reviews* 8, 307-351.

Fairbanks, R.G. (1989) A 17,000 year glacio-eustatic sea level record: influence of glacial melting rates on the Younger Dryas event and deep ocean circulation. *Nature* 342, 637-642.

Flather, R.A. and **Smith, J.** (1993) Recent progress with storm surge models – results from January and February 1993. In: *Proceedings of the MAFF Conference on River and Coastal Engineers*, University of Loughborough, pp. 6.2.1-6.2.16.

Godwin, H. (1959) Studies of the post-glacial history of British vegetation, XIV. Late-glacial deposits at Moss Lake, Liverpool. *Philosophical Transactions of the Royal Society* B 242, 127-149.

Gowen, R.J., Mills, D.K., Trimmer, M., and **Nedwell, D.B.** (2000) Production and its fate in two coastal regions of the Irish Sea: the influence of anthropogenic nutrients. *Marine Ecology Progress Series* 208, 51–64.

Gresswell, R.K. (1937) The Geomorphology of the South-West Lancashire coast-line. *Geographical Journal* 90, 335-348.

Gresswell, R.K. (1953) *Sandy shores in south Lancashire*. University Press, Liverpool.

Gresswell, R.K. (1957) Hillhouse coastal deposits in south Lancashire. Liverpool and Manchester *Geological Journal* 2, 60-78.

Gresswell, R.K. (1964) The origin of the Mersey and Dee estuaries. *Geological Journal* 4, 77–85.

Halliwell, A.R. (1973) Residual Drift near the Sea Bed in Liverpool Bay: an Observational Study. *Geophysical Journal International*, 32(4), 439-458.

Heaps, N.S. (1973) Three-dimensional numerical model of the Irish Sea. *Geophysical Journal of the Royal Astronomical Society* 35, 99-120.

Heaps, N.S. (1983) Storm surges, 1967-1982. *Geophysical Journal of the Royal Astronomical Society* 74, 331-376.

Heaps, N.S. and **Jones, J.E.** (1969) Density currents in the Irish Sea. *Geophysical Journal of the Royal Astronomical Society* 51, 393-429.

Heaps, N.S. and **Jones, J.E.** (1975) Storm surge computations for the Irish Sea using a three-dimensional numerical model. *Memoires de la Société Royale des Sciences Liége* 7, 289-333.

Heaps, N.S. and **Jones, J.E.** (1979) Recent storm surges in the Irish Sea. In: Nihoul, C.J. (ed.), *Marine Forecasting, Proceedings of the 10th International Liege Colloquium on Ocean Hydrodynamics, 1978*, Elsevier, Amsterdam, pp.285-319.

Hetherington, J.A. (1976) *Radioactivity in surface and coastal waters of the British Isles,* 1974. Rep. FRL-11. Ministry of Agriculture, Fisheries and Food, Lowestoft.

Hensley, R.T. (1996) A preliminary survey of benthos from the Nephrops norvegicus mud grounds in the north-western Irish Sea. *Estuarine, Coastal and Shelf Science* 42, 457–465.

Holt, J.T. and **Proctor, R.** (2003) The role of advection in determining the temperature structure of the Irish Sea. *Journal of Physical Oceanography* 33, 2288-2306.

Horsburgh, K.J. and **Hill, A.E.** (2003) A three-dimensional model of the density-driven circulation in the Irish Sea. *Journal of Physical Oceanography* 33, 343-365.

Howarth, M.J. (1984) Currents in the eastern Irish Sea. *Oceanography and Marine Biology: An Annual Review* 22, 11-53.

Howarth, M.J. and **Jones, J.E.** (1981) A comparison of numerical model and observed currents during a storm surge period. *Estuarine, Coastal and Shelf Science* 12, 655-663.

HR (1986) Wave Recording off Prestatyn. Report EX 1369, *Hydraulics Research*.

HRS (1958) *Radioactive Tracers for the study of Sand Movements.* Report on an experiment carried out in Liverpool Bay, Hydraulics Research Station, 1958.

HRS (1969a) *The Southwest Lancashire Coastline, A Report on the Sea Defences.* Hydraulics Research Station, Wallingford. Report no. EX 450, HMSO, 43.

HRS (1969b) *The Computation of Littoral Drift.* Hydraulics Research Station, Wallingford. Report no. EX 449, HMSO.

HRS (1977) *Sand winning at Southport, calculation of wave heights.* Report No. EX 785.

HR Wallingford (1990) *Joint Probability of Waves and Water Levels on the North Wales Coast.* Report Ex 2133, May 1990.

HR Wallingford (2000) *Sediment Transport and the Effects of Extraction.* Fylde Foreshore, Lancashire. Report EX 4152.

HR Wallingford (2002) *Crosby Marine Lake to Formby Point Defence Strategy Summary, Part I – Preliminary Studies.* Report EX 4496, August 2002.

HR Wallingford (2006a) *Crosby Marine Lake to Formby Point Coastal Defence Strategy, Part II: Interim Studies.* Report EX 5276, Release 3.0, November 2006.

HR Wallingford (2006b) *Crosby Marine Lake to Formby Point Coastal Defence Strategy, Part II: Interim Studies- Numerical Modelling.* Report EX 5273, Release 1.0, January 2006.

Huddart, D. (1992) Coastal environmental changes and morphostratigraphy in southwest Lancashire, England. *Proceedings of the Geologists' Association* 103, 217-236.

Huddart, D., Tooley, M.J. and **Carter, P.A.** (1977) *The coasts of north-west England. In: The Quaternary History of the Irish Sea,* Kidson, C. and Tooley, M.J. (eds.). Seel House Press, Liverpool, 119-154.

Huddart, D., Gonzalez, S. and **Roberts G.** (1999a) The archaeological record and mid-Holocene marginal coastal palaeoenvironments around Liverpool Bay. *Quaternary Proceedings* 7, 563-574.

Huddart, D., Roberts, G. and **Gonzalez, S.** (1999b) Holocene human and animal footprints and their relationships with coastal environmental change, Formby Point, NW England. *Quaternary International* 55, 29-41.

ICES (1988) *The status of current knowledge on anthropogenic influences in the Irish Sea.* In: Dickson, R.R. and Boelens, R.G.V. (eds.), ICES cooperative research report (88 pp.) Vol. 155. Palaegade 2–4, Copenhagen, Denmark.

Innes, J.B., Tooley, M.J. and **Tomlinson, P.R.** (1989) A comparison of the age and palaeoecology of some sub-Shirdley Hill sand peat deposits from Merseyside and south-west Lancashire. *Naturalist* 114, 65-69.

Jackson, D.I., Jackson, A.A., Evans D., Wingfield, R.T.R., Barnes, R.P. and **Arthur M.J.** (1995) *The Geology of the Irish Sea.* British Geological Survey, UK, Offshore Regional Report, HMSO, London, 123p.

Jay, H. (1998) *Dune Sediment Exchange and Morphodynamic Responses: Implications for Shoreline Management.* PhD thesis, University of Reading.

Johnson, L.R. (1983) The transport mechanisms of clay and fine silt in the North Irish Sea. *Marine Geology* 52, M33-M41,

Jones, J.E. (1980) Computed distributions of elevation and current for the major Irish Sea storm surges of November 1977. *Institute of Oceanographic Sciences, Report*, No 101, 145p.

Jones, J.E. and **Davies, A.M.** (1996) A high resolution three dimensional model of the M2, M4, M6, S2, N2, K1 and O1 tides in the eastern Irish Sea. *Estuarine, Coastal and Shelf Science* 42, 311-346.

Jones, J.E. and **Davies A.M.** (1997) Storm surge computations for the eastern Irish Sea including wave-current interaction. *Annales Geophysicae*, 15, Supplement II, C379.

Jones, J.E. and **Davies, A.M.** (1998) Storm surge computations for the Irish Sea using a three-dimensional numerical model including wave-current interaction. *Continental Shelf Research* 18, 201-251.

Jones, J.E. and **Davies, A.M.** (2001) Influence of wave –current interaction and high frequency forcing upon storm induced currents and elevations. *Estuarine, Coastal and Shelf Science* 53, 397-413.

Jones, J.E. and **Davies, A.M.** (2003a) Processes influencing storm-induced currents in the Irish Sea. *Journal of Physical Oceanography* 33, 88-104.

Jones, J.E. and **Davies, A.M.** (2003b) On combining current observations and models to investigate the wind induced circulation of the eastern Irish Sea. *Continental Shelf Research* 23, 415-434.

Jones, J.E. and **Davies, A.M.** (2004a) On the sensitivity of computed surges to open boundary formulation. *Ocean Dynamics* 54, 142-162.

Jones, J.E. and **Davies, A.M.** (2004b) Influence of wind field and open boundary input upon computed negative surges in the Irish Sea. *Continental Shelf Research* 24, 2045-2064.

Kershaw, P.J. (1986) Radiocarbon dating of Irish Sea sediments. *Estuarine, Coastal and Shelf Science* 23, 295-303.

Kershaw, P.J., Swift, D.J. and **Denoon, D.C.** (1988) Evidence of recent sedimentation in the eastern Irish Sea. *Marine Geology* 85(1), 1-14.

King, C.A.M. (1972) *Beaches and Coasts*, 2nd Edition. Edward Arnold, London, 570.

King, C.A.M. and **Barnes, F.A.** (1964) Changes in the configuration of the inter-tidal beach zone of part of the Lincolnshire coast since 1951. *Zeitschrift für Geomorphologie* 8, 105-126.

King, C.A.M. and **Williams, W.W.** (1949) The formation and movement of sand bars by wave action. *Geographical Journal* 113, 70-85.

Kirby, R., Parker, W.R., Pentreath, R.J. and **Lovett, M.B.** (1983) *Sedimentation studies relevant to low-level radioactive effluent dispersal in the Irish Sea. Part III. An evaluation of possible mechanisms for the incorporation of radionuclides into marine sediments.* Institute of Oceanographic Sciences Report 178, 63.

Kirby, R. (1987) Sediment Exchanges across the coastal margins of NW Europe. *Journal of the Geological Society* 144, 121-126.

Krivtsov, V., Howarth, M.J., Jones, S.E., Souza, A.J. and **Jago, C.F.** (2008) Monitoring and modelling of the Irish Sea and Liverpool Bay: An overview and an SPM case study. *Ecological Modelling* 212, 37-52.

Kroon, A. and **DeBoer, A.G.** (2001) Horizontal flow circulation on a mixed energy beach. *Proceedings of Coastal Dynamics '01*, ASCE, 548-557.

Kroon, A. and **Masselink, G.** (2002) Morphodynamics of intertidal bar morphology on a macrotidal beach under low-energy wave conditions, North Lincolnshire, England. *Marine Geology* 190, 591-608

Lambeck, K. (1991) Glacial rebound and sea-level change in the British Isles. *Terra Nova* 3(4), 379-389.

Lambeck, K. (1993) Glacial rebound of the British Isles I: preliminary model results. *Geophysical Journal International* 115(3), 941-959.

Lambeck, K. (1996) Glaciation and sea-level change for Ireland and the Irish Sea since late Devensian/Midlandian time. *Journal of the Geological Society of London* 153, 853-872.

Lambeck, K. and **Purcell, A.P.** (2001) Sea-level change in the Irish Sea since the last glacial maximum; constraints from isostatic modelling. *Journal of Quaternary Science* 16, 497-506.

Lamplugh, G.W. (1903) Geology of the Isle of Man, *Memoir Geological Survey of Great Britain*.

Lee, J.C. and **Davies, A.M.** (2001) Influence of data assimilation upon M2 tidal elevations and current profiles in the Irish Sea. *Journal of Geophysical Research* 106, 30961-30986.

Lennon, G.W. (1963) The identification of weather conditions associated with the generation of major storm surges along the west coast of the British Isles. *Quarterly Journal of the Royal Meteorological Society* 89, 381-394.

Liverpool Bay Coastal Group (1999) *Shoreline Management Plan, Sub-Cell 11a, Great Orme's Head to Formby Point. Stage 1, Volume 2*, Consulation Document. Shoreline Management Partnership, 309p.

Masselink, G. (1993) Simulating the effects of tides on beach morphodynamics. *Journal of Coastal Research Special Issue* 15, 180-197.

Masselink, G. (2004) Formation and evolution of multiple intertidal bars on macrotidal beaches: application of a morphodynamic model. *Coastal Engineering* 51, 713-730.

Masselink, G. and **Anthony, E.J.** (2001) Location and height of intertidal bars on macrotidal ridge and runnel beaches. *Earth Surface Processes and Landforms* 26, 759-774.

Masselink, G. and **Turner, I.L.** (1999) The effect of tides on beach morphodynamics. In: Short, A.D. (ed.), *Handbook of Beach and Shoreface Morphodynamics*, Wiley, Chichester, pp. 204-229.

Masselink, G., Kroon, A. and **Davidson-Arnott, R.G.D.** (2006) Morphodynamics of intertidal bars in wave-dominated coastal settings – a review. *Geomorphology* 73, 33-49.

Mauchline, J. (1980) Artificial radioisotopes in the marginal seas of north-western Europe. In: F.T. Banner, M.B. Collins and K.S. Massie (eds.), *The North-West European Shelf Seas: the Sea Bed and the Sea in Motion. II. Physical and Chemical Oceanography, and Physical Resources*. Elsevier, Amsterdam, pp. 517-542.

McCabe, A.M. (1997) Geological constraints on geophysical models of relative sea-level change during deglaciation of the western Irish Sea Basin. *Journal of the Geological Society of London* 154, 601-604.

McCabe, A.M. (2008) Comment: Postglacial relative sea-level observations from Ireland and their role in glacial rebound modelling. *Journal of Quaternary Science* 23(8), 817-820.

McCabe, A.M., Cooper, J.A.G. and **Kelley, J.T.** (2007) Relative sea level change from NE Ireland during the last glacial termination. *Journal of the Geological Society of London* 164, 1059-1063.

McCarroll, D. (2001) Deglaciation of the Irish Sea Basin: a critique of the glacimarine hypothesis. *Journal of Quaternary Science* 16, 393-340.

McCarroll, D. and **Harris, C.** (1992) The glacigenic deposits of western Lleyn, North Wales: terrestrial or marine? *Journal of Quaternary Science* 7, 19-29.

McDowell, D.M. and **O'Connor, B.A.** (1977) *The Hydraulic Behaviour of Estuaries*. Macmillan, London.

McLaren, P. (1989) *The sediment transport regime in Morecambe Bay and the Ribble estuary.* Report commissioned by North West Water from GeoSea Consulting Ltd, Cambridge, U.K.

Mitchell, G.F. (1972) The Pleistocene history of the Irish Sea: second approximation. *Scientific Proceedings of the Royal Dublin Society* 4A, 181-200

Murthy, T.K.S. and **Cook, J.** (1962) *Maximum wave heights in Liverpool Bay.* Vickers Armstrong, Department of design, Report No. V3031/HYDRO/04.

Neal, A. (1993) *Sedimentology and morphodynamics of a Holocene coastal dune barrier complex, Northwest England.* PhD thesis, Reading University.

Newton, M., Lymbery, G. and **Wisse, P.** (2008) *A Description of Time Series Spatial Data Held and Used Within the Sefton Coast Database.* Sefton Council Technical Services Department, Bootle, 21p.

O'Connor, B.A. (1987) Short and long term changes in estuary capacity. *Journal of the Geological Society of London* 144, 187-195.

Osuna, P., Souza, A.J., and **Wolf, J.** (2007) Effects of the deep-water wave breaking dissipation on the wind-wave modelling in the Irish Sea. *Journal of Marine Systems* 67, 59-72.

Pantin, H.M. (1977) Quaternary sediments of the northern Irish Sea. In: C. Kidson and M.J. Tooley (eds.), *The Quaternary History of the Irish Sea*. Seel House, Liverpool, pp. 27 54.

Pantin, H.M. (1978) Quaternary sediments from the northeast Irish Sea: Isle of Man to Cumbria. *Bulletin of the Geological Survey of Great Britain*, 64, 43.

Parker, W.R. (1971) *Aspects of the Marine Environment at Formby Point, Lancashire.* PhD thesis, University of Liverpool, two volumes.

Parker, W.R. (1974) Sand transport and coastal stability, Lancashire, UK. In: *Coastal Engineering 1974,* Proceedings of the 14th Coastal Engineering Conference, June 24-28th, Copenhagen, Denmark. American Society of Civil Engineers, New York, pp. 828-851.

Parker, W.R. (1975) *Sediment mobility and erosion on a multibarred foreshore (Southwest Lancashire, U.K.).* In: J. Hails, and A. Carr (eds.), Nearshore Sediment Dynamics and Sedimentation. London. John Wiley and Sons, pp. 151-179.

Peltier, W.R., Shennan, I., Drummond, R. and **Horton, B.** (2002) On the postglacial isostatic adjustment of the British Isles and the shallow viscoelastic structure of the Earth. *Geophysical Journal International* 148, 443-475.

Picton, J.A. (1849) The changes of sea-levels on the west coast of England during the historic period. (Abstract) *Proceedings of the Literary and Philosophical Society of Liverpool* 36th session 5, pp. 113-115.

Plater, A.J. (2004) Late Quaternary relative sea-level trends in the Irish Sea region. In: Chiverrell, R.C., Plater, A.J., and Thomas, G.S.P. (eds.), The Quaternary of the Isle of Man and North West England: Field Guide. *Quaternary Research Association*, London, 11-20.

Plater, A.J., Huddart, D., Innes, J.B., Pye, K., Smith, A.J. and **Tooley, M.J.** (1999) Coastal and sea-level changes. In: *The Sand Dunes of the Sefton Coast*, D. Atkinson and J.A. Houston (eds.). National Museums and Galleries on Merseyside in association with Sefton Metropolitan Borough Council, Liverpool, pp. 23-34.

Plater, A.J., Hodgson, D., Newton, M. and **Lymbery, G.** (This volume) Sefton South Shore: Understanding Coastal Evolution from Past Changes and Present Dynamics. In: *Sefton's Dynamic Coast: proceedings of the conference on coastal geomorphology, biogeography and management 2008* Worsley, A.T., Lymbery, G., Holden, V.J.C. and Newton, M. (eds.) Coastal Defence: Sefton MBC Technical Services.

POL (1995) Estimates of extreme sea conditions: Extreme sea-levels at the UK A-Class sites – optimal site-by-site analyses and spatial analyses for the east coast. Document 72, Proudman Oceanographic Laboratory.

POL (1997) Estimates of extreme sea conditions; Spatial analyses for the UK coast. Internal Document 112, Proudman Oceanographic Laboratory.

Prandle, D. (1984) A modelling study of the mixing of 137Cs in the seas of the European continental shelf. *Philosophical Transactions of the Royal Society of London* A 310, 407-436.

Price, W.A. and **Kendrick, M.P.** (1963) Field and model investigations into the reasons for siltation in the Mersey estuary. *Journal of the Institution of Civil Engineers* 24, 473-517.

Proctor, R. (1981) *Tides and residual circulation in the Irish Sea: a numerical modelling approach.* PhD thesis, University of Liverpool, 248.

Pye, K. (1977) *An analysis of coastal dynamics between the Ribble and Mersey estuaries, with particular reference to erosion at Formby Point.* BA dissertation, University of Oxford, 137.

Pye, K. (1990) *Physical and human influences on coastal dune development between the Ribble and Mersey estuaries, northwest England.* In: Nordstrom, K.F., Psuty, N.P. and Carter, R.W.G. (eds.), Coastal Dunes: Forms and Process. Chichester. Wiley, pp. 339-359.

Pye, K. and **Blott, S.J.** (2008) Decadal-scale variation in dune erosion and accretion rates: an investigation of the significance of changing storm tide frequency and magnitude on the Sefton coast, UK. *Geomorphology* 102(3-4), 652-666.

Pye, K. and **Blott, S.J.** (This volume) Geomorphology of the Sefton Coast sand dunes. In: *Sefton's Dynamic Coast: proceedings of the conference on coastal geomorphology, biogeography and management 2008* Worsley, A.T., Lymbery, G., Holden, V.J.C. and Newton, M. (eds) Coastal Defence: Sefton MBC Technical Services.

Pye, K. and **Neal, A.** (1993a) Late Holocene dune formation on the Sefton coast, northwest England. In: Pye, K. (ed.), *The Dynamics and Environmental Context of Aeolian Sedimentary Systems*, Geological Society of London, Special Publication 72, Geological Society Publishing House, Bath, pp. 201-217.

Pye, K. and **Neal, A.** (1993b) Stratigraphy and age structure of the Sefton dune complex: preliminary results of field drilling investigations. In: Atkinson, D. and Houston, J. (eds.), *The Sand Dunes of the Sefton Coast.* National Museums and Galleries on Merseyside in association with Sefton Metropolitan Borough Council, Liverpool, pp. 41-44.

Pye, K., and **Neal, A.** (1994) Coastal dune erosion at Formby Point, north Merseyside, England: Causes and Mechanisms. *Marine Geology* 119, 39-56.

Pye, K. and **Smith, A.J.** (1988) Beach and dune erosion and accretion on the Sefton Coast, Northwest England. In: Psuty, N.P. (ed.) Dune / Beach Interaction. *Journal of Coastal Research*, Special Issue 3, 33-36.

Pye, K., Blott, S.J., Short, B. and Whitton, S.J. (2006) *Preliminary investigation of sea bed sediment characteristics in Liverpool Bay.* Ken Pye Associates Ltd, External Research Report No.ER602, Crowthorne, Berkshire.

Ramster, J.W. and Hill, H.W. (1969) The current systems in the northern Irish Sea. *Nature* 224, 59-61.

Reade, T.M. (1871) The geology and physics of the post-glacial period, as shewn in the deposits and organic remains in Lancashire and Cheshire. *Proceedings of the Liverpool Geological Society* 2, 36-88.

Roberts, D.H., Chiverrell, R.C., Innes, J.B., Horton, B.P., Brooks, A.J., Thomas, G.S.P., Turner, S. and Gonzalez, S. (2006) Holocene sea levels, Last Glacial Maximum glaciomarine environments and geophysical models in the northern Irish Sea Basin, UK. *Marine Geology* 231, 113-128.

Russell, P.E. and Huntley, D.A. (1999) A cross-shore transport 'shape function' for high energy beaches. *Journal of Coastal Research* 15, 295-205.

Scourse, J.D. and Furze, M.F.A. (2001) A critical review of the glaciomarine model for Irish Sea deglaciation: evidence from southern Britain, the Celtic shelf and adjacent continental slope. *Journal of Quaternary Science* 16, 419-434.

Sefton MBC (2010) *Northwest Strategic Monitoring Programme: Data Report.* Sefton MBC, Bootle.

Sharaf El Din, S.H. (1970) Some oceanographic studies on the west coast of England. *Deep-Sea Research* 17, 647-654.

Sharples, J. (1992) *Time dependent stratification in regions of large horizontal gradient.* PhD thesis, University of Wales.

Sharples, J. and Simpson, J.H. (1995) Semi-diurnal and longer period stability cycles in the Liverpool bay region of freshwater influence. *Continental Shelf Research* 15, 295-313.

Shennan, I. and Horton, B.P. (2002) Holocene land- and sea-level changes in Great Britain. *Journal of Quaternary Science* 17(5-6), 511-526.

Shennan, I. Peltier, W.R., Drummond, R. and Horton, B.P. (2002) Global to local scale parameters determining relative sea-level changes and post-glacial isostatic adjustment of Great Britain. *Quaternary Science Reviews* 21, 397-408.

Shennan, I., Bradley, S., Milne, G., Brooks, A., Bassett, S. and Hamilton, S. (2006) Relative sea-level changes, glacial isostatic modelling and ice-sheet reconstructions from the British Isles since the Last Glacial Maximum. *Journal of Quaternary Science* 21(6), 585-599.

Short, A.D. and Hesp, P.A. (1982) Wave, beach and dune interactions in southeastern Australia. *Marine Geology* 48, 259-284.

Simpson, J.H. (1971) Density stratification and microstructure in western Irish Sea. *Deep-Sea Research* 18(3), 309-319.

Simpson, J.H. and Hunter, J.R. (1974) Fronts in the Irish Sea. *Nature* 250, 404-406.

Simpson, J.H., Brown, J. Matthews, J. and Allen, G. (1990) Tidal straining, density currents, and stirring in the control of estuarine stratification. *Estuaries* 13, 125-132.

Sly, P.G. (1966) *Marine geological studies in the Eastern Irish Sea and adjacent estuaries, with special reference to sedimentation in Liverpool Bay and the River Mersey.* PhD thesis, University of Liverpool.

Smith, A.J. (1982) *A Guide to the Sefton Coast Database.* Sefton Borough Council Engineer and Surveyor's Department, Bootle, 88p. + figures.

Stride, A.H. (Editor) (1982) *Offshore Tidal Sands.* Chapman and Hall, London, 222p.

Sunamura, T. and **Takeda, I.** (1984) Landward migration of inner bars. *Marine Geology* 60, 63-78.

Thomas, C.G., Spearman, J.R. and **Turnbull, M.J.** (2002) Historical morphological change in the Mersey estuary. *Continental Shelf Research* 22, 1775-1794

Thomas, G.S.P. (1985) The Quaternary of the northern Irish Sea Basin. In: Johnson, R.H. (ed.), *The Geomorphology of North-West England*, Manchester University Press, Manchester, pp.143-158.

Thomas, G.S.P. and **Dackombe, R.V.** (1985) Comment on 'Glaciomarine sediments of the Isle of Man as a key to late Pleistocene stratigraphic investigations in the Irish Sea Basin'. *Geology* 13, 445-447.

Thomas, G.S.P., Chiverrell, R.C. and **Huddart, D.** (2004) Ice-marginal depositional responses to probable Heinrich events in the Devensian deglaciation of the Isle of Man. *Quaternary Science Reviews* 23, 85-106.

Tooley, M.J. (1970) The peat beds of the southwest Lancashire coast. *Nature in Lancashire* 1, 19-26.

Tooley, M.J. (1974) Sea-level changes during the last 9000 years in northwest England. *Geographical Journal* 140, 18-42.

Tooley, M.J. (1976) Flandrian sea-level changes in west Lancashire and their implications for the 'Hillhouse Coastline'. *Geological Journal* 11, 137-152.

Tooley, M J. (1978a). *Sea-level changes in north west England during the Flandrian Stage*. Oxford Research Studies in Geography, Clarendon Press, Oxford.

Tooley, M.J. (1978b). Interpretation of Holocene sea-level changes. *Geologiska Föreningens i Stockholm Förhandlingar* 100, 203-212.

Tooley, M.J. (1980) Theories of coastal change in North-West England. In: Thompson, F.H. (ed.), *Archaeology and Coastal Change*. Society of Antiquaries of London, Occasional Papers New Series No.1, London, pp. 74-86.

Tooley, M.J. (1982) Sea-level changes in northern England. *Proceedings of the Geologists' Association* 93, 43-51.

Tooley, M.J. (1985) Sea-level changes and coastal morphology in northwest England. In: Johnson, R.H. (ed.), *The Geomorphology of Northwest England*. Manchester University Press, Manchester, pp. 94-121.

Tooley, M.J. (1990) The chronology of coastal dune development in the United Kingdom. In: Bakker, Th.W.M., Jungerius, P.D. and Klijn, J.A. (eds.), *Dunes of the European coasts*. Catena Supplement 18, 81-88.

Tooley, M.J. (1992) Recent sea-level changes. In: Allen, J.R.L. and Pye, K. (eds.), *Saltmarshes: Morphodynamics, Conservation and Engineering Significance*. Cambridge University Press, Cambridge, pp. 19-40.

Tooley, M.J. and **Kear, B.S.** (1977) Mere Sands Wood (Shirdley Hill Sand). In: Bowen, D.Q. (ed.), *The Isle of Man, Lancashire coast and Lake District*. INQUA X Congress. Geo Abstracts, Norwich, pp. 9-10.

Travis, C.B. (1926) The peat and forest bed of the south-west Lancashire coast. *Proceedings of the Liverpool Geological Society* 14, 263.

Trimmer, M., Gowen, R.J., Stewart, B.M., and **Nedwell, D.B.** (1999) The spring bloom and its impact on benthic mineralisation rates in western Irish Sea sediments. *Marine Ecology Progress Series* 185, 37-46.

Tucker, M.J. (1963) Analysis of records of sea waves. *Proceedings of the Institution of Civil Engineers* 26, 304-316.

Uehara, K., Scourse, J.D., Horsburgh, K.J., Lambeck, K. and Purcell, A.P. (2006) Tidal evolution of the northwest European shelf seas from the Last Glacial Maximum to the present. *Journal of Geophysical Research* 111, C09025, doi:10.1029/2006JC003531.

van der Wal, D., Pye, K. and Neal, A. (2002) Long-term morphological change in the Ribble estuary, northwest England. *Marine Geology* 189, 249-266.

van Landeghem, K.J.J., Uehara, K., Wheeler, A.J., Mitchell, N.C. and Scourse, J.D. (2009) Post-glacial sediment dynamics in the Irish Sea and sediment wave morphology: data-model comparisons. *Continental Shelf Research* 29, 1723-1736.

Williams, S.J., Kirby, R., Smith, T.J. and Parker, W.R. (1981) *Sedimentation studies relevant to low-level radioactive effluent dispersal in the Irish Sea. Part II. Sea bed morphology, sediments and shallow sub-bottom stratigraphy of the eastern Irish Sea.* Rep. Inst. Oceanogr. Sci., Wormley, No. 120, 50.

Wilson, P., Bateman, R.M. and Catt, J.A. (1981) Petrography, origin and environment of deposition of the Shirdley Hill Sand of southwest Lancashire, England. *Proceedings of the Geologists' Association* 92, 211-229.

Wingfield, R.T.R. (1992) Quaternary changes of sea level and climate. *The Irish Sea,* Irish Sea Forum Seminar Report, Global Warming and Climatic Change, Liverpool University Press, Liverpool, pp. 56-66.

Wingfield, R.T.R. (1995) A model of sea-level in the Irish and Celtic seas during the end-Pleistocene to Holocene transition. In: *Island Britain: a Quaternary Perspective*, R.C. Preece (ed.). Geological Society of London, Special Publication 96. Geological Society Publishing House, Bath, pp. 209-242.

Wolf, J., Wakelin, S.L. and Holt, J.T. (2002) A coupled model of waves and currents in the Irish Sea. *Proceedings of the Twelfth (2002) International Offshore and Polar Engineering Conference* Vol 3, 108-114.

Woodworth, P.L. and Blackman, D. L. (2002) Changes in extreme high waters at Liverpool since 1768. *International Journal of Climatology* 22(6), 697-714.

Wright, J.E., Hull, J.H., McQuillin, R. and Arnold, S.E. (1971) *Irish Sea investigations 1969-70.* Report of the Institute of Geological Sciences No. 71/19, 43p.

Wright, W.B. (1914) *The Quaternary Ice Age.* Macmillan, London.

Wright, P. (1976) *The morphology, sedimentary structures and processes of the foreshore at Ainsdale, Merseyside, England.* PhD thesis, University of Reading.

Xing, J. and Davies, A.M. (2001) A three-dimensional baroclinic model of the Irish Sea: formation of the thermal fronts and associated circulation. *Journal of Physical Oceanography* 31, 94-114.

The historic development of the north Sefton Coast

Vanessa J.C. Holden

Department of Natural, Geographical & Applied Sciences
Edge Hill University, Ormskirk, Lancashire, L39 4QP
holdenvj@edgehill.ac.uk

ABSTRACT

A vast archive of information exists concerning the north Sefton coast, collected and maintained by Sefton Metropolitan Borough Council. This paper gives an initial account of the information reviewed from within this archive, which dates from early navigation charts of the 1700s, through the Victorian era of controlling the environment, through to the present day, where data regarding the coastal environment continues to be collected and analysed. The form of the information reviewed ranges from maps, to historic books, to survey data, to 'grey literature' such as correspondence and meeting notes. The majority of the information has been collected by various Borough Engineers or Engineers employed by the Ribble and Mersey Navigation Companies. As a result, this paper is focused upon the three main topics of: (i) navigation, including the operation of the Port of Preston, and the dredging and sediment dynamics of both the Ribble and Mersey estuaries; (ii) reclamation of land, predominantly for agricultural purposes; and (iii) the urban development of Southport, particularly changes in the foreshore, development of salt marsh to the north, and changes in the Bog Hole Channel. Issues of data confidence of the archive are also briefly considered.

INTRODUCTION

As concerns grow about the impacts of climate change and the likely scenarios resulting from rising sea levels, so increasingly, research programmes based in coastal zones are attempting to unravel the nature of sedimentary and ecological changes and their relationships to anthropogenic activity (Smith et al., 1998; Adam, 2002; Crooks et al., 2002; Pethick, 2002; Thompson et al., 2002; Warren et al., 2002; Woodroffe, 2002; Cooper and Pilkey, 2004; Cooper, 2007). Planning for changes in coastal zones therefore requires an awareness of the nature of developments, both natural and anthropogenic, upon which to base present and future understanding. This paper, therefore, sets out to provide a concise account of the history of the north Sefton coast for the last 200 years.

The historic information available for the north Sefton coast, held by Sefton Metropolitan Borough Council, comprises a vast archive of knowledge, to an extent that allows a detailed picture of the development of the coastline to be constructed. Such an archive of detailed information is rare (Thomas et al., 2002). The information dates from the time that records and evidence of the landscape began to be collected in sufficient detail to accurately show its relatively natural and un-altered state; with the beginnings of human intervention in the nineteenth century; through the later 1800s, with the Victorian ethos of claiming and training the land; through the twentieth century with increasing awareness of the implications of human activities on natural processes; up to the present day situation with the need for sustainable strategic planning for the future. Table 1 shows the main types of secondary data that are available for the north Sefton coast held by Sefton Metropolitan Borough Council (SMBC).

Changes in the form of the Ribble estuary during the late 1800s, through into the twentieth century, led to an increasing awareness of the transitory nature of many estuarine features, and the implications of human interaction upon the natural processes. Much of the information held within the archive relates to either the training of the navigation channel; to the reclamations made on the southern side of the estuary; or to the Bog Hole Channel at Southport. This is largely due to the source and nature of the information collated, being from archives of the Borough Engineers, with much of

Table 1. Type and format of secondary data sources available within the Sefton Metropolitan Borough Council archives.

Data Type	Format	Date Range
Aerial Photographs	Digital and printed original	1945 to present
Ordnance Survey Maps	Digital and printed original	1843 to present
Navigation and Bathymetric Charts	Digital and printed original	1736 to 1937
Historic Photographs	Digital and printed original	c.1900 onwards
Unpublished Reports / 'Grey' Literature	Original	c.1880 onwards
Published Books	Original	1888 onwards
Vegetation Survey Data	Original	1998 to present
Profile Survey Data	Original	1913 to present

the literature being written by the Engineer and General Superintendent of the Ribble Navigation 1901-1933 (James Barron); Southport Borough Engineer, early 1900s (Mr A.E. Jackson); Engineer for Preston Corporation, 1890 until 1900, and later consulting engineer to Mersey Estuary (Mr Alexander F. Fowler); Ribble Engineer 1933 to mid 1900s (Mr A.H. Howarth, successor to James Barron); and consulting engineers (Mr L.F. Vernon-Harcourt, early 1900s, and Commander F.W. Jarrad, 1907).

It is for this reason that the evidence surrounding the development of the location is centred on the three main issues of navigation, reclamation, and the urban development of Southport. The main features discussed in the following text are overlain on the contemporary situation seen in the estuary (Plate 1).

NAVIGATION AND INDUSTRIAL ACTIVITY
Navigation of the Ribble Estuary

Table 2 shows the early stages of commercial developments prior to the opening of the Port of Preston.

From the chart of Fearon and Eyes (1736-7), (Figure 1), and from the range of charts that were produced during the 18th and 19th Centuries shown in Figures 2 to 7, it is clear that navigation of the channels of the estuary were important long before the opening of the Port of Preston in 1892. The pertinent points of each chart are discussed within the box in each figure.

Table 2. Developments prior to the opening of the Port of Preston. Based upon information from Wheeler (1893) and Barron (1938).

Year	Activity
1806	Company of Proprietors obtained Parliamentary Act to improve Ribble Channel between the Naze (SD 343200, 427200) and Preston. Company failed.
1837	New Company formed.
1853	Received power to dredge channel half a mile below Preston, construct 4.5 miles of training walls, build quay walls at Preston, and construct a lighthouse at St. Anne's.
1883	The Corporation of Preston obtained Parliamentary powers to purchase the rights of the Ribble Navigation for the sum of £72,500. Authorisation given to construct Preston docks (initially 36 acres, extended to 40 acres), to dredge between Lytham and Preston (even though it had previously been opposed by landowners), and to extend the training walls a further 3.5 miles (SD 335220, 426420 and 335000, 426073).
1892	The docks subsequently opened on 25th June 1892.

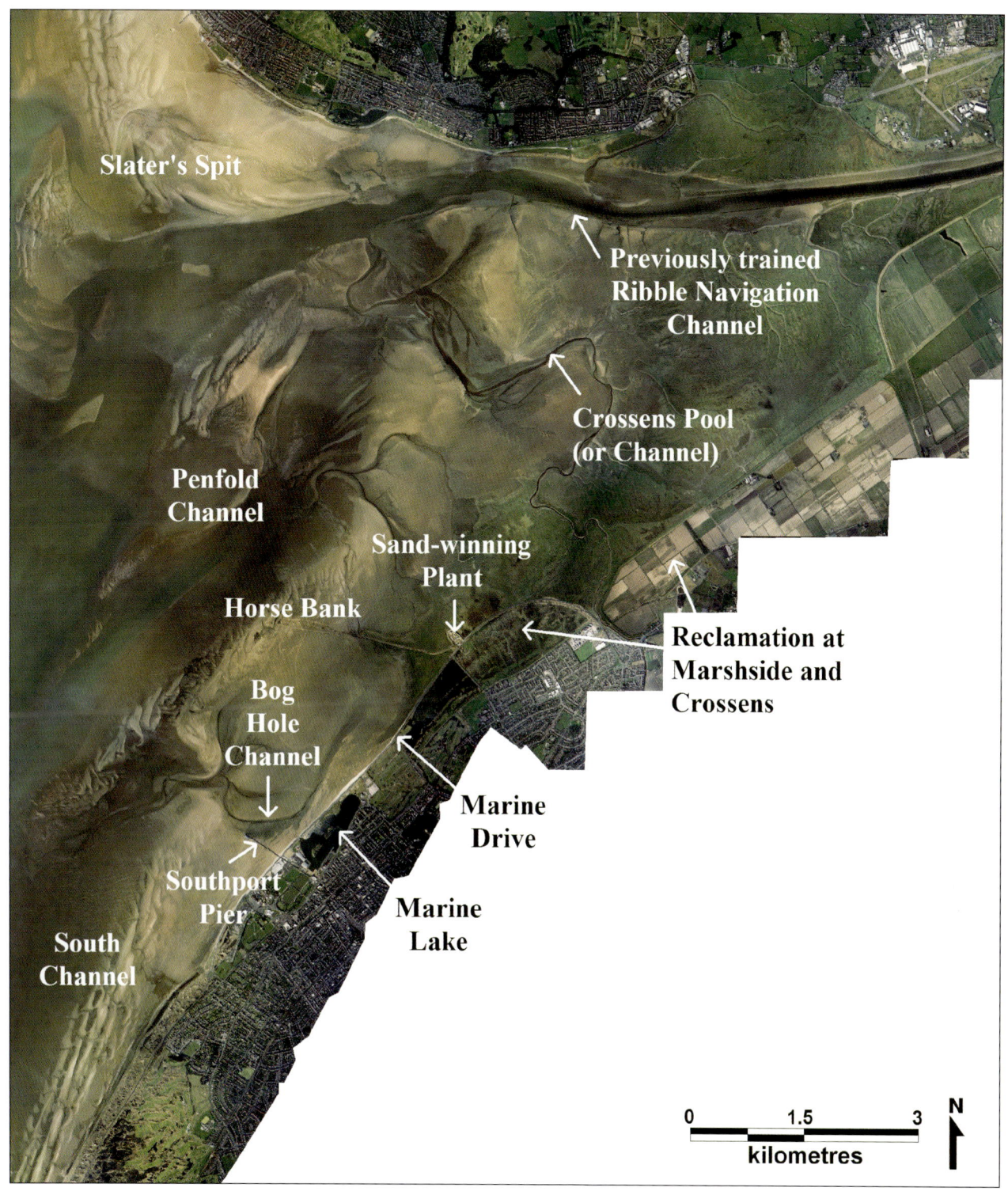

Plate 1. Annotated aerial photograph of present day conditions in the Ribble Estuary, 2002, with features discussed in detail in the text shown. Aerial photograph courtesy of SMBC © Crown copyright. All rights reserved. Sefton Council Licence number 100018192 2010.

The training of the navigable channel had implications for the natural evolution of the estuary and the movement of sediment within it. Between 1904 and 1929, James Barron produced five surveys of the Ribble estuary, (with an additional partial survey in 1935 by A. Howarth, and a subsequent complete survey in 1937), with selected charts shown in Figures 8 and 9. These charts demonstrate the movements of channels and sand banks within the estuary during the first part of the 20th Century. The natural tendency of an estuary to infill is heightened with the introduction of engineering structures such as training walls. The

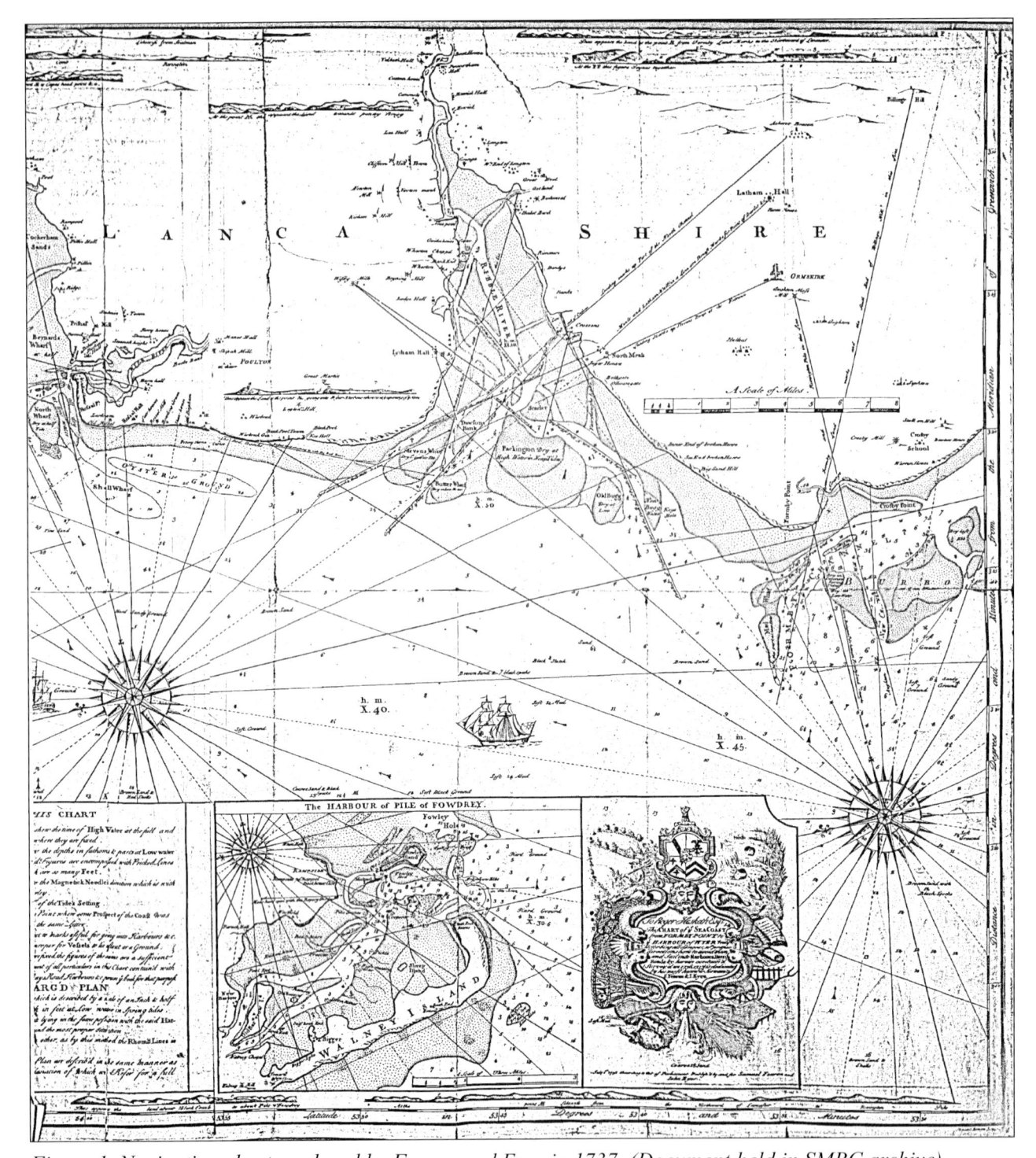

Figure 1. Navigation chart produced by Fearon and Eyes in 1737. (Document held in SMBC archive).

Ribble has a flood tide faster than its ebb with the faster flowing flood tide capable of transporting a greater sediment load than the lower energy ebb tide, thereby moving material into the estuary.

As the Navigation Channel was designed to concentrate the ebb flow to within the trained walls, the ebb flow was resultantly reduced outside the trained channels (Hydraulics Research Station, 1968; van der Wal et al., 2002). There was potentially also a strengthening of the flood tide outside of the main channel, leading to increased sedimentation inshore, subsequently accelerating the process of infilling in the estuary (Barron, 1938; Inglish and Kestner, 1958). Ashton (1920) noted the area of the Estuary covered by water at low tide reduced from 30% (in 1820) to only 5% (in 1910).

Dredging of the Ribble Channel

The earliest recorded dredgings of the Ribble were

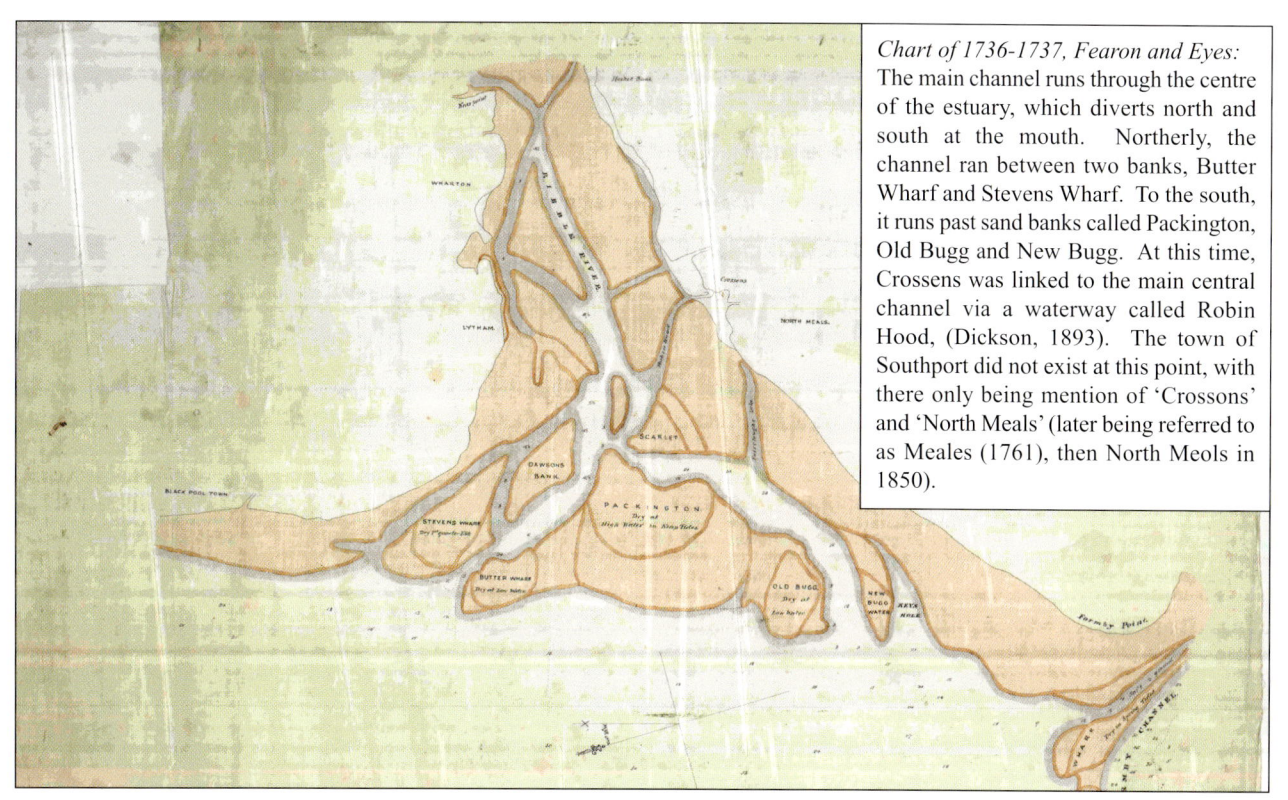

Chart of 1736-1737, Fearon and Eyes:
The main channel runs through the centre of the estuary, which diverts north and south at the mouth. Northerly, the channel ran between two banks, Butter Wharf and Stevens Wharf. To the south, it runs past sand banks called Packington, Old Bugg and New Bugg. At this time, Crossens was linked to the main central channel via a waterway called Robin Hood, (Dickson, 1893). The town of Southport did not exist at this point, with there only being mention of 'Crossons' and 'North Meals' (later being referred to as Meales (1761), then North Meols in 1850).

Figure 2. Chart of the Ribble Estuary, 1736-7, original by Fearon and Eyes. (Document held in SMBC archive).

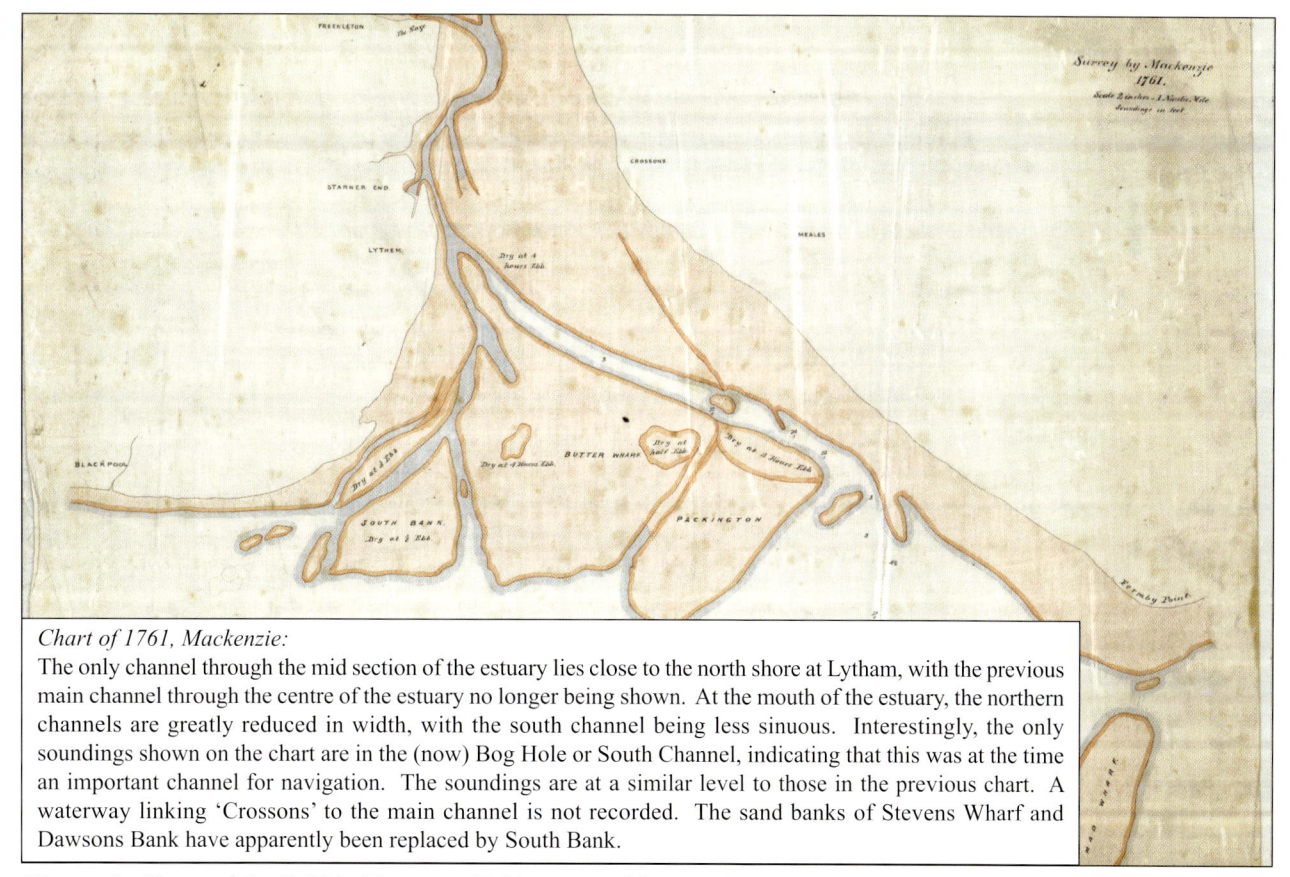

Chart of 1761, Mackenzie:
The only channel through the mid section of the estuary lies close to the north shore at Lytham, with the previous main channel through the centre of the estuary no longer being shown. At the mouth of the estuary, the northern channels are greatly reduced in width, with the south channel being less sinuous. Interestingly, the only soundings shown on the chart are in the (now) Bog Hole or South Channel, indicating that this was at the time an important channel for navigation. The soundings are at a similar level to those in the previous chart. A waterway linking 'Crossons' to the main channel is not recorded. The sand banks of Stevens Wharf and Dawsons Bank have apparently been replaced by South Bank.

Figure 3. Chart of the Ribble Estuary, 1761, original by Mackenzie. (Document held in SMBC archive).

Chart of 1820-1824, Brazier:
The main channel has migrated back to the centre of the estuary. The two channels that ran to the north of the mouth of the estuary now take a more central route to the sea. The previously straight south channel is now occupied by numerous sand banks, and diverges around the sand bank 'Butter Wharf', which has moved further south across the mouth of the estuary from the previous charts. Southport is shown, with a small, but presumably navigable channel leading to it. 'Crossands' once again has a major link with the sea, via the 'Southport Channel', but is no longer connected to the main channel in the mid section of the estuary, (Wheeler, 1893). The area that was previously Butter Wharf and South Bank are referred to as 'Horse Sand'.

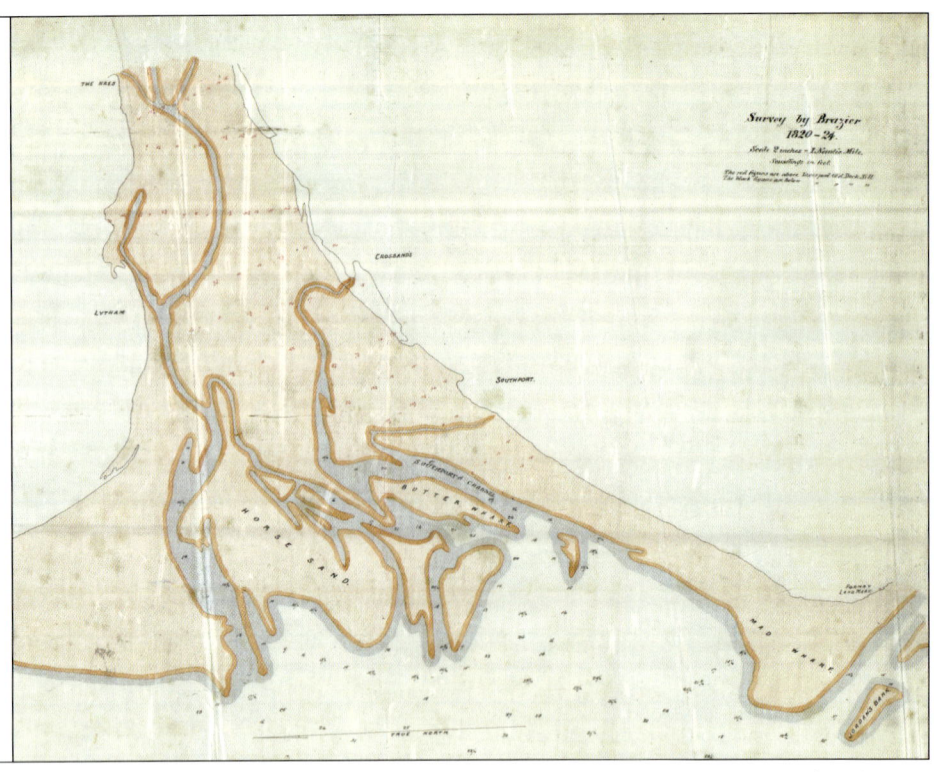

Figure 4. Chart of the Ribble Estuary, 1820-24, original by Brazier. (Document held in SMBC archive).

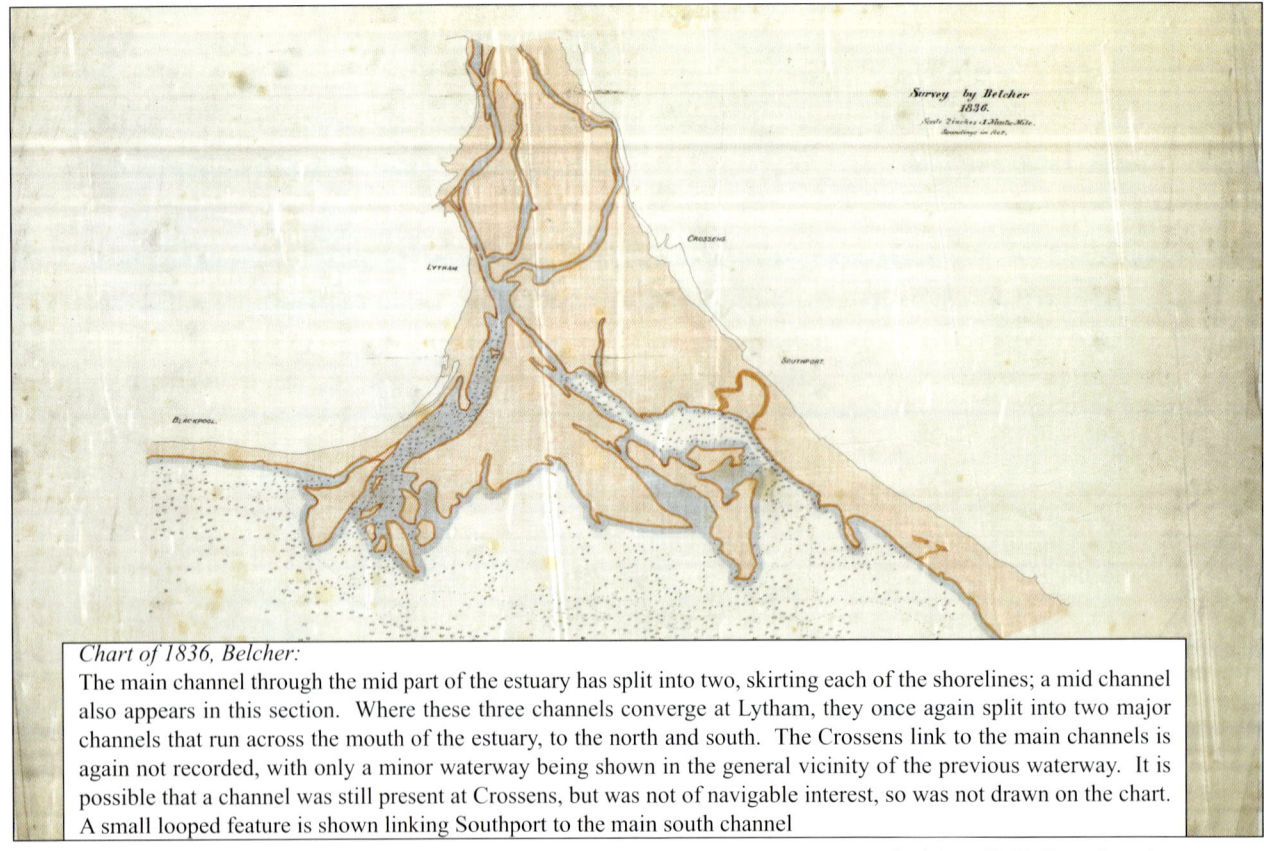

Chart of 1836, Belcher:
The main channel through the mid part of the estuary has split into two, skirting each of the shorelines; a mid channel also appears in this section. Where these three channels converge at Lytham, they once again split into two major channels that run across the mouth of the estuary, to the north and south. The Crossens link to the main channels is again not recorded, with only a minor waterway being shown in the general vicinity of the previous waterway. It is possible that a channel was still present at Crossens, but was not of navigable interest, so was not drawn on the chart. A small looped feature is shown linking Southport to the main south channel

Figure 5. Chart of the Ribble Estuary, 1836, original by Belcher. (Document held in SMBC archive).

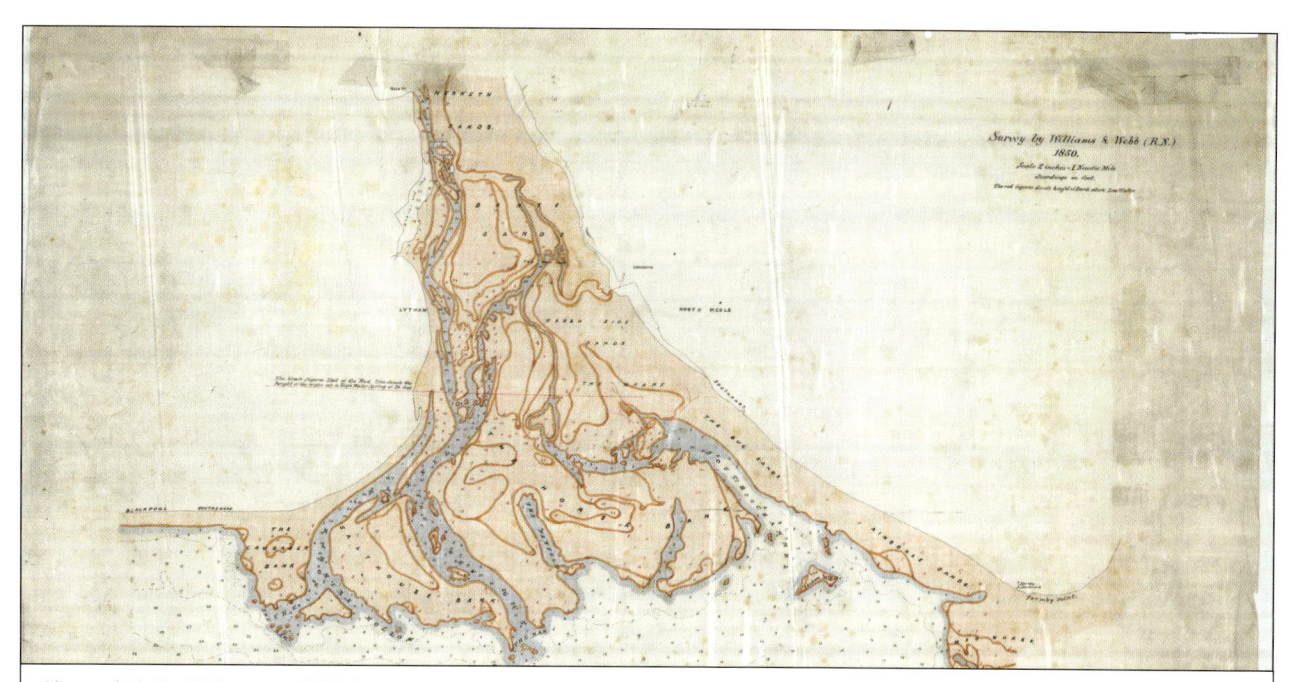

Chart of 1850, Williams and Webb:
One main channel drains the mid section of the estuary to the north shore, with the channel that followed the southern shore now becoming disconnected from the other channels toward the landward end. The South Channel that runs past Southport is still sizeable to this point, but narrows considerably further into the estuary past the town. Crossens is linked to the sea by minor channels that join the North and South Channels. 'The Horse Bank' is now a major sand bank across the mouth of the estuary, with 'Salthouse Bank' separating the 'North Channel' that runs very close to Blackpool, and the 'New Gut' that takes a more central route to the sea. 'The Penfold' channel makes an appearance, cutting midway across Horse Bank. Fronting Southport is now called the 'Bug Sands'.

Figure 6. Chart of the Ribble Estuary, 1850, original by Williams and Webb. (Document held in SMBC archive).

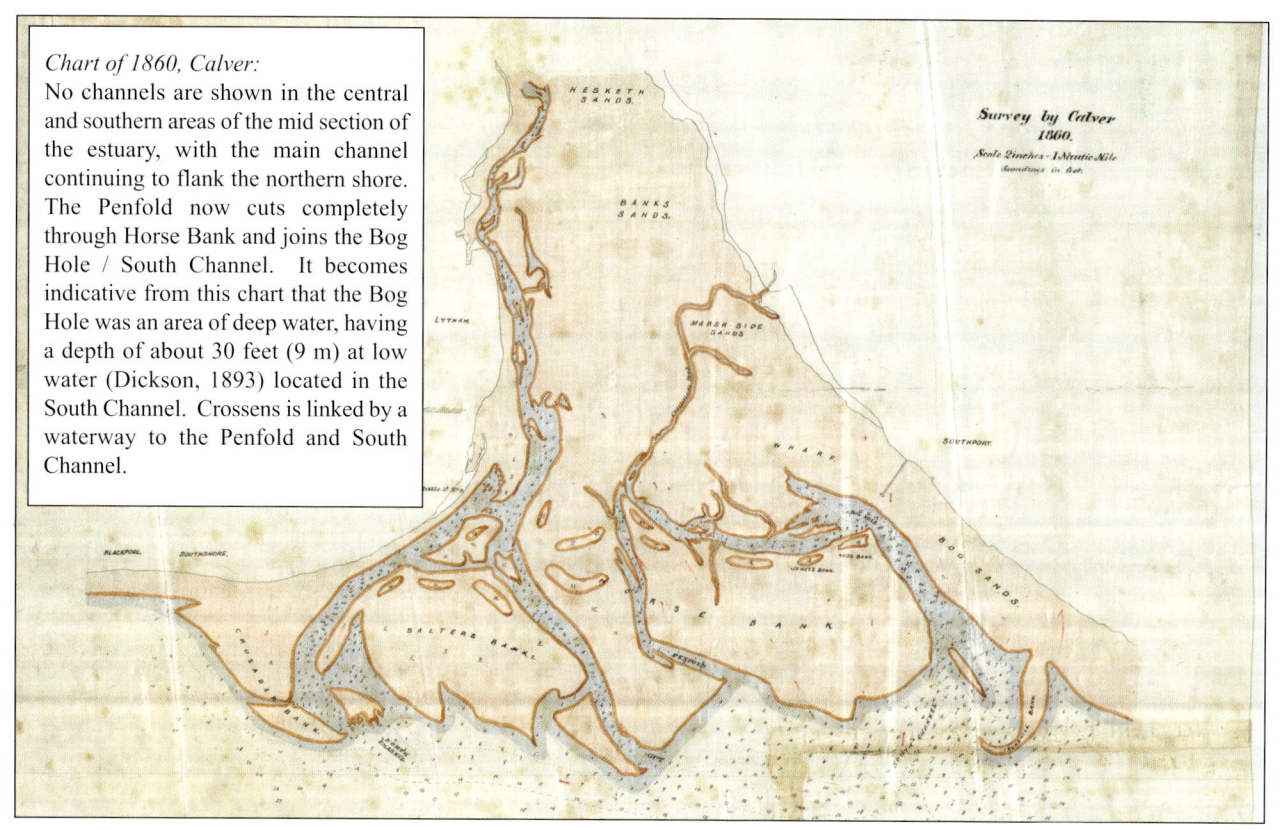

Chart of 1860, Calver:
No channels are shown in the central and southern areas of the mid section of the estuary, with the main channel continuing to flank the northern shore. The Penfold now cuts completely through Horse Bank and joins the Bog Hole / South Channel. It becomes indicative from this chart that the Bog Hole was an area of deep water, having a depth of about 30 feet (9 m) at low water (Dickson, 1893) located in the South Channel. Crossens is linked by a waterway to the Penfold and South Channel.

Figure 7. Chart of the Ribble Estuary, 1860, original by Calver. (Document held in SMBC archive).

made in the late 1800s and deposited at the southern end of the main channel training walls (as they were at that time), in an area extending between one and four miles westward of Lytham Pier (Messent, 1888; Anon, 1896). Following the Ribble Navigation Act 1905, Clause 8 specified all dredgings except that used for construction of the training walls, be deposited at sea "westward of a meridian drawn 3 degrees 9 minutes west of Greenwich", or deposited above the Spring High Water Mark, (Fowler, 1909b).

In the mid 1960s a number of investigations were undertaken by the Hydraulics Research Station (HRS) for the Corporation of Preston in order to propose improvements to the navigable Ribble Channel. In their report of 1965, they concluded that over the period 1949 to 1962, 1.6 million cubic yards of material had been

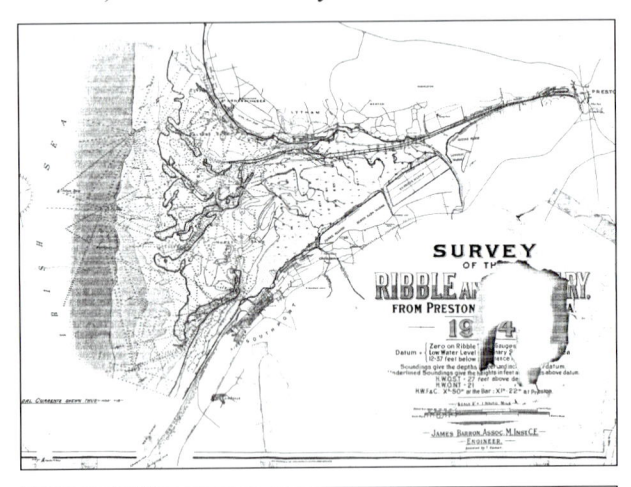

Figure 8. Charts of surveys of the Ribble Estuary, 1904 and 1912. Between 1904 and 1912 the extent of the additional training of the main navigation channel is apparent, extending to the seaward edge of Salter's Bank and the Old Gut Channel.

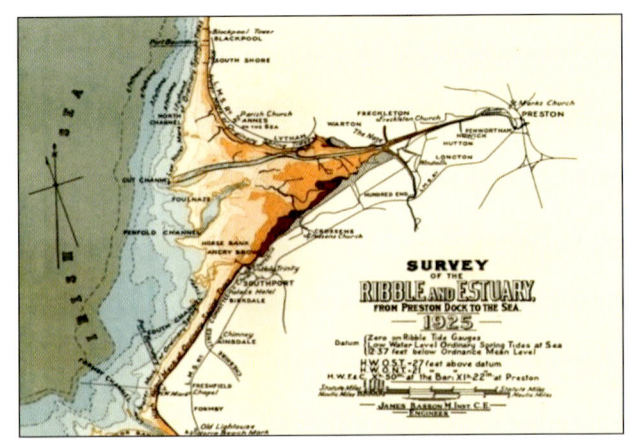

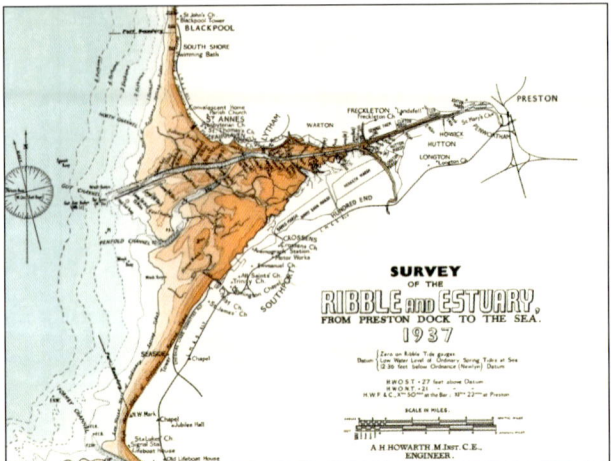

Figure 9. Charts of surveys of the Ribble Estuary, 1925 to 1937. By 1925 the infilling and disappearance of the Bog Hole at the end of the South Channel is particularly marked, as is the virtual disappearance of the New Gut Channel that previously dissected Salter's Bank to the north of the trained channel. The final chart of 1937 shows further changes around the end of the trained channel, and also the virtual disappearance of the Bog Hole Channel.

dredged per annum (HRS, 1965). At the same time, consulting engineers Rendel, Palmer and Tritton (1967) were in general agreement regarding the transport of material within the estuary, being that sediment was carried up the South (or New) Gut on a flood tide, and moved down the Navigation Channel on the ebb tide.

Evidence from the 1970s show dredgings from the upper tidal reaches were pumped into onshore lagoons to allow sand to settle out, with suspended finer material being returned to the channel (Fairhurst, c. 1974). Dredgings from further seaward were initially deposited to the north of the training wall, but following the formation of Salter's Spit, so they were deposited to the south (Fairhurst, c. 1974).

By 1982, the Port of Preston had closed to commercial traffic (Mamas et al., 1995), with a subsequent cessation of the dredging activities of the main channel. The maintenance of the training walls was also discontinued. Therefore, the sediment dynamics of the estuary can be anticipated to be undergoing further changes that may have implications for the future evolution of the area. The beginning of these anticipated changes are demonstrated by changes in the channels and sand banks in the outer estuary (Hansom et al., 1993), notably Salter's Spit now extending southwards beyond the end of the previous navigation channel since the cessation of the dredging (Rainfords, 2005) (see Plate 1). The implications of the channel no longer being dredged may include impeded land drainage of the low lying (largely reclaimed) land, and potentially flooding, should the channel accrete sufficiently to raise low flow levels and there be conditions of high freshwater flow into the channel, (Kestner, 1979).

Dredging of the Mersey Estuary

Dredging of the River Mersey began in 1890 (Jarrad, 1907; Fowler, 1909a). Jarrad (1907) recorded that the total amount dredged from the Mersey during 1905 was 9,119,100 tons, which were deposited (at that time) on the 'dividing line' between indraughts for the Mersey and Ribble estuaries (Figures 10 and 11). His investigations showed that the material was rarely deposited within the agreed area, being generally dumped 'anywhere within a mile westward' of the boundaries, as larger dredgers could not approach the area for some time before and after Low Water, nor in stormy weather. Jarrad (1907) further concluded that, between 1889 and 1904, there had been 'no appreciable diminution' of the depth of the water over the area, indicating that the 80 million tons of dredgings that had been deposited at the site during that period had been removed elsewhere. Following further investigations, Jarrad concluded that if dredgings were deposited slightly westward and northward of the prescribed area, then there would be an equal chance of the material returning on a flood tide to either the Mersey or Ribble indraught, however, during prevailing wind conditions of south-westerly, the likelihood was that material would have been carried into the Ribble indraught.

More recently, a study by the University of Liverpool in 1986 identified residual circulations were transporting material from the spoil ground at Jordan's Spit onto Taylor's Bank, then on to Formby, and subsequently into the drift feeding Horse Bank; hence the stability of Horse Bank may be linked to the dredging of the Mersey. Similarly, Pye and Neal (1994) determined that due to the dredging and subsequent spoil dumping of the Mersey, alterations were made to the wave regime around Formby Point, in part contributing to coastal erosion in this area, with much of the eroded material being moved towards the Ribble Estuary.

Dredged spoil continues to be dumped at 'Site Z', about 10 km due west of Formby Point, which in 1972 was receiving 15 million tonnes of dredged spoil per annum (Department of the Environment, 1972), which by 1992 had reduced to 3 million wet tonnes of sediment (Rees et al., 1992). According to Rees et al. (1992) much of the considerable decrease in the quantities of dredgings dumped is due to changes in dredging practices. Disposal at the 'new' Site Z, south west of the old site, commenced in 1982 due to shoaling at the 'old' Site Z (Rees et al., 1992).

Waste Disposal within Liverpool Bay

From 1896 to 1998, dumping of sewage sludge and industrial waste into Liverpool Bay has taken place (Allison, 1949; Murray and Norton, 1979; Norton et al., 1984; Jones et al., 1991; Jones, 2006). Many wastes and sludge were deposited at a site 30 km offshore (Norton et al., 1984). Between 1973 and 1980, between 15×10^{-5} and 20×10^{-5} tonnes of sewage sludge and industrial wastes were dumped at this site per year (Norton et al., 1984). Between 1880 and 1953, Liverpool Corporation also dumped large quantities of domestic refuse into Liverpool Bay (Department of the Environment, 1972; Jones et al., 1991).

During the 1970s, concerns were raised that wastes

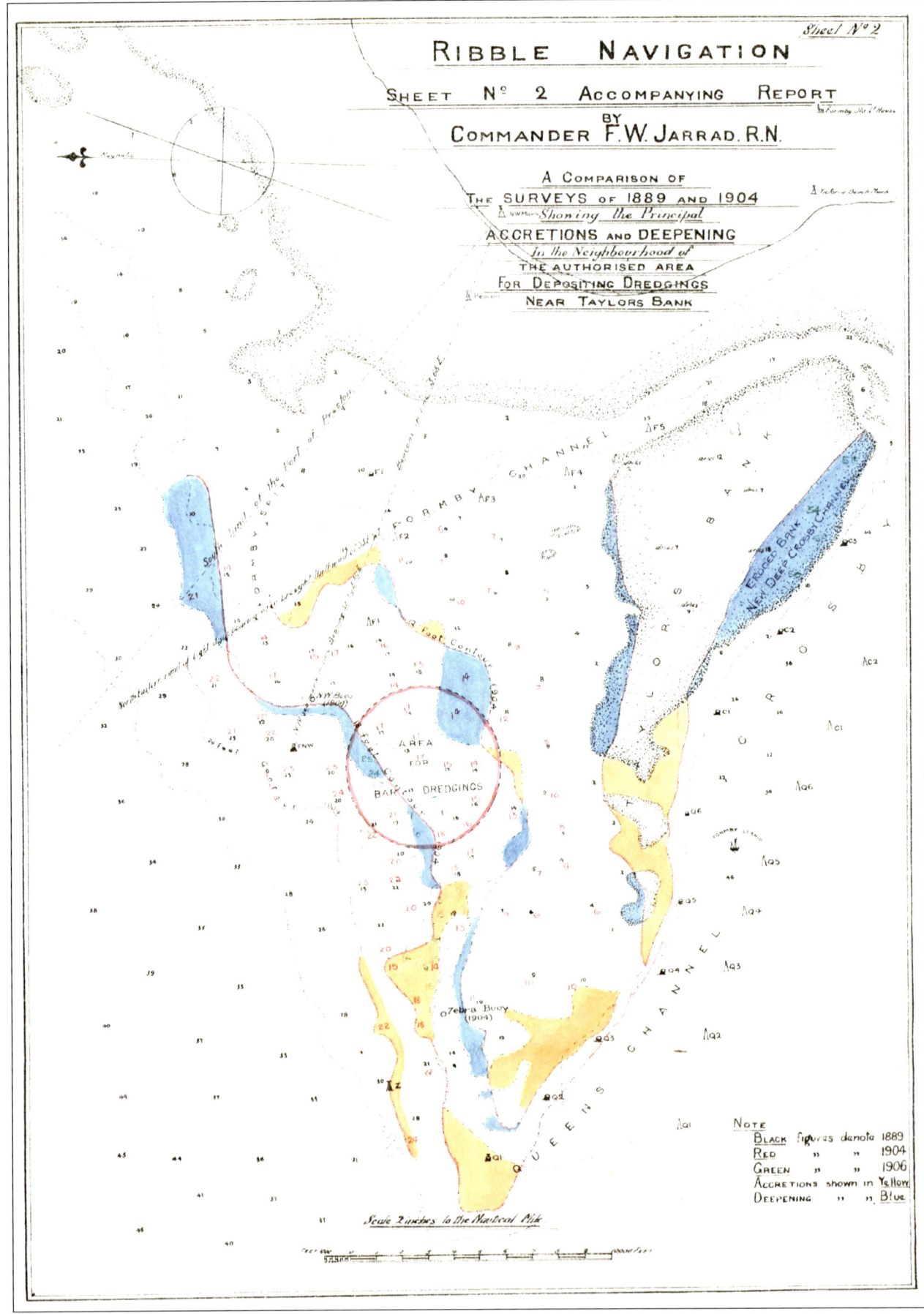

Figure 10. Position of the dumping ground for Mersey dredgings (1904), which at the time were close to the 'dividing line' between the Mersey and Ribble indraughts, from Jarrad, (1907). (Document held in SMBC archive).

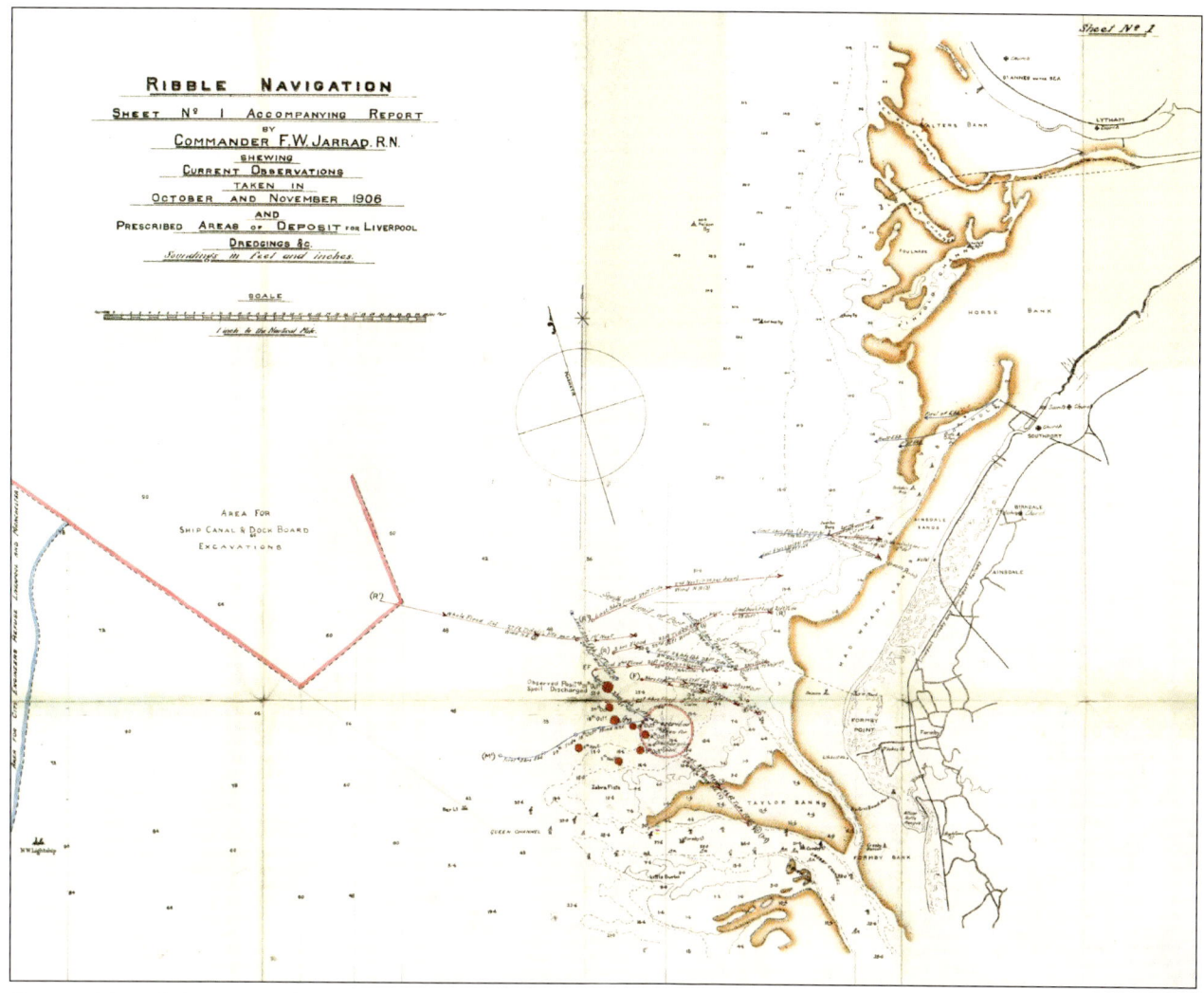

Figure 11. Tidal current observations around the Mersey dredging dumping site, drawn by Jarrad, 1906. (Document held in SMBC archive).

dumped in Liverpool Bay, due to the predominant currents, would return not only to the Mersey Estuary, but also to the Ribble. Ramster and Hill (1969) identified that 'fine suspended material in mid to surface waters tended to be more widely dispersed [away from the dumping site] with generally northwards residual movement' (Norton et al., 1984). The Department of the Environment (1972) reported that of the sludge dumped, much of the solids reach the bed of the sea after 12 hours, with the upper layers of the sea having a general drift northwards, whilst near bed waters drift south and eastwards.

The dumping of sewage and waste at sea was however, at the time, felt to be a feasible disposal option due to the 'tremendous diluting and treatment capacity of the sea' (Department of the Environment, 1972). Even with such tonnages being dumped, there was no formal routine monitoring in place regarding the effects of the dumping on the marine system. Ultimately, the concerns over pollution from dumping of sewage sludge and waste at sea led to the Food and Environment Protection Act of 1985, which since 1992 has prohibited the dumping of industrial waste at sea, and since 1998 has prohibited the dumping of sewage sludge (DEFRA, 2002).

Reclamation of Land

From the middle of the nineteenth century onwards, reclaimed salt marsh became deemed as valued agricultural land. In 1838 the Ribble Navigation Reclamation Plan was made (Figure 12) (Barron, 1938). The scale of the reclamation within the estuary was recorded by Gresswell (1953) when he calculated that between 1830 and 1880, over a mile in width of salt marsh had been reclaimed. More recently, it has been estimated that approximately 2,230 hectares of land has been reclaimed in the estuary since 1800 (Pye and French, 1993; Healy and Hickey, 2002). Plate 2 demonstrates the extent of the reclamation on the south side of the estuary.

In 1932, to further enable reclamation, 'saltings' or evenly spaced clumps of *Spartina townsendii* were planted to promote sedimentation and encourage new salt marsh formation at Marshside (Gresswell, 1953; Berry, 1967), (Plate 3). However, not all sedimentation seemed desirable, as only five years later an article from the Southport Visitor newspaper of February 1937 recorded that the Borough Engineer, Mr Jackson, stated that the build up of sand on the foreshore at Southport was due to the presence of "…the Ribble training walls, the diversion of the Crossens Channel from the South Channel into the Pinfold [Penfold] Channel, land reclamation and the growth of salt marshes, and the Mersey sand dredgings".

The evidence collated and analysed suggests that the salt marsh at Marshside has been expanding spatially and stratigraphically for at least the last 200 years. Until recently, much of this expansion has been actively encouraged by human intervention to enable reclamation of land.

Development and Urbanisation of Southport

The town of Southport, as exists today, did not come into being until the start of the Nineteenth century, when in 1798 a hotel was built by William Sutton. It is recorded that at the opening of the hotel, a Dr Barton declared the area be called Southport (Cameron, 1996). Prior to the town developing, there was believed to be very little habitation in the area that was then known as 'North Meols', with only isolated fishermen's cottages, which are now a considerable distance from what is now the foreshore (Tovey, 1965). The major activities regarding the development of Southport are given in Table 3.

Accretion of sediment along the Southport coast is evidently not a new occurrence. Dickson (1893) suggested that a prevalence of sand accumulating along the foreshore was due to the increasing numbers of buildings at Southport impeding the movement of wind blown sand inshore, and that the draining of the land under the town and culverting was allowing sand to dry out and be more susceptible to movement further up the foreshore.

In 1906, notes made by the Borough of Southport's Surveyor's Office records that:

"… although there had been erosion in many places the accretions facing the Borough exceed the erosion by many millions of cubic yards during the last 20 or 25 years. To the north-east of the Borough and higher up the Estuary enormous accretions have taken place… since the year 1839 above 5,000 acres have been grassed over." [Borough Surveyor's Office, Southport, 1906].

In addition to the sandy sediment referred to by Dickson (1893), the presence of muddy fine sediment along the foreshore to the north of the Pier is evidently also not a recent occurrence. Fowler (1910) reported it was almost impossible to walk to the Bog Hole Channel eastward of the Pier, with the whole of that area below the 9 feet (2.7 m) contour being in a "stagnant condition of wet sand and mud". Fowler summarises the condition of the foreshore by saying:

"The contrast between the muddy beach above the Pier and the hard clean Beach extending from the Pier to Formby and down to spring low water contour is very marked."[Fowler, Report to the Chairman and Members of the Improvement and Parks Committee, Southport, 1910]. Ashton (1920) also described the 'clay beds' near Crossens, and at low water on the landward side of the Bog Hole north of the Pier.

It is probable that the extensive reclamations within the estuary were an exacerbating factor in the accretion of sediment along the foreshore. The reclamation of land would reduce the tidal prism within an estuary (Long et al., 2006), and lower mean current velocities, resulting in sediment deposition (van der Wal et al., 2002). This deposition would have a consequential effect of establishing a positive feedback by further reducing the

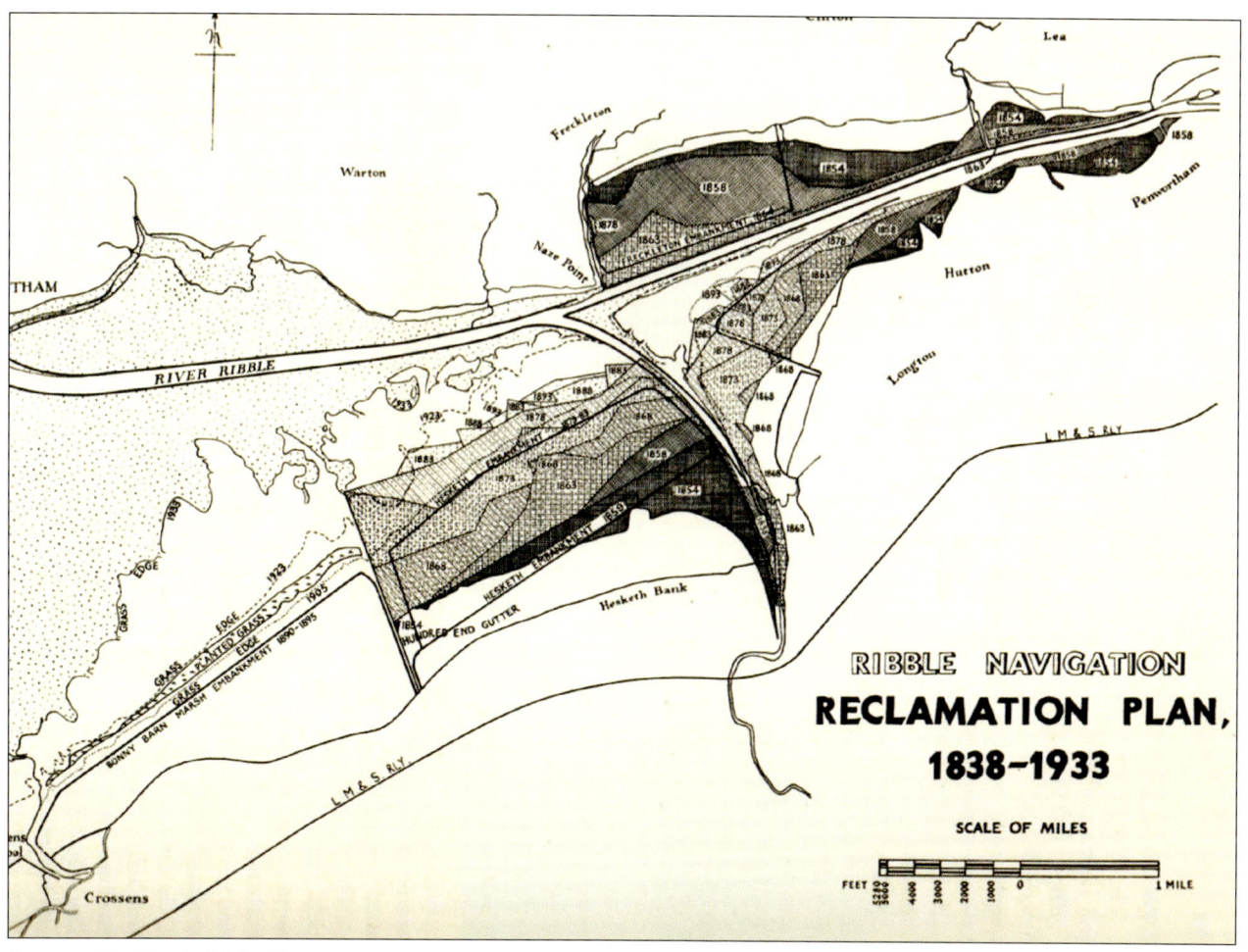

Figure 12. The Ribble Navigation Reclamation Plan. (Source: Barron, 1938, 341).

Plate 2. Aerial photograph showing field boundaries, highlighting the extent of reclamations on the south side of the Ribble Estuary, September 1984. (Image courtesy of SMBC).

Plate 3. Plate of the 'saltings' at Marshside, early twentieth century.
(Source: Gresswell, 1953, facing 72).

Plate 4. Aerial photograph of Southport foreshore looking north west, pre construction of the Coast Road and reclamation to create the Marine Lake extension, estimated mid 1950s. (Image courtesy of SMBC).

tidal prism, thereby promoting even more deposition, increasingly of finer sediment as the energy reduces (van der Wal et al., 2002).

Development of the Salt Marsh at Marshside

The coastal zone of Sefton has been surveyed at intervals by SMBC (or their predecessor, Southport Corporation) along a series of fixed transect profile lines, covering the entire Sefton coastline, in some places for almost 100 years, with all the survey data collected still available. This survey data can be used to assess changes in the topography and elevation over the coastal zone over time.

Since the start of the sand-winning operations from Horse Bank, with the plant situated on the foreshore of Marshside (SD 335300, 420640), localised alterations in the topography of the surrounding salt marsh have been apparent. Between 1968 and present, topographic surveys along a transect that runs to the south of the sand-winning plant, (known as 'Profile 28'), has shown a notable increase in surface elevation, averaging 0.5 metres increased height along its length (Figure 13). Along this profile between 1979 and 1996 surveys, a ridge developed at 250 metres chainage, approximately 0.4 metres high in 1996, increasing to 0.8 metres in 2003. This feature apparently relates to the man-made track constructed for vehicle access across the marsh,

Plate 5. Photograph of Marine Lake, looking west, showing dates of construction, c.1966.
(Image courtesy of SMBC).

leading from the sand-winning plant out to Horse Bank. Although a very localised effect, the presence of the slightly raised track at approximately 45 degrees to the High Water Mark may have been acting similar to a groyne, providing suitable conditions for enhanced salt marsh development. Although this cannot readily be proven, the presence of the salt marsh edge flanking the course of the track would suggest this to be the case. Figure 14 and Plate 6 demonstrate the changes in the location of the marsh edge at the study location over time.

In a report by R.J. Holliday of the University of Liverpool in 1976, the 'problem' of the spread of *Spartina* towards the Southport foreshore is considered. In which, reference is made to previous attempts to control the *Spartina* by spraying with Dalapon (a herbicide), but which had failed, assumed at the time to be due to insufficient application of the herbicide (Holliday, 1976). The report suggests all *Spartina* south of Marshside Road (near to the sand-winning plant) be eradicated by chemical control. The spread of salt marsh vegetation was subsequently controlled between 1977 and 1988 by the aerial spraying of Dalapon (Smith, 1982; Doody, 1984), with some more recent herbicide treatment in subsequent years (Bennett, 1996; Tomlinson, 1997).

In more recent years, the salt marsh has been allowed to develop naturally, as the value of salt marsh as a sea defence, and its importance locally, nationally and internationally as a specific habitat has been acknowledged.

At time of writing (autumn 2008), the sand-winning plant at Marshside / Horse Bank had ceased operations due to non-renewal of the extraction licence, a decision taken by the operator. Proposals for the site are to include extended facilities for the Royal Society for the Protection of Birds (RSPB), and reinstating a proportion of the site to the level of the surrounding salt marsh, to allow colonisation of the area by salt marsh vegetation.

Evolution of the Bog Hole Channel

Changes in the morphology of the Bog Hole Channel have been recorded in relative detail, probably due to its proximity to the foreshore at Southport and its presence providing the town with a constant marine

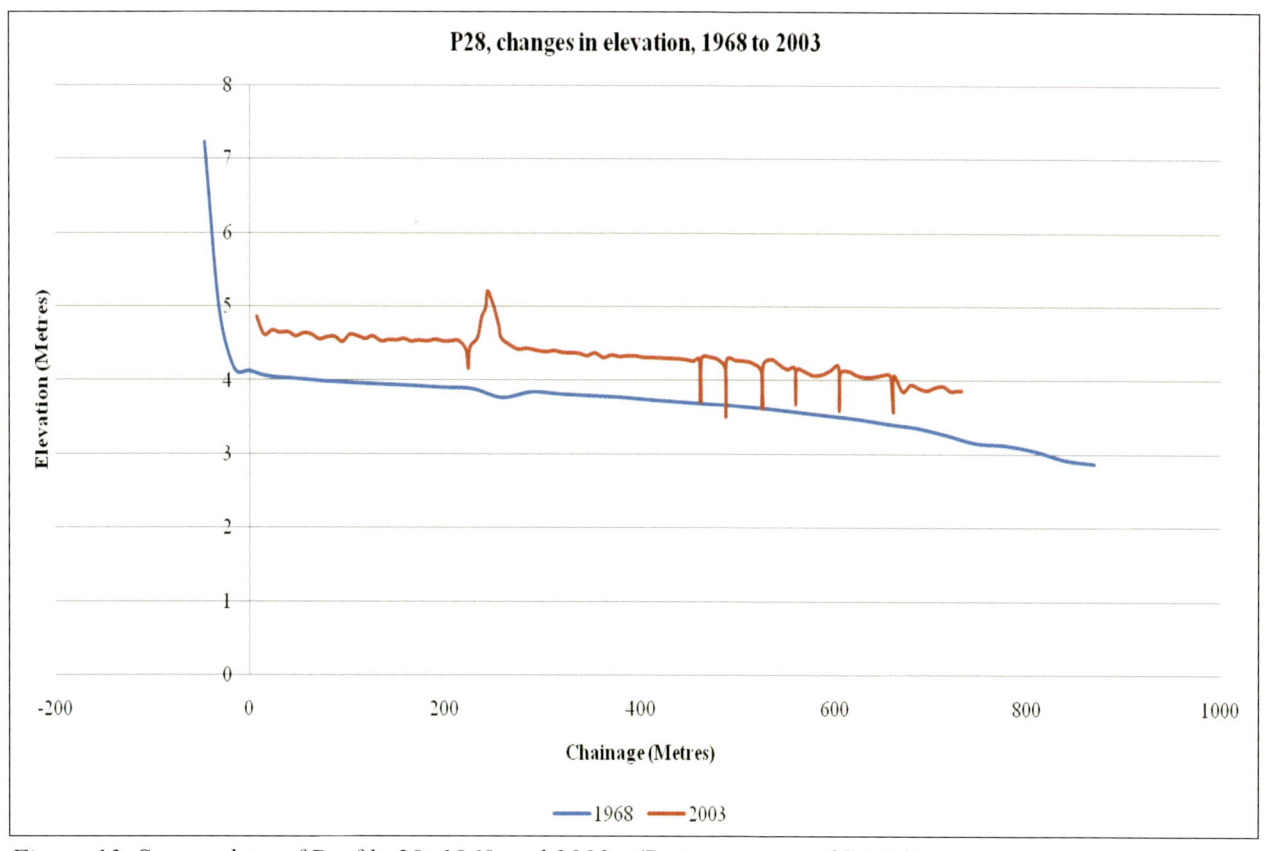

Figure 13. Survey data of Profile 28, 1968 and 2003. (Data courtesy of SMBC).

Table 3. Chronology of the development of the town of Southport.

Year	Activity	Reference
Early 1800s	Lord Street constructed.[1]	Tovey, 1965
1834	First Promenade built at level of Ordinary High Water Mark (SD 333600, 417700).	
1848	Railway line from Liverpool to Southport opened; 10,000 visitors a year by 1850.	Freeman et al., 1966; Gresswell and Lawton, 1964
1855	Railway line from Wigan to Southport opened.[2]	Freeman et al., 1966
1860	Pier[3] constructed, leading 1,100 metres out from the Promenade.	Tovey, 1965; Smith, 1982
1868	Pier extended to 1,335 metres.	Smith, 1982
1879 - 1881	Northern extension to Promenade, 45 acres reclaimed (SD 333820, 418050).	Borough Surveyor's Office, 1906; Gresswell, 1953
1887	South Marine Lake & Marine Park constructed, reclaiming 19.5 acres (SD 333230, 417500).	
1892	North Marine Lake constructed (SD 333640, 418000).	Borough Surveyor's Office, 1906
1895	Two lakes joined & Marine Drive (SD 333561, 418436) built, 59 acres enclosed.	Borough Surveyor's Office, 1906
1907	Bank End Sewage Treatment Works (SD 336948, 420736) built.	Tovey, 1965
1912	Sewage Treatment works extended.	Tovey, 1965
Mid 1960s	Construction of sea wall fronting Marine Lake, reclaiming 90 acres & extending the lake (333940, 418810) (Plates 4 & 5).	Tovey, 1965
Late 1960s	Extension to the sea wall (the 'Coast Road'[4]) as far as Marshside Road (SD 335220, 420440).	
1972	Sand-winning began on Horse Bank following a Town Planning Enquiry in 1968, operated by William Rainfords Ltd.[5]	Smith, 1982.
1974	Further extension to sea wall.[6]	

1. Lord Street was built with an average width of 240 feet (73 m) between building lines. The width of the street suited the propensity of the Victorians to 'promenade', though it was, however, built to such dimensions as at that time the ground in the middle of the road was still liable to flooding, hence the buildings were built on the slightly higher ground to either side, (the issue of flooding continued to some degree until the construction of the Esplanade Pumping Station in 1951) (Tovey, 1965).

2. The railway extension subsequently closed in 1952 (Jones et al., 1991).

3. It is worth noting that following the structural deterioration of the original pier, the current pier was constructed in 2001 on the same site, but was marginally shorter in length than its original counterpart. Therefore, any subsequent measurements referred to relating to distances from the Pier will relate to the original Pier (and extension), not the pier as exists today unless otherwise stated.

4. The entire length of the sea wall forming the road fronting Southport is correctly called 'Marine Drive'. However, the road (and, in particular, the most recent northerly section fronting Marshside and Crossens), is locally referred to as the 'Coast Road'. Therefore, when Coast Road is referred to in this text, it generally refers to the northern section of Marine Drive.

5. Between 1966 and 1994, 4.5×10^6 m^3 of sand was extracted for the glass industry (van der Wal et al., 2002). van der Wal et al. recorded that this volume of sand extracted was small compared to the changes in sediment movement seen across the estuary as a whole.

6. 1968 section, Hesketh Road to Marshside Road: SD 334506, 419404 to 335241, 420433. Northern 1974 section: SD 335241, 420433 to 337481, 420344

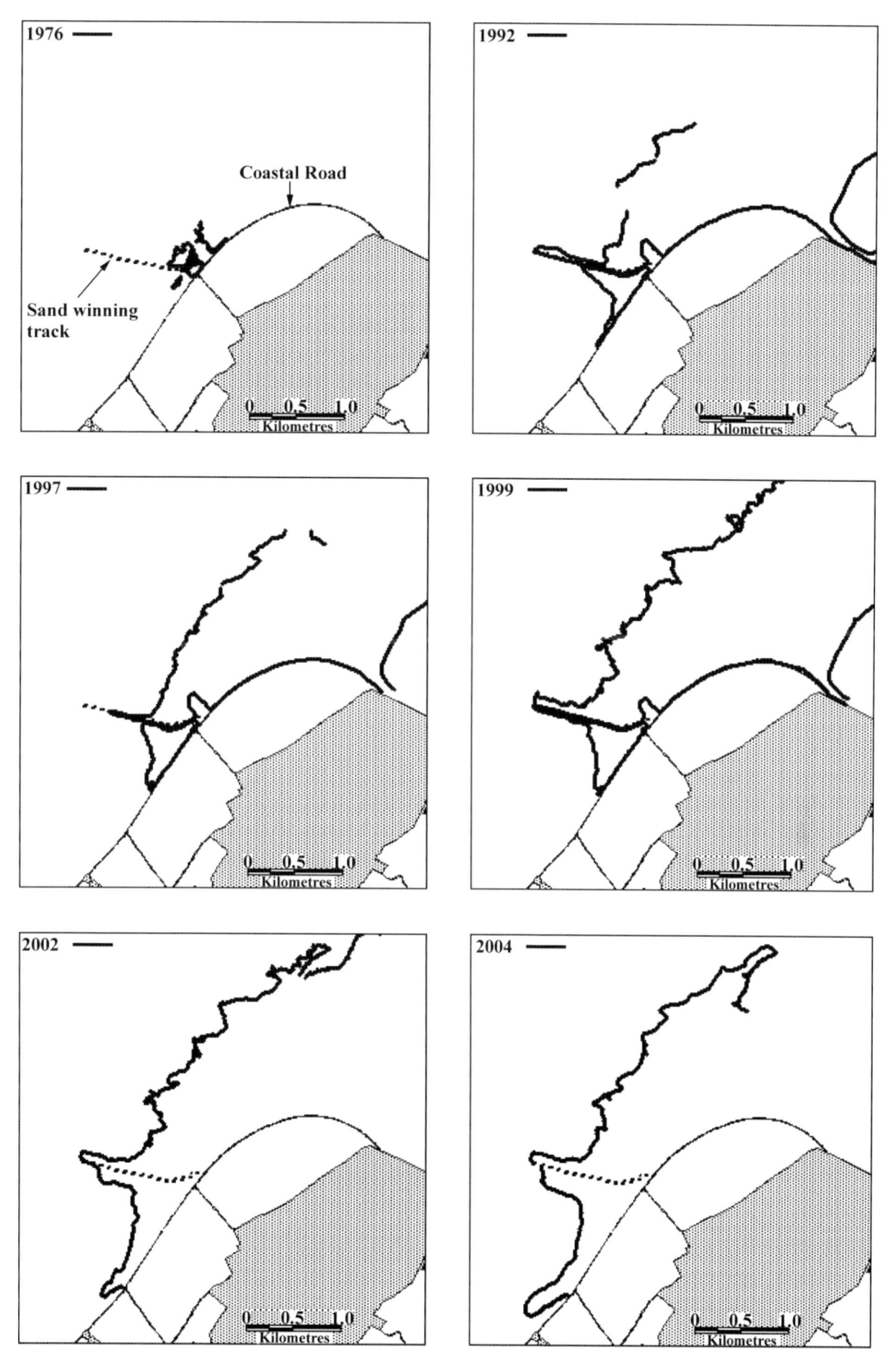

Figure 14. Expansion of salt marsh edge over time taken from aerial photography. (Illustration courtesy of SMBC).

Plate 6. Photographs of the same area south of the sand-winning plant (looking south, Coast Road to the left of both pictures) highlighting the expansion of the salt marsh at Marshside; a) approximately 1970s, courtesy of SMBC; b) 02/03/2007, V. Holden.

feature even at low tide. As Southport developed as a tourist destination during the nineteenth century, so too the depth of the Bog Hole increased (Table 4 and Figure 15). However, during the early part of the twentieth century the Channel reversed this trend and shallowed.

From the first proposal by the Corporation of Preston in the Ribble Navigation Act of 1883 to adopt the central channel in the estuary as the main navigable channel, the Borough of Southport opposed the Bill due to the potential impact upon the Bog Hole Channel, which at that time was of 'material interest' to the town of Southport (Anon, date unknown). A consulting engineer advising the Corporation of Southport at this time, Mr Vernon-Harcourt, advised that the Bog Hole (or South Channel) should be adopted as the main navigable approach to Preston, and not the central channel (Figure 16). Due to these conflicts of opinion, in the Act of 1889, a commission was established to investigate which channel should be adopted. Following the investigation, the central channel was however chosen as the main navigable channel (Anon, 1896; Fairhurst, c.1974).

Although previously in 1885, Southport Corporation had proposed a Parliamentary Bill to divert the flow of the Crossens Channel back into the Bog Hole, they were opposed by the riparian landowners whom claimed rights of accretion, with the Bill being dropped (Fairhurst, c.1974). It was subsequently suggested again at the beginning of the 20th century (e.g. Vernon-Harcourt (1905); Fowler (1910), (Figure 17)). Ashton, however, in his book 'The Battle of Land and Sea', quoted in Fowler (1910), disagreed with the proposal, believing the quantity of water flowing through the Crossens Channel to be insufficient to make a difference to the infilling of the Bog Hole. Potentially, in addition to the landowners' opposition to training the Crossens Channel, some reluctance to its training across the foreshore at Southport may have been due to the fact that prior to the Banks Sewage Treatment Works, untreated sewage was discharged via the Crossens Channel.

In reports of 1888 and 1893, Messent and Dickson, respectively, surmised that the existence of the Bog Hole was due to it no longer being connected to the other channels, so the flood tide causing an eddying motion that kept the depression scoured and hence relatively deep. The Bog Hole was recorded as continuing to deepen and narrow up to 1893, at which point it started to deteriorate, with infilling accelerated

Table 4. Changing depths of the Bog Hole channel, increasing through the Nineteenth Century, decreasing through the Twentieth Century.

Nineteenth Century (Data from Wheeler, 1893)			
Year	1824	1850	1860	1882
Depth (feet)	21.5	43 max/19 min	48	57
Depth (metres)	6.6	13.1 / 5.8	14.6	17.4
Twentieth Century (Data from Gresswell, 1953)			
Year	1909	1924	1928	1933
Depth (feet)	60	37	Dry at low water	Dry by several feet at low water
Depth (metres)	18.3	11.3	-	-

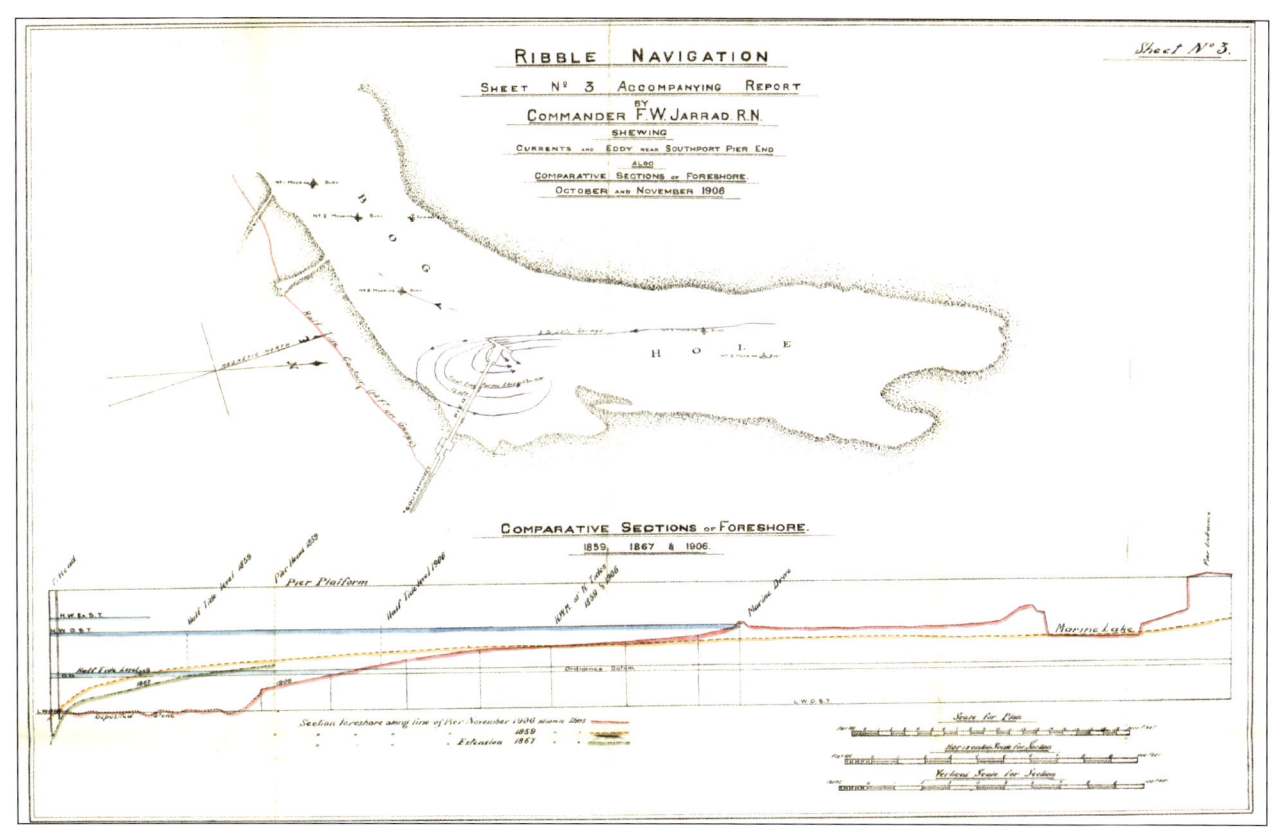

Figure 15. Plan showing the deepening of the Bog Hole Channel in the late Nineteenth Century, drawn by Jarrad, 1906. The comparative sections at the bottom show the deepening of the Bog Hole at the head of the Pier between 1859 (orange line), 1867 (green line), and 1906 (red line). (Document held in SMBC archive).

from 1904 (Fowler, 1909a; Fowler, 1910). Where there had previously been frequent steamer excursions from Southport Pier to Lytham across the estuary, by 1913, very few steamers were able to reach the Pier due to the accreting sand banks (Tovey, 1965). From the late 1920's, steamers were unable to sail from the Pier due to accretion of sand (Tovey, 1965), and to the infilling of the Bog Hole Channel (Figure 18 and Table 5).

It was not only the steamers that suffered because of the Bog Hole shallowing. Correspondence exists from the 'Southport and District Fishermen's Association', established in 1911 (e.g. Jackson, 1914; Marshall, 1914; Jackson, 1919), suggesting the number of local

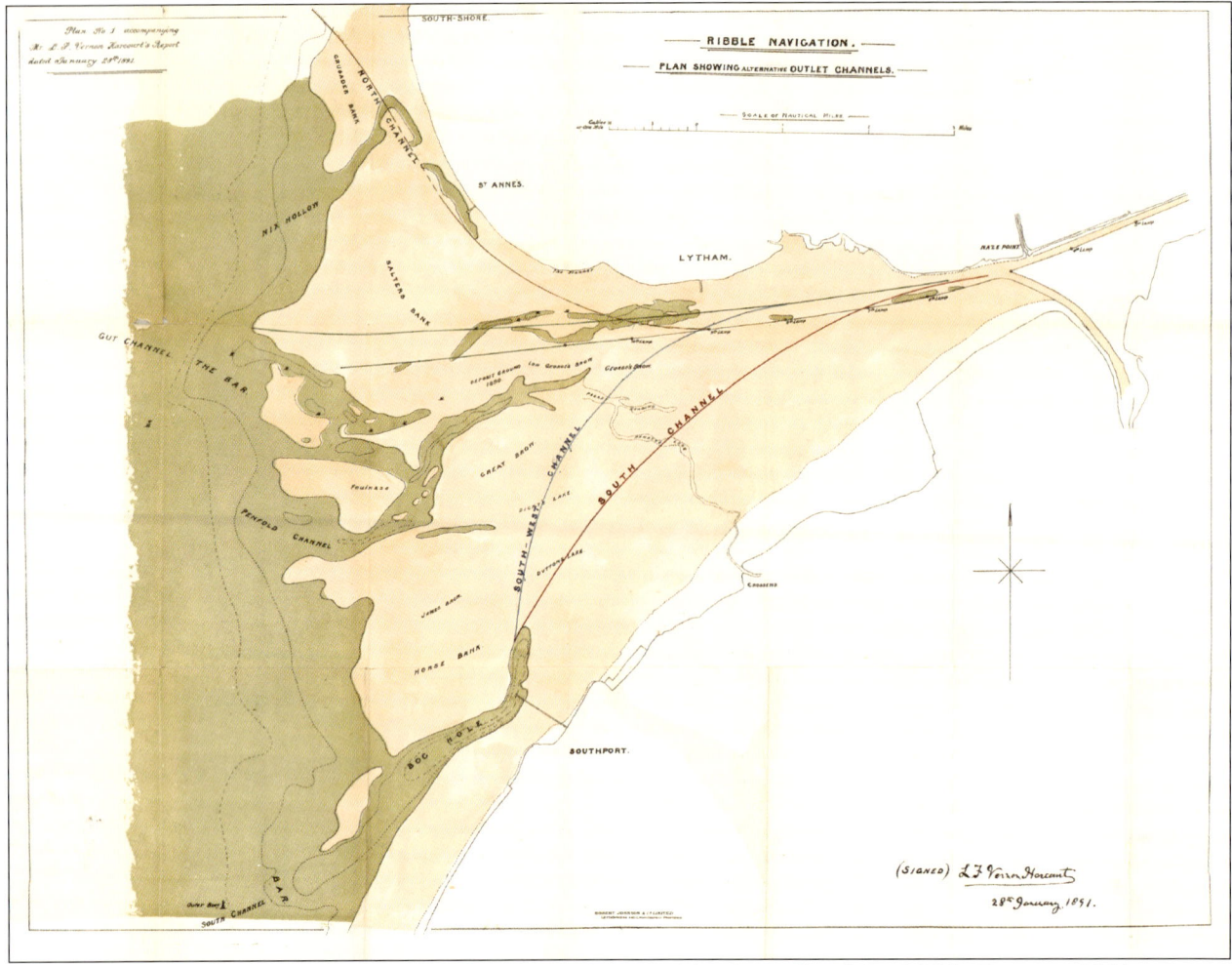

Figure 16. Plan by Vernon-Harcourt of the proposed navigable channels of the Ribble Estuary, 1891. (Document held in SMBC archive).

fishermen was significant; not only by the presence of an association, but because the correspondence regards the provision of leading lights to light the Southport (Bog Hole) Channel. In addition, Stammers (1999) recorded the building of boats at Shellfield Road in Marshside, which were moved by the local fishermen to be launched at Marine Drive, and records the presence of a shipyard at Crossens, possibly from the 18th or 19th century.

Jarrad (1907) suggested that it was the Mersey dredgings that were leading to the deterioration of the Bog Hole Channel, and recommended dredgings should be deposited in the area allocated for the 'City Engineers refuse Liverpool and Manchester'. It is evidence along these lines, along with the generally held view that the accumulation of sediment on the foreshore and the infilling of the Bog Hole Channel became marked in 1890, when the Mersey dredging began, that led people at the time to believe that it was, in part at least, the Mersey dredgings that were leading to the accretion at Southport.

In contrast to this, however, Fowler (1909a) recorded that the Mersey dredgings had been deposited in the "…tideway about six miles south-west of Jumbo Buoy" and did not believe that the deposition of the dredgings had a significant effect on the deterioration of the Bog Hole Channel, as the depth of the outer bar (at the southern end of the Channel) had remained constant. Ashton (1920) however stated that "The … highering up of the Ribble estuary banks… has its main source in the Mersey".

The general consensus during the early twentieth century was that the deterioration of the Bog Hole Channel was of great detriment to Southport, and that the main causing factors for the infilling were

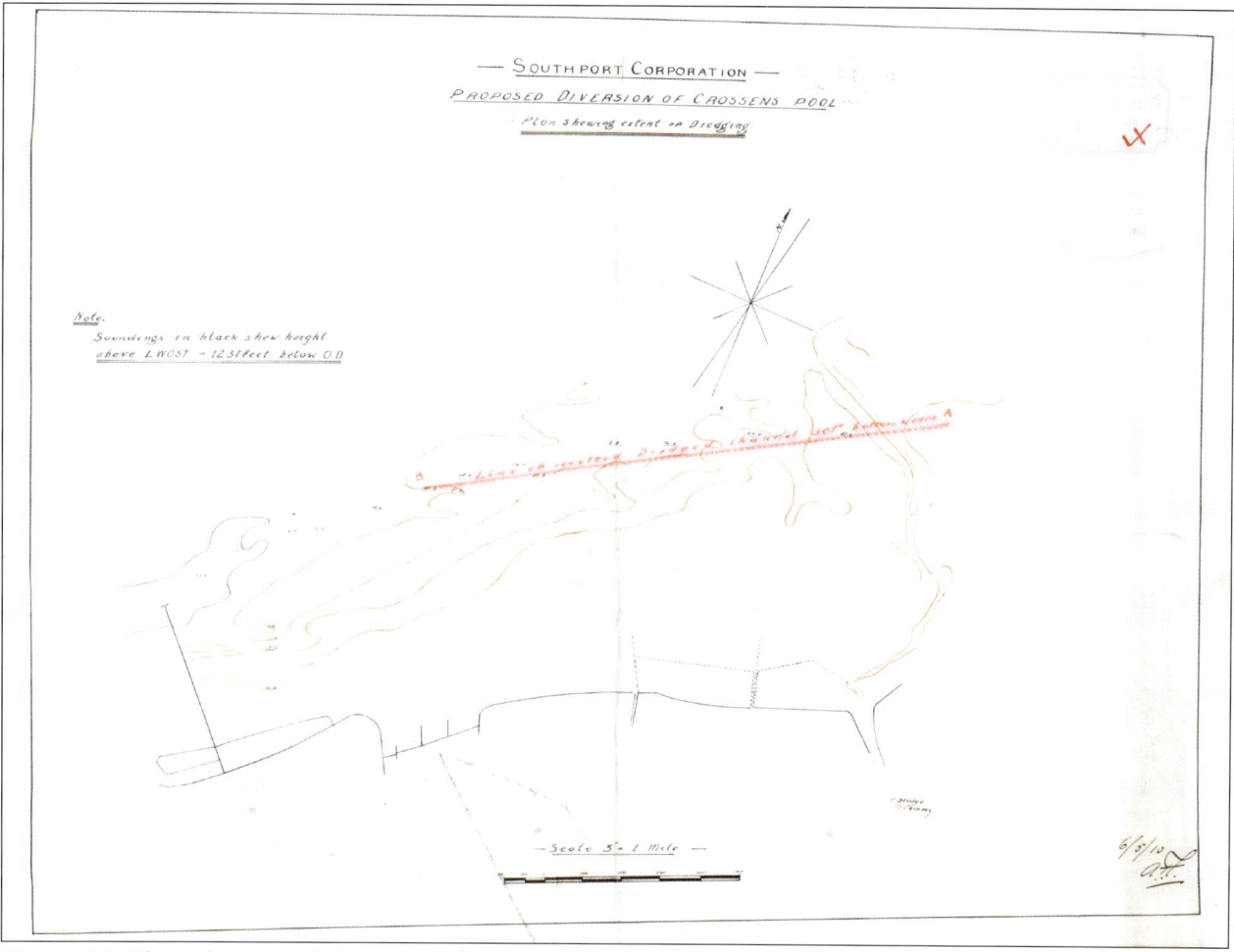

Figure 17. Plan of proposed diversion of Crossens Pool into the Bog Hole Channel, 1910, probably drawn by Fowler (Borough Surveyor). Text in red reads: "Line of revetted dredged channel 30ft bottom width". (Document held in SMBC archive).

predominantly; (i) the presence of the 1883 Ribble training walls (although this was always stated by engineers associated with Southport Corporation, and not Preston Corporation); (ii) to a lesser degree the reclamation of the foreshore on the southern side of the estuary, reducing the tidal capacity; and (iii) the effect of the deposition of Mersey dredgings since 1890.

More recently, topographic survey data along the SMBC survey profile nearest to the Pier that cuts across the Bog Hole Channel, shows a continued gradual increase in the foreshore level over time, (Figure 19). There is a noticeable concave section to the profile between 400 and 600 metres chainage, can be attributed to the remains of the Bog Hole Channel, with the profiles indicating a gradual landward shift of the deepest part of the channel.

Secondary Data Confidence

Historic accounts and data provide an invaluable record of the developments that have taken place along the coastline, without which, the significance and reasoning of many events would be lost, facts obscured, or simply forgotten. However, it is important to remember that most records are subject to the opinions and observations of the individuals making them, which can in some circumstances, give little indication of the accuracy of the data. Wherever possible, data was collected from original sources rather than subsequent accounts of the information. The accuracy of early charts may be questionable, particularly as they would potentially have been drawn up for specific purposes, hence certain channels or features that were not considered by the surveyor to be important may have been recorded in much less detail, or even omitted. However, as a certain level of accuracy would have been required for navigation purposes, the charts can

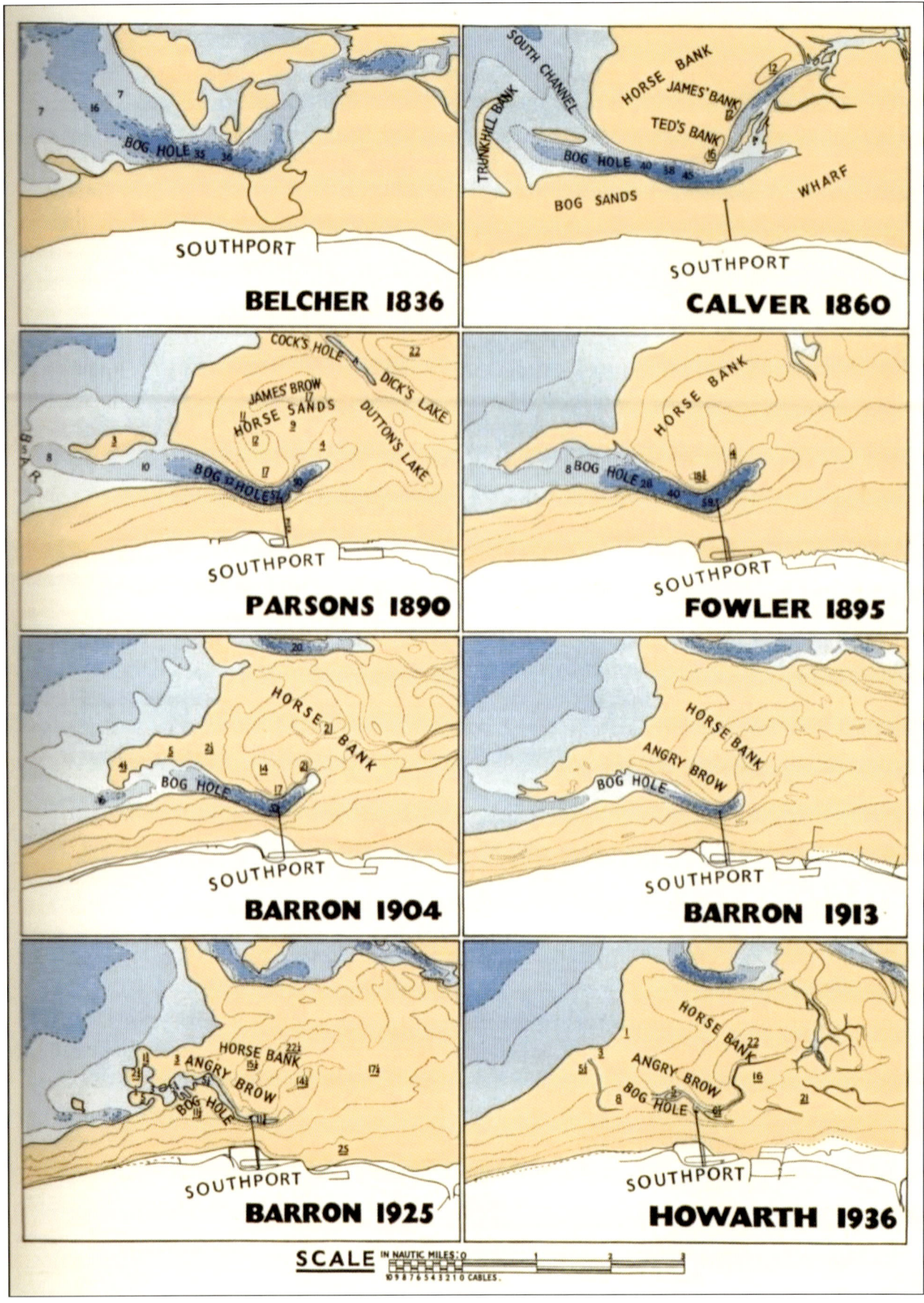

Figure 18. Successive charts showing the changes in the South (Bog Hole) Channel, redrawn by Barron (1938).

Table 5. Changing characteristics of the Bog Hole Channel, 1871 – 1919; (data taken from Ashton (1920), quoting data believed to have been collected by Fowler; and Fowler, 1910).

Characteristic	Year	Measurement	Measurement (metric conversion)
Height of water/sand 1 mile above Pier	1871	9 feet of water	3 m
	1905	9 feet of sand	3 m
Distance of head of channel from Pierhead (at Low Water)	1887	2700 yards	2469 m
	1903	1200 yards	1097 m
	1909	1000 yards	914 m
	1919	900 yards	823 m
Length of channel at 15 feet depth	1904	2 miles	3.2 km
	1910	1 mile 70 yards	1.67 km
Length of channel at 21 feet depth	1904	960 yards	878 m
	1910	730 yards	668 m
Area of channel above Pierhead	1890	127 acres	51.4 ha
	1913	47 acres	19.0 ha
Channel width, line from Pier to Horse Bank	1904	720 feet	220 m
	1910	660 feet	201 m
	1919	450 feet	137 m

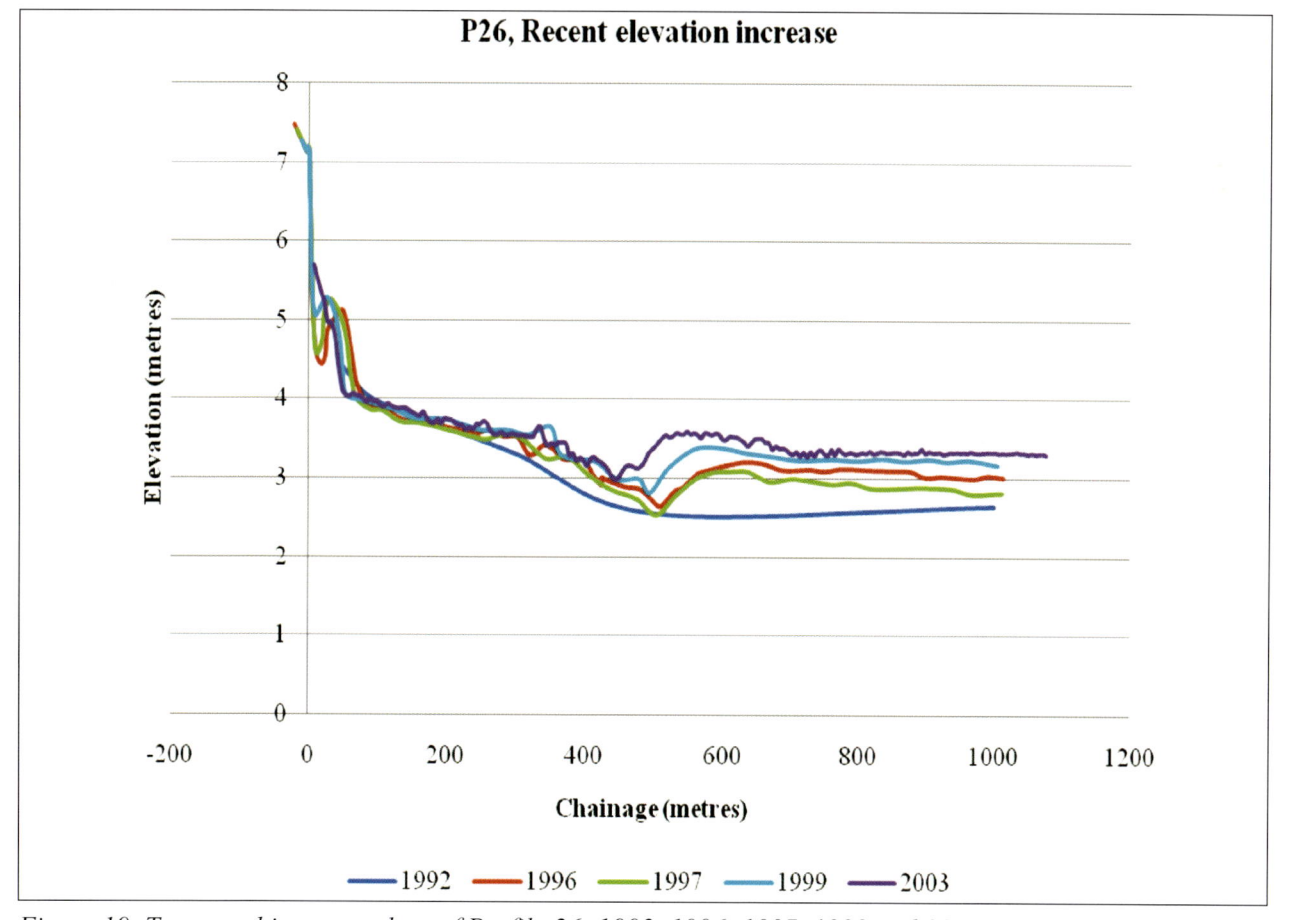

Figure 19. Topographic survey data of Profile 26, 1992, 1996, 1997, 1999 and 2003. (Data courtesy of SMBC).

be assumed to be representative of general episodes within the estuary. The charts shown by Mackenzie, Belcher, Williams and Webb, and Calver were made and published by the Admiralty (Barron, 1938; van der Wal and Pye, 2003), so some degree of data confidence can be attributed to their provenance.

Discrepancies between certain channel depths and low water marks may have potentially arisen between texts due to the variance between datums used for soundings. During the Nineteenth Century, the datums referred to are generally those of the 'Old Dock Sill' at Liverpool (from the first enclosed Liverpool Dock built in 1715 (Cashin, 1949)). From around 1844, Ordnance Datum referred to the mean sea level at Liverpool, which was 4.67 feet (1.42 m) above the Old Dock Sill. From around 1933, Ordnance Datum was based upon mean sea level at Newlyn, which was 14.54 feet (4.43 m) above the Old Dock Sill (Barron, 1938; Cashin, 1949; Gresswell, 1953). The level of the Old Dock Sill was believed to be 7ft. 9in. (2.36 m) above the level of Low Water at Southport (Jarrad, 1907), with the Ribble being calculated as 12.37 feet (3.77 m) below Ordnance Datum (Liverpool) in 1890 (Barron, 1938).

Similarly, early Ordnance maps based the High and Low Water Marks around surveys taken at times when the Admiralty informed them that there were expected to be 'ordinary' tides, which when combined with weather conditions, may produce misleading levels on flat stretches of coastline such as in the Ribble Estuary (Barron, 1938). Ordnance Survey maps, similarly, showed a low water mark of a 'Mean Spring Tide' in 1845, whereas later revisions show a low water mark of an 'Ordinary Tide', which is the actual low water level on a day between a spring and a neap tide (Gresswell, 1953).

SUMMARY

Evidence presented in this paper demonstrates how in an area significantly influenced by anthropogenic activities, it is not sufficient to merely understand the contemporary physical processes occurring when considering sustainable management of the coastal zone. It highlights the importance of appreciating the historical evolution and human influences in the area, both past and present. Trends observed regarding current conditions could be misleading without this knowledge, thereby making an awareness of the physical and anthropogenic development of the area an essential requirement for successful interpretation of contemporary data and subsequent strategic decisions.

Throughout the last 300 years the Ribble Estuary has evidently undergone continual change since the first available plans of 1737 illustrate. The estuary has experienced a net infilling, with evidence indicating that the natural processes of change have been exacerbated during the last 200 years by human activities (O'Connor, 1987) related to the development of the Port of Preston, and the evolution of the town of Southport; notably the construction of the Ribble training walls, land reclamation, and dredging of both the Ribble and Mersey Channels. During this time, human activities have been of noteworthy influence on the estuarine system compared to natural forcing factors, but with natural factors such as sea-level rise and climate change potentially becoming of increasing significance in the future (van der Wal et al., 2002).

Due to the closure of the Port of Preston and the subsequent cessation of dredging and maintenance of training walls within the estuary, it is probable that the sediment dynamics will undergo further alterations, requiring further investigation in order to effectively understand the processes operating in the estuary and to manage the future strategy for the area.

ACKNOWLEDGEMENTS

The author would like to thank Sefton Metropolitan Borough Council for providing access to the archives of data relating to the coast, and their help in interpreting them. In particular, Graham Lymbery, Paul Wisse, and Michelle Newton. Edge Hill University kindly provided funding through the Research Development Fund for the author to undertake a PhD, of which this research into the development of the Sefton coastline was an integral part.

The generations of engineers whose data has been

collated within in this paper are gratefully acknowledged. Without their hard work and conscientious recording of information, this reconstructive history would not have been possible and such valuable data would have been lost. The views expressed by the quoted authors are, of course, purely theirs alone, and do not necessarily reflect the views of the author, who has endeavoured to reflect their arguments as accurately as possible from the information available.

REFERENCES

Adam, P. (2002) Saltmarshes in a time of change. *Environmental Conservation*, 29, 39-61.

Allison, J.E. (1949) *The Mersey Estuary*, Liverpool, Liverpool University Press.

Anon. (1896) Minutes of Proceedings, Southport Corporation. *Unpublished archive document,* SMBC.

Anon. (Unknown) Note on the history of the Ribble Training Walls. *Unpublished archive document*, SMBC.

Ashton, W. (1920) *The Evolution of a Coastline*. Wm. Ashton & Sons, Southport.

Barron, J. (1938) *A History of the Ribble Navigation*. Preston, Guardian Press.

Bennett, C.M. (1996) Saltmarsh management on an amenity beach of international nature conservation interest at Southport, Merseyside. *Aspects of Applied Biology*, 44, 301-306.

Berry, W.G. (1967) Salt marsh development in the Ribble Estuary. IN Steel, R. W. & Lawton, R. (Eds.) *Liverpool Essays in Geography. A Jubilee Collection.* Liverpool, Liverpool University Press.

Borough Surveyor's Office. (1906) Royal Commission on Coast Erosion. Borough Surveyor's notes on questions framed by the Commission. *Unpublished archive document*, SMBC.

Cameron, K. (1996) *English Place Names*, London, B.T. Batsford, Ltd.

Cashin, J.A. (1949) Engineering works for the improvement of the estuary of the Mersey. *Journal of the Institution of Civil Engineers*, 32, 296-367.

Cooper, A. (2007) Temperate coastal environments. In Perry, C. & Taylor, K. (Eds.) *Environmental Sedimentology*. Oxford, Blackwell Publishing.

Cooper, J.A.G. and **Pilkey, O.H.** (2004) Sea-level rise and shoreline retreat: time to abandon the Bruun Rule. *Global and Planetary Change*, 43, 157-171.

Crooks, S., Schutten, J., Sheern, G.D., Pye, K. and **Davy, A.J.** (2002) Drainage and elevation as factors in the restoration of salt marsh in Britain. *Restoration Ecology*, 10, 591-602.

DEFRA. (2002) Safeguarding Our Seas. A strategy for the conservation and sustainable development of our marine environment. London, Department for Environment, Food and Rural Affairs.

Department Of The Environment. (1972) Out of Sight. Out of Mind. Report of a Working Party on the Disposal of Sludge in Liverpool Bay. Vol.1: Main Report. London, HMSO.

Dickson, E. (1893) The history of the Ribble Valley and Estuary. *The Southport Society of Natural Science*, Fourth Session, 3-13.

Doody, J.P.E. (1984) *Spartina anglica* in Great Britain: a report of a meeting held at Liverpool University on 10th Nov. 1982. Focus on Nature Conservation No.5. Nature Conservancy Council.

Fairhurst, G.J. (c.1974) *A study into changes in the estuary of the River Ribble, Lancashire.* Unpublished BSc Project Report, Liverpool Polytechnic.

Fowler, A.F. (1909 (a)) Report to the Chairman and Members of the Improvement and Parks Committee, Southport Corporation on the Bog Hole Channel. *Unpublished archive document*, SMBC.

Fowler, A.F. (1909(b)) Letter to J.E. Jarrat (Town Clerk, Southport) regarding the Bog Hole Channel. *Unpublished archive document*, SMBC.

Fowler, A.F. (1910) Report to the Improvement and Parks Committee, Southport. *Unpublished archive document*, SMBC.

Freeman, T.W., Rodgers, H.B. and **Kinvig, R.H.** (1966) *Lancashire, Cheshire and the Isle of Man*. London, Thomas Nelson and Sons Ltd.

Gresswell, R.K. (1953) *Sandy Shores of South Lancashire*. Liverpool, Liverpool University Press.

Gresswell, R.K. and **Lawton, R.** (1964) *Merseyside*. Sheffield, The Geographical Association.

Hansom, J.D., Comber, D.P.M. and **Fahy, F.M.** (1993) Estuaries Management Plans, Coastal Processes and Conservation. Ribble Estuary. Report for English Nature.

Healy, M.G. and **Hickey, K.R.** (2002) Historic land reclamation in the intertidal wetlands of the Shannon Estuary, western Ireland. *Journal of Coastal Research Special Issue 36*, 363-373.

Holliday, R.J. (1976) An assessment of the problem of *Spartina* control in the Marshside region of the Southport shore. Unpublished report to Metropolitan Borough of Sefton., University of Liverpool, Department of Botany.

Hydraulics Research Station (1965) An investigation of sand movements in the Ribble Estuary using radioactive tracers. Report No. Ex.280.

Hydraulics Research Station (1968) Note on Engineering Works to Reduce Dredging in the Ribble Estuary. Report No. Ex.391.

Inglish, C.C. and **Kestner, F.J.T.** (1958) The long term effects of training walls, reclamation and dredging on estuaries. *Proceedings of the Institution of Civil Engineers*, 9, 193-216.

Jackson, A.E. (1914) Letter to Mr Marshall, Southport and District Fishermen's Association, re: Leading Lights, Southport Channel. *Unpublished archive document*, SMBC.

Jackson, A.E. (1919) Letter to Mr Black, Borough Electrical Engineer, re: Leading Lights, Southport Channel. *Unpublished archive document*, SMBC.

Jackson, A.E. (1937) Southport Foreshore. *The Southport Visitor.*

Jarrad, R.N. (1907) Report on the South (Bog Hole) Channel. *Unpublished archive document*, SMBC.

Jones, C.R., Houston, J.A. and **Bateman, D.** (1991) A history of human influence on the coastal landscape. In Atkinson, D. & Houston, J. (Eds.) *The Sand Dunes of the Sefton Coast. Proceedings of the Sefton Coast Research Seminar, Liverpool*. Liverpool, National Museums & Galleries on Merseyside.

Jones, P.D. (2006) Water quality and fisheries in the Mersey Estuary, England: A historical perspective. *Marine Pollution Bulletin*, 53, 144-154.

Kestner, F.J.T. (1979) Loose boundary hydraulics and land reclamation. In Knights, B. and Phillips, A. J. (Eds.) *Estuarine and Coastal Land Reclamation and Water Storage*. Farnborough, Saxon House.

Long, A.J., Waller, M.P. and **Stupples, P.** (2006) Driving mechanisms of coastal change: Peat compaction and the destruction of late Holocene coastal wetlands. *Marine Geology*, 225, 63-84.

Mamas, C.J.V., Earwaker, L.G., Sokhi, R.S., Randle, K., Beresford-Hartwell, P.R. and **West, J.R.** (1995) An estimation of sedimentation rates along the Ribble Estuary, Lancashire, UK, based on radiocaesium profiles preserved in intertidal sediments. *Environment International*, 21, 151-165.

Marshall, J. (1914) Various letters to Mr Jackson, Southport Borough Engineer, from Southport and District Fishermen's Association, re: Lighting of Southport Channel. *Unpublished archive document*, SMBC.

Messent, P.J. (1888) Effect of the works now being executed in the Ribble by the Preston Corporation, a report to the Southport Corporation and the Lords of the Manor of North Meols. *Unpublished archive document*, SMBC.

Murray, L.A. and **Norton, M.G.** (1979) The composition of dredged spoils dumped at sea from England and Wales. Ministry of Agriculture, Fisheries & Food, Directorate of Fisheries Research, Technical Report No.52. Lowestoft, MAFF.

Norton, M.G., Rowlatt, S.M. and **Nunny, R.S.** (1984) Sewage sludge dumping and contamination of Liverpool Bay sediments. *Estuarine, Coastal and Shelf Science*, 19, 69-87.

O'Connor, B.A. (1987) Short and long term changes in estuary capacity. *Journal of the Geological Society*, London, 144, 187-195.

Pethick, J.S. (2002) Estuarine and tidal wetland restoration in the United Kingdom: Policy versus practice. *Restoration Ecology*, 10, 431-437.

Pye, K. and **French, P.W.** (1993) Erosion and accretion processes on British saltmarshes. Vol II: Database of British Saltmarshes. Final report to MAFF, MAFF.

Pye, K. and **Neal, A.** (1994) Coastal dune erosion at Formby Point, north Merseyside, England: Causes and mechanisms. *Marine Geology*, 119, 39-56.

Rainfords Ltd (2005) Ribble Estuary Sandwinning - Horse Bank, Southport. Aerial inspection monitoring summary report 1993 to 2001.

Ramster, J.W. and **Hill, H.W.** (1969) Current system in the northern Irish sea. *Nature*, 224, 59-61.

Rees, H.L., Rowlatt, S.M., Limpenny, D.S., Rees, E.I.S. and **Rolfe, M.S.** (1992) Benthic studies at dredged material sites in Liverpool Bay. IN MAFF Directorate of Fisheries Research, *Aquatic Monitoring Report No. 28*, MAFF.

Rendel, Palmer and **Triton** (1967) Ribble Estuary Navigation Channel. Report to Port of Preston Authority. *Unpublished archive document*, SMBC.

Smith, A.J. (1982) Guide to the Sefton Coast Database. Southport, Sefton MBC.

Smith, G.M., Spencer, T., Murray, A.L. and **French, J.R.** (1998) Assessing seasonal vegetation change in coastal wetlands with airborne remote sensing: an outline methodology. *Mangroves and Salt Marshes*, 2, 15-28.

Stammers, M. (1999) Shipways and steamchests: the archaeology of 18th- and 19th- Century wooden merchant shipyards in the United Kingdom. *The International Journal of Nautical Archaeology*, 28, 253-264.

Thomas, C.G., Spearman, J.R. and **Turnbull, M.J.** (2002) Historical morphological change in the Mersey Estuary. *Continental Shelf Research*, 22, 1775-1794.

Thompson, R.C., Crowe, T.P. and **Hawkins, S.J.** (2002) Rocky intertidal communities: past environmental changes, present status and predictions for the next 25 years. *Environmental Conservation,* 29, 168-191.

Tomlinson, R. (1997) Survey of beach vegetation, Southport. Nov 96 - March 97.

Tovey, N.E. (1965) Municipal works in Southport. *17th Annual Conference of the Institute of Works and Highways Superintendents.* Southport.

University of Liverpool (1986) Sefton Coastal Study. Stage VII. Volume I. *Unpublished archive document,* SMBC.

van der Wal, D. and **Pye, K.** (2003) The use of historical bathymetric charts in a GIS to assess morphological change in estuaries. *Geographical Journal,* 169, 21-31.

van der Wal, D., Pye, K. and **Neal, A.** (2002) Long-term morphological change in the Ribble Estuary, northwest England. *Marine Geology,* 189, 249-266.

Vernon-Harcourt, L. F. (1905) Report on the extension of the Ribble training walls proposed in the Ribble Navigation Bill of 1905. *Unpublished archive document,* SMBC.

Warren, R.S., Fell, P.E., Rozsa, R., Brawley, A.H., Orsted, A.C., Olson, E.T., Swamy, V. and **Niering, W.A.** (2002) Salt marsh restoration in Connecticut: 20 years of science and management. *Restoration Ecology,* 10, 497-513.

Wheeler, W.H. (1893) *Tidal Rivers.* London, Green & Co.

Woodroffe, C.D. (2002) *Coasts: Form, Process and Evolution.* Cambridge, Cambridge University Press.

Sefton South Shore: Understanding coastal evolution from past changes and present dynamics

Andrew J. Plater+, David Hodgson*, Michelle Newton and Graham Lymbery

+Department of Geography, University of Liverpool, Roxby Building, Chatham Street, Liverpool, Merseyside, L69 7ZT, UK

*Department of Earth and Ocean Sciences, University of Liverpool, 4 Brownlow Street, Liverpool, Merseyside, L69 3GP, UK

Sefton Council – Coastal Defence, Ainsdale Discovery Centre, The Promenade, Shore Road, Ainsdale-on-Sea, Southport, PR8 2QB, UK

ABSTRACT

The evolution, morphology and character of the Sefton South Shore (from Formby Point to Bootle) are considered in relation to changes in relative sea-level, patterns of sediment erosion, transport and deposition, and human activity from early settlement to present-day engineering and management actions. The morphology of the shoreline plays an important role in directing sediment along and away from the coast at Formby Point towards the inter- and sub-tidal sand banks that characterise the outer Mersey estuary. Equally, the foreshore is an important source of sand for the Sefton dunes, which show a high degree of susceptibility to changes in sediment supply and episodic storms. About 6,000 years ago the Sefton South Shore became the location for a coastal dune barrier complex and a barrier estuary in the vicinity of the River Alt. The variety of wetland environments in the back-barrier lowland proved an attractive area for human settlement and activity from about 5000 to 3000 years ago, which was probably brought to a close by a phase of dune mobility and encroachment. Human activity in the last 150 years or so, in the form of engineering works and dredging to maintain navigation and, more recently, for coastal defence, has had an important impact on sedimentation at the interface between Liverpool Bay and the River Mersey, leading to significant changes in nearshore sedimentation and the morphology and location of the sand banks and main drainage channels. Throughout its development, the Sefton South Shore has been characterised by dynamic response to changing conditions, whether brought about by climate, sea level or humans. It is critical that this is taken into consideration in future coastal management.

INTRODUCTION

Set within our understanding of long-term (post-glacial, the last c.20,000 years) coastal evolution and the present-day dynamics of Irish Sea-Liverpool Bay-Sefton Coast continuum (Plater and Grenville, This volume), this chapter aims to first describe the morphology of the Sefton South Shore (from Formby Point to Bootle) (Figure 1). The nature and drivers of coastal change are then considered – from the link between Holocene (the last 10,000 years) sea-level change and the development of coastal wetland environments, to the impact of changes in sediment budget and storm events on the integrity of the dune coast in recent times. From this evidence of past and recent change, we are therefore equipped to consider how the Sefton South Shore will respond to climate change in the near future, and the challenges that face the people who manage, work on, and live at the coast. The focus of the chapter is very much on the results of recent research and monitoring, within the context of previous work, but it also considers new developments in research that aim to improve knowledge more widely than that of local and regional significance.

The dynamics of the Sefton South Shore are closely tied to the southeasterly drift of sediment from Liverpool Bay, and indeed alongshore from Formby Point, and the outflows from the rivers Mersey and Alt (Figure 1). Consequently, it lies at a crossing point or interface between a littoral drift cell (an area of coastal sediment erosion, transport and deposition) and cyclic tidal flushing of the estuaries. It thus exhibits a morphology

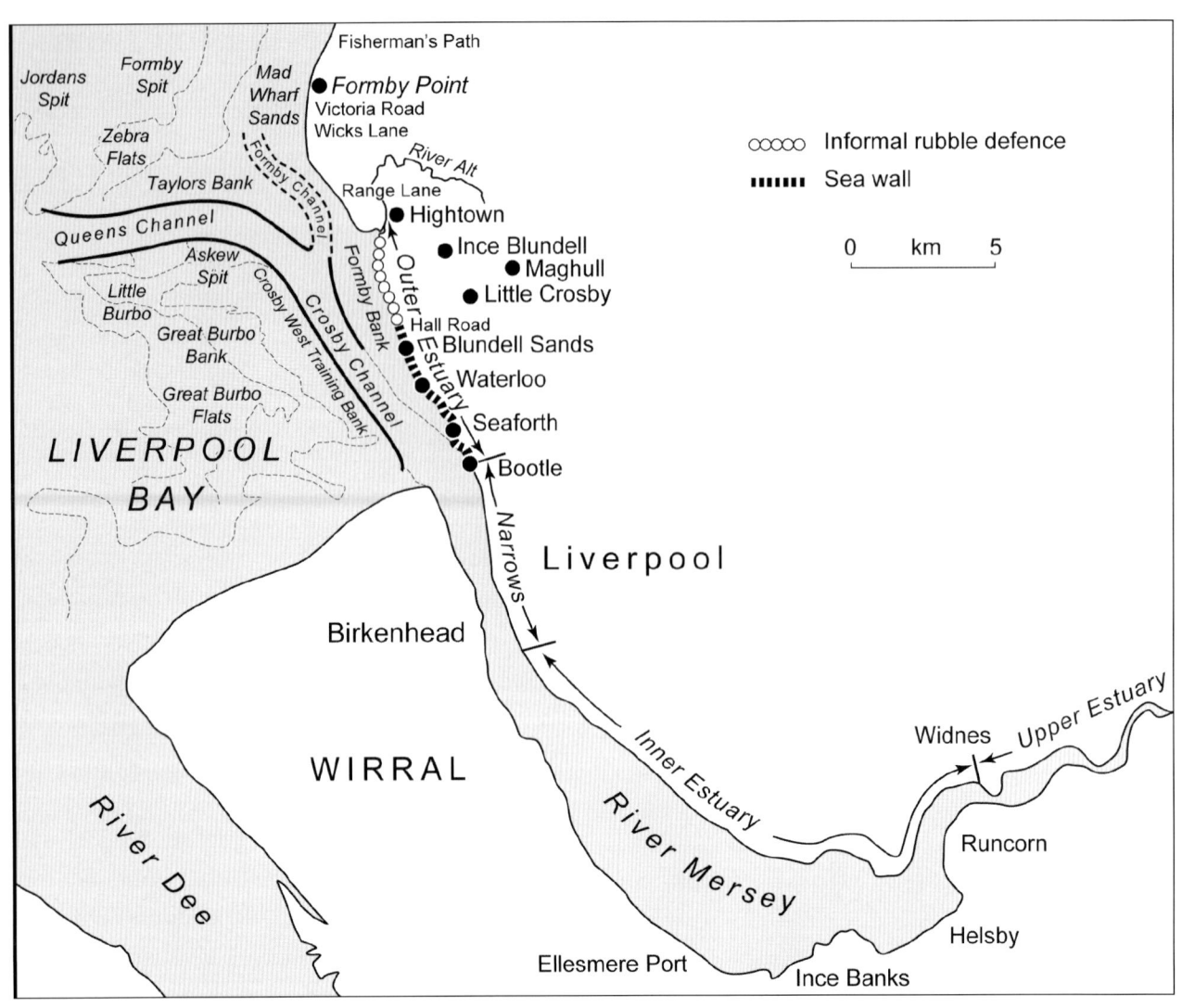

Figure 1. Location map of the Sefton South Shore and the Mersey estuary, showing main features and location of sites considered in the text. (After Blott et al. 2006).

and degree of morphological change that is highly responsive to a range of external forcing factors and internal controlling mechanisms. In addition, the southern part of this shore is characterised by Seaforth Docks, Crosby Marina, the Marine Lake at Waterloo, and a series of hard defences and drainage works, and thus has been significantly affected by humans during the last c.150 years.

MORPHOLOGY

Foreshore geomorphology

The Sefton coast has a wide gently-sloping shore, characterised by a series of low-amplitude ridges made up of sand and mud (Parker, 1971, 1975). A range of dynamic conditions occur along the coastline, from accretion to the north and south forming a foreshore with long continuous ridges and troughs and regularly spaced rip channels (channels that cut through the ridges thus enabling offshore drainage), to a retreating sand dune coast in the region of Formby Point (Figure 2). On the retreating part of the coast, erosion operates on a series of timescales and varies spatially (Parker, 1974), where recorded trends correspond qualitatively with theoretical calculations of sand flux (HRS, 1969) (Figure 3). Southward, a wider intertidal area marks the onset of shoreline accretion. The transition from one state to the other was the focus of sediment budget calculations undertaken by the Institute of Oceanographic Sciences, Taunton, in the 1970s (Parker, 1974). Previous work had shown that the primary modes of sediment transport on the foreshore were: (i) sediment movement mainly as bedload via bedforms; (ii) sediment transport concentrated in zones – the troughs and the upper foreshore are dominated by longshore transport, and beach ridges are zones where

Table 1. Sand transport on a low-amplitude ridge foreshore. (After Parker, 1971, 1975).

Tidal state	Predominant geomorphic process
Low Water	Currents draining the trough (runnel) transport sand alongshore.
Flood - early	Sand is swept landward by swash over the ridge (bar) crest and into the trough which is filling with water.
Flood - mid	Breaker zone operates over the ridge crest. Sand is moved landward into the trough which is the site of a longshore current. Sediment bedforms (megaripples and ripples) form in the trough.
Flood – late	Breaker zone decays, longshore currents diminish. Limited sand movement.
High Water	Low currents, little sediment movement.
Ebb – early	Breaker zone develops over ridge. Longshore currents are generated. Sand is moved along the trough, as well as being swept over the ridge into the longshore current.
Ebb – mid	Sand is swept landward by swash into the trough draining alongshore. Breaker zone develops over the next ridge to seaward. Eventually the ridge becomes exposed.
Ebb – late	Trough drains. Currents move sand alongshore.

transport directions are more at right-angles to the shoreline; and (iii) the timing and landward vectors of sand movement on the beach bars are short. On both flood and ebb tides, sand transport on the bar takes place when longshore currents are most active. Here, the sand swept over the bar crest and into the adjacent trough is moved alongshore by currents (see Table 1). Hence, the longshore and rip currents in the topographic lows play an important role in sand movement on the foreshore. However, because of the relative orientation of the coast and the foreshore bars, the sand in the troughs has a net offshore transport direction and the topography generally causes sand to move away from the high water mark. Sediment is therefore largely recycled on the foreshore by this return flow, and the onshore transfer of sediment between the beach and the adjacent sand dunes is limited.

Parker (1971) demonstrated that sand moves predominantly in bedforms as ripples and megaripples. Megaripple bedforms or dunes (not to be confused with the large-scale aeolian dunes at the coast) provide useful indicators of the direction and extent of sediment transport over the foreshore in this context. Bedforms and field experiments show a pattern of divergence of bedload transport away from the centre of the erosion front at Wick's Lane, concentrated in the troughs where the main transport direction is away from the high water

Figure 2. Aerial photograph montage showing the character and morphology of the Sefton South Shore. (Source: Sefton Council). © Crown copyright. All rights reserved. Sefton Council Licence number 100018192 2010.

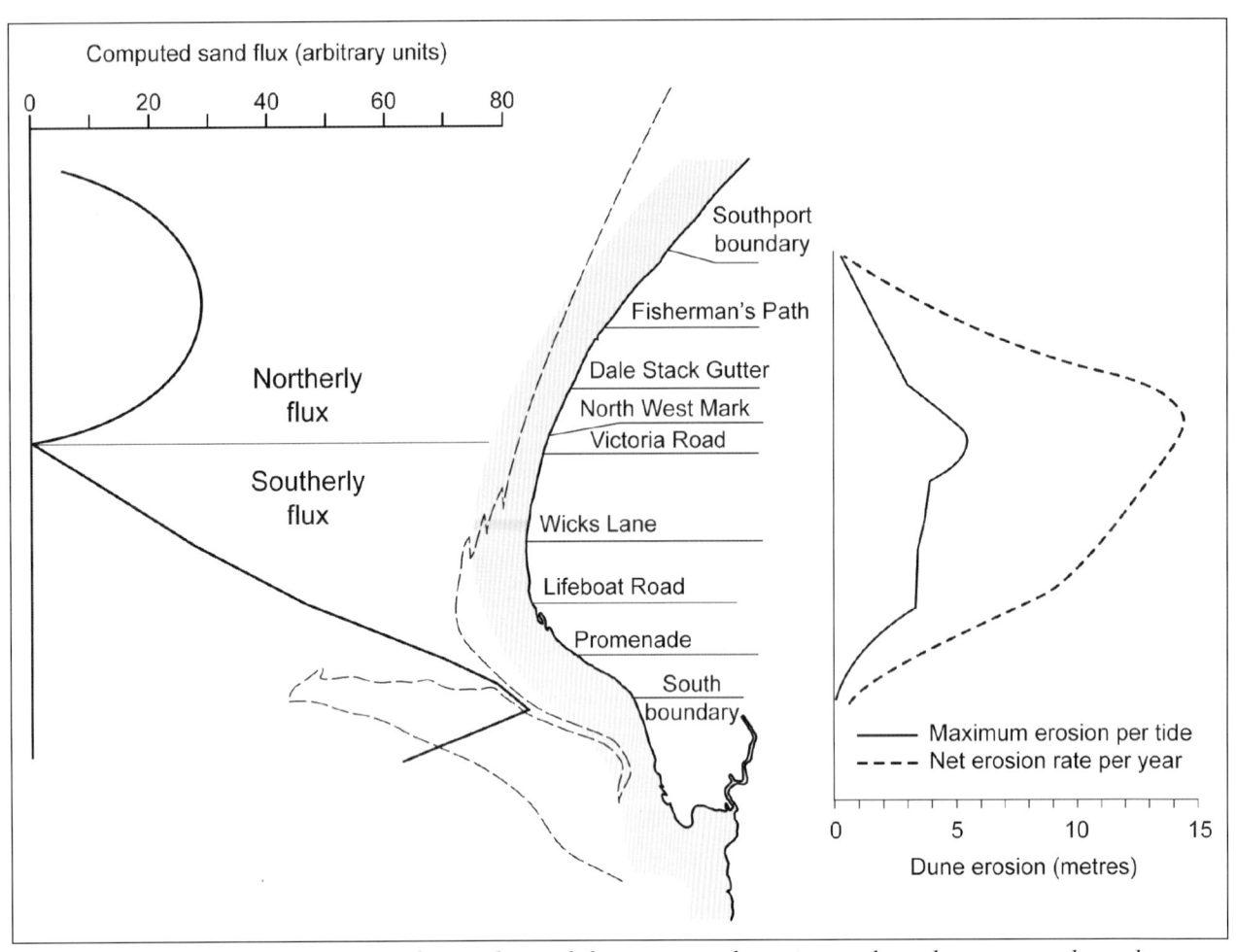

Figure 3. Theoretical calculations of spatial variability in coastal erosion and sand transport along the Sefton shore in the vicinity of Formby Point. (After Parker, 1974).

mark. This offshore component is reinforced by seaward discharges of sediment-laden water through rip channels. Only transport over the upper foreshore planar area maintains a supply along the coast without taking it offshore.

To the south of Formby Point, there is an interaction of the southerly sand flux and sedimentation processes within Formby Channel (Parker 1974), comprising the sediment being transported by the intertidal troughs and that taking place on the planar upper foreshore between the most landward bar and the sand dune frontage. During the ebb tide and at low water, the troughs discharge sand into Formby Channel, leading to significant sedimentation and emergence in some parts. Parker (1975) notes that this sediment feed has maintained two large deltaic systems at the exits from the troughs into the main tidal drainage channel. Through time, this process leads to changes in the shape and migration of the channel (see Nearshore bathymetry page 93).

Booth et al. (2008) recently undertook detailed levelling, dGPS and granulometric analysis of the low-amplitude ridges on the foreshore at Crosby where Antony Gormley's 'Another Place' is installed. Here, the statues form reference points against which to examine the short-term dynamics of the beach morphology and grain size. Particle size distributions were largely unimodal and well sorted, medium to coarse sands, with a slight fining trend towards the sea. More significantly, the beach height data revealed no overall pattern of accretion and/or erosion, but daily changes in elevation of the order of 10 cm compared with annual changes of the order of 100 cm. This highlights that nested within the long-term persistence of ridges on the foreshore, there is considerable diurnal variability in beach morphology that cannot easily be linked to tidal phase but rather is indicative of high sensitivity to the prevailing state of tide-wave interaction (Plater and Grenville, This volume).

Beach-dune interaction

Aeolian sediment transport across beaches is complex and controlled by a number of factors. Sediment flux at any given point and time is dependent upon wind stress, wind field variability, fetch, transported grain size, available sand supply, moisture content and grain size distribution of the source sand surface, bed roughness, beach slope, vegetation cover and surface heterogeneity (Bauer et al., 2008). Aeolian sediment transport over the Sefton foreshore is largely ineffective despite the onshore westerly winds. The troughs tend to remain wet at low water whilst wet areas are also found on the upper foreshore due the impermeable nature of the underlying Holocene stratigraphy (Parker, 1975). These areas are then unable to liberate significant volumes of sediment through aeolian transport unless they have sufficient time during the tidal cycle to become dry. The planar section of the upper foreshore is, therefore, the only potential source of sand for aeolian transport.

As a further consideration, mud sedimentation increases the natural cohesion of the sediment surface, and hence bed stability. Fine sediment in suspension becomes flocculated and deposited, quickly forming a cohesive surface through dewatering and evaporation. Parker (1975) suggests a surface cohesion capable of withstanding currents of the order of 0.3-0.4 m/s in 0.3 m water depth, yet the muds possess sufficiently low shear strengths as to be deformed by wave-induced currents. Whilst the supply of mud to the foreshore ensures continued mud deposition (Halliwell and O'Connor, 1966) over much of the foreshore (Parker 1973, 1975), its longevity and preservation are limited to those areas protected from ebb breakers, i.e. in troughs between ridges. These veneers of mud then act to limit sediment mobility.

The gradient and width of the beach also have a profound effect on the potential rate of landward aeolian sediment transport; the flatter the surface, the less the wind velocity fluctuates and the greater the potential for sand transport (Hsu, 1977). Here, changes in beach topography induce changes in the velocity fields resulting in flow separation, thus reducing velocity gradients, surface shear stress and the volume of sand transported (Hesp, 1982). The Sefton shore exhibits the morphology of a dissipative shoreline (Short and Hesp, 1982), even though the prevailing wave climate (see Plater and Grenville, This volume) is more akin to reflective surf zone dynamics. Dissipative beaches are characterised by a wide low-gradient beach face and subdued shore-parallel bars and troughs. Such shorelines generally show low mobility with infrequent erosion which is continuous alongshore and causes parallel backbeach-foredune scarping. This morphology induces minimum disturbance of the wind flow across these beaches and the potential sand transport onshore by winds is at a maximum, particularly at low tide. However, because of the low wave energy in the case of the Sefton South Shore, beach surfzone gradients tend to be lower and the intertidal beach width greater (Short and Hesp, 1982). Sediment is, therefore, not as readily transported to the beach over time, and aeolian sediment transport is reduced – especially as the water table is very close to the beach surface (Short and Hesp, 1982).

Sherman and Bauer (1993) designed a conceptual framework, based on that of Valentin (1952), for considering beach-dune interaction in the context of relative sea-level rise (Figure 4). Two end-member states can be identified as: (i) unconditionally advancing coasts, where depositional processes and/or emergent trends create a seaward migration; and (ii) unconditionally retreating coasts, where erosion or relative submergence cause shoreline retreat. Within this framework, it is easy to envisage a system switch from an advancing to a retreating coast (bottom left sector of Figure 4) due to a subtle increase in the rate of relative sea-level rise or a reduction in the extent of net deposition.

Rivers at the coastal interface

The main fluvial drainage on the Sefton South Shore is the River Alt (Figure 1). This rises to the east, flows northward across the lowland coastal plain towards Formby before turning southward and entering the sea at Hightown. The Alt is 17 km in length, with a

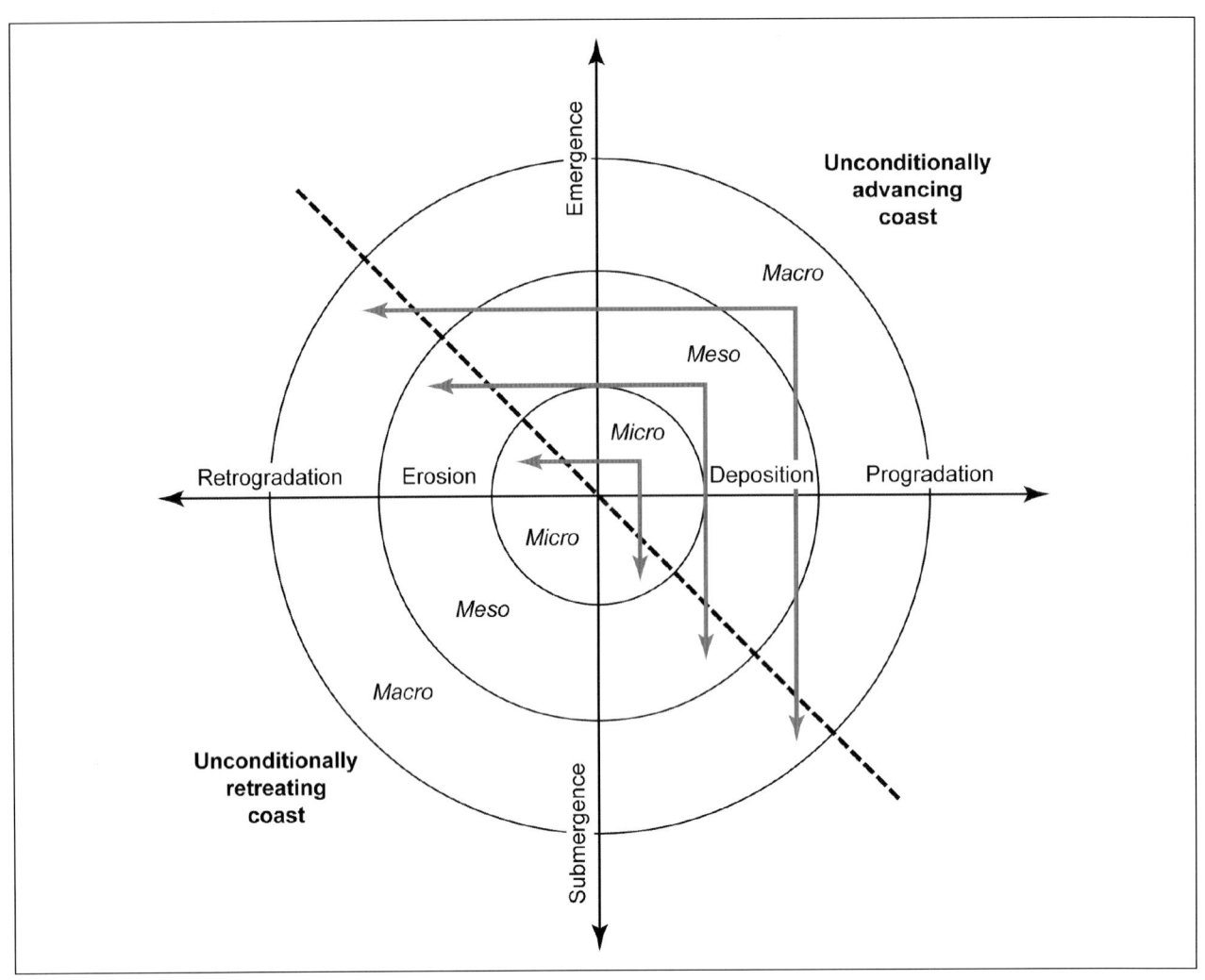

Figure 4. Beach classification scheme showing the influence of relative sea level and net sedimentation on coastal response (after Valentin, 1952, and Sherman and Bauer, 1993). The rate of relative sea-level change (emergence vs submergence) is expressed on the vertical axis, and net sediment accretion vs erosion on the horizontal axis. In addition, tidal regime is indicated by the concentric circles. Advancing and retreating coasts are separated by the dashed line. The arrows pointing from right to left indicate the likely coastal response to reduced sediment supply, whilst those pointing downwards indicate the consequence of enhanced sea-level rise.

catchment covering 240 km² (Mersey and Weaver River Authority, 1971). The lower Alt from Maghull to Hightown is particularly low-lying, mainly agricultural land at considerable risk from flooding. Indeed, the Alt lowlands have a history of flooding as the seaward gradient is very low and it is bordered by wide alluvial flats. It has, therefore, been managed from at least the C13th and certainly from the C17th by Dutch engineers (Newton et al., 2007b). Tidal gates were built in the C18th at the mouth of the river and at Ince Blundell to prevent tidal flooding, and were improved in 1830. This has not, however, prevented flooding from river discharge and runoff during extreme weather (Payne, 1976).

The adjacent Mersey estuary (Figure 1) has a macrotidal regime, with a mean spring tidal range of 8.4 m at Liverpool (Admiralty, 2005). Tidal current velocities reach a maximum of 1 m/s on the flood tide in the outer estuary, at the entrance to Queen's Channel, and c.2.2 m/s on both the flood and ebb tides in the narrows between Liverpool and Birkenhead (Blott et al., 2006). The estuary is partially mixed; a mean freshwater flow of 66 m³/s (Shaw, 1975) contrasts with an average spring tidal influx through the narrows of c.2000 m³/s (McDowell and O'Connor, 1977).

Depending on river inflow, average salinity values range from 4 to 11 g/l (Dyer, 1997). Although the largest waves are from the west (Sly, 1966; Pye and Neal, 1994), with c.52% of all waves approaching from the SW to NW, wave action in the estuary is limited due to the constriction at the narrows. Indeed, despite significant waves with heights of 5 m being observed in Liverpool Bay during winter (McDowell and O'Connor, 1977), the mean annual wave height in the estuary is 0.8 m. Morphologically, the Mersey estuary may be divided into: (i) the wide outer estuary adjoining Liverpool Bay (Figure 1), characterised by extensive sand banks and maintained navigation channels; (ii) the narrows, where the estuary becomes restricted between Permo-Triassic rocky shores to c.1.5 km in width near Liverpool; (iii) the inner estuary, where the estuary widens and experiences sediment deposition in the form of sand bars and fringing tidal flats and saltmarshes; and (iv) the upper estuary, extending from Runcorn to the tidal limit at Warrington.

CHANGES IN MORPHOLOGY

Long-term evolution of the Sefton South Shore

The buried peat deposits and intercalated sands and muds of the Mersey and the wider Merseyside region have long been known for their contained evidence of environmental change (Binney and Talbot, 1843; Picton, 1849; Morton, 1887, 1888, 1891; de Rance, 1869, 1872, 1877; Reade, 1871, 1872, 1881, 1908; Blackburn (in Cope, 1939)) and vegetation history (Travis, 1908, 1926, 1929; Travis, 1922; Erdtman, 1928). Indeed, the combined works of R. Kay Gresswell (1937, 1953, 1957) and Michael Tooley (1969, 1970, 1974, 1976, 1977, 1978a, 1978b, 1980, 1982, 1985a, 1985b, 1990, 1992; Huddart et al., 1977; Tooley and Kear, 1977; Innes et al., 1989) have provided a strong stratigraphical, geomorphological and palaeoecological framework for studies of coastal evolution and sea-level change during the Holocene period, and have been added to in recent years by works on the Sefton Coast (Wilson et al., 1981; Innes et al., 1989; Huddart, 1992; Innes and Tooley, 1993; Neal, 1993; Pye and Neal, 1993a, 1993b; Pye et al., 1995; Gonzalez et al., 1997; Roberts et al., 1996; Zong and Tooley, 1996, Holden, This volume) and the Wirral Shore (Kenna, 1976; Innes et al., 1990; Bedlington, 1995).

The general stratigraphic sequence of the Sefton South Shore comprises glacial deposits overlain by a sand, the Shirdley Hill Sand, of predominantly late Glacial aeolian origin but with considerable evidence of fluvial and especially marine reworking. This gives way to a marine silt with frequent Phragmites rhizomes, the Downholland Silt, being indicative of an extensive tidal flat and estuarine system giving way to saltmarsh and perimarine wetland environments during the mid-Holocene. This grades into extensive peat beds inland, whilst at the coast the thin peats are overlain by a dune sand complex. Erdtman (1928) completed pollen analysis of the peat at Hightown, revealing evidence of a bog or fen deposit. The peat in the Alt valley averages about 1-1.5 m in thickness where it overlies Downholland Silt and, to the east, Shirdley Hill Sand. The peat contains large tree stumps or 'stocks' of ash, oak and birch. Near Hightown, the peat, or forest bed, is up to 2 m thick but thins away from the Alt mouth and towards Hall Road to between 0.2-1.2 m (Wray and Cope, 1948). This has been described in detail by Travis (1926) and Tooley (1977, 1982), who dated the transition from marine to freshwater conditions to 4545+/-90 ^{14}C years BP. Peat continues beneath dune sands southward along the coast, being overlain by about 2 m, and as much as 4 m, of blown sand between Hall Road station and Little Crosby. The peat attains a thickness of as much as 9 m at Seaforth station, but is generally c.1 m thick in the region of Bootle. At Sniggery Wood near Little Crosby, the peat immediately beneath the dune sand yielded an age of 4510+/-50 years BP, and a later date of 3910+/-50 years BP was obtained for sand overblowing the peat at Mount Pleasant, Waterloo (Innes and Tooley, 1993). This phase of dune building between 4600 and 4000 years BP is closely equivalent to the Older Dunes in the Netherlands (Jelgersma et al., 1970), and probably dates from a period of increased storminess and/or enhanced wind climate. Sand blow continued to consume the South Sefton coastal hinterland beyond this date, as evidenced by a thin peat bed resting on glacial till at Church Road, Waterloo, which was buried by sand after 3200+/-60 years BP, and a peat beneath 5 m of sand is dated to 2680+/-50 years BP at Murat Street, Waterloo (Innes and Tooley, 1993).

The Holocene stratigraphy in the region of the River Alt was investigated by Huddart (1992). This included a sequence of late Glacial or early Holocene nearshore sands exposed in drainage ditches to the north-west of Lady Green, near Ince Blundell, potentially derived from sandy flats bordering the former shoreline (Jones et al., 1938; Wray and Cope, 1948). These were subsequently stabilised through soil development and colonisation by vegetation, before being inundated by high tidal levels and forming a lower to upper marsh environment. Sections were also exposed in a northern meander cut of the Alt at Alt Bridge and Abraham's Bridge, where more than 2 m of laminated fine sand and silt is indicative of an estuarine and tidal facies incised into a saltmarsh during a negative tendency in relative sea-level, with a clear connection developed to the open sea. It was concluded that the estuary and connection to the sea from the Holocene back-barrier environments of Downholland Moss were to the south-west through the present location of the River Alt outfall, emphasising some degree of inheritance in the present-day drainage pattern. This is supported by aerial photographic evidence that the southern part of the back-barrier wetland was drained by creeks flowing towards the River Alt (Pye and Neal, 1993b). Tooley (1992) suggested that the relict tidal creeks mapped by Huddart (1992) and later by Neal (1993) were associated with the most recent phase of tidal flat development c.6000-5600 ^{14}C years BP.

Holocene coastal sediments in the region also provide evidence of groundwater and sea-level change which have been important controls on the nature and development of past coastal environments and, hence, prehistoric human activity (Cowell and Innes, 1994; Roberts et al., 1996; Gonzalez et al., 1997; Huddart et al., 1999a, b; Gonzalez and Huddart, 2002a, b; Gonzalez and Cowell, 2004; Adams and Harthen, 2007; Cowell, 2008; Roberts and Worsley, 2008). At Formby, Tooley (1970) noted 'cattle' hoofprints in the foreshore silts which were subsequently recorded in Hale (1985). These were found beneath a woody detrital peat dated to 2334+/-120 ^{14}C years BP. Subsequently, large numbers of prehistoric human and animal footprints (aurochs, red deer, roe deer, dog) and animal bones (aurochs, red deer and dog) preserved in the beach sediments of Formby Point have been recorded during the last 30 years (Cowell et al., 1993; Pye and Neal, 1994; Roberts et al., 1996; Gonzalez et al., 1997; Huddart et al., 1999a). Gordon Roberts completed a systematic survey of the range of animal, bird and indeed human imprints, in which they have largely been recorded in two separate levels (Figure 5). The upper series is found in intertidal sediments, which overlie a pre-existing sandy barrier facies, that date from 3650-3250 years BP (Gonzalez et al., 1997; Roberts et al., 1996). The period during which the imprints were made is considered to be during the later Neolithic to early Bronze Age (c.4000-3600 years BP). Relationships between the animal and human footprints have been interpreted as evidence of some degree of husbandry, with the coast being used for access as inland was either too wet or too heavily wooded (Huddart et al., 1999b).

On the Sefton South Shore, large numbers of mammalian remains have been obtained from Hightown, mainly from the intertidal peat bed exposed at the Alt Mouth. Reade (1881) and Moore (1881) described the remains of horse, boar, red deer, small ox, sheep and dog associated with a Phragmites peat dating from 4545+/-90 ^{14}C years BP. These have been added to with later findings of aurochs, red deer, horse and wild pig (Huddart et al., 1999b). Here, a series of Neolithic artefacts include an arrowhead and three axeheads (Cowell and Innes, 1994), as well as flint scatters around Ince Blundell and Lady Green. Most recently, Gonzalez and Cowell (2004) described a prehistoric trackway just south of the main Hightown intertidal peat exposure. This is a timber brushwood feature, 2.0 m in length, 1.4 m wide and 0.3 m in depth, radiocarbon dated to 5020+/-60 and 4910+/-60 ^{14}C years BP. In addition, a facetted point, driven vertically into the underlying sediments, dates from 4430+/-80 ^{14}C years BP. Foraminiferal data show that the

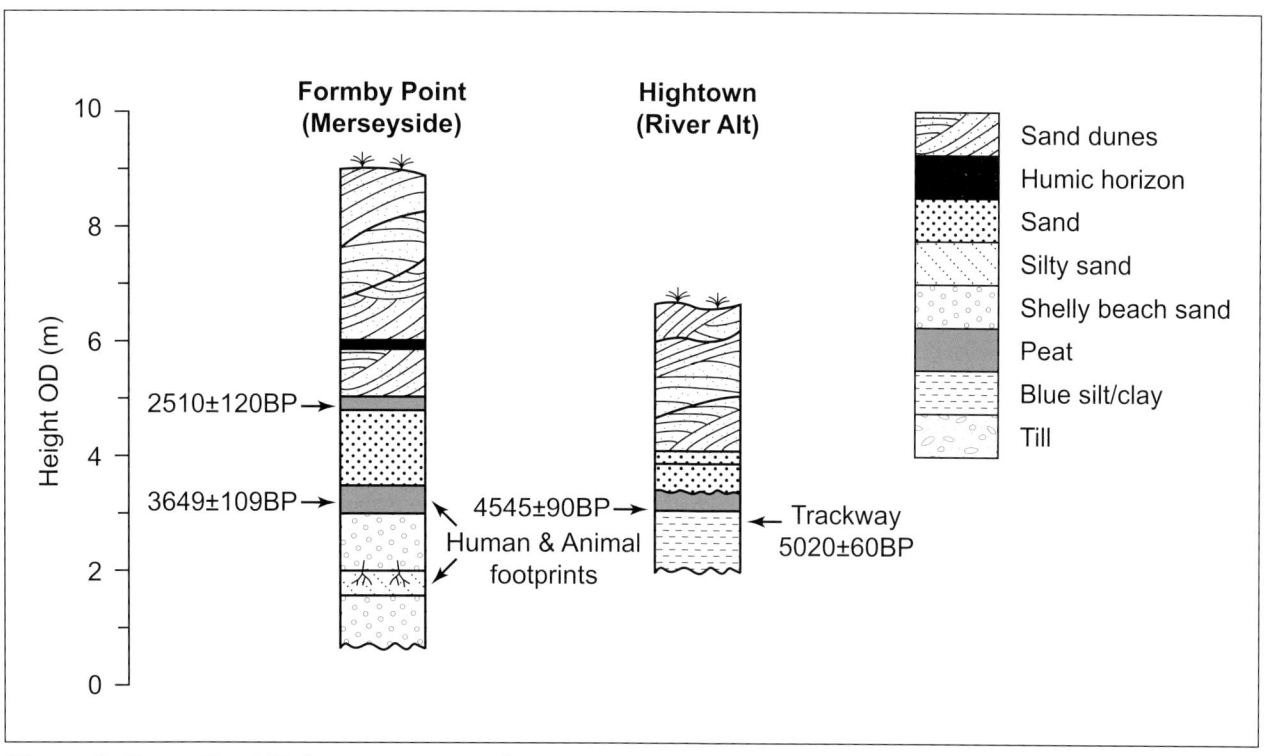

Figure 5. Generalised Holocene stratigraphic sequence on the Sefton South Shore. (Edited and redrawn from Chiverrell et al., 2004).

trackway was constructed across a saltmarsh and Phragmites reed swamp.

In terms of the coastal palaeoenvironments of the Sefton South Shore, active saltmarsh may have existed on the margins of the early Alt estuary during the Neolithic with more freshwater marsh and lagoonal conditions prevailing behind the coastal barrier complex. The breaks in the silt exposures at the coast provide evidence of tidal creeks or drainage outlets from this back-barrier wetland. The intertidal environment appears to have been overcome relatively rapidly by sand dunes soon after 3250 ^{14}C years BP (Neal, 1993), which were subsequently stabilised with the formation of peat horizons c.2500-2250 ^{14}C years BP. A similar phase of dune development is also recorded at Waterloo, Crosby (Innes and Tooley, 1993). Cowell and Innes (1994) suggested that isolated flint production sites may have been present on the coast fringe during the Mesolithic. Indeed, the lower series of human footprints may well date from the Later Mesolithic c.6800-4000 BC (Huddart et al., 1999a). Here, the dynamic coastal environments of the Mesolithic gave way to a more stable barrier estuary wetland complex during the Neolithic, and then a more established dune coast during the Bronze and Iron Ages.

According to Gonzalez and Cowell (2004) the Alt estuary archaeological sites all either lie on the eastern edge of the perimarine/intertidal wetland transition or on small islands within it. In addition, the sites identified, both wetland edge and further inland, are small and discrete, suggesting some degree of continuity from the Mesolithic to the Neolithic. In contrast, evidence from Woodham Knoll, Little Crosby (Cowell and Innes, 1994) is indicative of the fens here being used for summer pasture for cattle and associated settlements during the Neolithic. There is also evidence for Neolithic cultivation on the Sefton South Shore from Mount Pleasant, Waterloo (Innes and Tooley, 1993). Hence, the further consolidation of the dune coast led to a less opportunistic, more sedentary coastal community.

More recent research on Holocene coastal evolution has focussed on the Mersey estuary proper, especially the inner estuary in the region of Ince Banks and Helsby Marshes. Here, Wilson (2004) and Wilson et al. (2004, 2005a, b) have investigated the potential for using isotopic data to reveal changes in relative sea-level during the Holocene. Wilson et al. (2005a, b) measured $\delta^{13}C$ and C/N ratios of C3 saltmarsh plants and suspended estuarine Particulate Organic Carbon (POC); the two main sources of organic carbon in intertidal sediments. The two sources of organic carbon in high saltmarsh and sub-tidal sediments resulted in differing bulk sediment $\delta^{13}C$ values (-27.8‰ and -22.8‰, respectively) and C/N ratios (11.6 and 9.3, respectively). It was also shown that surface elevation within the tidal frame, i.e. relative to the high tide level, was a key determinant in controlling bulk surface $\delta^{13}C$ and C/N ratios. Here, supratidal and high saltmarsh sediments consisted almost entirely of organic carbon derived from overlying C3 vegetation, whilst the intertidal and sub-tidal flats were almost exclusively characterised by tidally-derived POC. Wilson (2004) and Wilson et al. (2005b) also present the results from two long sediment cores from Helsby and Ince Marshes that reveal the nature of coastal evolution and relative sea-level change since approximately 8200 cal. yr BP. An intercalated sediment sequence from the inner estuary shows periodic switching between tidal flat, saltmarsh and perimarine wetland environments under the influence of sea-level rise, even extending to the existence of mixed oak–hazel woodland during the early to mid-Holocene. Peat layers from the more landward Helsby Marsh core date from c.8300-7600 and c.6800-3400 cal. yr BP, and an organic clay in the 'seaward' Ince Banks core indicates the development of saltmarsh conditions around 7400-7200 cal. yr BP. In these cores, a broad agreement is found between the $\delta^{13}C$ and C/N ratios and more established palaeoenvironmental indicators, i.e. pollen and diatoms. In general, Holocene super-tidal and perimarine environments may be distinguished from inter- and sub-tidal sediments on the basis of more negative $\delta^{13}C$ and higher C/N ratios.

RECENT COASTAL CHANGE

Dune erosion

Coastal dunes often form an important element of coastal defences. Hence, understanding their response to storm and inter-storm processes is an essential component of any coastal management strategy. Indeed, the Middle Age settlement of Argameols had disappeared by 1503, probably washed away by the sea during a storm (Smith, 1999), and there are further accounts of severe storms affecting the coast during the early C18th. Recognising the vulnerability of the coast, measures were already in place from the C17th to prevent the dunes blowing inland by planting and managing dune grasslands, and succeeded by tree plantations in the early C18th.

Between 1845 and 1906, 220 m of accretion was recorded in the region of Victoria Road at Formby Point (Smith, 1999). However, erosion has been progressing at Formby Point since 1906, with as much as 400 m of dunes being lost during the C20th. This has been countered by over 100 m of dune accretion at Range Lane between 1958 and 1991.

Forcing factors of dune erosion and accretion include changes in mean sea level, wind and wave climate, storm surge magnitude/frequency and human endeavour (Saye and Pye, 2007; Pye et al., 2007; Esteves et al., 2009). In relation to dune erosion under storm conditions, van de Graaff (1986, 1994) identified maximum surge level, significant wave height during the maximum surge, grain size of the dune sediment, initial profile shape and dune height, storm surge duration, and the occurrence of squalls and gusts during the storm as important determinants, whilst Edelman (1968) and van der Meulen and Gourlay (1968) previously recognised that the distance of dune front retreat is inversely correlated with dune height. Indeed, the height and extent of foredune development relative to storm surge elevation is a primary control on the response of barrier islands to extreme storm events (Theiler and Young, 1991; Morton, 2002). Parker (1975) stated that combined marine and wind erosion are responsible for dune cliff retreat at Formby Point, particularly where blow-through paths are present. Marine erosion is achieved by undercutting and soaking. Studies of dune erosion (Parker, 1971, 1975) and comparison with model output (Edelman, 1968, 1973) and the findings of van der Meulen and Gourlay (1968) indicate that storm surges are primarily responsible for enhanced erosion when the water level exceeds +9.6 m above local chart datum (given as -4.42 m OD for Liverpool Bay Datum in Parker (1975), thus converting equivalent altitude of +5.18 m OD). Undercutting by waves becomes supplemented during surge conditions by high water levels which cause saturation and slumping. The sand is then easily dispersed by waves. Locally, a water level of approximately +5.2 m OD is therefore sufficient to cause soaking and, thus greatly enhance the rate of cliff retreat.

Over the long term, variations in the magnitude and frequency of storms, changes in nearshore sediment budget, sea level and engineering structures are likely to influence dune erosion and accretion patterns (Psuty, 1988, 1993; Anthony, 2000). Indeed, Houser et al. (2008) argue that sediment is returned to the beachface through nearshore bar migration following the storm, and hence the ability of coastal dunes to recover and the morphological consequences of the next storm are largely dependent upon the availability of sediment from the beachface (Psuty, 1992). This is coupled with the frequency of damaging storms.

Whilst there is a general pattern of dune retreat focussed on Formby Point, the greatest and perhaps most widespread phases of erosion results from storm events. For example, the storm surge of 25th-29th February 1990 produced a water level that was above the +5.2 m OD threshold for soaking plus undercutting for more than 6 hours. This resulted in as much as 13.6 m of dune being eroded at Victoria Road, declining in extent southward to 5.9 m between Albert Road and Alexandra Road and 4.0 m at Range Lane. Erosion during this event also extended southward, affecting the area between Hightown and Crosby. Pye and Blott (2008) show that the frequency and magnitude of storm surges and high tides have had a detectable effect on the rate of dune erosion and accretion along the Sefton coast. Records of dune toe erosion/accretion since 1958 and various field surveys (Parker, 1971, 1975; Pye, 1977, 1991; Jay, 1998) have shown that storms which generate positive surges on successive tides typically cause 2-6 m of dune erosion, and even as much as 14 m. However, on the more accretionary shorelines, individual storms rarely cause more than 3 m of erosion and the damage is generally recovered over a period of days to months (Pye and Blott, 2008).

Nearshore bathymetry

Morphological and bathymetric changes over the historical period along the Sefton South Shore have been investigated by Spearman et al. (2000), Thomas (2002), Thomas et al. (2002) and, more recently, Blott et al. (2006). Here, changes at the interface of the Mersey estuary and Liverpool Bay have been investigated using Historical Trend Analysis (HTA) and Expert Geomorphological Assessment (EGA) (cf. Pye and van der Wal, 2000a, b) based on a series of digitized Admiralty charts dating from 1912, 1949, 1988 and 2002, Ordnance Survey maps, aerial photographic surveys and CASI (Compact Airborne Spectrographic Imager) imagery. A broad view of changing morphology in the outer estuary can be obtained from historical charts dating back to that of Fearon and Eyes in 1738 (see Figure 3 in Blott et al. (2006)). This chart identifies Formby Channel, running offshore and parallel to the Sefton shoreline; Mad Wharf Sands, a wide accretionary shore extending from the River Alt northwards to Southport; Formby Bank, a narrow accretionary shore in the vicinity of Crosby; Rock Channel, running offshore and parallel to the north Wirral shoreline; and Great Burbo Bank, an extensive complex of offshore sandbanks immediately west from Crosby (Figure 6). In Williamson's chart of 1766 (in Allison, 1949), the low water tidal channel of the River Alt that had extended south towards the Mersey had disappeared. This is further evidenced by the 1833 chart, which also shows a greatly reduced area in the region of Mad Wharf Sands, continued existence of Formby Channel - although no longer directly linked to the main Mersey low tide channel, and a southward migration of Great Burbo Bank. The 1873 chart documents an increase in the area of sand flats to the west of Formby Point (e.g. Mad Wharf Sands, Jordan's Bank, Taylor's Bank and Askew Spit), with Formby Channel extending northward from the main Crosby Channel navigation and separating the sandbanks. Great Burbo Bank and Flats continue to expand in area, as evidenced by the 1912 chart which shows northward migration of Crosby Channel, reducing the extent of Taylor's Bank which remains separated from the Formby Point foreshore by Formby Channel. By 1949, a series of training walls that had been constructed along Crosby Channel (Cashin, 1949; Agar and McDowell, 1971) constraining the Queen's Channel between the southwestern margin of Taylor's Bank, and the north of Great Burbo Bank. Whilst Formby Channel remains evident, shifting to the west at its northern extent, Rock Channel along the north Wirral shore has largely disappeared. The 1988 chart shows considerable narrowing of Great Burbo Bank into Askew Spit extending westward concurrently with Queen's Channel, with almost complete infilling of Formby Channel, and the connecting of Taylor's Bank to the Formby foreshore; trends that are further developed in the 2002 chart (Blott et al., 2006).

According to Blott et al. (2006) bathymetric changes over the periods 1912-1949, 1949-1988, and 1988-2002 are characterised by accretion in the sub-tidal zone for all three time periods, especially during 1912-1949 at an average sedimentation rate of c.6 mm/yr, and a net loss of sediment in the intertidal zone for the first two periods followed by only minor accretion during 1988-2002. Sediment volume calculations show that the exposed intertidal flat areas in the west decreased in volume by 36% between 1912 and 1988, whilst those in the north and east decreased by c.16% between 1912 and 1949. Both Queen's and Crosby channels experienced a reduction in sediment volumes between 1912 and 1949 (the latter showing considerable deepening after the turn of the century) due to training wall construction and dredging. Outer estuary cross-sections further evidence the changes described above, particularly the migration and 'flattening' of Taylor's Bank to the west and Great Burbo Bank to the east, the deepening of Queen's Channel and the infilling of Formby Channel (Blott et al., 2006).

Blott et al. (2006) identify the outer estuary as being particularly susceptible to morphological change because it is largely open to marine influences, with the Sefton South Shore in the region of Crosby consisting of erodible unconsolidated Quaternary deposits, and date the main morphological changes to the latter part of the C19th and the early C20th before reaching a new condition of equilibrium and reduced inshore sediment movement after about 1950, and especially since the late 1970s. On balance, the available evidence relating to sediment transport, tidal dynamics, wave regime, sea level and storm surges over the corresponding period very much points to dredging, training wall construction and dredge spoil disposal as the most likely factors to have affected the morphology and sediment patterns in the outer estuary during the last 100 years. The construction of the training wall along the margins of the Crosby Channel from 1909 and subsequent extension into the Queen's Channel and towards the River Mersey between 1910 and 1957 reduced the need for dredging for navigation due to enhanced natural flushing (dredged volumes reduced from 25×10^6 tonnes in 1924 to $5-9 \times 10^6$ tonnes after WWII) but limited the extent of natural scour by ebb tidal currents in the Rock and Formby channels (Price and Kendrick, 1963; McDowell and O'Connor, 1977). According to Blott et al. (2006), this reduced ebb flow also gave a strengthened flood tide a longer time window to move sediment inshore, causing eastward migration of Taylor's Bank, eventually amalgamating with Formby Bank, and Great Burbo Bank. Both dredging and the dumping of dredge spoil also impacted on bathymetry, particularly increasing water depths at the Mersey Bar and in Crosby Channel and decreasing water depths over Jordan's Spit and infilling Formby Channel (Pye and Neal, 1994). This shallowing probably had a major effect on the wave regime at Formby Point and changed frontal dune accretion to erosion after about 1900. Of the resulting southward drift of sediment, much of this has become incorporated into the beach-dune complex between Hightown and Seaforth. In essence, navigation and dredging works have concentrated the ebb flow to the main channel in the outer estuary, producing flood-dominated zones outside the trained areas where sediment, including dredge spoil, has infilled the adjacent channels.

COASTAL DEFENCES AND LAND CLAIM

Hard defences on the Sefton South Shore are located

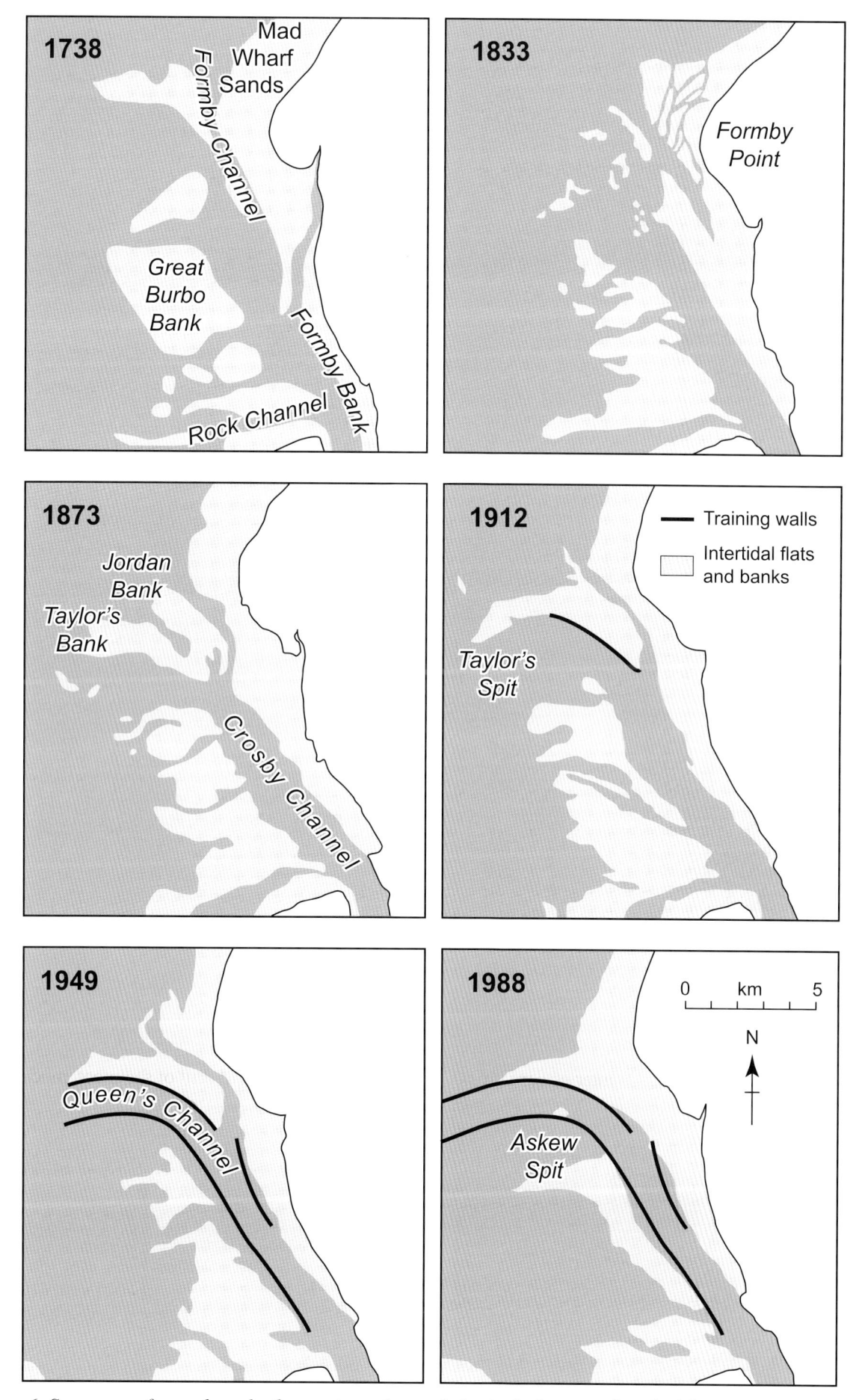

Figure 6. Summary of nearshore bathymetric and morphological changes identified from selected historical charts of the outer Mersey estuary (redrawn from Blott et al., 2006).

between Crosby and the River Alt (Figure 1) (Newton et al., 2007a). Structures were installed in this area for both defence and land claim, with defence being mainly due to problems linked to the River Alt outfall. Present coastal problems are linked to increased exposure seaward of mean high water (e.g. Hall Road West), where wind and waves are both frequent and strong, and sand accumulation landward of mean high water (e.g. Mariners Road to Crosby Marine Lake) which causes considerable problems from sand blow across the promenade and dune migration.

As noted above, the main driver for the defence works north of Hall Road has been the southerly and easterly movement of the River Alt, being constrained close to the shore by sand drifting southward from Formby Point. Gresswell (1937) noted that coastal erosion has been most active at the mouth of the Alt, and has been accentuated by the southward migration of the main tidal channel. Reade (1908) estimated the rate of land loss at Great Crosby for approximately 1866-1906 to be about 1 metre per year. In comparison, Travis (1920) noted an acceleration in this retreat during the early C20th to approximately 7.5 metres per year, and as much as 10 metres per year. The maps of both Saxton (C16th) and Speed (C17th) show the River Alt running out to sea due west, during which time the Alt was supposedly navigable for over three miles upstream. Newton et al. (2007b) completed a review of the lower coastal Alt, documenting the historical changes that have taken place during the period of cartographic and aerial photographic record (1945-2005). Significantly, dredging and navigation works in Crosby Channel, as well as the alongshore drift of sediment from Formby Point, has caused the tidal channel of the River Alt to migrate southwards and landwards. This has contributed to the erosion of the base of the dunes along the adjacent coast (Land Reclamation Development Consultants Limited, 1984), particularly so at Blundellsands from 1910 (Smith, 1982). In August 1929, a scheme to divert the Alt by blasting a channel proved unsuccessful in the short term. The tidal gates were extended in 1933 in an attempt to reduce siltation (which subsequently led to the construction of a pumping station at Altmouth in 1972), but the main works were undertaken in 1936 when a training wall was built to divert the channel away from the shore. This has had to be repaired in 1949 and again in 1969, and is currently barely higher than the adjoining shore. The training wall had the effect of moving the Alt channel away from the shore between 1951 and 1980, but by 1997 it had moved back to the training wall (Figure 7). The migration pattern of the channel to the north of the training wall seems to be progressive landward movement between 1945 and 2005.

DISCUSSION

The long-term evolution of the Sefton South Shore is linked to post-glacial inundation of a sand-dominated coastal plain, the establishment of a highly-dynamic barrier and back-barrier lagoonal environment and, eventually, a stable barrier and back-barrier perimarine wetland complex. Neolithic to early Bronze Age (c.6,000-3,000 years ago) people, in particular, took advantage of this hospitable back-barrier environment, showing evidence of animal husbandry in addition to a highly opportunistic lifestyle. Intermittent coastal barrier mobility characterises the late Holocene, with sand over-blowing the back-barrier peatlands as early as 4,000 years ago along the more exposed coast and as recent as 2,500 years ago further into the Mersey estuary. Throughout this period of coastal change, and indeed through to the present-day, the Alt estuary shows a considerable degree of persistence in terms of its function as a back-barrier drainage outlet.

Sediment movement of the foreshore is largely controlled by the interaction between beach morphology (ridges and troughs) and the interaction of prevailing tidal and wave dynamics, with the troughs between the ridges encouraging a net southward drift of sediment and large-scale recycling between the nearshore and foreshore. These ridges may appear static, but they actually exhibit a high degree of variance in terms of location and grain size over the short-term, and show evidence of alternate states under storm versus inter-storm conditions, i.e. onshore versus offshore migration. The spatial patterns and timescales of sediment cycling across the nearshore-

Andrew J. Plater, David Hodgson, Michelle Newton and Graham Lymbery

Sefton South Shore: Understanding coastal evolution from past changes and present dynamics

Figure 7. Changing positions of the River Alt from Altmouth to the sea, 1945-2005. (Source: Sefton Council). © Crown copyright. All rights reserved. Sefton Council Licence number 100018192 2010.

foreshore continuum are one of the key parameters in determining the response of the Sefton South Shore to future climate and environmental change.

Sediment delivery from the foreshore to the coastal dunes is inhibited by beach morphology (present and fossil), as well as the moisture content and sediment grain size. Contemporary sediment flux from the beach to the coastal dunes is limited, thus making it highly sensitive to changes in sea level, storm magnitude/frequency and sediment supply – all of which show a future trajectory, consequent upon climate change and sustainable catchment/coast management, that is likely to tip the beach-dune system into an unconditionally retreating regime (Figure 4).

During at least the last 200 years or so, and probably dating back to early evidence of coastal change c.1500 AD, the dune coast has witnessed a southward movement of sediment resulting from erosion of Formby Point. This erosion is especially prevalent during storm events when the rate of retreat is enhanced by a combination of undercutting and soaking. Again, this sediment is transported both along- and off-shore during storms, but may eventually make its way back to the foreshore through onshore bar migration. However, there is evidence to suggest that increased storm magnitude/frequency may lead to enhanced dune retreat, especially if the rate of sediment cycling across the foreshore is unable to keep pace. The future sedimentary response of the Sefton coast to climate change may therefore be limited by the increased frequency of extreme levels and an associated increase in net off-shore sediment transport.

Whilst the rivers may have a negligible impact on sediment delivery in comparison with tidal sediment flux, the location of the main low-water channels have had a marked impact on nearshore bathymetry and to some extent the pattern of coastal erosion. Furthermore, navigation works in the Mersey estuary during recent times have greatly impacted on nearshore bathymetry and morphology as a consequence of dredging, the release of dredge spoil and engineering works (enhancing the background sediment flux). Blott et al.

(2006) conclude that this strong anthropogenic imprint on the dynamics of the outer estuary has left it sensitive to natural environmental change in the future. Acceleration in the rate of sea-level rise would lead to increased water depths, tidal prism and nearshore current velocities, thus increasing potential for sediment reworking.

The Sefton South Shore and the Mersey estuary continue to be important areas for research, not necessarily by obtaining new information on long- and short-term evolutionary trends but through previous work providing a firm foundation for developing new techniques. This is especially true of the stable isotope (Wilson 2005a, b) and ecological transfer-function approaches to sea-level reconstruction (e.g. Horton et al., 1999; Edwards and Horton, 2000; Gehrels, 2000), and in providing a firm evidence base for testing and validating models of coastal response to extreme events (see: http://www.geog.plym.ac.uk/CoFEE/index.html).

Critically, continued economic development and sustainable management of the Sefton South Shore is highly dependent on the wider recognition of the coast's dynamic resilience – that is, long-term stability is achieved by active response to environmental change through erosion and deposition, on a variety of timescales and spatial patterns. Whilst the late C19th witnessed a period of stability, mainly due to an enhanced sediment budget, the C21st is likely to see accelerated coastal retreat as a result of limited (or at least finite) sediment supply, sea-level rise and enhanced storm magnitude/frequency. Communities need to acknowledge that this is the true nature of a coastal existence, and that we have become accustomed to shorelines that are either relatively static or at least easy to engineer and manage. New ways of living on a dynamic coast will emerge, many of which have the potential to enhance quality of life through new economic opportunities (e.g. Plater and Kirby, 2006).

ACKNOWLEDGEMENTS

The authors should like to thank Jason Kirby for his comments on the initial draft, John Grenville for the provision of data and information, Annie Worsley and

Vanessa Holden for their sterling work in reviewing earlier versions of the manuscript, and a wide variety of colleagues who have undertaken research on the Sefton coast and on beach-dune systems worldwide.

REFERENCES

Admiralty. (2005) Admiralty Tide Tables. *Volume 1, 2006. Admiralty Charts and Publications, Chart NP201*, UK Hydrographic Office, Taunton.

Adams, M. and **Harthen, D.** (2007) *An Archaeological Assessment of the Sefton Coast, Merseyside, Parts 1 and 2*. National Museums Liverpool Field Archaeological Unit, Liverpool, 132.

Agar, M. and **McDowell, D.M.** (1971) The sea approaches to the Port of Liverpool. *Proceedings of the Institution of Civil Engineers* 49, 145-156.

Allison, J.E. (1949) *The Mersey Estuary*. Liverpool University Press, Liverpool.

Anthony, E.J. (2000) Marine sand supply and Holocene coastal sedimentation in northern France between the Somme estuary and Belgium. In: Pye, K. and Allen, J.R.L. (eds.), *Coastal and Estuarine Environments: Sedimentology, Geomorphology and Geoarchaeology*. Geological Society Special Publication 175, Geological Society, London, 87-97.

Bauer, B.O., Davidson-Arnott, R.G.D., Hesp, P.A., Namikas, S.L., Ollerhead, J. and **Walker, I.J.** (2008) Aeolian sediment transport on a beach: surface moisture, wind fetch, and mean transport. *Geomorphology* in press.

Bedlington, D. (1995) *Holocene Sea-level Changes and Crustal Movements in North Wales and Wirral*. Unpublished Ph.D. thesis, University of Durham, 244p.

Binney, E.W. and **Talbot, J.H.** (1843) On the petroleum found in the Downholland Moss, near Ormskirk. Paper read at the Fifth Annual General Meeting of the Manchester Geological Society, 6th October 1843. *Transactions of the Manchester Geological Society* 7, 41-48.

Blott, S.J., Pye, K., van der Wal, D. and **Neal, A.** (2006) Long-term morphological change and its causes in the Mersey Estuary, NW England. *Geomorphology* 81, 185-206.

Booth, M., Hodgson, D.M. and **Kavanagh, J.P.** (2008) Taking beach morphology analysis to 'Another Place': high resolution investigation of coastal dynamics and sediment budget in a low-lying coastal setting utilizing status plinths as a fixed datum. Poster, STRATgroup, Department of Earth and Ocean Sciences, University of Liverpool.

Cashin, J.A. (1949) Engineering works for the improvement of the estuary of the Mersey. *Journal of the Institution of Civil Engineers* 32, 355-396.

Chiverrell, R.C., Plater, A.J. and **Thomas, G.S.P.** (eds) (2004) *The Quaternary of the Isle of Man and North West England, Field Guide*. Quaternary Research Association, London. 252.

Cope, F.W. (1939) Oil occurrences in South-West Lancashire. *Bulletin of the Geological Survey of Great Britain* 2, 18.

Cowell, R.W. (2008) Coastal Sefton in the Prehistoric Period. In: Lewis, J.M. and Stanistreet, J.E. (eds.), *Sand and Sea – Sefton's Coastal Heritage*. Sefton Council Leisure Services Department, Bootle, 182p.

Cowell, R.W. and **Innes, J.B.** (1994) *The Wetlands of Merseyside, North West Wetlands Survey* 1. Lancaster. National Museums and Galleries on Merseyside/Lancaster University Archaeological Unit.

Cowell, R.W., Milles, A. and **Roberts, G.** (1993) Prehistoric footprints on Formby Point beach, Merseyside. *North West Wetlands Annual Report 1993*, 43-48.

De Rance, C.E. (1869) *Geology of the country between Liverpool and Southport*. Memoir of the Geological Survey of Great Britain. HMSO, London.

De Rance, C.E. (1872) *Geology of the country around Southport, Lytham and South Shore.* Memoir of the Geological Survey of Great Britain. HMSO, London.

De Rance, C.E. (1877) *The superficial geology of the country adjoining the coasts of south-west Lancashire.* Memoir of the Geological Survey of Great Britain. HMSO, London.

Dyer, K. (1997) *A Physical Introduction to Estuaries.* Wiley, London.

Edelman, T. (1968) Dune erosion during storm conditions. *Proceedings of the 11th Conference on Coastal Engineering*, Volume 2, London, 719-722.

Edelman, T. (1973) Dune erosion during storm conditions. *Proceedings of the 12th Conference on Coastal Engineering*, 1305-1312.

Edwards, R.J. and **Horton, B.P.** (2000) Reconstructing relative sea-level change using UK saltmarsh foraminifera. *Marine Geology* 169(1-2), 41-56.

Erdtman, G. (1928) Studies of the Post-Arctic History of the Forests of North Western Europe I. Investigations in the British Isles. *Geologiska Föreningens i Stockholm Förhandlingar* March-April, 133p.

Esteves, L.S., Williams, J.J., Nock, A. and **Lymbery, G.** (2009) Quantifying shoreline changes along the Sefton coast (UK) and the implications for research-informed coastal management. *Journal of Coastal Research* 56, 602-606.

Gehrels, W.R. (2000) Using foraminiferal transfer functions to produce high-resolution sea-level records from salt-marsh deposits, Maine, USA. *The Holocene* 10(3), 367-376.

Gonzalez, S. and **Cowell, R.** (2004) Formby Point Foreshore, Merseyside: Palaeoenvironment and Archaeology. In: Chiverrell, R.C., Plater, A. and Thomas, G. (eds.), *The Quaternary of the Isle of Man and North West England, Field Guide*, Quaternary Research Association, UK, pp. 206-216.

Gonzalez, S. and **Huddart, D.** (2002a) Hightown. In: Huddart, D. and Glasser, N.F. (eds.), *The Quaternary of Northern England.* Geological Conservation Review Series, No.25, Joint Nature Conservation Committee, Peterborough, pp. 582-588.

Gonzalez, S. and **Huddart, D.** (2002b) Formby Point. In: Huddart, D. and Glasser, N.F. (eds.), *The Quaternary of Northern England.* Geological Conservation Review Series, No.25, Joint Nature Conservation Committee, Peterborough, pp. 569-582.

Gonzalez, S., Huddart, D. and **Roberts, G.** (1997) Holocene development of the Sefton Coast: a multidisciplinary approach to understanding the archaeology. In: Sinclair, A., Slater, E. and Gowlett, J. (eds.), *Archaeological Science 1995*. Oxbow Monograph 64, Oxford, pp. 289-299.

Gresswell, R.K. (1937) The Geomorphology of the South-West Lancashire coast-line. *Geographical Journal* 90, 335-348.

Gresswell, R.K. (1953) *Sandy shores in south Lancashire.* University Press, Liverpool.

Gresswell, R.K. (1957) Hillhouse coastal deposits in south Lancashire. *Liverpool and Manchester Geological Journal* 2, 60-78.

Hale, W.G. (1985) *Martin Mere: its History and Natural History.* The Wildfowl Trust, Martin Mere, Burscough, 24.

Halliwell, A.R. and **O'Connor, B.A.** (1966) Suspended sediment in a tidal estuary (Mersey). *Proceedings of the 10th Conference of Coastal Engineering, Tokyo,* Vol.1, 687-706.

Hesp, P.A. (1982) *Dynamics and morphology of foredunes in S.E. Australia.* PhD thesis, University of Sydney.

Holden V.J.C. (This volume) The historic development of the North Sefton Coast In: *Sefton's Dynamic Coast: proceedings of the conference on coastal geomorphology, biogeography and management 2008* Worsley, A.T., Lymbery, G., Holden, V.J.C. and Newton, M. (eds) Coastal Defence: Sefton MBC Technical Services.

Horton, B.P., Edwards, R.J. and **Lloyd, J.M.** (1999) A foraminiferal-based transfer function: implications for sea-level studies. *Journal of Foraminiferal Research* 29(2), 117-129.

Houser, C., Hapke, C. and **Hamilton, S.,** (2008) Controls on coastal dune morphology, shoreline erosion and barrier island response to extreme storms. *Geomorphology* 100 (3-4), 223-240.

HRS. (1969) The Southwest Lancashire Coastline, A Report on the Sea Defences. Hydraulics Research Station, Wallingford. Report no. EX 450, HMSO, 43p.

Hsu, S.A. (1977) Boundary-layer meteorological research in the coastal zone. In: Walker, H.J. (ed.), *Geoscience and Man, Vol. 18; Research Techniques in Coastal Environments*. School of Geoscience, Louisiana State University, pp. 99-111.

Huddart, D. (1992) Coastal environmental changes and morphostratigraphy in southwest Lancashire, England. *Proceedings of the Geologists' Association* 103, 217-236.

Huddart, D., Tooley, M.J. and **Carter, P.A.** (1977) The coasts of north-west England. In: *The Quaternary History of the Irish Sea,* Kidson, C. and Tooley, M.J. (eds.). Seel House Press, Liverpool, 119-154.

Huddart, D., Gonzalez, S. and **Roberts, G.** (1999a) The archaeological record and mid-Holocene marginal coastal palaeoenvironments around Liverpool Bay. *Quaternary Proceedings* 7, 563-574.

Huddart, D., Roberts, G. and **Gonzalez, S.** (1999b) Holocene human and animal footprints and their relationships with coastal environmental change, Formby Point, NW England. *Quaternary International* 55, 29-41.

Innes, J., and **Tooley, M.J.** (1993) The age and vegetational history of the Sefton Coast dunes. In: Atkinson, D. and Houston, J. (eds), *The Sand Dunes of the Sefton Coast.* National Museums and Galleries on Merseyside/Sefton Metropolitan Borough Council, Liverpool, pp. 35-40.

Innes, J.B., Tooley, M.J. and **Tomlinson, P.R.** (1989) A comparison of the age and palaeoecology of some sub-Shirdley Hill sand peat deposits from Merseyside and south-west Lancashire. *Naturalist* 114, 65-69.

Innes, J.B., Bedlington, D.J., Kenna, R.J.B. and **Cowell, R.W.** (1990) A preliminary investigation of coastal deposits at Newton Carr, Wirral, Merseyside. *Quaternary Newsletter* 62, 5-12.

Jay, H. (1998) *Beach-dune sediment exchange and morphodynamic responses: implications for shoreline management, the Sefton Coast, NW England.* PhD thesis, University of Reading.

Jelgersma, S., deJong, J., Zagwijn, W.H. and **van Regteren Altena, J.F.** (1970) The coastal dunes of the western Netherlands; geology, vegetational history and archaeology. *Mededelingen Rijks Geologische Dienst* NS 21, 93-167.

Jones, R.C.B., Tonks, L.H. and **Wright, W.B.** (1938) *The Geology of Wigan.* Memoir of the Geological Survey of Great Britain, HMSO, London.

Kenna, J.B. (1976) The Flandrian sequence of North Wirral, N.W. England. *Geographical Journal* 21, 1-27.

McDowell, D.M. and **O'Connor, B.A.** (1977) *Hydraulic Behaviour of Estuaries.* Macmillan, London.

Mersey and Weaver River Authority. (1971) *River Alt Pumping Station*. Mersey and Weaver Authority.

Moore, T.J. (1881) Notes on the mammalian remains from Hightown. *Proceedings of the Liverpool Geological Society* 4, 277.

Morton, G.H. (1887) Stanlow, Ince and Frodsham Marshes. *Proceedings of the Liverpool Geological Society* 5, 349-351.

Morton, G.H. (1888) Further notes on the Stanlow, Frodsham and Ince Marshes. *Proceedings of the Liverpool Geological Society* 6, 50-55.

Morton, G.H. (1891) *Geology of the Country around Liverpool, including the North of Flintshire*, George Philip and Son, London.

Morton, R.A. (2002) Factors controlling storm impacts on coastal barrier and beaches – a preliminary basis for near real-time forecasting. *Journal of Coastal Research* 18, 486-501.

Neal, A. (1993) *Sedimentology and morphodynamics of a Holocene coastal dune barrier complex, Northwest England*. PhD thesis, Reading University.

Newton, M., Lymbery, G. and **Wisse, P.** (2007a) *Report on Hard Sea Defences on the Sefton Coast*. Sefton Council Technical Services, Bootle. 48p.

Newton, M., Lymbery, G. and **Wisse, P.** (2007b) *Report on the Changing Morphology of the Lower Alt, from Altmouth Pumping Station to the Sea*. Sefton Council Technical Services, Bootle. 18.

Parker, W.R. (1971) *Aspects of the Marine Environment at Formby Point, Lancashire*. PhD thesis, University of Liverpool, two volumes.

Parker, W.R. (1974) Sand transport and coastal stability, Lancashire, UK. In: *Coastal Engineering 1974*, Proceedings of the 14th Coastal Engineering Conference, June 24-28th, Copenhagen, Denmark. American Society of Civil Engineers, New York, pp. 828-851.

Parker, W.R. (1975) Sediment mobility and erosion on a multibarred foreshore (Southwest Lancashire, U.K.). In: J. Hails and A. Carr (eds.), *Nearshore Sediment Dynamics and Sedimentation*. London. John Wiley and Sons, pp. 151-179.

Payne, G.E. (1976) *Investigation into the recent siltation of the tidal reaches of the River Alt. (Lancashire)*. Anthony Brown and Partners.

Picton, J.A. (1849) The changes of sea-levels on the west coast of England during the historic period. (Abstract) *Proceedings of the Literary and Philosophical Society of Liverpool* 36th session 5, 113-115.

Plater, A.J. and **Grenville, J.** (This volume) Liverpool Bay: Linking the Eastern Irish Sea to the Sefton Coast. In: *Sefton's Dynamic Coast: proceedings of the conference on coastal geomorphology, biogeography and management 2008* Worsley, A.T., Lymbery, G., Holden, V.J.C. and Newton, M. (eds) Coastal Defence: Sefton MBC Technical Services. pp. 27-54.

Plater, A.J. and **Kirby, J.R.** (2006) The potential for perimarine wetlands as an ecohydrological and phytotechnological management tool in the Guadiana estuary, Portugal. *Estuarine, Coastal and Shelf Science* 70, 98-108.

Price, W.A. and **Kendrick, M.P.** (1963) Field and model investigation into the reasons for siltation in the Mersey Estuary. *Proceedings of the Institution for Civil Engineering* 24, 473-518.

Psuty, N.P. (1988) Sediment budget and dune/beach interaction. *Journal of Coastal Research* special issue 3, 1-4.

Psuty, N.P. (1992) Spatial variation in coastal foredune development. In; Carter, R.W.G., Curtis, T.G.F. and Sheehy-Skeffington, M.J. (eds.), *Coastal dunes: geomorphology, ecology and management for conservation.* Balkema, Rotterdam, pp. 3-13.

Psuty, N.P. (1993) Foredune morphology and sediment budget, Perdido Key, Florida, USA. In: Pye, K. (ed.), *The Dynamics and Environmental Context of Aeolian Sedimentary Systems.* Geological Society Special Publication 72, Geological Society Publishing House, Bath, pp. 145-157.

Pye, K. (1977) *An analysis of coastal dynamics between the Ribble and Mersey estuaries, with particular reference to erosion at Formby Point.* BA dissertation, University of Oxford, 137p.

Pye, K. (1991) Beach deflation and backshore dune formation following erosion under storm surge conditions: an example from northwest England. *Acta Mechanica* supplement 2, 171-181.

Pye, K. and **Blott, S.J.** (2008) Decadal-scale variation in dune erosion and accretion rates: an investigation of the significance of changing storm tide frequency and magnitude on the Sefton coast, UK. *Geomorphology* 102(3-4), 652-666.

Pye, K. and **Neal, A.** (1993a) Late Holocene dune formation on the Sefton coast, northwest England. In: Pye, K. (ed.), *The Dynamics and Environmental Context of Aeolian Sedimentary Systems,* Geological Society of London, Special Publication 72, Geological Society Publishing House, Bath, pp. 201-217.

Pye, K. and **Neal, A.** (1993b) Stratigraphy and age structure of the Sefton dune complex: preliminary results of field drilling investigations. In: Atkinson, D. and Houston, J. (eds.), *The Sand Dunes of the Sefton Coast.* National Museums and Galleries on Merseyside in association with Sefton Metropolitan Borough Council, Liverpool, pp. 41-44.

Pye, K., and **Neal, A.** (1994) Coastal dune erosion at Formby Point, north Merseyside, England: Causes and Mechanisms. *Marine Geology* 119, 39-56.

Pye, K. and **van der Wal, D.** (2000a) Historical Trend Analysis (HTA) as a tool for long-term morphological prediction in estuaries. EMPHASYS *Consortium, Estuaries Research Programme Phase 1, MAFF contract CSA 4938, Modelling Estuary Morphology and Process,* Final Report, TR111, pp. 89-96.

Pye, K. and **van der Wal, D.** (2000b) Expert Geomorphological Assessment (EGA) as a tool for long-term morphological prediction in estuaries. EMPHASYS *Consortium, Estuaries Research Programme Phase 1, MAFF contract CSA 4938, Modelling Estuary Morphology and Process,* Final Report, TR111, pp.97-102.

Pye, K., Stokes, S. and **Neal, A.** (1995) Optical dating of aeolian sediments from the Sefton coast, northwest England. *Proceedings of the Geologists' Association* 106, 281-292.

Pye, K., Saye, S.E. and **Blott, S.J.** (2007) S*and Dune Processes and Management for Flood and Coastal Defence.* Parts 1-5. Joint Defra/EA Flood and Coastal Erosion Risk Management RandD Technical Report FD1302/TR. Defra/EA, London.

Reade, T.M. (1871) The geology and physics of the post-glacial period, as shewn in the deposits and organic remains in Lancashire and Cheshire. *Proceedings of the Liverpool Geological Society* 2, 36-88.

Reade, T.M. (1872) The post-glacial geology and physiography of west Lancashire and the Mersey estuary. *Geological Magazine* 9, 111-119.

Reade, T.M. (1881) On a section of the Formby and Leasowe marine beds and superior peat bed, disclosed by the cuttings for the outlet sewer at Hightown. *Proceedings of the Liverpool Geological Society* 4, 269-277.

Reade, T.M. (1908) Post-glacial beds at Great Crosby, as disclosed by the New Outfall Sewer. *Proceedings of the Liverpool Geological Society* 10, 249.

Roberts, G. and **Worsley, A.** (2008) Evidence for human activity in the mid-Holocene coastal palaeoenvironments of Formby, North West England. In: Lewis, J.M. and Stanistreet, J.E. (eds.), *Sand and Sea – Sefton's Coastal Heritage.* Sefton Council Leisure Services Department, Bootle, pp. 28-43.

Roberts, G., Gonzalez, S., and **Huddart, D.** (1996) Intertidal Holocene footprints and their archaeological significance. *Antiquity* 70, 647-651.

Saye, S.E. and **Pye, K.** (2007) Implications of sea level rise for coastal dune habitat conservation in Wales, UK. *Journal of Coastal Conservation* 11, 31-52.

Shaw, D.F. (1975) Water inputs to Liverpool Bay. In: Liverpool Bay Study Group (ed.), *Liverpool Bay: An Assessment of Present Knowledge.* Natural Environmental Research Council, Series C, 14, London, pp. 11-12.

Sherman, D.J. and **Bauer, B.O.** (1993) Dynamics of beach-dune systems. *Progress in Physical Geography* 17(4), 413-447.

Short, A.D. and **Hesp, P.A.** (1982) Wave, beach and dune interactions in southeastern Australia. *Marine Geology* 48, 259-284.

Sly, P.G. (1966) *Marine geological studies in the Eastern Irish Sea and adjacent estuaries, with special reference to sedimentation in Liverpool Bay and the River Mersey.* PhD thesis, University of Liverpool.

Smith, A.J. (1982) Guide to the Sefton Coast Data Base. Sefton Metropolitan Borough Council and University of Liverpool, Liverpool.

Smith, P.H. (1999) *The Sands of Time: an introduction to the sand dunes of the Sefton coast.* National Museums and Galleries on Merseyside/Sefton Metropolitan Borough Council, 196p.

Spearman, J., Turnbull, M., Thomas, C., and **Cooper, A.** (2000) *Two and three dimensional modelling of sediment transport mechanisms in the Mersey Estuary.* Estuaries Research Programme, Phase 1B, Modelling Estuary Morphology and Process. Final Report prepared by the EMPHASYS Consortium for MAFF, Contract CSA 4938, December 2000.

Theiler, E.R. and **Young, R.S.** (1991) Quantitative evaluation of coastal geomorphological changes in South Carolina after Hurricane Hugo. *Journal of Coastal Research* 8, 187-200.

Thomas, C. (2002) *The Application of Historical Data and Computational Methods for Investigating Causes of Long-Term Morphological Change in Estuaries: A Case Study of the Mersey Estuary, UK.* PhD thesis. Oxford Brookes University and HR Wallingford.

Tooley, M.J. (1969) *Sea-level changes and the development of coastal plant communities during the Flandrian in Lancashire and adjacent areas.* Unpublished PhD thesis, University of Lancaster.

Tooley, M.J. (1970) The peat beds of the southwest Lancashire coast. *Nature in Lancashire* 1, 19-26.

Tooley, M.J. (1974) Sea-level changes during the last 9000 years in northwest England. *Geographical Journal* 140, 18-42.

Tooley, M.J. (1976) Flandrian sea-level changes in west Lancashire and their implications for the 'Hillhouse Coastline'. *Geological Journal* 11, 137-152.

Tooley, M.J. (1977) The Altmouth. In: M.J. Tooley (ed), The Isle of Man, Lancashire Coast, and Lake District. 10th INQUA Excursion Guide. Norwich. Geoabstracts 8, 32.

Tooley, M.J. (1978a) *Sea-level changes in north west England during the Flandrian Stage.* Oxford Research Studies in Geography, Clarendon Press, Oxford.

Tooley, M.J. (1978b) Interpretation of Holocene sea-level changes. *Geologiska Föreningens i Stockholm Förhandlingar* 100, 203-212.

Tooley, M.J. (1980) Theories of coastal change in North-West England. In: Thompson, F.H. (ed.), *Archaeology and Coastal Change.* Society of Antiquaries of London, Occasional Papers New Series No.1, London, pp. 74-86.

Tooley, M.J. (1982) Sea-level changes in northern England. *Proceedings of the Geologists' Association* 93, 43-51.

Tooley, M.J. (1985a) Sea-level changes and coastal morphology in northwest England. In: Johnson, R.H. (ed.), *The Geomorphology of Northwest England.* Manchester University Press, Manchester, pp. 94-121.

Tooley, M.J. (1985b) Climate, sea level and coastal changes. In: Tooley, M. J. and Sheail, G. (eds.) *The Climatic Scene: Essays in honour of Gordon Manley.* George Allen and Unwin, London, pp. 206-234.

Tooley, M.J. (1990) The chronology of coastal dune development in the United Kingdom. In: Bakker, Th.W.M., Jungerius, P.D. and Klijn, J.A. (eds), *Dunes of the European coasts.* Catena Supplement 18, 81-88.

Tooley, M.J. (1992) Recent sea-level changes. In: Allen, J.R.L. and Pye, K. (eds.), *Saltmarshes: Morphodynamics, Conservation and Engineering Significance.* Cambridge University Press, Cambridge, pp. 19-40.

Tooley, M.J. and **Kear, B.S.** (1977) Mere Sands Wood (Shirdley Hill Sand). In: Bowen, D.Q. (ed.), *The Isle of Man, Lancashire coast and Lake District.* INQUA X Congress. Geo Abstracts, Norwich, pp. 9-10.

Travis, C.B. (1908) On some borings near Hightown. *Proceedings of the Liverpool Geological Association*, New Series 3, 11.

Travis, C.B. (1920) Coastal Changes at the Alt Mouth. *Proceedings of the Liverpool Geological Society* 33.

Travis, C.B. (1926) The peat and forest bed of the south-west Lancashire coast. *Proceedings of the Liverpool Geological Society* 14, 263.

Travis, C.B. (1929) The peat and forest bed of the south-west Lancashire coast. *Proceedings of the Liverpool Geological Society* 15, 157-178.

Travis, W.G. (1922) On peaty beds in the Wallasey Sand-Hills. *Proceedings of the Liverpool Botanical Society* 1, 47-52.

Valentin, H. (1952) Die kusten der Erde. *Petermanns Geografisches Mitteilungen, Erganzungsheft* 246.

van de Graaff, J. (1986) Probabilistic design of dunes, an example from The Netherlands. *Coastal Engineering* 9, 479-500.

van de Graaff, J. (1994) Coastal dune erosion under extreme conditions. In: Finkl, C.W. (ed.), Coastal Hazards – Perception, Susceptibility and Mitigation. *Journal of Coastal Research*, special issue 12, 253-262.

van der Meulen, T. and **Gourlay, M.R.** (1968) *Beach and dune erosion tests.* Proceedings of the 11th conference on Coastal Engineering, London, Volume 1, pp. 710-707.

Wilson, G.P. (2004) *Microfossil and isotopic evidence for Holocene sea-level change in the Mersey Estuary.* Unpublished PhD thesis, Liverpool John Moores University.

Wilson, G., Lamb, A., Leng, M., Gonzalez, S. and **Huddart, D.** (2004) The potential of carbon isotope ratios as indicators of coastal palaeoenvironmental change; preliminary results from the Mersey Estuary, North West England. In: Chiverrell, R.C., Plater, A.J. and Thomas, G.S.P. (eds.), *The Quaternary of the Isle of Man and North West England, Field Guide,* Quaternary Research Association, London, pp. 217-225.

Wilson, G.P., Lamb, A.L., Leng, M.J., Gonzalez, S. and **Huddart, D.** (2005a) Variability of organic $\delta^{13}C$ and C/N in the Mersey Estuary, UK and its implications for sea-level reconstruction studies. *Estuarine, Coastal and Shelf Science* 64, 685-698.

Wilson, G.P., Lamb, A.L., Leng, M.J., Gonzalez, S. and **Huddart, D.** (2005b) $\delta^{13}C$ and C/N as potential coastal palaeoenvironment indicators in the Mersey Estuary, UK. *Quaternary Science Reviews* 25, 2015-2029.

Wilson, P., Bateman, R.M. and **Catt, J.A.** (1981) Petrography, origin and environment of deposition of the Shirdley Hill Sand of southwest Lancashire, England. *Proceedings of the Geologists' Association* 92, 211-229.

Wray, D.A. and **Cope, F.W.** (1948) *Geology of Southport and Formby.* Memoir of the Geological Survey of Great Britain, one-inch Geological Sheets 74 and 83 New Series, HMSO, London 54p.

Zong, Y. and **Tooley, M.J.** (1996) Holocene sea-level changes and crustal movements in Morecambe Bay, northwest England. *Journal of Quaternary Science* 11, 43-58.

Effects of climate change on wet slacks in the dune system at Ainsdale

Derek Clarke*, Ruth K. Abbott and Sarinya Sanitwong Na Ayutthaya

School of Civil Engineering and the Environment, University of Southampton.
*Corresponding Author : Dr Derek Clarke, School of Civil Engineering and the Environment, University of Southampton, Southampton SO17 1BJ.
dc@soton.ac.uk

ABSTRACT

Slack floors in dune systems are important for biodiversity, but are dynamic in their moisture characteristics, both spatially and in time. Slack floors can be characterised as dry, partly wet or flooded, depending on the water table levels and in each case they support differing vegetation and animal species. This paper examines the effects of changes in water table levels on the extent of wet slack floors in the Ainsdale Sand Dunes National Nature Reserve on the Sefton Coastline near Liverpool. A groundwater model was used to calculate the water table levels on a monthly basis based on climatic, hydrological and vegetation conditions. Detailed topographic survey data at 2m grid intervals from LIDAR imagery were used to represent the ground surface in the slack floors. The expected slack type (dry, damp, flooded) was generated by intersecting the LIDAR digital terrain model with the predicted water table levels. The model has been used to calculate the distribution and frequency of occurrence of flooded slack floors between 1972 and 2004. UKCIP'02 climate change predictions have been used to estimate the future extent and size of flooded slack floors at Ainsdale up to the year 2100.

INTRODUCTION

The dune system of the Sefton Coastline in North West England near Liverpool is an example of a fragile system where the dunes perform a coastal defence role and support an exceptionally rich bio-habitat. (Sefton Coast Partnership, http://www.seftoncoast.org.uk/). This is due in part to the presence of a shallow unconfined water table in the sand aquifer. When the water table is close to the dune slack floors, rich assemblages of flora exist (Jones et al, 2006). In wet winters, these slacks often flood to a depth of 10-30cm providing breeding grounds for amphibians such as the Natterjack Toad. There are significant seasonal and annual changes in the water table levels resulting in changes in the distribution and frequency of occurrence of wet slack floors. Prior to 1970 the water table was relatively high and many of the dune slacks were flooded in winter. In the early 1970's, the water table fell by 0.8m and many slacks dried up (Clarke and Pegg, 1992). Little was known of the impact of factors such as the development of the pumped agricultural drainage system inland of the dune system and the expansion of nearby areas of urbanisation with associated road drainage systems, which intercept water which would naturally drain to the water table. Clarke (1980) demonstrated that the changes could be explained by variations in rainfall patterns. Post 1977, the water table levels rose again, however there are still sequences of years with low water levels, causing difficulty in managing the ecology in the dune floors (Davy et al., 2006).

A well tube monitoring network was installed in part of the dune system at Ainsdale Sand Dunes National Nature Reserve in 1972 and water table levels were measured every month (Figure 1). The water table rises from sea level (4m OD) to a maximum +10.5mOD approximately 2km inland from the sea in the vicinity of the Liverpool-Southport railway line (Figure 2). This indicates that groundwater flows through the dunes towards the sea west of the railway and inland east of the railway. Long term monitoring of the wells over the last 35 years has provided a comprehensive data set to understand the influences of seasonality, changes in climate, coastal erosion and land use management on groundwater levels in this area.

Modelling changes in water table levels

Clarke (1980) described a conceptual water balance

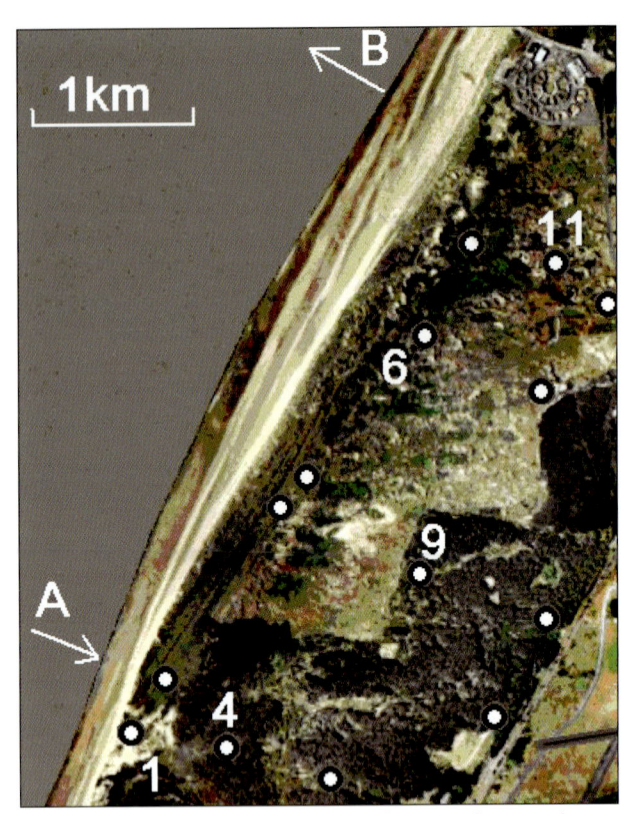

Figure 1. Well tube network and sites of coastal erosion (A) and accretion (B).

model for the dune system, based on the principles of catchment hydrology. This was later improved to include soil moisture deficit simulation and tree interception (Clarke and Sanitwong Na Ayutthaya, 2007; Sanitwong Na Ayutthaya, 2008). The improved model predicts the water table level at the end of each month by simulating a daily water balance in the dune system (Figure 3). The following section is an overview of the water balance model and its sub models, more detail can be found in Clarke and Sanitwong Na Ayutthaya (2010).

The soil moisture deficit in the vegetation root zone is calculated based on daily potential evapotranspiration, rainfall and a root zone model. The latter includes data for root depth, soil water holding capacity and a sub-model that calculates actual evapotranspiration based on the previous day's soil moisture deficit. Groundwater flow is calculated using the Dupuit equation, based on the saturated sand permeability and the hydraulic

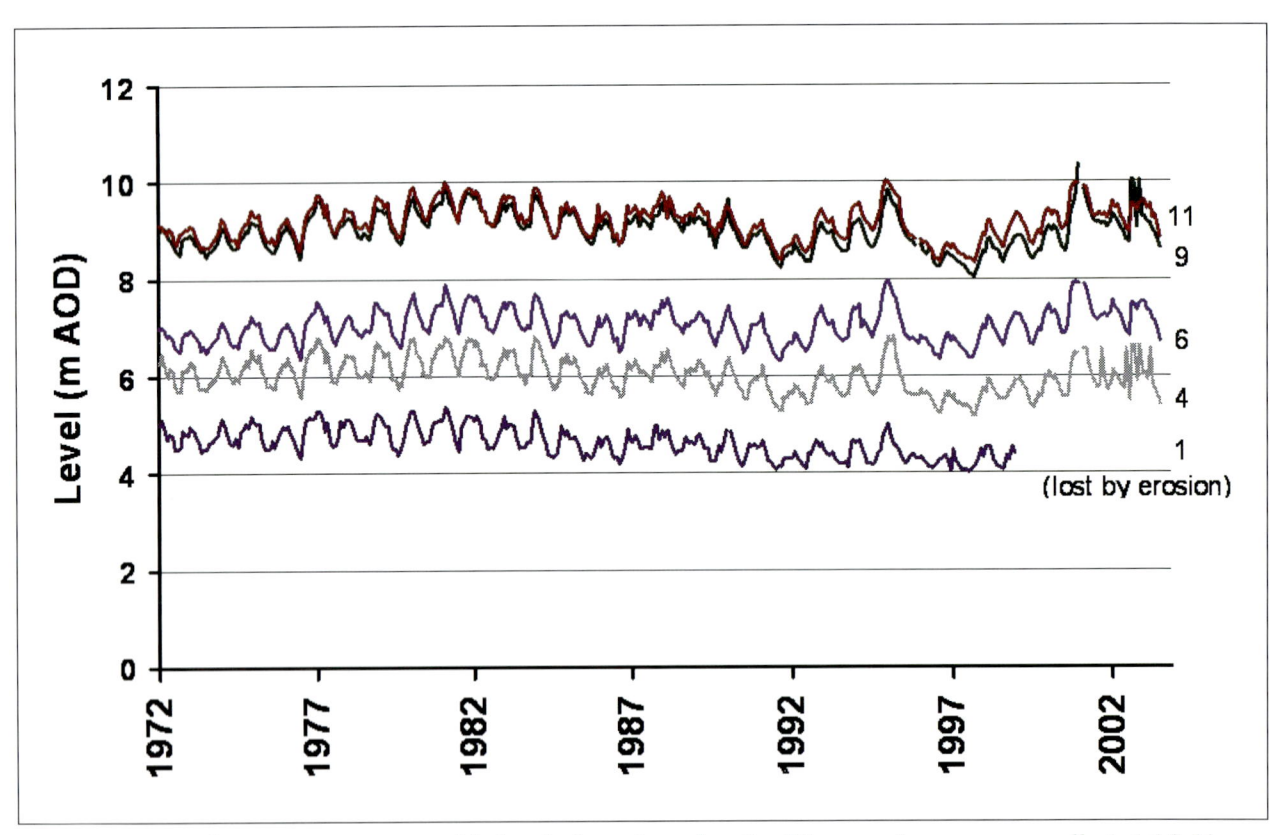

Figure 2. Typical variation in water table levels for selected wells. Distance from coast : wells 1,4,6,9,11 <25m, 400m, 500m, 950m, 1500m. Well 1 was abandoned due to coastal erosion. See Figure 1 for location of wells.

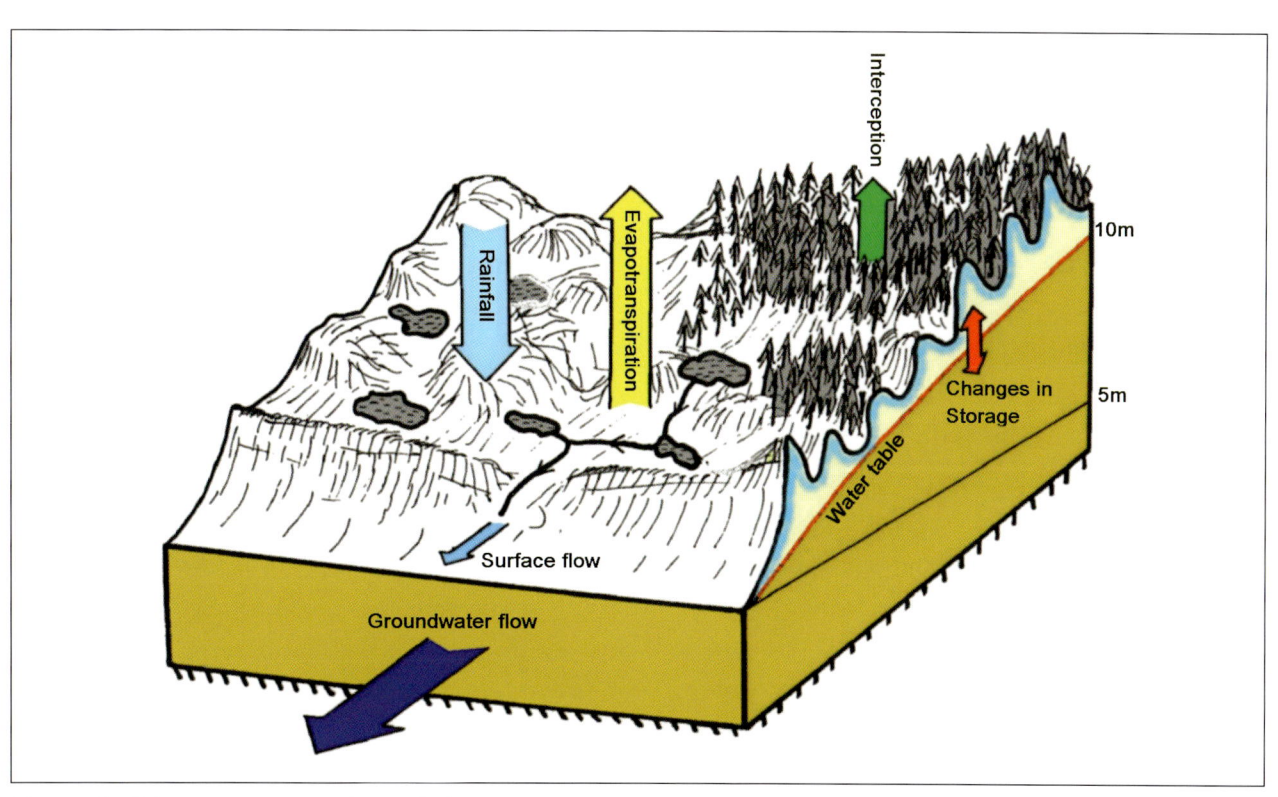

Figure 3. Schematic section through the dune system showing components of the water balance model. Flooded slacks occur where the water table intercepts the dune slack floors.

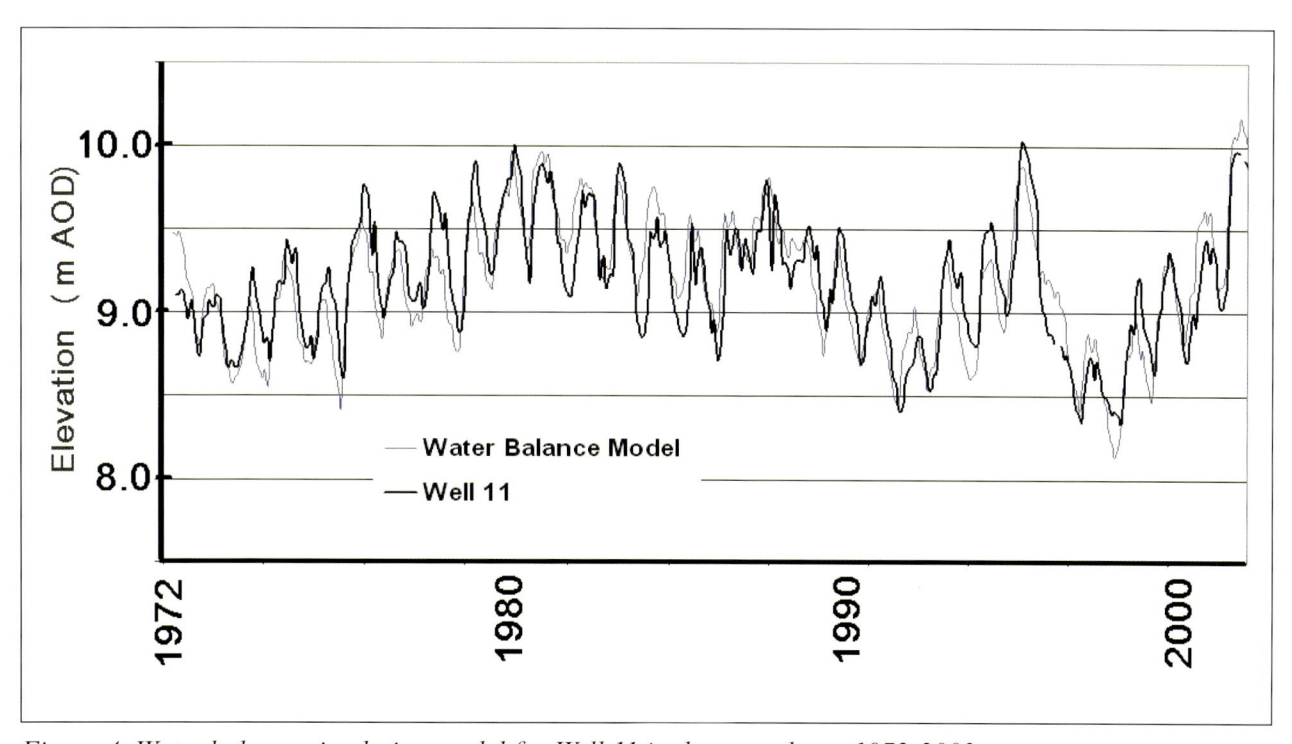

Figure 4. Water balance simulation model for Well 11 in the open dunes 1972-2002.

gradient between the current water table height and sea level (Todd and Mays, 2005). If the surface layer of the soil is at field capacity or near saturation, any excess water is assumed to drain to the water table where it saturates the sand and raises the water table. The change in water storage is calculated from the balance between water that drains vertically into the water table minus the groundwater flow. The water storage change is then converted into a water table elevation change using a depth dependent root zone porosity function. At depths greater than 1m the effective porosity is set to 28% a laboratory figure derived for clean dune sand (Clarke,

1980), in the root zone (0 to 0.5m) it rises to 35% due to the presence of roots and organic materials. At or above ground level, the porosity is assumed to be 100% but the regional porosity is adjusted to cater for the fact that a proportion of the water table is in open air (flooded dune slacks) whilst the majority of the water table is under the dune ridges (saturated sand). In exceptionally wet conditions with a very high water table level a series of heuristic rules are applied to simulate the removal of surface water by drainage ditches which cross some parts of the dune system.

The model was calibrated using physically realistic parameter optimization and has been tested against four of the eleven wells in the Nature Reserve. It is able to describe the changes in water table levels with an accuracy of about +/- 0.2m over the 30 year period of available data. A typical run of the model for well 11 located in the open dunes in the northern part of the area is shown in Figure 4.

Effect of vegetation type on water table levels.

The area comprises a mixture of coniferous woodland, deciduous scrub, fixed dune and dune slack. Pine plantations were established starting in the 1900's in the southern part of the Nature Reserve. The forested areas coincide with an area of water table levels that are approximately 50cm lower than the rest of the system. To investigate the effect of pine trees on the water system, the water balance model for the open dunes was modified to include interception losses calculated using models suggested by Calder (2005). The revised model shows that pine trees intercept and re-evaporate about 150mm of rain each year, equivalent to lowering the water table levels under the pine trees by 0.5-0.6m. In 1992, 4.5ha of trees were removed as part of a management plan to increase biodiversity in the dune system. A further 16ha were removed in 1997 (Atkins, 2004). Following the tree removal, groundwater levels rose by approximately 50cm in two years. These observations and model results agree with observed water table levels and support other work that suggests that pine trees have a significant effect in reducing aquifer recharge in temperate climates (Minderman and Leeflang, 1968; Calder et al., 2002).

Impacts of climate change on water table levels.

To investigate the likely effects of climate change on the water table levels in the Sefton dunes, climate predictions from the UK Climate Impact Programme (Hulme et al., 2002) were incorporated into the water balance model. The prediction for climate change

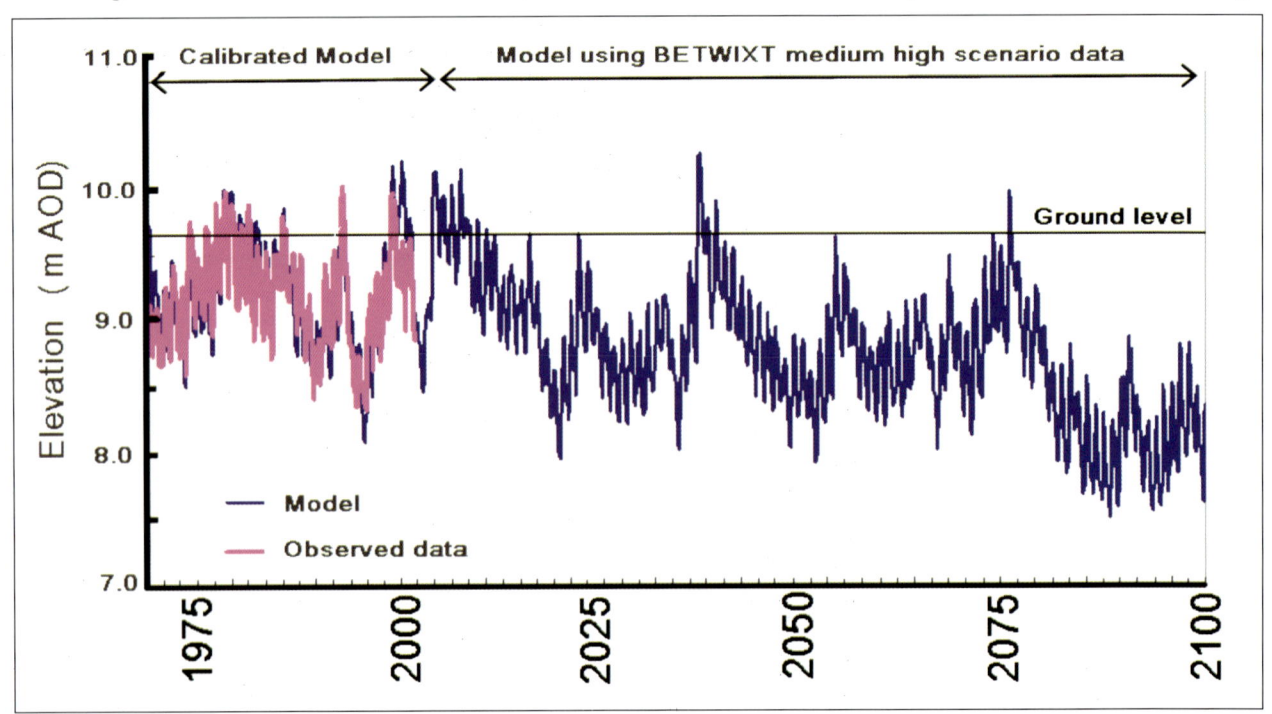

Figure 5. One simulation of the impacts of climate change on water table levels at well 11 (1972-2095).

under a medium high emission scenario in 2100 for this area is an increase in mean air temperature of about +3°C, together with a small decrease in rainfall (3-5%). However rainfall patterns are expected to change so that summers become drier and winters are wetter. Wetter winters would result in more recharge into the water table in winter and the predicted higher summer temperatures would result in higher potential evapotranspiration. The balance between these opposing changes was investigated with the water balance model. The model was run from 2005 to 2100 using the UKCIP'02 medium high emissions scenario by applying the stochastically generated time series of daily climatic data published by the BETWIXT project (BETWIXT, 2005). These data embody the expected average and temporal variability in temperature, humidity, wind speed, sunshine and rainfall over the next 100 years at Ringway airport near Manchester, modified to take into account the distance between Ainsdale and Ringway (approx 60km).

The soil moisture deficit calculation in the water balance model showed that despite expected higher potential evapotranspiration, actual evapotranspiration from the open dunes will be limited by soil drying in summer. It should be noted that drier summers may stress vegetation and result in long term change of vegetation types, although this was not accounted for in the model formulation. An example output from the model for Well 11 in the open dunes between 1972-2100 is shown in Figure 5. The model suggests an overall fall in water table levels of between 0.8-1.2m. There are short sequences (3-4 years) of exceptionally wet years, similar to current wet conditions, separating long sequences of years with low water table levels.

The modelled water table levels were sensitive to the timing and magnitude of rain events and the BETWIXT data contains only one of many possible data sequences of rainfall patterns. The model was run 500 times using a random sampling approach to create differing sequences of rainfall events. In each run the total amount of rainfall remained the same and each run of the model was equally likely to happen. The ranges of predicted water table levels from 1972 – 2080's for wet conditions in March are shown in Figure 6. The overall pattern is a lowering of the water table levels with time, but also greater variability in these levels. The average end of winter water table level at Well 11 (sited in the open dunes, about 1.5km from the sea) may fall from

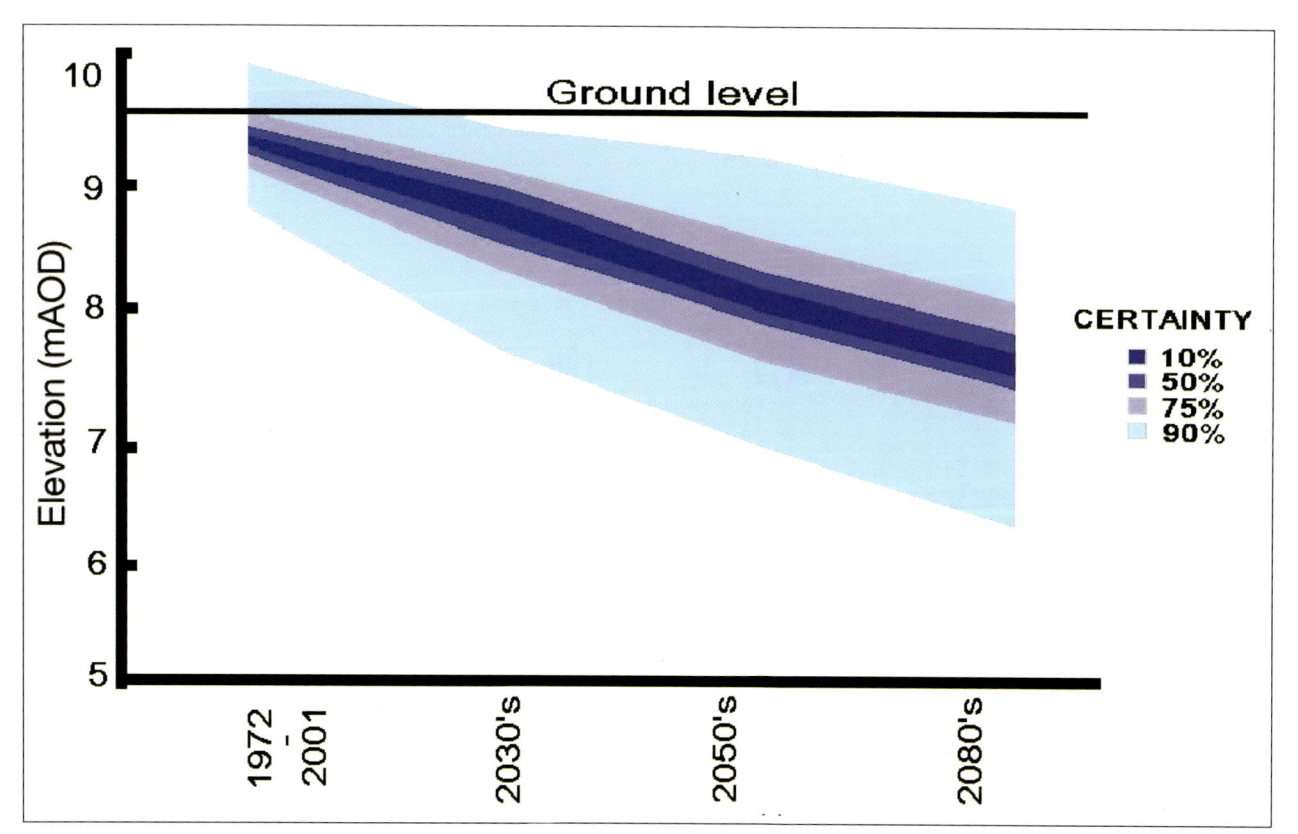

Figure 6. Range of likely springtime water levels in the open dunes (mAOD).

a current average level of +9.5mAOD to around +8.0mAOD by the 2080's. This will impact both the vegetation and the wildlife in the sand dune system, particularly the current wet and damp slack floor areas which are of high biodiversity (Davy et al., 2006). It is expected that many slack floors will dry out, suggesting that species dependent on standing water for breeding (such as the Natterjack Toad) will become threatened.

Coastal Erosion and Sea Level Rise

Undefended coastal zones are vulnerable to erosion and sea level rise. A rising sea level may cause permanent inundation of wetlands and other lowlands, increase of the salinity of rivers, bays and accelerate coastal erosion (Holman et. al, 2002). Rising sea levels may also increase the salinity of groundwater through salt water intrusion. Sea level rise in this region of the UK is expected to be of the order of +2 to +7mm/y (UKCIP, 2005; DEFRA, 2007). The dune coastline at Formby is currently being eroded at up to 4.5m/year (Lymbery et al., 2007) and the beach at Southport has widened by over 200m (Edmondson et al, 1999). Erosion will steepen the hydraulic gradient between the water table and the sea and increase groundwater flow, whereas accretion will decrease the groundwater flow. Additionally, sea level rise will reduce the hydraulic gradient between the groundwater and the sea, reducing groundwater flow.

Simulations of the effects of changing the position of the coastline and scenarios of sea level rise were carried out. The effects were modelled at different distances inland, at Well 7 (750m from the coastline) and Well 11 (1550m from the coastline). The water balance model was run by varying the coastline position by -500m to +500m from its present position and average sea levels were varied from -3m to +3m. Results indicate that coastal erosion will lower water table levels by 0.18m for every 100 metres of coastline set back and will increase water table levels by 0.14m for every 100m of accretion. The expected sea level rise by 2100 is +0.6m (UKCIP 2005). The model suggests that this rise will cause water table levels to rise by about 0.2m within 500m of the coastline and by 0.1m 1km inland.

Importance of wet dune slack habitats

Climate change will be a significant determinant on the future state of the dune slacks and the need to consider this factor is recognised in conservation strategies. Of particular importance are the Natterjack Toads which are supported in the flooded dune slacks. Approximately 40% of Britain's total population have been reported in the Sefton Dunes (North West Biodiversity Forum, 2008). Natterjacks rely on the numerous slack areas found within the reserve for breeding pools during March to June (Bebee and

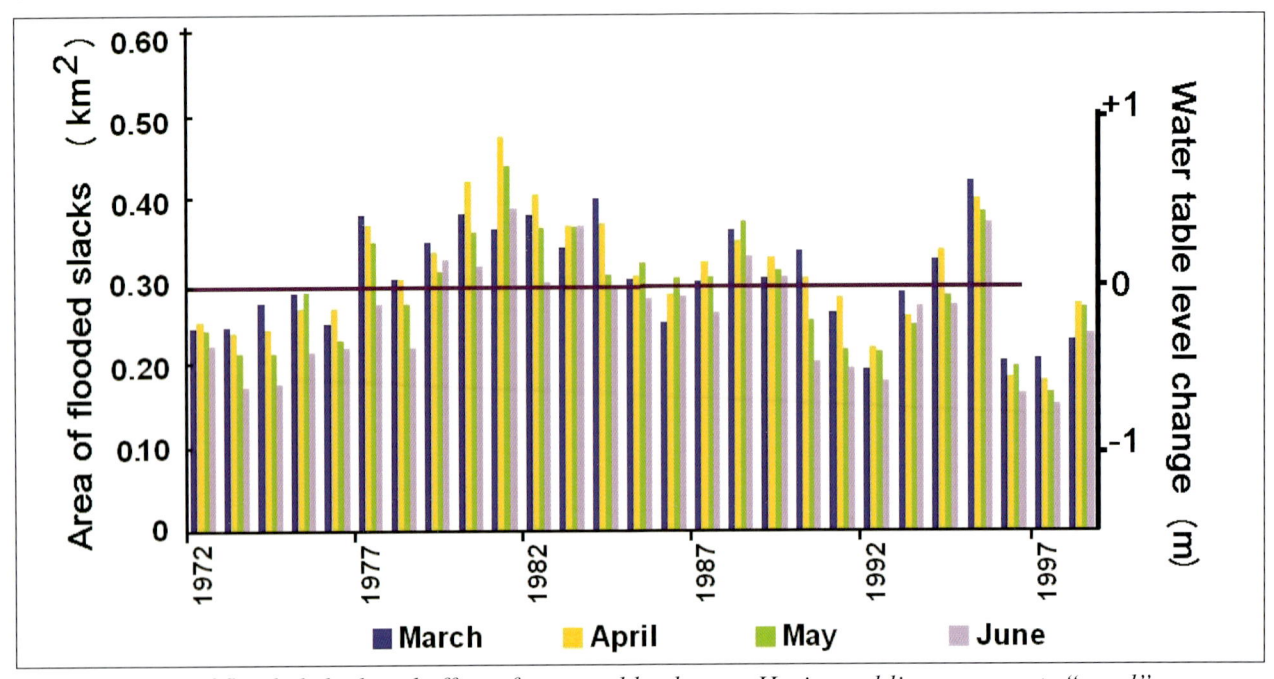

Figure 7. Area of flooded slack and effect of water table change. Horizontal line represents "good" conditions for Natterjack Toad breeding.

Denton, 1996; Natural England, 2007). Conserving, enhancing and restoring the toad population and dune slack habitat is included in many land management plans, including the Ainsdale Sand Dunes NNR Management Plan (Gee, 1998), the UK Biodiversity Action Plan (1995) and the North Merseyside Biodiversity Action Plan (1994). However, with the exception of the present study, little is known about how climate change may impact on slack ecology, therefore no confident management strategy can exist (Downey, pers comm., 2007).

Future water table levels & flooded areas

Wet dune slacks occur where the water table is close to or above the ground level. The water balance model outputs were combined with a Digital Elevation Model (DEM) of the open dune areas of the Nature Reserve. The DEM was generated from LIDAR data provided by the Environment Agency at a spatial resolution of 2m. Grids of predicted water table levels were produced by fitting a second order (parabolic) surface to the well data. This surface was combined with the ground level DEM to produce monthly maps of flooded slack floors (Abbott, 2007). A strong relationship was found between water table level and the modelled flooded slack area in the open dunes ($r^2 = 0.978$). This relationship was used to estimate the changes in total wet slack floor area over the next 100 years using results from the water balance model.

The water table model suggested that the average water table levels near Well 7 in the open dunes during March-April are likely to fall from 7.23mAOD at present to 6.21mAOD by 2100. Discussions with the staff at Ainsdale Sand Dunes NNR indicate that a "good" year for Natterjack breeding occurred in years that coincide with a modelled flooded slack area greater than or equal to 0.3km^2 (Downey, pers comm.). The predicted fall in water table levels during the 21st century suggests a reduction in the springtime wet slack area in the open dunes from an average of 0.32km^2 (1972-2000) to 0.012km^2 by the year 2100. This fall in the area of flooded and wet slacks is far below the current Biodiversity Action Plan target (shown as the horizontal line in Figure 7). The inter-annual variability of predicted water table levels means that there may be some rare years with relatively abundant slack areas, but these are likely to be interspersed by sequences of 5-10 years of low water table levels. Smith (2006, cited in Davy et al., 2006) stated that the number of species in a given slack type can be correlated with the slack area. The predicted dry periods will result in a loss of the continuous breeding environment for species such as the Natterjack toad and a gradual change in flora habitat from flooded to dry slack floor. Therefore the predicted decrease in flooded slack area raises serious concern for the future abundance and diversity of species in wet or flooded slacks.

SUMMARY

- The key environmental factors that affect groundwater levels in a coastal sand dune system in North West England have been evaluated up to the year 2100.
- Using UKCIP'02 estimates of climate change, with a medium-high emissions scenario, it is expected that water table levels may fall by approximately 0.8m, having the effect of drying out the inland slack floors and potentially reducing the biodiversity of these areas.
- The presence of pine trees appears to have the effect of lowering water table levels by 0.5-0.8m.
- Coastal erosion will increase the hydraulic gradient and lower water tables by about 0.15m per 100m of coastline recession. These effects will be larger closer to the coast.
- In areas of coastal accretion, water table levels will rise and new areas will develop that offer wet or flooded slack conditions.
- In order to assess the likely scale of impact of climate change on the wet dune habitats of the Sefton Coastline, there is an urgent need to extend the modelling work described in this paper to the whole dune system.

ACKNOWLEDGEMENTS

We wish to thank the staff of Natural England for the long term data used in this study, and the Environment Agency for access to the LIDAR data. Work for this paper was carried out with funds provided by the

School of Civil Engineering and the Environment in the University of Southampton, the Government of Thailand and NERC.

REFERENCES

Abbott, R. (2007) *A model for estimating wet slack areas in sand dunes.* Unpublished MSc thesis, University of Southampton.

Atkins. (2004) *Ainsdale Sand Dunes National Nature Reserve: Environmental impact assessment of options for management of seawards areas.* WS Atkins Consultants Limited report.

Beebee, T. and **Denton, J.** (1996) *Natterjack toad conservation handbook.* English Nature. ISBN 1 85716 220 X G.

BETWIXT. (2005) *The Built Environment: Weather scenarios for investigation of impacts and extremes.* Climatic Research Unit, University of East Anglia, UK. http://www.cru.uea.ac.uk/projects/betwixt/

Calder, I.R. (2005) *Blue revolution: integrated land and water resources management.* 2nd edition. Earthscan.

Calder, I.R., Reid, I., Nisbet, T., Armstrong, A., Green, J.C. and **Parkin, G.G.** (2002) *Study of the Potential Impacts on Water Resources of Proposed Afforestation.* Department for Environment, Food and Rural Affairs, UK.
http://www.defra.gov.uk/environment/water/resources/research/pdf/tadpole_execsumm.pdf

Clarke, D. (1980) *The groundwater balance of a coastal sand dune system.* Unpublished PhD thesis, University of Liverpool, UK.

Clarke, D. and **Pegg, R.** (1992) Hydrological investigations in the Ainsdale Sand Dunes National Nature Reserve. In Atkinson, D. and Houston, J. (eds) *Sand dunes of the Sefton Coast.* Eaton Press 0 096367 62 X, 55-58.

Clarke, D. and **Sanitwong Na Ayutthaya, S.** (2007) A probabilistic assessment of future coastal groundwater levels in a dune system in England. International conference on management and restoration of coastal dunes. ICCD, *Santander.* pp. 38-39.

Clarke, D. and **Sanitwong Na Ayutthaya, S.** (2010) Predicted effects of climate change, vegetation and tree cover on dune slack habitats at Ainsdale on the Sefton Coast, UK. *Journal of Coastal Conservation.* DOI 10.1007/s11852-009-0066-7.

Davy, A.J., Grootjans, A.P., Hiscock, K. and **Petersen, J.** (2006) Development of eco-hydrological guidelines for dune habitats – Phase 1, English Nature Research Report, No 696.

DEFRA (2007) Sea level rise at selected sites:1850-2004.
http://www.defra.gov.uk/environment/statistics/globatmos/kf/gakf14.htm

Downey, M. (2007) *Personal communication.* Ainsdale Sand Dunes National Nature Reserve, Merseyside, UK.

Edmonson, S.E., Traynor, H. and **McKinnnell, S.** (1999) The development of a green beach on the Sefton coast, Merseyside, UK. In *Coastal Dune Management* Houston J., Edmonson S.E. and Rooney, P.J. (eds) pp. 48-58. Liverpool University Press. ISBN 9 780853 23854 5.

Gee, M. (1998) *Ainsdale Sand Dunes National Nature Reserve Management Plan 1998-2003.* Unpublished report, English Nature, Wigan.

Holman, I.P., Loveland, P.J., Nicholls, R.J., Shackley, S., Berry, P.M., Rounsevell, M.D.A., Audsley, E., Harrison, P.A. and **Wood, R.** (2002) REGIS: Regional climate change impact and response studies in East Anglia and north west England. Environment Agency & English Nature, 20.

Hulme, M., Jenkins, G., Lu, X. et al. UKCIP (2002). Climate change scenarios for the United Kingdom: the UKCIP02 scientific report. Tyndall Centre for Climate Change Research. School of Environmental Sciences. University of East Anglia.

Jones, M.L.M., Reynolds, B., Brittain, S.A., Norris, D.A., Rhind, P.M. and **Jones, R.E.** (2006) Complex hydrological controls on wet dune slacks: the importance of local variability. *Science of the Total Environment*, 372 266-277.

Lymbery, G., Wisse, P., and **Newton, M.** (2007) Report on Coastal Erosion predictions for Formby Point, Formby, Merseyside. http://www.sefton.gov.uk/pdf/TS_cdef_predict2.pdf

Minderman, G. and **Leeflang, K.W.F.** (1968) The amounts of drainage water and solutes from lysimeters planted with either oak, pine or natural dune vegetation or without any vegetation cover. *Plant Soil* 28, 61-80.

Natural England (2007) Advice for Land Managers: European protected species: What you need to know about Natterjack toads.
www.naturalengland.org.uk/conservation/wildlife-management-licensing/docs/Guidanceland-managers-natterjacktoad.pdf

North Merseyside Biodiversity Action Plan (1994) http://www.merseysidebiodiversity.org.uk

North West Biodiversity Forum (2008) http://www.biodiversitynw.org.uk/

Sanitwong Na Ayutthaya, S. (2008) *Impacts of Climate Change on Groundwater in Coastal Aquifers*. Unpublished PhD thesis, University of Southampton.

Sefton Coast Partnership web site (2007) http://www.seftoncoast.org.uk/

Todd, D.K. and **Mays, W.L.** (2005) *Groundwater Hydrology*, John Wiley & Sons. ISBN 0-471-05937-4. 148.

UK Biodiversity Action Plan. (1995) Biodiversity: The UK Steering Group Report - Volume II: Action Plans (December 1995, Tranche 1, Vol 2, 114). http://www.ukbap.org.uk/UKPlans.aspx?ID=173

UKCIP (2005) Updates to regional net sea-level change estimates for Great Britain. (www.ukcip.org.uk)

Distinguishing dune environments based on topsoil characteristics: a case study on the Sefton Coast

Jennifer A. Millington[1]*, Colin A. Booth[1], Mike A. Fullen[2], Ian C. Trueman[2] and Annie T. Worsley[3].

[1]SEBE, University of Wolverhampton WV1 1SB, United Kingdom.
[2]SAS, University of Wolverhampton WV1 1SB, United Kingdom.
[3]NGAS, Edge Hill University L39 4QP, United Kingdom.

*Corresponding author: Miss Jennifer Millington

Jennifer.Millington@wlv.ac.uk

ABSTRACT

It is important to understand the effects of coastal change on the migration of coastal dune environments and their associated imprint on soil processes, for both environmental and ecological motives. Geographical Information Systems (GIS) have been applied to investigate soil spatial patterns and their controlling influences on the Sefton dunes. To verify relationships between plant communities and soil types, ground-truthing of existing vegetation maps has been achieved through analysis of representative, geo-referenced, topsoil (0-5 cm) samples (n = 115), from classified dune environments (n = 10), for the purpose of distinguishing dune environments from their soil characteristics. Samples were analysed for pH, organic matter content, particle size, total soil organic carbon and total soil nitrogen, geochemical composition and magnetic susceptibility. Significant differences (p <0.05) are apparent for the suite of soil characteristics collated, indicating individual dune environments are associated with specific soil properties. Therefore, identification and mapping of dune soil habitats can provide baseline information for conservation management.

INTRODUCTION

Successful sand dune management plans require accurate baseline maps of land cover. The biological diversity of these systems can only be maintained when all dune and vegetation successional environments are encouraged. Over-stabilization of the dunes is potentially as great a threat as erosion and can result in an acceleration of succession from a natural fixed dune habitat towards dune heath climax, resulting in loss of biological diversity (Doody, 1989; Kutiel et al., 2000). In an ecologically significant dune system, such as the Sefton coast, bare sand, mobile and fixed dune environments have ideal conditions for successionally young slack habitats. Identification and mapping of both favourable (sandy) and unfavourable (scrub-rich) habitats for such slack developments provide baseline information for conservation management (Lucas et al., 2002). Gibson and Looney (1992) states that effective coastal conservation programmes should be established using baseline maps of relative proportions of sand and vegetation, to monitor stages in dune succession.

Effective use of baseline maps can be taken one step further to distinguish dune soil environments, subsequent patterns in soil development and to understand current environmental processes in operation on the dune landscape. These assist plans to minimize potential future loss of natural resources. Presently, the Sefton coast soil landscape is represented on the 1:250,000 Soil Survey of England and Wales (1984) map as the Sandwich Association Soil sub-group 361, a typical sand-pararendzina (very thin organic layer overlying sandy parent material) and has not been differentiated further. However, any small variations in soil development indicate phases of landscape stability and vegetation succession; therefore relationships can be attributed between both plant communities and soil type. Existing vegetation descriptions for this part of the coast (e.g. Smith, 1978; Payne, 1983; Rothwell, 1985; Edmondson et al., 1993) suggest great variability in the existing vegetation communities, the patterns of which can be recognized as either occurring along topographic transects or related to individual landforms within just a few metres of each other. This, in turn, suggests great variability within soil environments.

This paper distinguishes soil characteristics in all natural and semi-natural dune environments on the Sefton coast. The main objective is to apply a selection of physio-chemical data from sand dune topsoils (0-5 cm) to the ArcView Geographical Information System' (GIS) to map spatial patterns of soil characteristics.

METHODOLOGIES

National Vegetation Classification survey (2003/2004)

Two phases of survey and map development of the sand dune vegetation resource of the Sefton coast, covering 1970 ha, were carried out by Sefton Metropolitan Borough Council (SMBC), initially in 1988/89 and, more recently, between April 2003 and July 2004. These applied both CORINE and National Vegetation classification (NVC) definitions. The vegetation boundaries, mapped in the field and verified by aerial photographs, were digitized in MapInfo v 6.5 GIS. Across the 5,369 GIS polygons, 2,513 quadrats were recorded, resulting in 563 vegetation categories, corresponding to 61 broad NVC community types.

Development of the dune environment map

Conversion of existing vegetation datasets and aerial photographs from MapInfo v 6.5 GIS (courtesy of SMBC) into ArcView GIS, allowed classification of existing vegetation types (n = 61) into feasible parcels of land associated with both dune morphology and vegetation environments (n = 13). If not already done so, the initial CORINE classification vegetation map of the area based on habitat categories was modified to the NVC, a classification of vegetation type, not habitat, enabling the vegetation data to be more readily utilized. The 13 dune environments represent areas of blown sand remaining outside residential development; amenity grassland (golf clubs), arable, bare sand (beach, blowouts), dune heath, dune pasture, embryo dune, fixed dune, mobile dune, plantations (coniferous), salt marsh communities, scrub, slacks and woodland (deciduous).

The embryo dune environment was incorporated into the bare sand category, as embryo dunes represent a geomorphological feature within the landscape, rather than a classified parcel of land that can be associated with soil development. Areas not associated with natural dune succession (amenity grassland, arable and salt marsh communities) were excluded from further analyses to prevent anomalies in the data. However, although not distinguished through vegetation analysis, the existence of an area of felled pine is apparent in the field, and therefore was added to the overall dune landscape profile. As the felled area is integrated into the fixed dune vegetation on the dune environment map it is represented on an overlay in GIS.

Field, laboratory and desk-based techniques

A ground-truth exercise assessed the accuracy and effectiveness of the dune environment map, and therefore to confirm the assumption that soil characteristics and vegetation type are related. Representative topsoil (0-5 cm) samples were collected from each of the chosen 10 classified dune environments (total n = 113) for soil property determinations. Similar ground-truth testing of methodological approaches is described in Rühl et al. (2007), where ground data was used to link virtual GIS derived data with field data for accuracy assessment of the applied methodology.

Qualitative information of soil type were recorded, both during the survey and in the laboratory to be used along side numerical data to model typical distributions of physio-chemical soil properties in GIS. Similar methodology descriptions can be found in Costantini et al. (2007) and Repe (2007). Parameters measured were: distance from mean high water (MHW), pH (using 1:2.5 soil:distilled-water), soil organic matter (SOM) content (percentage loss of soil weight after ignition at 375°C for 16 hours) and geochemical composition (analyzed by X-ray fluorescence spectroscopy) (van Reeuwijk, 2002), total soil organic carbon and total soil nitrogen (determined by the dry combustion method) (Kundu et al., 2007), particle size (using Laser Diffraction) (Booth et al., 2003) and magnetic susceptibility (χLF) (the ratio of the induced magnetization of a sample to a magnetic field) (Mullins, 1977).

After digitization, ArcView GIS was used to automatically assigned interpolation quantitative values to unsampled sites within the area covered by geo-referenced sample points, to display patterns of soil characteristics. As the sample points are representative, rather than transectional or from a grid, interpolation was achieved by nearest neighbour analysis (Andrews et al., 2002), as it is possible to calculate diagonal values accurately. Grid cell size, or resolution, of 10 m was determined following Burrough and McDonnell (1998).

All remaining data analyses were performed using MINITAB PC (version 15). To establish the significance of differences between each soil parameter measured for individual dune environments, a series of independent Mann-Whitney U tests (a non-parametric test for assessing whether two samples of observations come from the same distribution) were conducted.

RESULTS
Spatial variations in dune environments

A complete, 'textbook', successional series of plant communities across the full width of blown sand is not evident on the dune environments map (Figure 1). The order of dune environments described is based on a proposed path of natural dune succession, initiating on the backshore and proceeding inland. However, it is important to note that this is not necessarily a complete and accurate indication of development over time, especially when considering mean distance from mean high water (MHW). Some environments have been

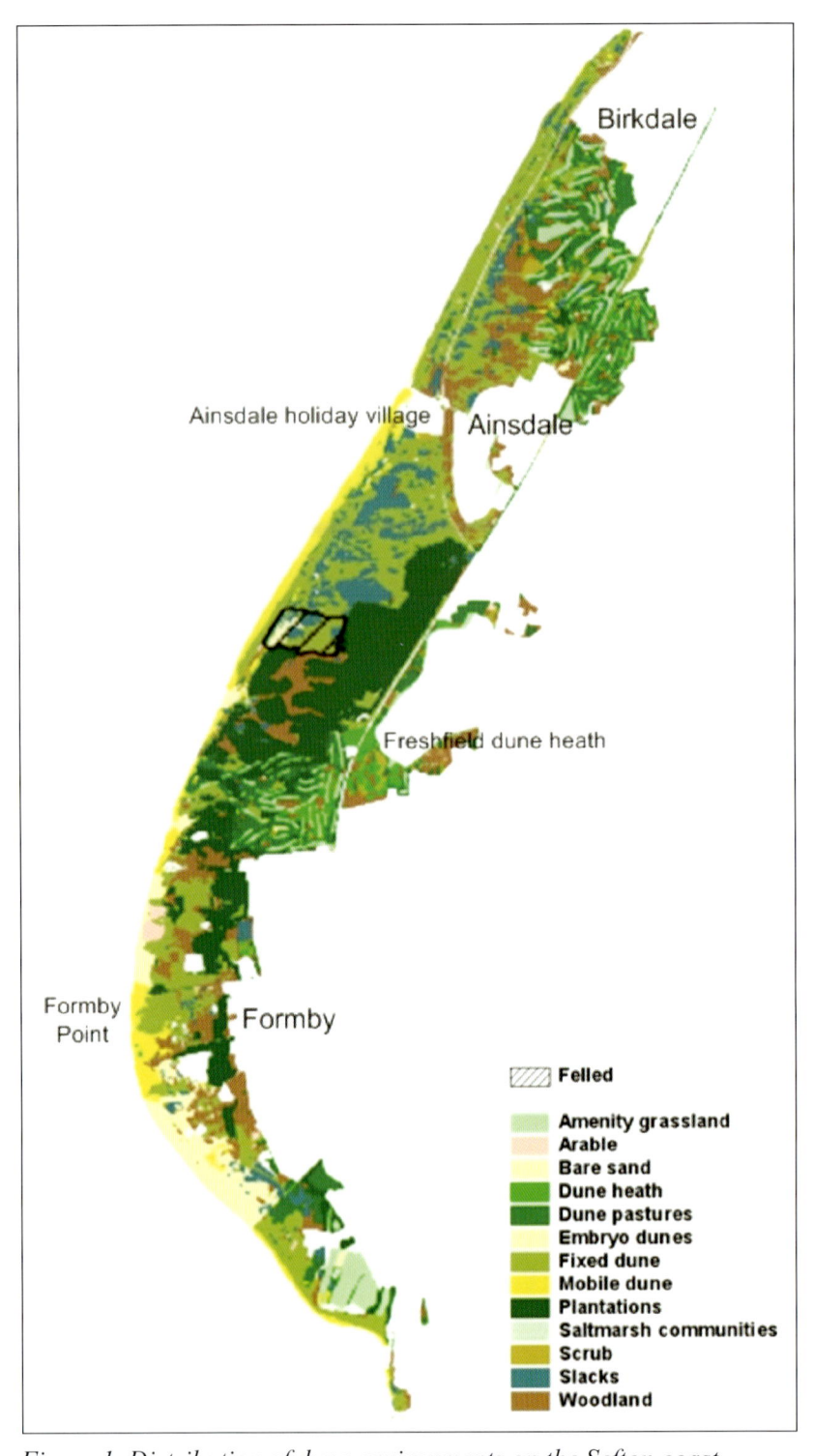

Figure 1. Distribution of dune environments on the Sefton coast.

both altered by natural coastal processes and anthropogenic modifications. Attempts have been made to explain spatial variations of dune environments in an environmental processes context.

Bare sand, or unvegetated advancing dunes (mean (n = 9) distance from MHW = 203 m), is represented at the seaward margin of the eroding Formby Point, where Pye (1990) suggested recreational pressure has caused the formation of large blowouts and subsequent transgressive sand sheets at beach access points. Embryo dunes are present where the coast is accreting both north and south of the erosion front. Mobile dunes (mean (n = 12) distance from MHW = 226 m) occur behind the embryo dunes and are represented by ridges parallel to the coast. They are clearly visible running almost the entire length of the project area, except for the disturbed northern extremity and where they are truncated by erosion around Formby Point.

The most common and widespread environmental community is fixed dune (mean (n = 18) distance from MHW = 832 m). This landscape, interspersed with slack habitats, is most likely to be the result of an irregular series of hummock dunes with incipient blowouts and parabolic dunes, which Pye (1990) suggested were prograding seaward pre-1800. The slack environments (mean (n = 11) distance from MHW = 439 m) have successfully been distinguished by vegetation analyses. However, it is evident from the polygon shapes on the dune environments map that geomorphological aspects can identify at least two stages of slack development. The older, secondary slacks at Birkdale and Ainsdale have a mainly east-west orientation suggesting an origin from large blowouts. The younger, primary dune slacks, occurring at Ainsdale, are associated with a later phase of dune development where successive dune ridges, parallel to the coast, have cut-off former beach surfaces. Smith (1978) pointed out that natural dune slacks south of Formby Point have been largely eliminated by extensive modification for asparagus agriculture, represented on this map by arable environments.

Dune heath (mean (n = 9) distance from MHW = 1724 m) is the most fragmented of all the dune environments, occupying the eastern fringes of the dune system, where most building development has occurred. According to Edmondson et al. (1993), Freshfield dune heath was formerly a golf course, and therefore cannot be classed as a natural dune environment. In fact, all areas of heath communities appear to occur on golf courses, often interspersed with dune pastures, probably occupying former slack areas (Edmondson et al., 1993). Although,

dune grasslands are usually the next successional stage developing from both fixed dune and slack communities, on this coast the pastures (mean (n = 10) distance from MHW = 661 m) appear either on the edges of larger slacks or totally colonizing older slack hollows (Edmondson et al., 1993).

Scrub areas (mean (n = 10) distance from MHW = 484 m) are extensive, although sparse just north of the central areas of the dune system. In the far northern part of the dune landscape, succession appears to be attributable to initial phases of slack development and has progressed to scrub following a generalized trend of increased stability. According to Edmondson et al. (1993) only limited areas of this scrub are natural, or have been colonized by native species. The main areas of deciduous woodland (mean (n = 14) distance from MHW = 635 m) are also planted (Edmondson et al., 1993) and occur on the eastern fringes of the southern half of the map, often surrounded by pine plantations. An area of deciduous woodland, in the centre of the map, has probably developed due to sheltering properties of the pine plantations from coastal erosion and deposition of aeolian derived sand. These pine plantations (mean (n = 10) distance from MHW = 1001 m) occupy large areas of the central dune system. Areas of felled coniferous woodland (mean (n = 10) distance from MHW = 476 m) found on the edge of the pine plantations, is represented by vegetation associated with fixed dune communities, despite only being felled during winter 1995/96 (Purdie, 2002).

Spatial variations in soil properties

Figures 2, 3 and 4 show spatial variations in physical and chemical soil properties. Table 1 displays descriptive statistics for all variables measured. Statistically significant differences in successional soil properties, of varying degrees, are found between all dune environments using Mann-Whitney U tests (Table 2).

Seaward margins, including bare sand and mobile dunes, are pH neutral to slightly alkaline along the entire length of the coast (Figure 2a) and are rich in silicon (Si) (Figure 3a). Large areas of neutral to

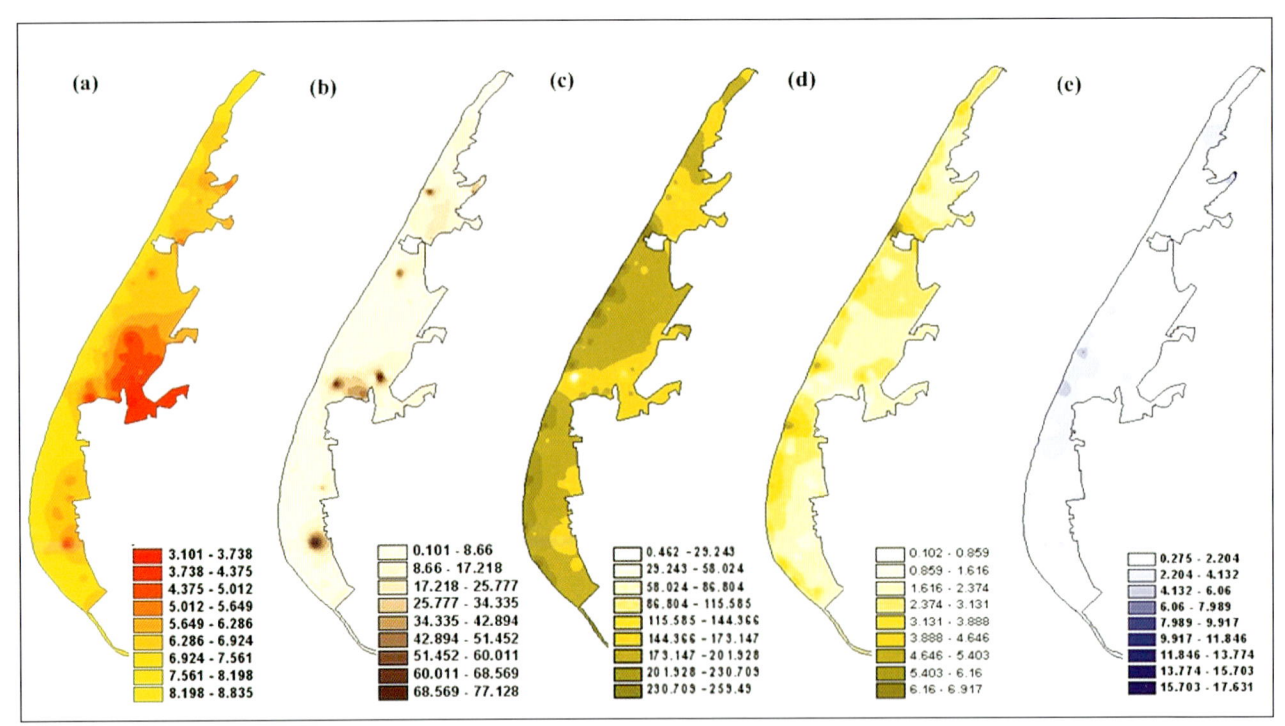

Figure 2. Spatial distribution of (a) pH, (b) % SOM, (c) mean particle size (μm), (d) % clay and (e) χLF on the Sefton Coast.

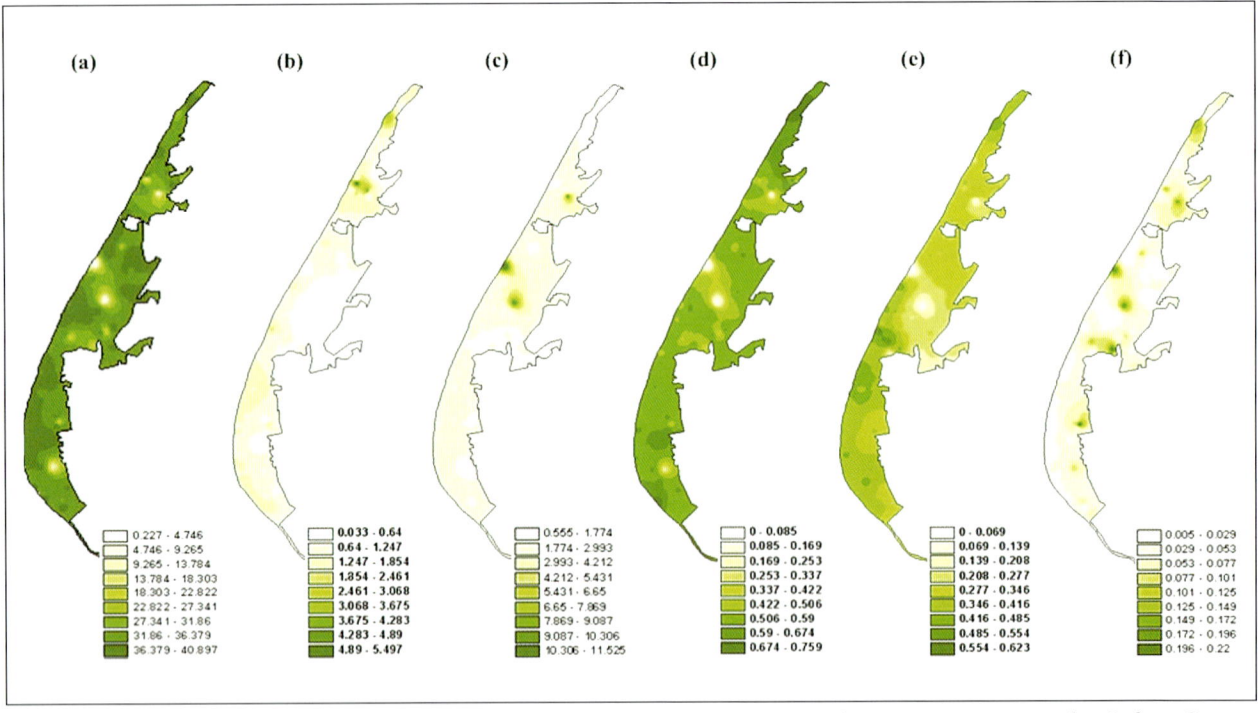

Figure 3. Spatial distribution of (a) Si, (b) Ca, (c) Na, (d) K, (e) Mg and (f) P percentages on the Sefton Coast.

alkaline pH appear directly behind the erosional point and continue to the edge of the urban area of Formby. There are very low levels of soil % carbon (C) (Figure 4a) and soil % nitrogen (N) (Figure 4b) calculated for these habitats, which are significantly different ($p < 0.01$) to all other environments. Mean ($n = 113$) particle size of the dune system ranges from 200-260 μm (medium sand) along the coastal margin, to 115-175 μm (fine sand) inland on the eastern fringes (Figure 2c). Mean ($n = 21$) particle size of bare sand and mobile dunes is similar ($p = 0.696$), but these are significantly different ($p < 0.05$) to all other remaining dune environments.

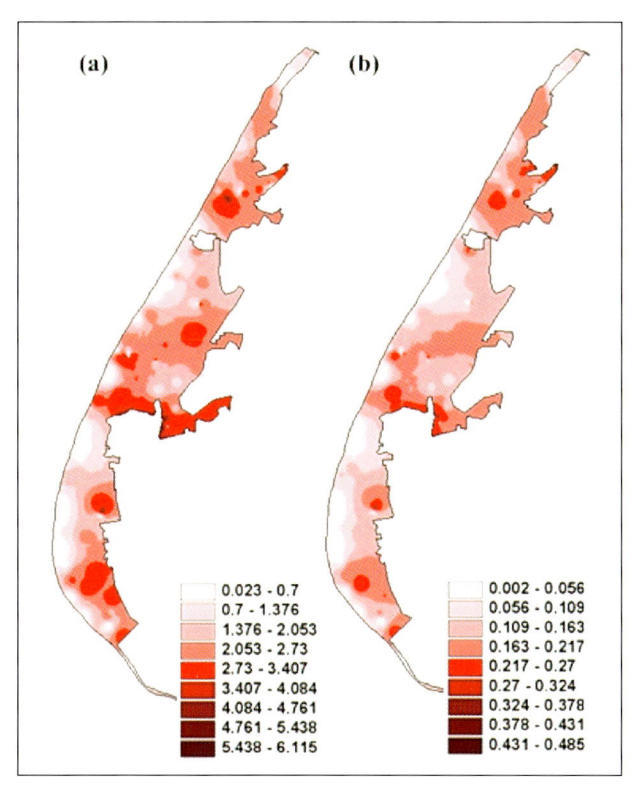

Figure 4. Spatial distribution of (a) total soil organic C and (b) total soil N on the Sefton Coast.

Across most of the dune landscape, the clay fraction only contributes ~2% of the total, with the exception of a small area on the mobile dunes by Ainsdale Holiday Village, where clays contribute 6% of the total (Figure 2d). Apart from this anomaly, the densest distribution of clays is on the fixed dunes, associated with higher concentrations of calcium (Ca) (Figure 3b), sodium (Na) (Figure 3c), potassium (K) (Figure 3d) and magnesium (Mg) (Figure 3e). Percentages of Mg in the fixed dunes are significantly different ($p < 0.001$) to bare sand environments, and percentages of K are different ($p < 0.05$) to those of mobile dunes. Clay values of slack soils are significantly different ($p < 0.05$) to bare sand and mobile dunes, and also have low χLF values (Figure 2e), compared to other dry dune environments ($p < 0.05$). These soils have a much higher SOM content than both comparable-aged dry fixed dune soils ($p < 0.01$), and soils associated with dune pasture and felled areas ($p < 0.05$) (Figure 2b). Some SOM accumulation is associated with scrub and woodland

Table 1. Statistical descriptions of (a) Sand, (b) Mobile dune, (c) Fixed dune, (d) Heath, (e) Slack, (f) Pasture, (g) Scrub, (h) Woodland, (i) Plantations and (j) Felled.

		(a) Sand (n=9)	(b) Mobile (n=12)	(c) Fixed (n=18)	(d) Heath (n=9)	(e) Slack (n=11)	(f) Pasture (n=10)	(g) Scrub (n=10)	(h) Wood (n=14)	(i) Pine (n=10)	(j) Felled (n=10)
Distance (m)	mean	202.778	226.042	794.444	1723.611	438.636	661.250	483.750	634.821	1056.250	476.250
	SD	69.253	266.951	339.826	137.137	148.477	423.668	350.993	268.148	276.464	216.703
	CoV	34.152	118.098	42.775	7.956	33.850	64.071	72.557	42.240	26.174	45.502
	Min	75.000	50.000	150.000	1587.500	250.000	150.000	175.000	212.500	575.000	225.000
	Max	300.000	1025.000	1475.000	1950.000	675.000	1212.500	1175.000	1075.000	1525.000	862.500
pH	mean	8.088	7.717	6.391	3.863	6.733	6.845	6.831	6.138	5.155	6.344
	SD	0.382	0.868	0.831	0.367	0.665	0.629	0.962	1.078	0.879	0.788
	CoV	4.723	11.248	13.003	9.500	9.877	9.189	14.083	17.563	17.051	12.421
	Min	7.620	6.490	5.010	3.100	5.260	5.450	5.140	3.890	4.060	5.410
	Max	8.680	8.840	7.830	4.310	7.740	7.730	7.700	7.370	6.660	7.490
SOM (%)	mean	0.210	2.148	4.426	6.814	18.709	4.775	8.494	12.977	33.785	3.917
	SD	0.067	6.043	6.013	3.952	20.431	8.698	11.637	20.576	27.500	3.678
	CoV	31.905	281.331	135.856	57.998	109.204	182.157	137.003	158.557	81.397	93.898
	Min	0.108	0.101	0.406	1.826	0.940	0.335	0.105	2.357	2.916	0.642
	Max	0.325	21.289	24.508	13.770	61.244	29.363	31.544	77.165	76.923	13.422
Mean size (μm)	mean	236.029	220.342	173.907	166.010	153.385	178.181	181.707	174.022	156.869	185.821
	SD	13.401	32.877	19.818	7.186	29.749	19.149	19.860	18.462	28.006	13.051
	CoV	5.678	14.921	11.396	4.329	19.395	10.747	10.930	10.609	17.853	7.023
	Min	210.570	168.459	124.294	159.083	98.924	146.636	142.813	135.338	95.701	161.875
	Max	249.134	259.500	203.174	177.322	195.689	212.017	201.977	199.237	182.822	205.688
Clay (%)	mean	4.321	3.582	2.676	1.761	2.285	2.115	2.156	1.765	1.881	2.052
	SD	0.859	1.296	0.880	0.398	0.933	0.942	0.655	0.695	0.717	0.534
	CoV	19.880	36.181	32.885	22.601	40.832	44.539	30.380	39.377	38.118	26.023
	Min	3.525	1.279	1.715	1.331	1.312	0.000	1.450	1.152	1.141	1.594
	Max	5.716	6.918	4.739	2.462	3.831	3.338	3.393	3.955	3.562	3.314

Table 1. Statistical descriptions of (a) Sand, (b) Mobile dune, (c) Fixed dune, (d) Heath, (e) Slack, (f) Pasture, (g) Scrub, (h) Woodland, (i) Plantations and (j) Felled.

		(a) Sand (n=9)	(b) Mobile (n=12)	(c) Fixed (n=18)	(d) Heath (n=9)	(e) Slack (n=11)	(f) Pasture (n=10)	(g) Scrub (n=10)	(h) Wood (n=14)	(i) Pine (n=10)	(j) Felled (n=10)
Xlf	mean	1.302	1.047	1.191	1.202	1.264	2.000	3.847	2.178	1.760	2.741
	SD	1.127	0.746	0.629	0.658	0.599	0.904	5.168	1.253	0.669	2.366
	CoV	86.559	71.125	52.813	54.742	47.389	45.200	134.338	57.530	38.011	86.319
	Min	0.336	0.275	0.419	0.418	0.692	0.693	0.618	0.853	1.073	0.601
	Max	3.853	2.731	3.153	2.222	2.735	3.331	17.718	6.065	3.530	6.934
Carbon (%)	mean	0.297	0.506	1.647	3.421	2.448	2.340	2.118	2.324	3.184	3.214
	SD	0.529	0.916	1.416	1.560	1.613	1.466	1.168	1.241	1.387	1.743
	CoV	178.114	181.028	85.974	45.601	65.891	62.650	55.146	53.399	43.562	36.537
	Min	0.018	0.021	0.063	1.212	0.379	0.085	0.144	0.230	1.007	0.058
	Max	1.665	3.170	4.437	6.136	4.656	5.095	3.385	3.816	5.096	5.767
Nitrogen (%)	mean	0.021	0.040	0.114	0.223	0.163	0.180	0.186	0.166	0.229	0.207
	SD	0.035	0.075	0.106	0.087	0.101	0.132	0.102	0.092	0.096	0.096
	CoV	166.667	187.500	92.982	39.013	61.963	73.333	54.839	55.422	41.921	46.377
	Min	0.002	0.003	0.005	0.098	0.032	0.009	0.011	0.023	0.095	0.004
	Max	0.112	0.262	0.390	0.351	0.288	0.501	0.305	0.324	0.370	0.312
Sodium (%)	mean	2.034	2.895	2.039	1.752	1.682	1.976	1.868	2.446	2.241	1.900
	SD	0.175	2.726	0.210	0.236	0.471	0.252	0.258	2.303	2.699	0.263
	CoV	8.604	94.162	10.299	13.470	28.002	12.753	13.812	94.154	120.437	13.842
	Min	1.790	1.810	1.570	1.350	0.551	1.500	1.440	0.660	0.890	1.540
	Max	2.410	11.530	2.290	2.190	2.190	2.440	2.270	10.330	9.820	2.230
Magnesium (%)	mean	0.432	0.338	0.314	0.189	0.335	0.327	0.335	0.341	0.297	0.305
	SD	0.078	0.126	0.039	0.047	0.071	0.051	0.033	0.122	0.146	0.081
	CoV	18.056	37.278	12.420	24.868	21.119	15.596	9.851	35.777	49.158	26.557
	Min	0.359	0.001	0.262	0.111	0.241	0.262	0.297	0.001	0.001	0.197
	Max	0.625	0.485	0.395	0.277	0.476	0.442	0.389	0.475	0.540	0.422
Silicon (%)	mean	37.930	34.700	38.438	35.697	30.960	36.224	35.414	31.508	24.420	37.332
	SD	1.396	11.317	2.913	1.906	7.474	4.598	4.165	12.114	11.787	2.027
	CoV	3.680	32.614	7.578	5.339	24.141	12.693	11.761	38.447	48.268	5.430
	Min	35.950	0.210	30.340	33.760	13.170	23.890	28.300	0.213	0.223	32.360
	Max	40.540	40.860	40.910	38.930	39.500	40.110	39.790	38.910	39.220	39.730
Phosporus (%)	mean	0.017	0.038	0.027	0.047	0.072	0.039	0.047	0.057	0.111	0.028
	SD	0.006	0.054	0.024	0.016	0.047	0.032	0.034	0.041	0.075	0.019
	CoV	35.294	142.105	88.889	34.043	65.278	82.051	72.340	71.930	67.568	67.857
	Min	0.011	0.005	0.007	0.017	0.017	0.019	0.011	0.021	0.020	0.013
	Max	0.032	0.188	0.105	0.067	0.148	0.126	0.099	0.181	0.221	0.068
Potassium (%)	mean	0.590	0.507	0.599	0.555	0.530	0.582	0.559	0.512	0.420	0.558
	SD	0.082	0.169	0.043	0.030	0.086	0.053	0.041	0.179	0.173	0.045
	CoV	13.898	33.333	7.179	5.405	16.226	9.107	7.335	34.961	41.190	8.065
	Min	0.519	0.001	0.527	0.508	0.367	0.498	0.494	0.001	0.001	0.492
	Max	0.759	0.647	0.691	0.596	0.706	0.667	0.636	0.627	0.600	0.616
Calcium (%)	mean	1.611	1.028	0.620	0.092	2.361	0.800	0.969	0.685	0.676	0.598
	SD	0.426	0.435	0.361	0.017	2.419	0.288	0.426	0.497	0.613	0.429
	CoV	26.443	42.315	58.226	18.478	102.457	36.000	43.963	72.555	90.680	71.739
	Min	1.074	0.033	0.123	0.073	0.361	0.411	0.255	0.039	0.043	0.154
	Max	2.579	1.742	1.210	0.117	8.148	1.262	1.667	1.447	1.961	1.298

communities, but sites displaying extremely high values of SOM content are associated with edges of slack environments.

Areas on the central-eastern fringes of the dune landscape, associated with pine plantations and heathland, are the most acidic (min pH = 3.10) and display significantly different (p <0.01) pH and Ca values to all remaining dune environments. Concentrations of phosphorus (P) (>0.2%) (Figure 3f) and N on the heathland are significantly higher (p <0.01) than in the fixed dunes, whereas percentages of

Table 2. Mann-Whitney U test p-value matrixes of (a) pH, (b) SOM, (c) mean particle size (PS) μm, (d) clay %, (e) χLF, (f) Si, (g) Ca, (h) Na, (i) K, (j) Mg, (k) P, (l) total soil organic C and (m) total soil N on the Sefton coast. Data represents p-values of Mann-Whitney U test. Bold text signifies a statistically significant difference, whereas regular text signifies no difference. Sand, Mobile dune, Fixed dune, Heath, Slack, Pasture, Scrub, Woodland, Plantations, Felled.

(a) pH	Sand	Mobile	Fixed	Heath	Slack	Pasture	Scrub	Woodland	Plantations
Mobile	0.337								
Fixed	**<0.001**	**0.002**							
Heath	**<0.001**	**<0.001**	**<0.001**						
Slack	**<0.001**	**0.018**	0.234	**<0.001**					
Pasture	**0.001**	**0.032**	0.131	**<0.001**	0.833				
Scrub	**<0.001**	**0.038**	0.222	**<0.001**	0.647	0.791			
Woodland	**<0.001**	**0.001**	0.649	**<0.001**	0.218	0.121	0.084		
Plantations	**<0.001**	**<0.001**	**0.004**	**0.001**	**0.001**	**0.001**	**0.003**	**0.024**	
Felled	**<0.001**	**0.003**	0.886	**<0.001**	0.307	0.241	0.151	0.747	**0.013**

(b) LOI	Sand	Mobile	Fixed	Heath	Slack	Pasture	Scrub	Woodland	Plantations
Mobile	0.374								
Fixed	**<0.001**	**0.001**							
Heath	**<0.001**	**0.002**	**0.048**						
Slack	**<0.001**	**0.001**	**0.008**	0.362					
Pasture	**<0.001**	**0.004**	0.905	**0.016**	**0.022**				
Scrub	**0.004**	**0.011**	0.549	0.307	0.098	0.910			
Woodland	**<0.001**	**<0.001**	**0.018**	0.729	0.396	**0.013**	0.107		
Plantations	**<0.001**	**<0.001**	**0.001**	0.079	0.170	**0.001**	**0.007**	**0.038**	
Felled	**<0.001**	**0.002**	0.518	0.066	0.032	0.427	0.910	0.057	**0.004**

(c) Mean	Sand	Mobile	Fixed	Heath	Slack	Pasture	Scrub	Woodland	Plantations
Mobile	0.696								
Fixed	**<0.001**	**0.001**							
Heath	**<0.001**	**0.001**	**0.029**						
Slack	**<0.001**	**<0.001**	**0.014**	0.543					
Pasture	**<0.001**	**0.005**	0.981	0.111	**0.038**				
Scrub	**<0.001**	**0.011**	0.204	**0.037**	**0.018**	0.521			
Woodland	**<0.001**	**0.001**	0.985	0.055	**0.023**	0.838	0.230		
Plantations	**<0.001**	**0.005**	0.981	0.111	**0.038**	1.000	0.521	0.838	
Felled	**<0.001**	**0.017**	0.095	**0.002**	**0.004**	0.391	0.713	0.197	0.391

(d) Clay	Sand	Mobile	Fixed	Heath	Slack	Pasture	Scrub	Woodland	Plantations
Mobile	0.051								
Fixed	**<0.001**	**0.009**							
Heath	**<0.001**	**0.002**	**0.003**						
Slack	**0.002**	**0.029**	0.178	0.288					
Pasture	**<0.001**	**0.003**	0.222	0.111	0.805				
Scrub	**<0.001**	**0.004**	0.089	0.153	0.972	0.678			
Woodland	**<0.001**	**0.002**	**<0.001**	0.729	0.163	0.057	0.065		
Plantations	**0.001**	**0.003**	**0.008**	0.903	0.379	0.212	0.273	0.619	
Felled	**<0.001**	**0.003**	**0.029**	0.216	1.000	0.438	1.000	**0.022**	0.307

(e) Xlf	Sand	Mobile	Fixed	Heath	Slack	Pasture	Scrub	Woodland	Plantations
Mobile	0.644								
Fixed	0.662	0.320							
Heath	0.724	0.414	0.980						
Slack	0.447	0.282	0.770	0.704					
Pasture	0.079	**0.011**	**0.026**	0.066	0.085				
Scrub	0.066	**0.023**	0.052	0.111	0.073	0.678			
Woodland	**0.035**	**0.003**	**0.004**	**0.022**	**0.017**	0.884	0.930		
Plantations	0.094	**0.027**	**0.006**	0.079	**0.032**	0.623	0.910	0.169	
Felled	0.153	**0.044**	0.131	0.236	0.218	1.000	0.791	0.747	0.970

Table 2. continued

(f) Silicon	Sand	Mobile	Fixed	Heath	Slack	Pasture	Scrub	Woodland	Plantations
Mobile	0.644								
Fixed	0.138	0.241							
Heath	**0.027**	0.095	**0.011**						
Slack	**0.012**	**0.025**	**0.001**	0.068					
Pasture	0.540	0.410	**0.042**	0.153	0.062				
Scrub	0.270	0.277	**0.026**	0.596	0.170	0.623			
Woodland	**0.025**	0.057	**0.002**	0.592	0.286	0.121	0.501		
Plantations	**0.006**	0.013	**0.001**	0.055	0.170	**0.014**	**0.026**	0.107	
Felled	0.967	0.621	**0.031**	0.111	**0.012**	0.762	0.385	**0.050**	**0.007**

(g) Calcium	Sand	Mobile	Fixed	Heath	Slack	Pasture	Scrub	Woodland	Plantations
Mobile	**0.008**								
Fixed	**<0.001**	**0.014**							
Heath	**<0.001**	**0.002**	**<0.001**						
Slack	0.704	0.230	**0.004**	**<0.001**					
Pasture	**0.001**	0.060	0.183	**<0.001**	**0.032**				
Scrub	**0.007**	0.621	**0.047**	**<0.001**	0.130	0.273			
Woodland	**0.001**	0.157	0.827	**0.001**	**0.015**	0.429	0.188		
Plantations	**0.004**	0.156	0.900	**0.008**	**0.022**	0.308	0.162	0.838	
Felled	**0.001**	0.060	0.821	**<0.001**	**0.008**	0.241	0.141	0.578	0.910

(h) Sodium	Sand	Mobile	Fixed	Heath	Slack	Pasture	Scrub	Woodland	Plantations
Mobile	0.356								
Fixed	0.535	0.465							
Heath	**0.013**	**0.003**	**0.011**						
Slack	**0.044**	**0.005**	**0.009**	0.732					
Pasture	0.624	0.147	0.315	0.066	0.105				
Scrub	0.178	**0.019**	0.071	0.270	0.481	0.308			
Woodland	0.329	**0.045**	0.147	0.108	0.239	0.619	0.578		
Plantations	**0.025**	**0.008**	**0.008**	0.391	0.379	0.059	0.121	0.107	
Felled	0.488	0.070	0.160	0.289	0.275	0.762	0.850	1.000	0.112

(i) Potassium	Sand	Mobile	Fixed	Heath	Slack	Pasture	Scrub	Woodland	Plantations
Mobile	0.414								
Fixed	0.389	**0.028**							
Heath	0.659	0.915	**0.015**						
Slack	0.197	0.644	**0.007**	0.323					
Pasture	0.838	0.249	0.407	0.236	0.098				
Scrub	0.596	0.531	**0.022**	1.000	0.275	0.212			
Woodland	0.729	0.440	0.117	0.508	0.286	0.396	0.539		
Plantations	**0.025**	0.075	**0.001**	**0.037**	0.149	**0.009**	**0.045**	**0.047**	
Felled	0.488	0.621	**0.047**	0.775	0.245	0.345	1.000	0.704	**0.038**

(j) Magnesium	Sand	Mobile	Fixed	Heath	Slack	Pasture	Scrub	Woodland	Plantations
Mobile	0.082								
Fixed	**<0.001**	0.106							
Heath	**<0.001**	**0.002**	**<0.001**						
Slack	**0.008**	0.712	0.589	**0.001**					
Pasture	**0.003**	0.489	0.514	**<0.001**	0.916				
Scrub	**0.001**	0.644	0.183	**<0.001**	0.805	0.545			
Woodland	0.078	0.918	0.052	**0.002**	0.494	0.279	0.320		
Plantations	**0.020**	0.291	0.498	**0.025**	0.379	0.345	0.427	0.242	
Felled	**0.008**	0.138	0.482	**0.004**	0.481	0.385	0.308	0.188	0.970

(k) Phosphorus	Sand	Mobile	Fixed	Heath	Slack	Pasture	Scrub	Woodland	Plantations
Mobile	0.456								
Fixed	0.360	0.740							
Heath	**0.002**	**0.043**	**0.011**						
Slack	**0.001**	**0.013**	**0.002**	0.494					
Pasture	**0.004**	0.060	0.093	0.079	0.085				
Scrub	**0.014**	0.081	0.075	0.540	0.139	0.706			
Woodland	**<0.001**	**0.006**	**0.001**	0.925	0.494	0.057	0.306		
Plantations	**0.001**	**0.007**	**0.002**	0.066	0.307	0.064	0.089	0.169	
Felled	0.391	1.000	0.940	**0.031**	**0.012**	0.186	0.241	**0.021**	**0.005**

Table 2. continued

(l) Carbon Bare Sand	Sand	Mobile	Fixed	Heath	Slack	Pasture	Scrub	Woodland	Plantations
Mobile	0.303								
Fixed	**0.004**	**0.012**							
Heath	**<0.001**	**<0.001**	0.014						
Slack	**0.003**	**0.004**	0.270	0.244					
Pasture	**0.002**	**0.004**	0.257	0.206	0.807				
Scrub	**0.002**	**0.006**	0.280	**0.046**	0.961	0.850			
Woodland	**<0.001**	**<0.001**	0.154	0.062	1.000	0.733	0.515		
Plantations	**0.003**	**0.003**	**0.043**	0.680	0.284	0.303	0.093	0.148	
Felled	**0.003**	**0.005**	**0.045**	0.791	0.341	0.206	0.348	0.350	0.860
(m) Nitrogen Bare Sand	Sand	Mobile	Fixed	Heath	Slack	Pasture	Scrub	Woodland	Plantations
Mobile	0.394								
Fixed	**0.006**	**0.022**							
Heath	**<0.001**	**<0.001**	0.012						
Slack	**0.002**	**0.004**	0.229	0.341					
Pasture	**0.002**	**0.004**	0.163	0.206	0.961				
Scrub	**0.001**	**0.002**	0.069	0.540	0.807	0.385			
Woodland	**<0.001**	**<0.001**	0.072	0.217	0.968	0.852	0.556		
Plantations	**0.003**	**0.004**	**0.025**	1.000	0.284	0.303	0.704	0.254	
Felled	**0.003**	**0.006**	**0.039**	0.930	0.397	0.348	0.653	0.350	0.814

K are significantly lower ($p <0.001$), as well as C and N ($p < 0.05$), under pine plantations than under newly developed fixed dune soils. While Na and Ca are generally the most abundant elements, Ca values are low (0-2%). There are considerable variations between all element concentrations in all dune environments. Values of χLF are extremely low and relatively consistent throughout the entire dune landscape, apart from an area on the seaward side, north of Formby Point. SOM and P values from the clear-felled sites topsoil are statistically different ($p <0.01$) to those in the pine plantations.

DISCUSSION

Distinct soil characteristics are apparent within the Sefton dune landscape and can be confidently associated with individual dune environments by digitizing reliable data into a GIS and conducting Mann-Whitney U tests. Frontal dunes have relatively high alkalinity, Ca and Si contents, due to the abundance of calcareous bare sand reflecting low shell content and high quartz concentrations. Marine influences are also important for Na and Mg. Spatial variations in both geochemical and particle size characteristics of these coastal sediments provide indicators of sediment source and depositional conditions (Knight et al., 2002; Saye and Pye, 2006). For example, similarities in cations and mean particle size of bare sand and mobile dunes suggests these areas are responding to local variations in wave and wind energy regimes, which are strongly dependent on exposure (James et al., 1986; Saye and Pye, 2006), rather than soil processes. Strong prevailing south-westerly winds (James and Wharfe, 1989) are capable of transporting a wide range of particle sizes great distances. Varying pH values within this depositional zone are likely to be the result of differential rates of erosion and deposition, ranging from aeolian derived calcareous sand being transported from the frontal dunes, leaving large blowouts, to be subsequently deposited inland causing, in effect, slower rates of dune roll-back. This is reflected in lower levels of soil %N and soil %C in bare sand environments than in mobile dune environments. Although these environments are not significantly different at the 95 % level, higher values can be associated with increased cover of *Ammophila arenaria* (Jones et al., 2004) differentiating between frontal dunes and unstable blowouts. Soil development on the frontal dunes is limited to the incorporation of some SOM in the top few centimetres, therefore classifying the soil in these areas as raw terrestrial soil (Avery, 1980), differing little from the recently formed and well-drained sandy parent material.

Sediment accumulation north of Formby Point has resulted in stabilization of the northern fixed dune landscape, represented by neutral to slightly acidic topsoil. These fixed dune environments are associated with high concentrations of Na, K, Ca and Mg, therefore increasing soil fertility, appearing to develop from a terrestrial raw soil into a lithomorphic soil or sand-pararendzina (Avery, 1980). Again, bare sand and mobile dune environments can be categorized together by differing percentages of Mg and K compared with the fixed dunes, indicating that even the shallowest topsoils are becoming more productive than bare sand alone. Wet slacks are interspersed within the fixed dune environment. The immediate noticeable characteristic of dune slacks is that the SOM content is much higher than comparable-aged dry fixed dune soils, dune pasture and felled areas, similar to findings by Ranwell (1959). However, they have very similar values to heath, scrub, woodland and plantations, suggesting that they are at a much more advanced stage of development. Ranwell (1972) highlighted that in most dune systems, the well-developed wet slacks overlie impermeable clay, which account for the higher clay contents found in Ainsdale National Nature Reserve. It is widely known that particle size and soil moisture characteristics are related (e.g. Williams and Cooper, 1990). The low χLF values, compared to other dry dune environments, are evidence of waterlogged sites, resulting from dissolution of the strongly magnetic minerals under reducing conditions (Mullins, 1977; Wang et al., 2008). These characteristics are associated with ground water gley soils (Avery, 1980), which develop in areas where drainage is poor and the water table is high.

The semi-natural dune heath environment conditions are extremely acidic, although Salisbury (1925) considered that it takes 200 years for the pH of the soils on this coast to fall as low as 6.4 from its original value of 8.2 at the top of the beach, and Payne (1983) noted that acidity is slow to develop on these dunes, as the parent material is highly alkaline. According to Ranwell (1972), these trends depend upon a slowing down of the mobility of the dune system as vegetation colonizes, providing a source for SOM accumulation, and increasing exposure of soil to leaching. Evidence from other dune systems with initial calcium carbonate concentrations (<5 %) (Salisbury, 1952; Ranwell, 1959) show carbonate is lost from the topsoil within 300-400 years. As expected, there is also a decrease in mean particle size in the soils of the dune heathland, as coarser sand derived from aeolian processes is not capable of reaching this far inland. However, it would be expected that finer particle fractions would accumulate with time as a result of in situ weathering and increased levels of airborne dust, increasing the exchange capacity of the soil. But an overall decrease in percentage of clay is evident, suggesting that eluviation of clays proceeds with age and subsequent development of a micro-podzol soil (Avery, 1980).

There are considerable variations between all elemental concentrations across the entire dune landscape, due to varying leaching rates, translocation rates, atmospheric inputs, evapotranspiration, plant uptake and SOM content. K appears to vary more than the other cations due to close links with uptake and storage by a variable vegetation cover (James et al., 1986). Differences in soil C are not significant among individual dune environments inland from fixed dune communities, apart from a slight increase in semi-natural dune heath environments, similar to findings by Ordóñez et al. (2007). For most parameters, a seperation can be made between 'frontal dune' environments (bare sand, mobile, and to some extent, fixed dunes) and hind dune environments (heath, pasture, scrub, woodland and plantations). It is evident that temporal increases in SOM, and subsequent increases in the capacity of the soil to retain nutrients (Ranwell, 1972), support the large and diverse floral communities, such as successional scrub and deciduous woodland. These environments support deep, well-drained fertile brown soils (Avery, 1980) and therefore, any cations not stored in the soil or biomass are lost to groundwater at this stage (James et al., 1986). Coastal erosion processes, active on Formby point, have destroyed mobile dune barrier environments, consequently exposing these areas of previous inland woody habitats on the foreshore. Subsequent burial of these woodland communities is evident by equivalent values of K across all bare sand, scrub and woodland environments, as increased storminess has resulted in deposition of large, K-deprived, sand sheets inland burying existing topsoils. This is also reflected by χLF values for scrub being similar to those of bare sand. Higher values of χLF are associated with both mature scrub and woodland communities, now exposed on the coastal edge, due to a combination of the well-drained nature of these areas and, when considered alongside increases in particle size data, deposition of freshly eroded sand (Williams and Cooper, 1990).

Afforestation of the sand dunes as a stabilizing technique from 1887-1960 (Sturgess, 1992; Wheeler et al., 1993) has imposed a distinctive microclimate, ecological and soil environment on this part of the dunes (James and Wharfe, 1989), affecting the nature and transformation of SOM and associated nutrients (Chen et al., 2007). Soil C and N stocks have limited variability under open fixed dune and plantation ecosystems, possibly as a result of increased biomass in the latter depleting excess available C and N in the soil (Jones et al., 2004). This suggests that afforestation caused little change in total soil organic C and total soil N, similar to findings by McKinley et al.(2008). Topsoil characteristics have been altered substantially through deposition of a thick acidic pine needle litter layer, evident of micro-podzol development (Avery, 1980). Guo et al. (2007) suggest that the slower turnover of litter plays a role in the reduction in soil C, as under such rapidly developing acidic conditions, litter breakdown is also inhibiting SOM accumulation. Berendse (1998) suggests that during succession, the litter of woody species contains increased phenols and organic compounds, slowing decomposition rates. Enhanced mineralization of this SOM and associated P appears to have consequently improved P-availability in topsoil. The diamagnetic characteristic of increasing SOM may account for decreased values of χLF in this habitat.

Detrimental ecological changes, including loss of flora and fauna, soil changes, lowering of the water table and seeding into surrounding duneland, resulted after afforestation of the sand dunes (Atkinson, 1988; Sturgess, 1992, 1993). In response, dune managers attempted to restore fixed dune habitats, of greater conservation value, by removing areas of pine woodland. Due to the potentially large quantity of nutrients removed, it has generally been considered that clear felling can lead to decreased soil-available reserves of limiting nutrients (Olsson et al., 2000; Merino et al., 2004). This is evident by percentages of K under pine plantations being significantly lower than under newly developed fixed dune soils. However, following felling on the Sefton coast, promoted decomposition and leaching has resulted in increases in the movement of some minerals, facilitating plant colonization (Atkinson and Sturgess, 1991). The high SOM content retains water much better than bare sand, and this may be an important factor in the rapid establishment of fixed dune vegetatation on clear-felled sites (Sturgess, 1992, 1993). Colonization of such plants will have begun to recycle nutrients, accounting for losses in SOM and P from the clear-felled sites topsoil, compared with the pine plantations. Total N and C percentages remain the same under clear-felled areas as under the plantations, which according to Staaf and Berg (1982) is not normal as C would be expected to be lost at a higher rate through respiration than N. Although, Merino et al. (2004) propose that changes in total soil C and N cannot be reliably detected after such a short time period between felling in 1995/96 and today. The pH values of the plantations are different from felled areas, highlighting that during the last decade, since felling began, the micro-podzol soils are reverting back to fixed dune sand-pararendzina characteristics. The lack of vegetation differences proves that successful growth of dune plants, characteristic of young, calcareous dunes, has not been prevented.

CONCLUSIONS

Individual dune environments can be distinct from each other and their distinctiveness can be calculated using Mann-Whitney U tests. The pH is the the most appropriate soil characteristic, followed by mean particle size, Ca and Mg. For most parameters, a separation can be made between 'frontal dune' environments (bare sand, mobile, and to some extent, fixed dunes) and 'hind dune' environments (heath, pasture, scrub, woodland and plantations). χLF values can be used to differentiate well-drained and poorly-drained soils, therefore slack and dry dune environments. It seems that to be able to distinguish dune environments using soil characteristics, it is necessary to group the dune areas into fewer classified environments. But this runs the risk of losing some dune soil environments that do appear to have very distinct soil habitats.

ACKNOWLEDGEMENTS

Thanks to David Luckhurst (University of Wolverhampton) and Graham Lymbery (SMBC). The first author gratefully acknowledges receipt of a Ph.D. studentship, partly funded by Sefton MBC. Thanks to the Vivekananda Institute of Hill Agriculture, India, for total soil organic carbon and total soil nitrogen determinations.

REFERENCES

Andrews, B.D., Gares, P.A. and **Colby, J.D.** (2002) Techniques for GIS modelling of coastal dunes. *Geomorphology*, 48, 289-308.

Atkinson, D. (1988) The effects of afforestation on a sand dune grassland. *British Ecological Society Bulletin*, 29(2), 99-101.

Atkinson, D. and **Sturgess, P.W.** (1991) *Restoration of sand-dune communities following deforestation. Perturbation and recovery of terrestrial and aquatic ecosystems.* Ellis Horwood, London.

Avery, B.W. (1980) *Soil Classification for England and Wales (Higher Categories)*, Soil Survey Technical Monograph 14. Soil Survey of England and Wales, Harpenden.

Berendse, F. (1998) Effects of dominant plant species on soils during succession in nutrient-poor ecosystems. *Biogeochemistry*, 42, 73-88.

Booth, C.A., Fullen, M.A., Jankauskas, B. and **Jankauskiene, G.** (2003) International calibration of the textural properties of Lithuanian Eutric Albeluvisols. *Soil Science and Agrochemistry. Journal of the Lithuanian Academy of Sciences*, 4, 3-10.

Burrough, P.A. and **McDonnell, R.A.** (1998) *Principles of Geographical Information Systems*. Clarendon Press, Oxford.

Chen, C.R., Condron, L.M. and **Xu, Z.H.** (2008) Impacts of grassland afforestation with coniferous trees on soil phosphorus dynamics and associated microbial processes: A review. *Forest Ecology and Management,* 255, 396-409.

Costantini, E.A.C., Fantappiè, M., Bocci, M., Marzocchi, S., Sacchi, G., Paolanti, M., Riveccio, R., Giorgianni, A. and **Perciabosco, M.** (2007) Monitoring soil erosion and landslides in Sicily by an integrated physiographic and geomorphological GIS. In: C. Dazzi (Ed.) *Changing Soils in a Changing World: the Soils of Tomorrow.* Book of Abstracts. 5th International Congress of the European Society for Soil Conservation. Palermo, Italy.

Doody, J.P. (1989) Management for nature conservation. *Proceedings of the Royal Society of Edinburgh,* 247-265.

Edmondson, S.E., Gately, P.S., Rooney, P. and **Sturgess, P.W.** (1993) Plant communities and succession. In: D. Atkinson, and J. Houston (Eds.). *The Sand Dunes of the Sefton Coast.* Eaton Press Ltd, England.

Gibson, D.J. and **Looney, P.B.** (1992) Seasonal variation in vegetation classification on Perdido Key, a barrier island off the coast of the Florida Panhandle. *Journal of Coastal Research,* 8(4), 943-956.

Guo, L.B., Cowie, A.L., Montagu, K.D. and **Gifford, R.M.** (2008) Carbon and nitrogen stocks in a native pasture and an adjacent 16-year-old Pinus radiate D.Don. plantation in Australia. *Agriculture, Ecosystems & Environment*, 124, 205-218.

James, P.A. and **Wharfe, A.J.** (1989) Timescales of soil development in a coastal sand dune system, Ainsdale, north-west England. In: F. van der Meulen, P.D. Jungerius and J.H. Visser (Eds.) *Perspectives in coastal dune management.* The Hague, The Netherlands.

James, P.A., Wharfe, A.J., Pegg, R.K. and **Clarke, D.** (1986) A cation budget analysis for a coastal dune system in north-west England. *Catena*, 13, 1-10.

Jones, M.L.M., Wallace, H.L., Norris, D., Brittain, S.A., Haria, S., Jones, R.E., Rhind, P.M., Reynolds, B.R. and **Emmett, B.A.** (2004) Changes in vegetation and soil characteristics in coastal sand dunes along a gradient of atmospheric nitrogen deposition. *Plant Biology*, 6, 1-8.

Knight, J., Orford, J.D., Wilson, P. and **Braley, S.M.** (2002) Assessment of temporal changes in coastal sand dune environments using the log-hyperbolic grain-size method. *Sedimentology*, 49, 1229-1252.

Kundu, S., Bhattacharyya, R., Prakash, V., Ghosh, B.N. and **Gupta, H.S.** (2007) Carbon sequestration and relationship between carbon addition and storage under rainfed soybean–wheat rotation in a sandy loam soil of the Indian Himalayas. *Soil and Tillage Research*, 92 (1-2), 87-95.

Kutiel, P., Peled, Y. and **Geffen, E.** (2000) The effect of removing shrub cover on annual plants and small mammals in a coastal sand dune ecosystem. *Biological Conservation*, 94(2), 235-242.

Lucas, N.S., Shanmugam, S. and **Barnsley, M.** (2002) Sub-pixel habitat mapping of a coastal dune ecosystem. *Applied Geography*, 22, pp. 253-270.

McKinley, D.C., Rice, C.W. and **Blair, J.M.** (2008) Conversion of grassland to coniferous woodland has limited effects on soil nitrogen cycle processes. *Soil Biology & Biochemistry,* 40, 2627-2633.

Merino, A., Fernández-López, A., Solla-Gullón, F. and **Edeso, J.M.** (2004) Soil changes and tree growth in intensively managed Pinus radiata in northern Spain. *Forest Ecology and Management*, 196, 393-404.

Mullins, C.E. (1977) Magnetic susceptibility of the soil and its significance in soil science – a review. *Journal of Soil Science*, 28, 223-246.

Olsson, B.A., Lundkvist, H. and **Staaf, H.** (2000) Nutrient status in needles of Norway spruce and Scots pine following harvesting of logging residues. *Plant Soil*, 23, 161-173.

Ordóñez, J.A.B., Jong, B.H.J., García-Oliva, F., Aviña, F.L., Guerrero, P.G., Martínez, R. and **Masera, O.** (2008) Carbon content in vegetation, litter, and soil under 10 different land-use and land-cover classes in the Central Highlands of Michoacan, Mexico. *Forest Ecology and Management*, 255, 2074-2084.

Payne, K.R. (1983) *The vegetation of Ainsdale dunes.* Unpublished report NCC. ASD.

Purdie, J. (2002) *A pH survey of Ainsdale Sand Dunes National Nature Reserve, Merseyside.* Unpublished Report conducted on behalf of English Nature. Liverpool Hope University College.

Pye, K. (1990) Physical and human influences on coastal dune development between the Ribble and Mersey estuaries, northwest England. In: K.F. Nordstrom, N. Psuty and R.W.G. Carter (Eds.). *Coastal Dunes: form and process.* John Wiley & Sons, Chichester.

Ranwell, D.S. (1959) Newborough Warren, Anglesey. I. The dune system and dune slack habitat. *Journal of Ecology,* 47, 571-601.

Ranwell, D.S. (1972) *Ecology of salt marshes and sand dunes.* Chapman & Hall Ltd.

Repe, B. (2007) Geographical soil map – a different soil determination approach in Slovenia. In: C. Dazzi (Ed.) *Changing Soils in a Changing World: the Soils of Tomorrow.* Book of Abstracts. 5th International Congress of the European Society for Soil Conservation. Palermo, Italy.

Rothwell, P. (1985) Management problems on Ainsdale Sand Dunes National Nature Reserve. In: P. Doody. Focus on Nature Conservation No 13. *Sand Dunes and their Management.* Peterborough.

Rühl, J., Cullotta, S., Minacapelli, M. and **La Mantia, T.** (2007) Analysis of renaturation processes using LANDSAT 5 and 7 imagery. In: C. Dazzi (Ed.) *Changing Soils in a Changing World: the Soils of Tomorrow.* Book of Abstracts. 5th International Congress of the European Society for Soil Conservation. Palermo, Italy.

Salisbury, E.J. (1925) Note on the edaphic succession in some dune soils with special reference to the time factor. *Journal of Ecology,* 13, 322-328.

Salisbury, E.J. (1952) *Downs and Dunes. Their Plant Life and Its Environment.* G. Bell & Sons, London.

Saye, S.E. and **Pye, K.** (2006) Variations in chemical composition and particle size of dune sediments along the west coast of Jutland, Denmark. *Sedimentary Geology*, 183, 217-242.

Smith, P.H. (1978) The Ecological Evaluation of Wetland Habitats in the North-Merseyside Sand-dune System. Report to Merseyside County Council. JCAS database 075 Da 071.

Staaf, H. and **Berg, B.** (1982) Accumulation and release of plant nutrients in decomposing Scots Pine needle litter. Long term decomposition in a Scots Pine forest II. *Canadian Journal of Botany*, 60, 1561-1568.

Sturgess, P. (1992) Clear-felling dune plantations: Studies in vegetation recovery. In: R.W.G. Carter, T.G.F. Curtis and M.J. Sheehy-Skeffington (Eds.). *Coastal Dunes.* Balkema, Rotterdam.

Sturgess, P.W. (1993) Clear-felling dune plantations: Studies in vegetation recovery on the Sefton coast. In: D. Atkinson, and J. Houston (Eds.). *The Sand Dunes of the Sefton Coast.* Eaton Press Ltd, England.

van Reeuwijk, L.P. (2002) *Procedures For Soil Analysis, Sixth Edition.* International Soil Reference and Information Centre.

Wang, J-S., Grimley, D.A., Xu, C. and **Dawson, J.O.** (2008) Soil magnetic susceptibility reflects soil moisture regimes and the adaptability of tree species to these regimes. *Forest Ecology and Management*, In Press.

Wheeler, D.J., Simpson, D.E. and **Houston, J.A.** (1993) Dune use and management. In: D. Atkinson, and J. Houston (Eds.). *The Sand Dunes of the Sefton Coast.* Eaton Press Ltd, England.

Williams, R.D. and **Cooper, J.R.** (1990) Locating soil boundaries using magnetic susceptibility. *Soil Science*, 150(6), 889-895.

Geomorphology of the Sefton Coast sand dunes

Kenneth Pye* and Simon J. Blott

Kenneth Pye Associates Ltd, Crowthtorne Enterprise Centre, Old Wokingham Road, Crowthorne, Berkshire RG45 6AW, UK
*Corresponding author: e-mail: k.pye@kpal.co.uk

ABSTRACT

The Sefton coast contains the largest area of windblown sand in England, consisting of dunes, sandsheets and deflation basins. Most of the blown sand area is now stabilized by vegetation, significant areas have been built on, and much of the remainder has been significantly modified by a range of human activities including sand mining, creation of asparagus fields, waste dumping and dune re-profiling as part of woodland management and habitat restoration programmes. Nevertheless, the dune system retains numerous features of intrinsic geomorphological as well as ecological interest. The dunes form an important defence against tidal flooding of the hinterland and are also of major nature conservation and recreation value. For all these reasons, the geomorphological character, sedimentological composition and evolutionary history of the Sefton dune systems have been the subject of extensive research in recent decades. This paper provides an overview of the geomorphological character of the dunes, their evolutionary history and the relationship between dune morphology and shoreline changes in the historically recent past. New information relating to the frontal dune morphology, sand volumes and implications for coastal flood defence, derived from analysis of lidar data obtained in 2008, is also presented together with an update regarding frontal dune erosion and accretion on annual to centennial timescales.

Abbreviations:

OD = Ordnance Datum;
DTM = Digital Terrain Model;
NAO = North Atlantic Oscillation;
HAT = Highest Astronomical Tide;
MHW = Mean High Water;
AD = Anno Domini;
BP = Before Present.

INTRODUCTION

Coastal dunes front much of the Sefton coast between Southport and Crosby (Figure 1a). They provide an important defence against marine flooding and are also of high nature conservation and recreational importance (George, 1997; Smith, 1999; Hopkins and Radley, 2001; Rooney, 2001). A considerable amount of research has been undertaken in relation to the geomorphological evolution of the dunes, and a review of work undertaken before 1992 was presented by Plater et al. (1993). The present paper summarises some of the more important previous work but gives particular attention to research undertaken in the past 15 years.

Geological framework and coastal evolution

The overall form and orientation of the Sefton coast is controlled by the underlying geological framework and the history of weathering and erosion over the past few million years. The bedrock surface in West Lancashire and north Merseyside is characterised by a series of NW-SE trending buried valleys which broadly follow the regional pattern of faulting and which were over-deepened by glacial and fluvio-glacial action during the Pleistocene Period (Wray et al., 1948; Shackleton, 1953; Gresswell, 1953a,b, 1964; Howell, 1973; Johnson and Musk, 1974; Johnson, 1985). Glacial sediments up to 30 m thick, including till and outwash sands and gravels, were deposited in an irregular pattern across this surface. Towards the end of the last glacial stage, aeolian reworking of fluvio-glacial sediments led to the formation of coversands and low dunes (represented by the Shirdley Hill Sand) which extends as far inland as the Billinge - Ashurst ridge, east of Ormskirk (Figure 1b).

Between approximately 9,000 and 5,000 years ago, in the early to mid Holocene epoch (also often referred to

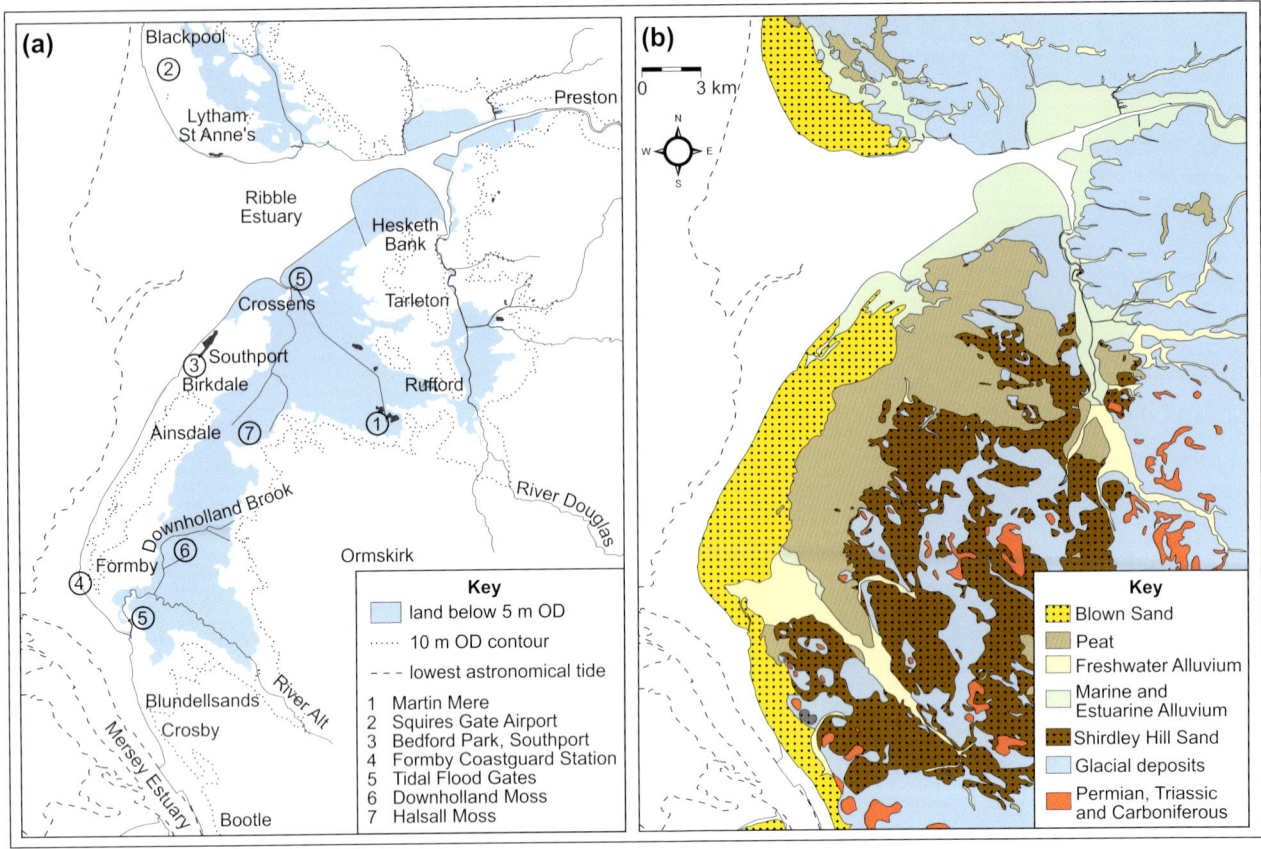

Figure 1. (a) Location map showing major localities mentioned in the text and areas below 5 m OD at risk of tidal flooding; (b) simplified geological map of Sefton and adjoining areas (data sources: Ordnance Survey and British Geological Survey).

in the UK as the Flandrian stage), rising sea level flooded lower ground in the western part of the area and a sequence of marine, estuarine and lagoonal sands, silts and clays (represented by the Downholland Silt) was deposited on top of the post-glacial surface. In some areas, including Orrell Hill Wood near Hightown, the Shirdley Hill Sand was reworked by marine and aeolian processes (Pye et al., 1995), and in the Mere Sands Wood area near Burscough there appears to have been some aeolian reworking of Shirdley Hill Sands from the shores of Martin Mere (Wilson, 1985). However, Gresswell's (1953c) suggestion that the main body of Shirdley Hill Sands were derived from marine beaches of early to mid-Holocene age has not been supported by later work.

Radiocarbon dating and palaeoecological studies suggest that sea level in this area rose from c. -16.0 m OD around 9000 yr BP to approximately its present level (+/- 1m) by 5000 years BP (Tooley, 1969, 1974, 1976, 1978, 1985a,b). The maximum inland extent of the postglacial marine transgression varied in time and space, but at c. 5000 yr BP marine influence extended up the River Alt valley and across Downholland Moss and Halsall Moss (Figure 1a). Within the tidally influenced area a thickness of several metres of predominantly silty sediments was deposited (the Downholland Silt). Borehole information suggests that Downholland Silt is not present in the Martin Mere area (Tooley, 1978, 1985a; Hale 2005a,b), and prior to drainage at the end of the 17th century Martin Mere was a large freshwater lake, approximately 6 km long, 3 km wide, and up to 2 m deep (Brodrick, 1902, 1903; Tooley, 1985a; Hale, 2005a). The lake formed at the end of the last glacial period in a shallow depression in the glacial till and Shirdley Hill Sand surface (Hale 2005a,b). Marine transgression into the lake basin was probably prevented by a low ridge of till (2.4 - 3.5 m OD) which extends between the higher ground at Banks and a bedrock outcrop near Brown Edge (National Grid Reference SD363147). Early maps (e.g. Saxton 1577, Speed, 1610, Blaeu, 1662) show that in the 16th and 17th centuries Martin Mere had an outlet eastwards into the River Douglas near Rufford, although at times of high lake level it may also have overflowed westwards into a stream flowing towards Crossens (Tooley, 1985a; Hale, 2005b). Drainage of the lake was begun by Thomas Fleetwood in 1692 and involved digging a new cut to the sea at Crossens (Hale, 2005c).

The present coastal dune barrier is underlain by up to 12 m of marine sands and silty sands (Tooley, 1974; Neal, 1993; Pye and Neal, 1993). Borehole evidence suggests that Downholland Silt is not present beneath the main dune belt at Formby Point and Ainsdale (Neal, 1993; Pye and Neal, 1993). The accumulation of silts and organic-rich deposits in the Downholland Moss area indicates low energy conditions and suggests that some form of barrier existed to the west throughout much of the early to mid Holocene. Environmental conditions in the back-barrier varied spatially and over time, leading to the formation of a range of sedimentary environments including sandflats, mudflats, saltmarshes, brackish reed marshes and open water lagoons (Tooley, 1978). The area was drained by two principal tidal creek systems, the largest of which entered the Alt estuary the other which entered the Ribble estuary at Crossens (Huddart, 1992; Neal, 1993). Progressive sedimentation in the back-barrier area, and possible short periods of slight relative sea level fall, led to the development of freshwater reed marsh and deciduous forest between about 4500 and 3500 yr BP. Forest beds with in situ tree stumps and fallen logs dated to this period are exposed on the foreshore near Blundellsands Sailing Club and have been encountered in many dock excavations as far south as Liverpool (Reade, 1881; Travis, 1926; Wray et al., 1948). Peat deposits containing oak and other tree stumps have been recorded in the Bog Hole Channel off Southport (Ashton, 1909, 1920), and in many boreholes to the east of the dune belt between Ainsdale and Crossens (Wray et al., 1948).

In several places along the eastern margin of the blown sand belt, radiocarbon dates of c. 4500 - 4000 14C yr BP have been obtained from peat deposits immediately beneath windblown sand (Innes and Tooley, 1993). The extent of mobile sand, and the nature and distribution of dune vegetation communities at this time, are uncertain.

Drilling investigations in the early 1990's (Neal, 1993; Pye and Neal, 1993; Pye et al., 1995) revealed relatively few fossil soil layers within the dune sand sequence itself, possibly due to the effects of later wind erosion and dune movement, and no palaeo-ecological analysis was undertaken on the soil horizons which were identified in this work. Subsequent palaeo-ecological work has been restricted mainly to inter-tidal peat and silt beds of archaeological interest at Formby Point and Hightown (Roberts et al., 1996; Huddart et al., 1999a,b; Gonzalez et al., 1996; Gonzalez and Huddart, 2002; Adams and Harthen, 2007; Roberts and Worsley, 2008; Cowell, 2008). The occurrence of these deposits on the present foreshore suggests the former existence of coastal barriers or a very wide expanse of intertidal flats to seawards of the present shoreline during much of the later Holocene.

Natural environmental and human influences on dune processes and morphology

The Sefton coast is influenced by processes in the Irish Sea to the west, the Ribble estuary to the north, the Mersey estuary to the south, and the West Lancashire coastal plain to the east (Figure 2). The eastern Irish Sea experiences a semi-diurnal tidal regime with a mean spring tidal range of more than 8.0 m at Formby, resulting in the exposure of a wide foreshore at low tide. The more exposed parts of the Sefton Coast between Birkdale and Crosby experience moderate wave energy, with dominant waves from the W - WNW, corresponding with the direction of maximum fetch (Gresswell, 1953c; Sly, 1966; Parker, 1975). The combination of a large tidal range and moderate wave energy has resulted in the formation of a sand-dominated, multi-barred foreshore (Parker, 1971, 1975; Wright, 1976).

The prevailing winds blow from the SW but there is considerable variation in local wind direction and strength along the coast due to variable shoreline orientation, topography and vegetation cover. Wind records for Squires Gate airport and Crosby (Figure 1a) show dominant winds from the SW and W, but at Formby Point and Southport winds from the WNW and

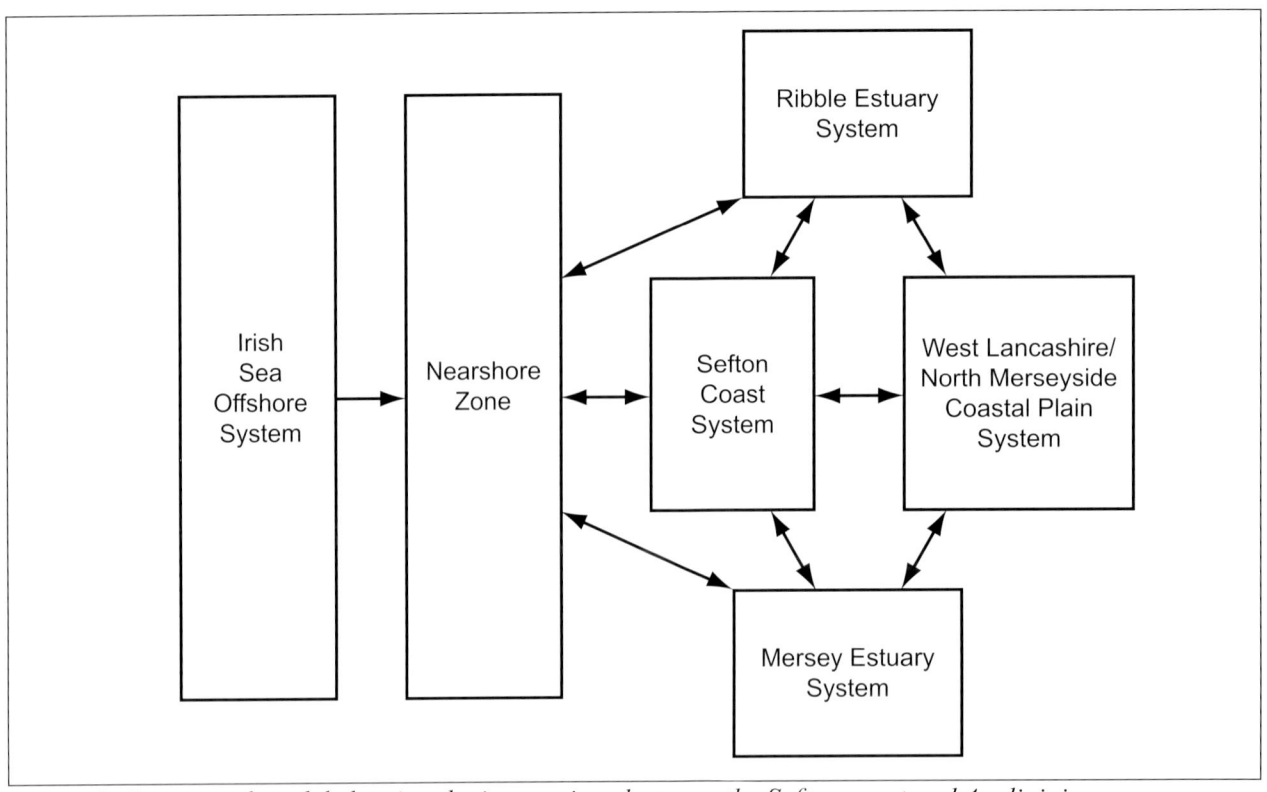

Figure 2. Conceptual model showing the interactions between the Sefton coast and 4 adjoining areas.

NW assume greater importance (Jay, 1998).

The main immediate sources of beach and dune sand on the Sefton coast are the adjoining nearshore and offshore zones. Sands with localised pockets of gravel and / or mud occur throughout Liverpool Bay and the eastern Irish Sea (Sly, 1966; Jackson et al., 1995; Pye et al., 2006). Most of the sand is derived from marine reworking of glacial deposits. The Ribble and Mersey estuaries have acted as net sinks for this reworked sediment throughout the Holocene (van der Wal et al., 2002; Thomas et al., 2002; Blott et al., 2006). Flood tidal streams diverge off Formby Point, and landward residual currents dominate in Liverpool Bay, favouring the movement of sea bed sediment towards the Mersey and Ribble estuaries (Sly, 1966; Halliwell, 1973).

The most significant waves in the eastern Irish Sea are almost entirely generated by winds blowing from the southwest, west and northwest. Under most conditions, waves approach the shoreline at an oblique angle, giving rise to littoral sediment drift. A sediment parting zone exists off Formby Point, with the relative magnitudes of inter-tidal zone sand transport to the north and south dependent on wave height and approach angle (HRS 1969a,b; Parker, 1971, 1975). Winds which approach Formby Point from SW to NW directions also contribute to significant longshore transport of windblown sand. Figure 3 presents a conceptual model showing the main sediment sources, sinks and transport pathways in the area.

Infilling of the Ribble estuary with sediment was enhanced in the 18th and 19th centuries by embanking, land reclamation, training wall construction and dredging (Barron, 1938; Pye, 1977; Smith, 1982; van der Wal et. al., 2002). Embanking of the marshes along the southern shore of the Ribble estuary probably began in the Middle Ages but most was undertaken after 1810. Southport began to develop as a seaside resort at the end of the 18th century and the town experienced rapid growth after 1830. Progradation of the shoreline between Birkdale and Hesketh Bank was encouraged by building of embankments and erection of groynes. The present sea wall, promenade and coastal road between Weld Road, Birkdale and Marshside, dates from the 1970's (Lymbery, 2008).

Embanking and reclamation in the Mersey estuary has been on a smaller scale than in the Ribble, but has still

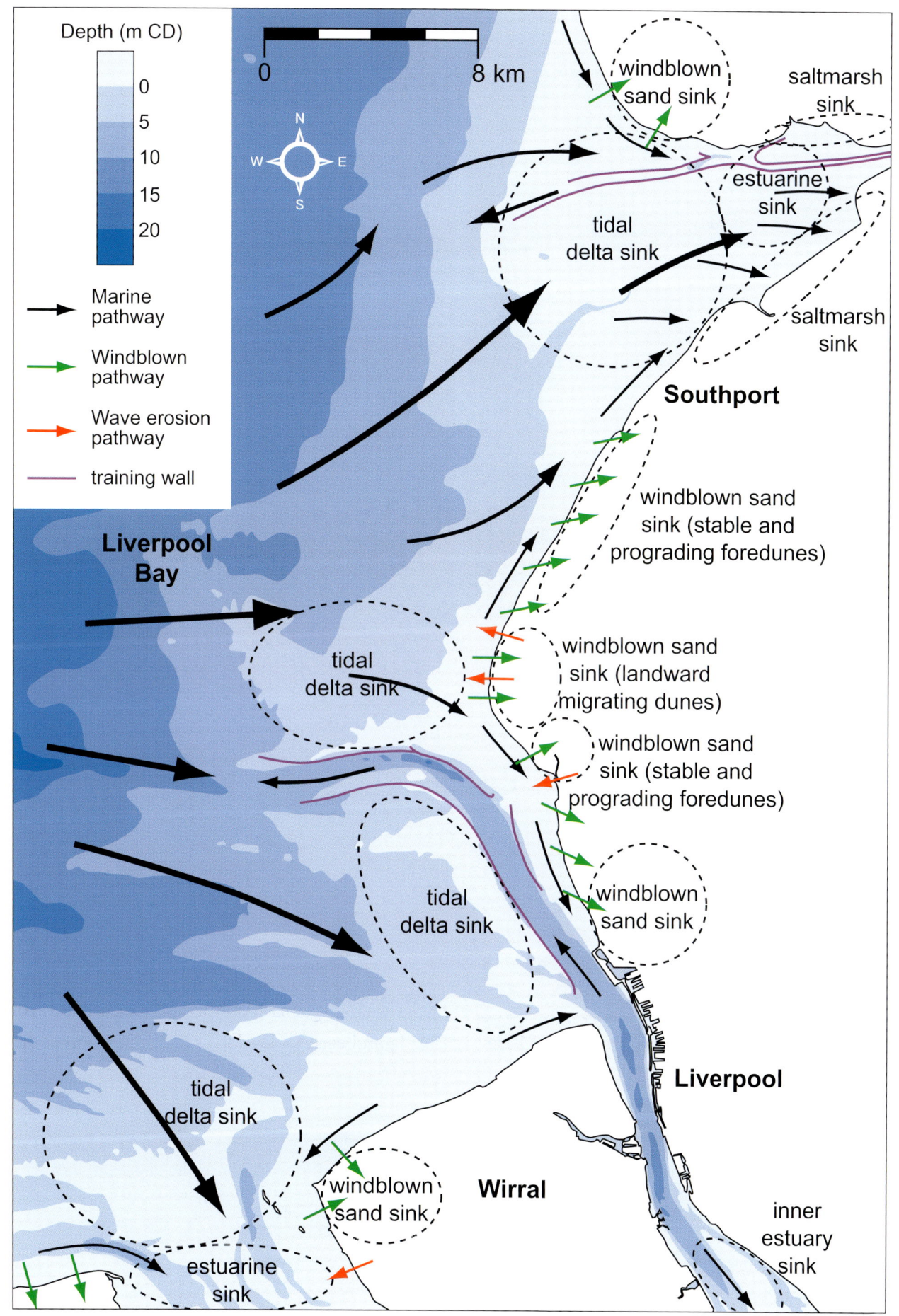

Figure 3. Map showing generalised sediment transport pathways and sediment sinks around Liverpool Bay, based on information from various sources.

Plate 1. Oblique aerial photographs showing (a) multiple foredune ridges formed within a shallow embayment, backed by a broad sand plain affected by previous sand mining activity, near Range Lane, Altcar; (b) the transition between accreting and eroding frontal dunes near the Ainsdale - Formby boundary; (c) eroding dune cliffs and blowouts near Victoria Road, Freshfield; (d) areas of recent dune accretion near the mouth of the River Alt at Hightown. Photographs taken by the Environment Agency on 17th March 2008.

been significant, especially in the Upper estuary above Runcorn, on the south side of the Inner estuary at Frodsham and Ince, and between Liverpool and Seaforth where reclamation for dock development took place after 1715 (Milne, 2005). Dredging and training wall construction in the outer estuary also had an important effect on the hydrodynamic regime, sediment transport and morphology of the estuary (Price and Kendrick, 1963; Smith, 1982; Thomas et al., 2002; Blott et al., 2006). Reclamation and engineering works in the Mersey and Ribble estuaries, combined with dredging and dredge spoil disposal, had a significant effect on the general pattern of sediment transport and sea bed morphology in Liverpool Bay. This in turn, had an impact on wave processes and patterns of sediment accretion and erosion on the Sefton coast (Pye, 1977, 1990; Smith, 1982; Pye and Neal, 1994).

The present geomorphological character of the frontal dunes between Southport and Crosby reflects alongshore differences in the balance between the beach and frontal dune sediment budgets, the rate of historical shoreline progradation / recession, the intensity of visitor pressure, and the history of dune management (Pye, 1990; Jay, 1998; Saye et al., 2005). In areas of shoreline progradation to the north and south of Formby Point, semi-continuous shore-parallel foredune ridges have formed (Plate 1a and b). An inverse relationship exists between the maximum crest height of the frontal dunes and the rate of seaward movement of the shoreline (Pye, 1990; Saye et al., 2005; Figure 4). Where seaward movement of the shoreline has been most rapid, and where the upper beach sediment budget is strongly positive but aeolian transport is limited by exposure to wind energy or by a high beach water table (e.g. at Birkdale), a series of relatively low foredune ridges has formed. In areas of slower shoreline progradation, where the beach sediment budget is less

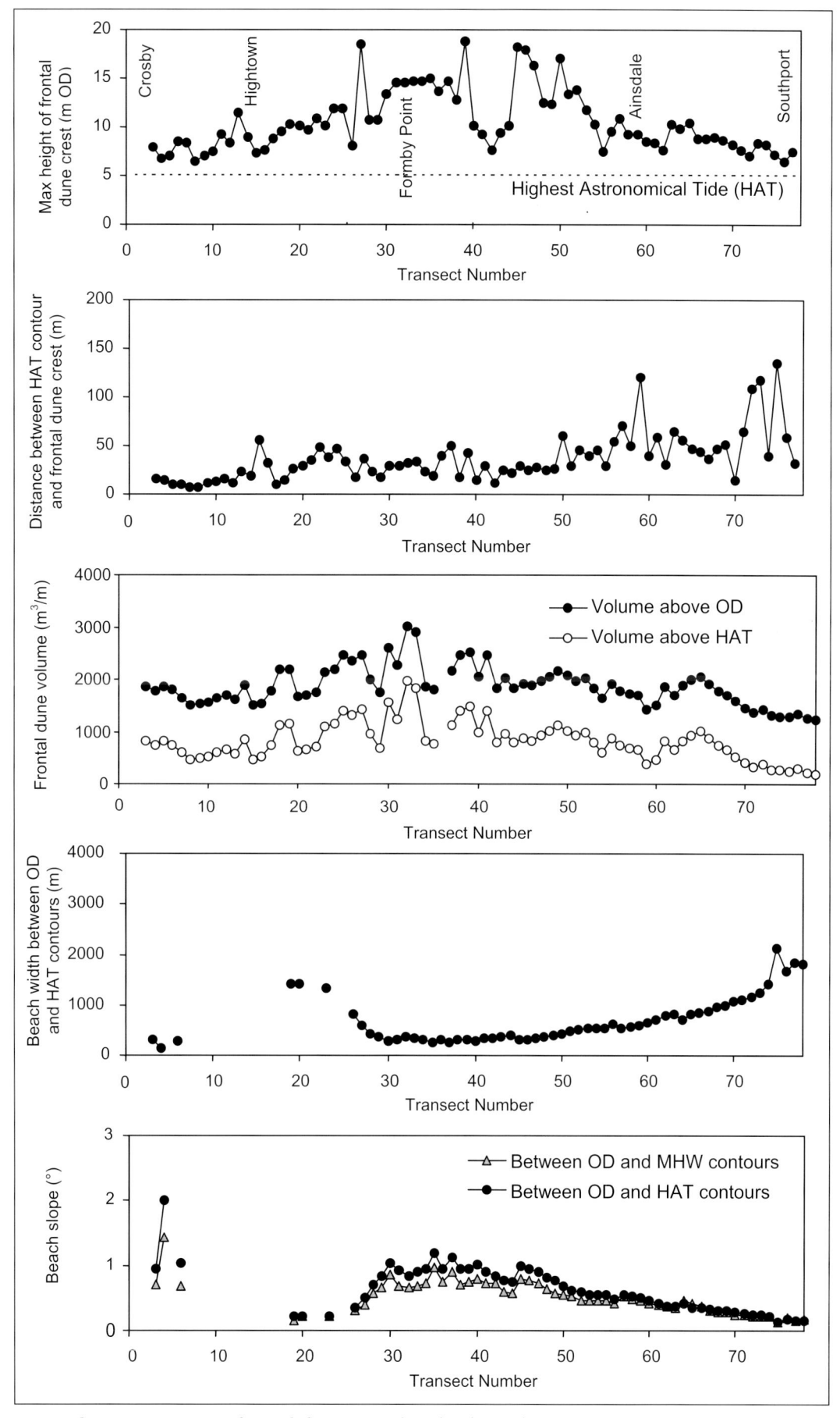

Figure 4. Longshore variation in frontal dune crest height, frontal dune sand volume and beach characteristics between Crosby (Transect 1) and Southport (Transect 78), derived from a lidar survey in 1999 (data from Saye et al., 2005).

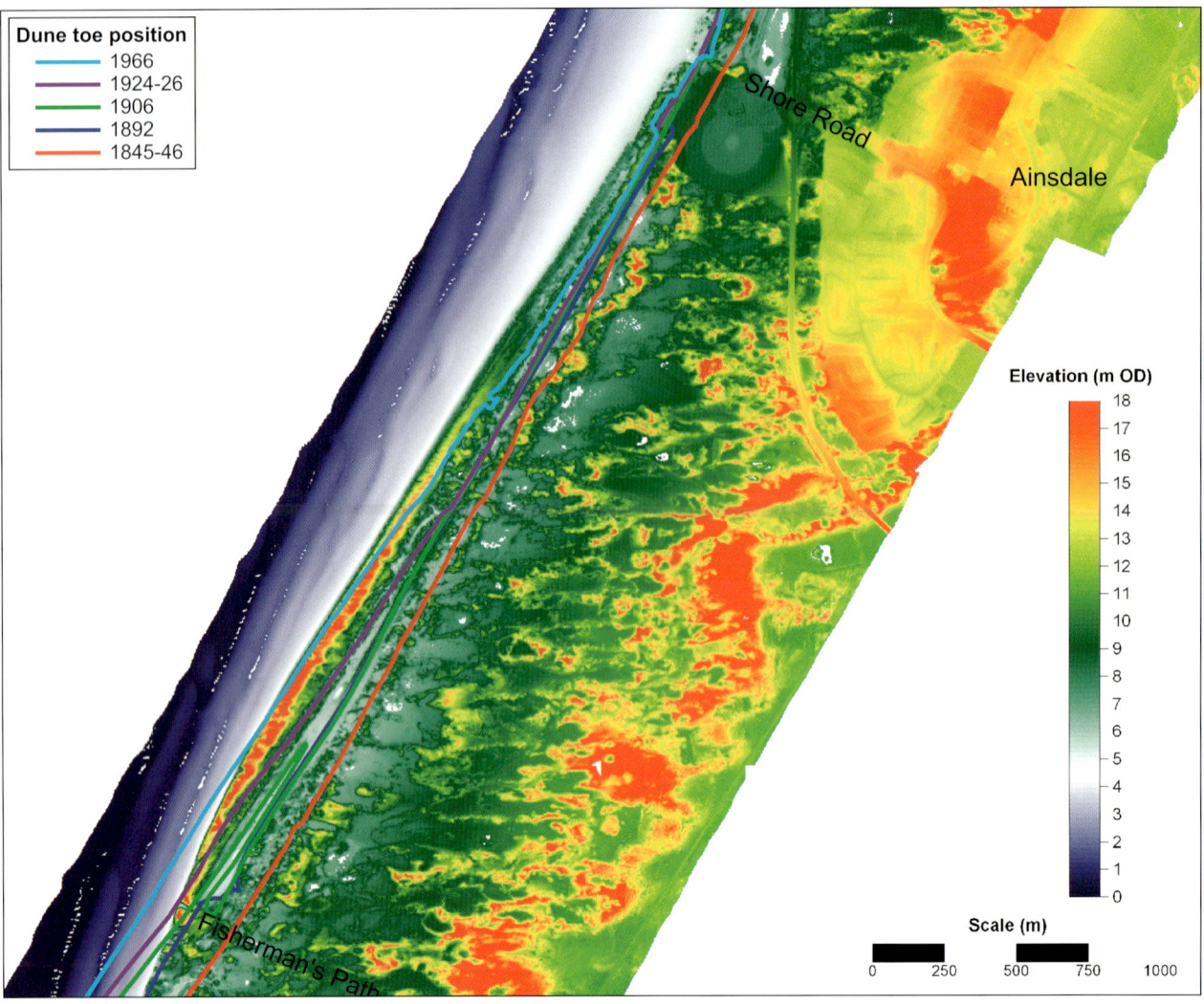

Figure 5. Digital terrain model of the area between Fisherman's Path and Shore Road, Ainsdale, based on filtered lidar data surveyed in March 2008, with superimposed historical dune toe positions determined from Ordnance Survey Six Inch maps. © Crown copyright. All rights reserved. Sefton Council Licence number 10001819 2010.

positive and/ or there is greater potential for aeolian transport due to greater wind exposure or a steeper, better drained beach (as at Ainsdale south), only one or two parallel dune ridges are present but the crest heights are higher than at Birkdale (Plate 1b). On the eroding shore between the Ainsdale / Formby Boundary and Lifeboat Road, where the upper beach sediment budget is negative but there is relatively high exposure to onshore winds, a single high frontal dune ridge is present behind the beach; this rolls landwards and gains sand volume as the high water mark moves landward. In places this ridge is disrupted by blowouts and transgressive sand sheets which transfer sand into the hind dune area (Plate 1c).

The more inland areas are characterised by complex hummocky topography composed of deflation troughs and sand mounds. Well-defined parabolic dune forms are rare, although the dune terrain in the Ainsdale area shows a general SW-NE 'grain' which reflects the dominance of winds from the WSW and SW (Figure 5). Between Ravenmeols and Freshfield (Figure 6) there is a ridge of relatively high dunes ('the fifty foot ridge') which runs in a gentle N-S trending arc 800 - 1200 m inland from the present shore. This ridge appears to have formed mainly in the early to mid 18th century during a period of extensive sand-blowing and subsequent widespread planting of marram to arrest the eastward sand drift. Planting of trees on this ridge by the Weld-Blundell and Formby estates, mainly after 1885, led to further stabilization of the sand (Yorke and Yorke, 2008a). To the west of this ridge lies a

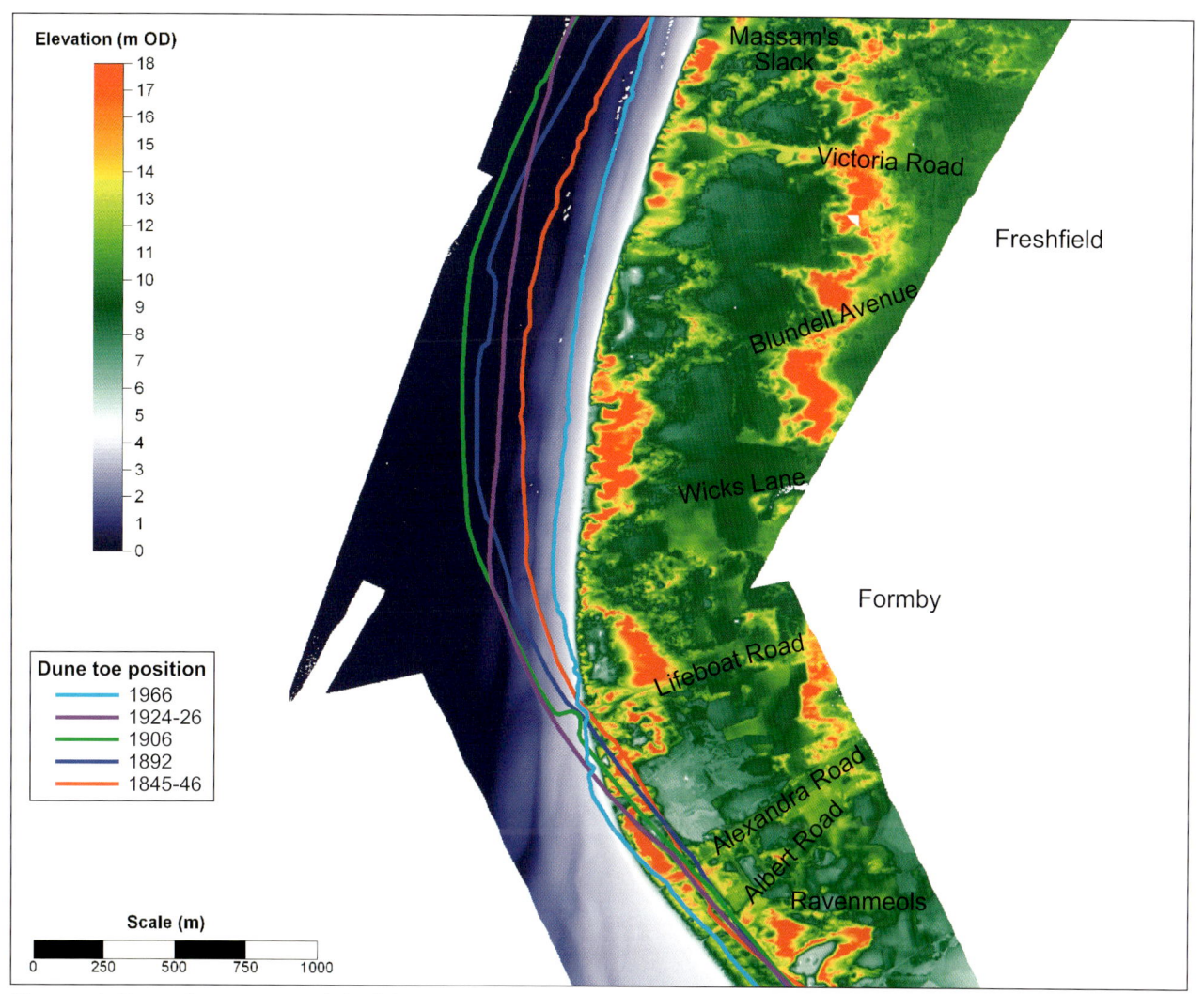

Figure 6. Digital terrain model of the Formby Point area based on filtered lidar data surveyed in March 2008, with superimposed historical dune toe positions determined from Ordnance Survey Six Inch maps. © Crown copyright. All rights reserved. Sefton Council Licence number 10001819 2010.

deflationary sand plain which is partly covered by low hummock dunes and artificial sand mounds. This area has been affected by a number of human activities including sand extraction, construction of rifle ranges and asparagus fields, and re-profiling of the most seawards dunes as a management measure during the 1980's (Houston and Jones, 1987; Jones et al., 1993).

Extensive sand extraction has also taken place in the Cabin Hill area, south of Ravenmeols (Figure 7), where an extensive low-lying sand plain now lies behind shore-parallel dune ridges which have accumulated within a former shallow embayment over the past 150 years (Figure 7). Altcar Rifle Range lies on a former inter-tidal sand flat (Balling's Wharf) which was reclaimed from the sea during the 19th century (Cook, 1989). Much of the area has been levelled to create firing ranges, but near the beach there is a relatively narrow belt of shore-parallel ridges (Plate 1d). During the Balling's Wharf reclamation the River Alt was deflected eastwards and since that time its low water channel has migrated across the upper inter-tidal zone between Hightown and Blundellsands, causing periodic erosion of the frontal dunes in this area (Figure 8).

Figure 9 shows the locations of several cross-sectional profiles where sand volumes within 200m and 500m of the high water mark have been calculated using the digital terrain model constructed using the Environment Agency 2008 lidar survey data. The cross-sections are shown in Figure 10 and the sand volumes summarised in Table 1. At Freshfield and Formby the surface dune forms sit on an underlying sand ramp which rises well above present sea level. This ramp is not present between Freshfield and Ainsdale or between Ravenmeols and Altcar. However, along the whole coast between Ainsdale and Ravenmeols there is a relatively large volume of sand within 200m and 500m of the shore, and the dunes form an effective flood defence with a wide 'buffer zone' against erosion. To the south of Ravenmeols the frontal dunes are much narrower, lower, and contain a much smaller sand volume. Until the end of the 18th century this area formed part of the estuary of the River Alt, and high dunes have never existed in the area. During the 20th century large quantities of sand were removed from the low dunes which occupied the area. Consequently these dunes form a less effective flood defence and are potentially more at risk from future coastal erosion. An artificial flood bank was constructed at Cabin Hill in 1973 (Figure 7).

Cross-section H in Figure 10 intersects a large wind deflation basin (the Devil's Hole) which developed after the Second World War. This feature was initiated by vegetation disturbance some distance inland from the shoreline and continued to grow during the 1950's and 1960's even though the shoreline was prograding seawards (Figure 11). The present feature and its associated dune has a composite form which has developed as a result of the coalescence of three smaller blowouts and associated dunes (Neal et al., 2001).

The extent of blowouts and sand mobility in other parts of the Sefton dune system has also varied considerably since 1945. Figure 12 shows changes in the extent of mobile sand around Lifeboat Road at Formby Point between 1945 and 1999. Recession of the shore in this area began around 1940 and accelerated until the late

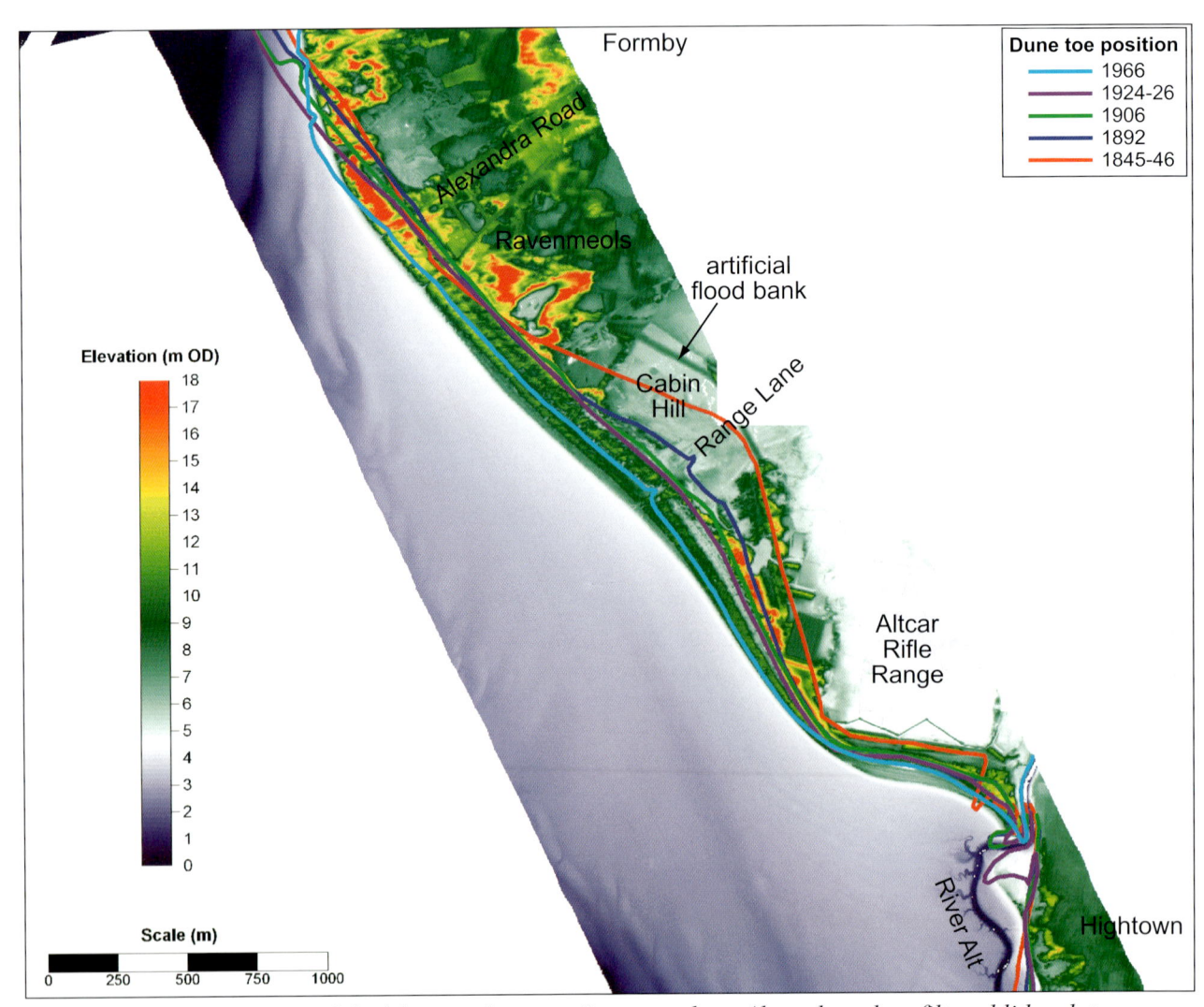

Figure 7. Digital terrain model of the area between Ravenmeols to Altcar based on filtered lidar data surveyed in March 2008, with superimposed historical dune toe positions determined from Ordnance Survey Six Inch maps. © Crown copyright. All rights reserved. Sefton Council Licence number 10001819 2010.

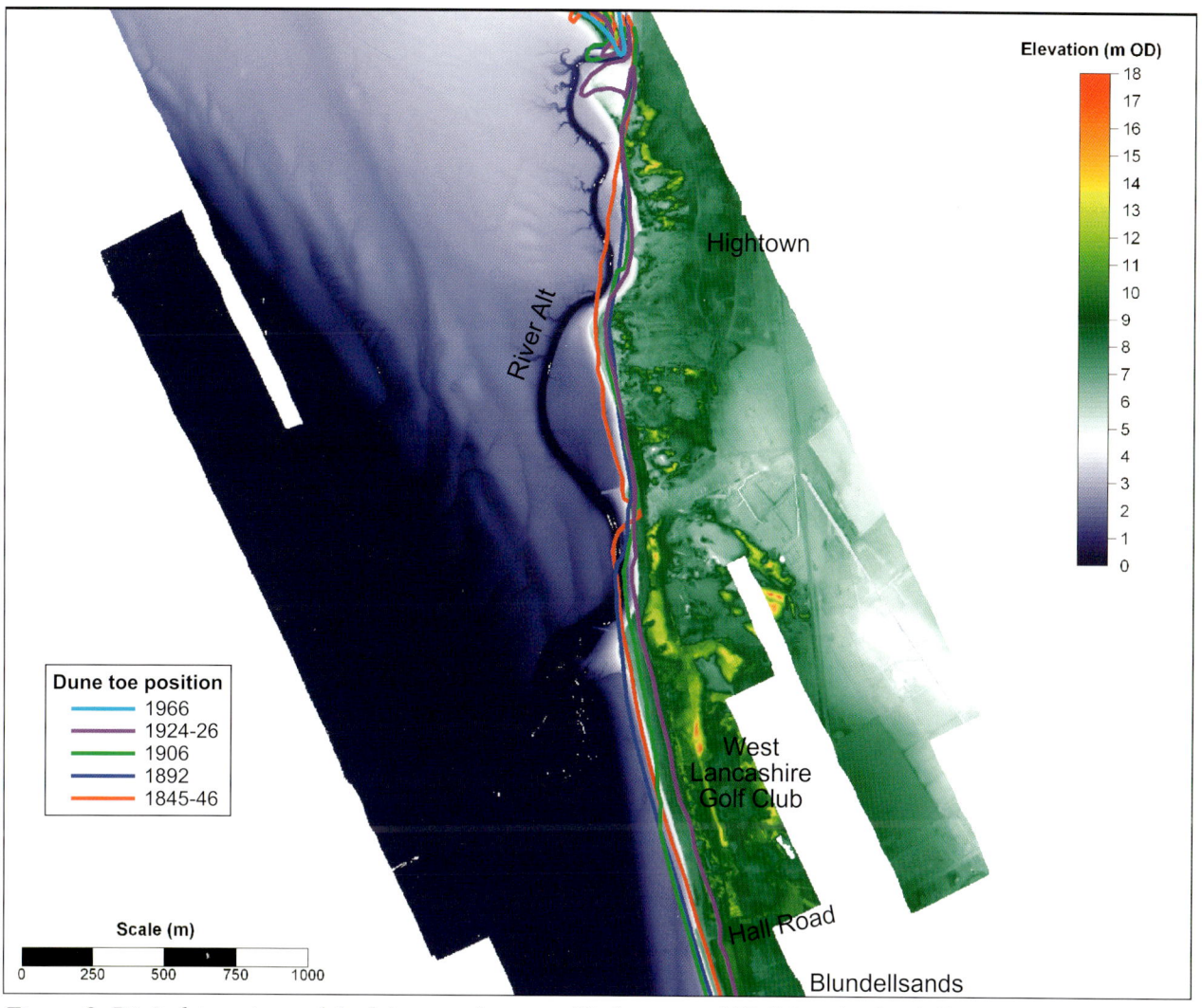

Figure 8. Digital terrain model of the area between Hightown and Hall Road, Blundellsands, based on filtered lidar data surveyed in March 2008, with superimposed historical dune toe positions determined from Ordnance Survey Six Inch maps. © Crown copyright. All rights reserved. Sefton Council Licence number 10001819 2010.

1960's, accompanied by the development of an extensive mobile sand sheet behind the beach. The former lifeboat station was closed in 1918 but was used as a tea room until erosion forced its demolition in 1965. The nearby lifeboat keeper's cottage was demolished in 1970. Following the establishment of the Sefton Coast Management Scheme in 1978, measures were taken to stabilize the mobile sand in this area, including re-profiling of a frontal dune ridge using bulldozers, control of visitor access using sand fences and boardwalks, placement of brashings to stabilize the surface sand, and marram planting (Houston and Jones, 1987; Wheeler et al., 1993). By 1999 much of the former bare sand had developed surface crusts and vegetation cover (Figure 12).

Sedimentological characteristics of the dunes

The beach and dune sediments of the Sefton coast consist predominantly of fine to medium, well-sorted to very-well sorted sands which show only limited longshore and cross-shore variation. (Kavanagh, 1997; Jay; 1998; Saye, 2003; Blott and Pye, 2003; Saye and Pye, 2003; Table 2). The mean sand size of the frontal dunes is similar to that of the adjacent upper beaches, since wind transport is not size-selective when the sand source consists of well-sorted medium and fine sands (Pye and Tsoar, 2009). There is very little variation in the modal size (most frequent size class) of the frontal dunes between Seaforth and Shore Road, Ainsdale, but north of this point the modal size becomes slightly finer

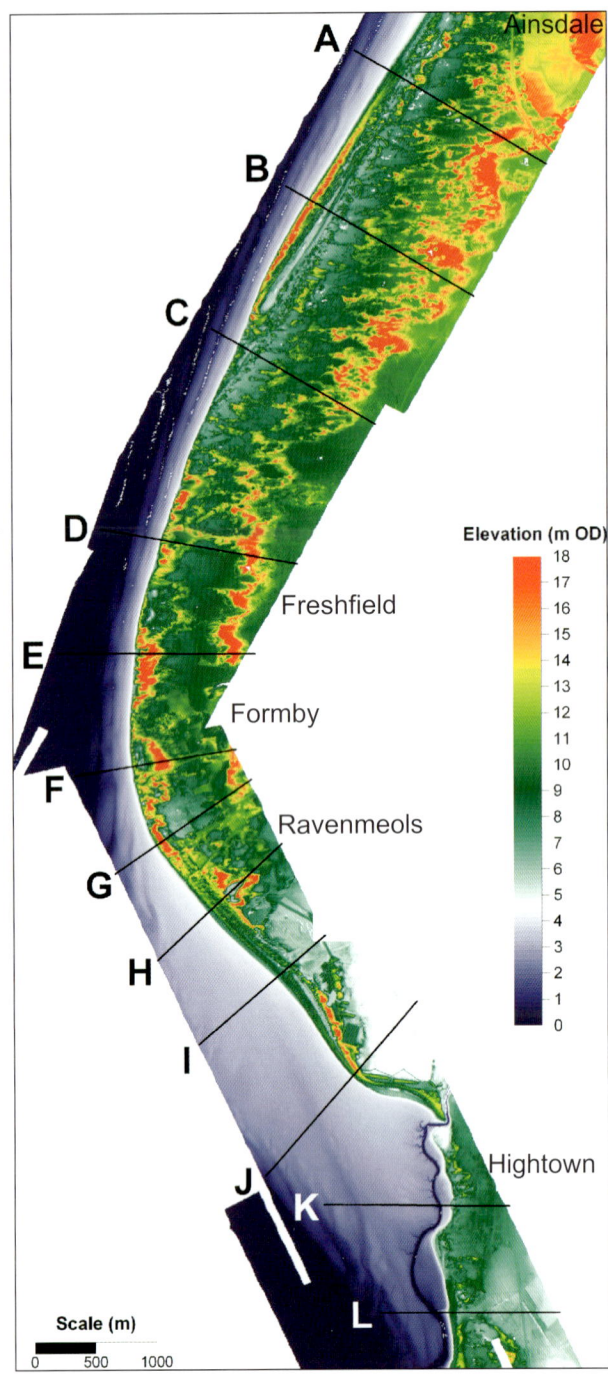

Figure 9. Locations of cross-sections through the dunes determined from the 2008 lidar digital terrain model.

(Figure 14). By contrast, the mean size shows a slight increase from about 225 μm at Hightown to >250 μm near Southport Pier, reflecting the presence of a significant broken shell component in the more northern area.

Mobile dunes further inland also typically consist of medium, well-sorted sands with a low silt and clay content (< 1%), but stabilized dune sands affected by soil development typically contain 2 - 4% percent silt and clay.

The dune sands consist predominantly of silica, on average containing more than 90% SiO_2 (Table 3; Kavanagh, 1997; Saye, 2003). Quartz is the dominant mineral present, with subsidiary plagioclase feldspar, potassium feldspar, calcite and heavy minerals. The calcium oxide (CaO) content, which can be taken to be a proxy for calcium carbonate ($CaCO_3$) content, is relatively low (overall average 2.36%), although there is a significant difference between the younger dunes nearer the beach, which have a higher CaO content, and older, stabilized, dunes further landward which have been partially leached. Differences in the calcium carbonate content are significant in terms of dune vegetation community development. Dune heath and some dune grass communities are essentially restricted to the areas of carbonate-poor sand where the pH of the dune soils is more acid.

Frontal dune erosion and accretion trends
Longer-term (decades to centuries)

There is clear sedimentological, geomorphological, palaeoecological and archaeological evidence of significant changes in the shoreline position and degree of dune stability / instability on the Sefton coast during the past few thousand years. Silts and silty sands which are periodically exposed on the foreshore at Formby Point were deposited between c. 5000 and 3000 years ago in a relatively low energy environment, probably protected from the open sea by a low barrier (Pye and Neal, 1993). These deposits were buried by windblown sand between 3200 and 2700 years ago, after which time there was a period of surface stability and peaty soil development which lasted until at least 2300 yr BP, when there was renewed aeolian sedimentation (Tooley, 1970; Pye and Neal, 1993; Pye et al., 1995). However, the causes of these changes, and their geographical extent, remain a matter of debate (Tooley, 1980, 1985a,b).

The period from Saxon times until the early Middle Ages appears to have been one of predominant stability in the dune landscape, but between AD 1300 and

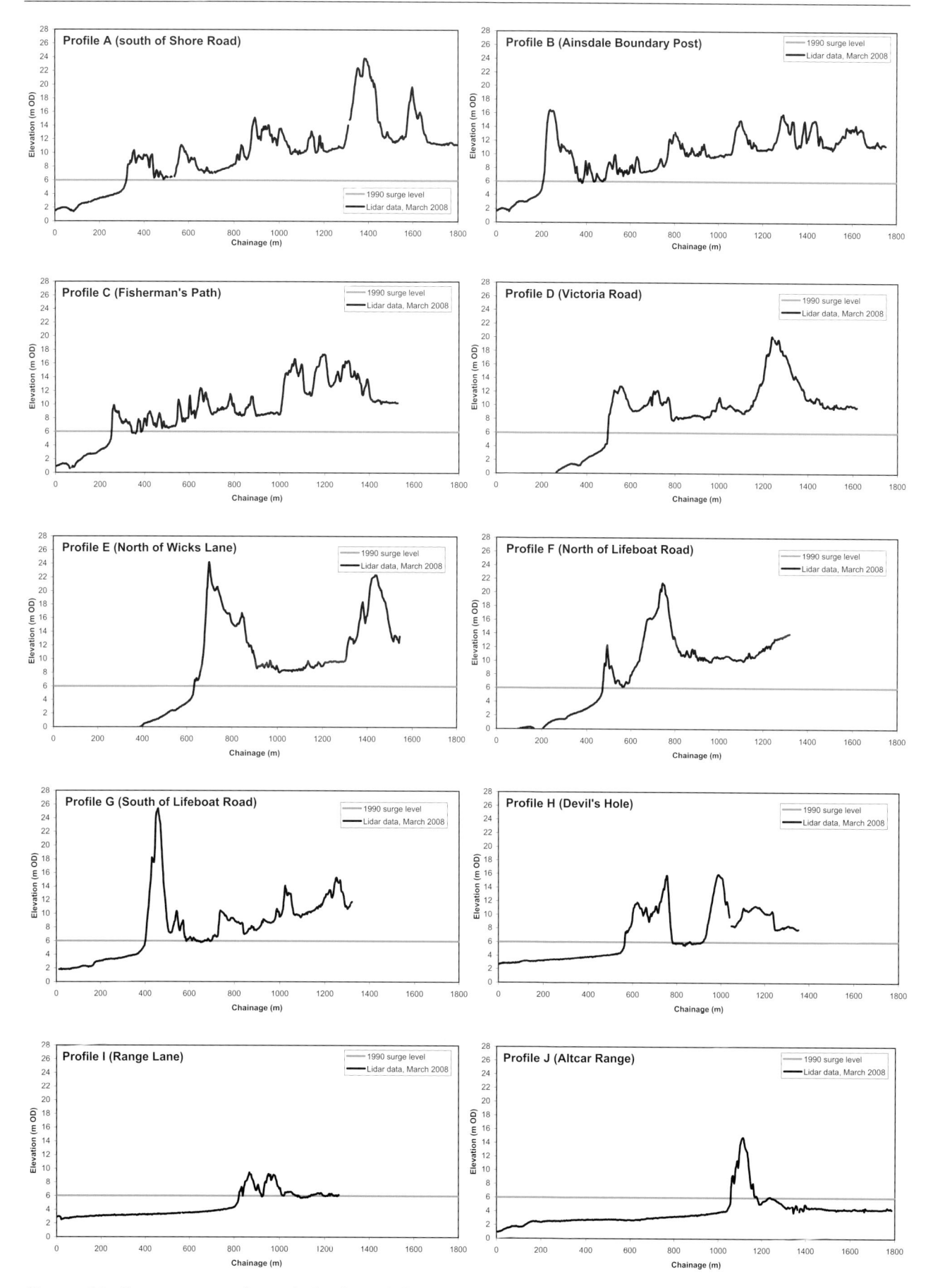

Figure 10. Cross-sections through the frontal dunes deterived from the 2008 lidar DTM; the horizontal line indicates the maximum water level likely to be reached by a surge tide with approximately a 1 in 10 year recurrence interval. Profile locations are shown on Figure 9.

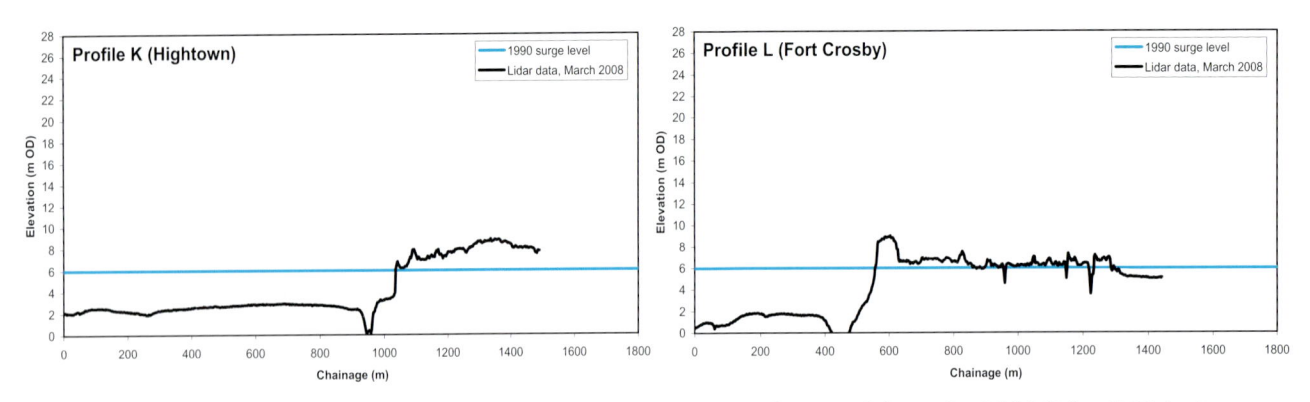

Figure 10 (continued). Cross-sections through the frontal dunes deterived from the 2008 lidar DTM; the horizontal line indicates the maximum water level likely to be reached by a surge tide with approximately a 1 in 10 year recurrence interval. Profile locations are shown on Figure 9.

Figure 11. Sequence of vertical air photographs showing the development of the Devil's Hole blowout south of Albert Road between 1945 and 1999. Former dune toe positions indicated by sequential Ordnance Survey map editions are also shown. © Crown copyright. All rights reserved. Sefton Council Licence number 10001819 2010.

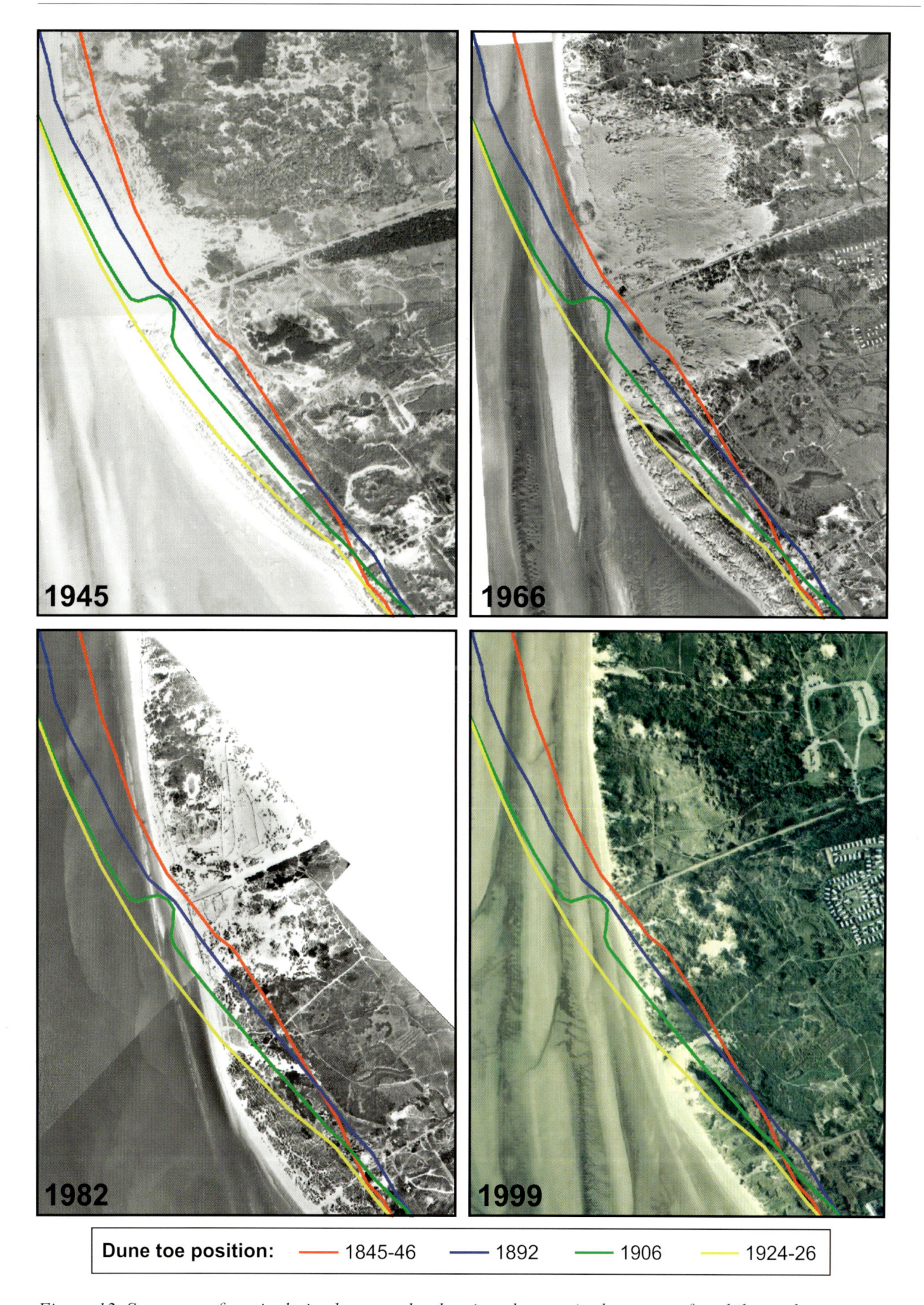

Figure 12. Sequence of vertical air photographs showing changes in the extent of mobile sand near Lifeboat Road between 1945 and 1999. Former shoreline positions indicated by Ordnance Survey maps are also shown. © Crown copyright. All rights reserved. Sefton Council Licence number 10001819 2010.

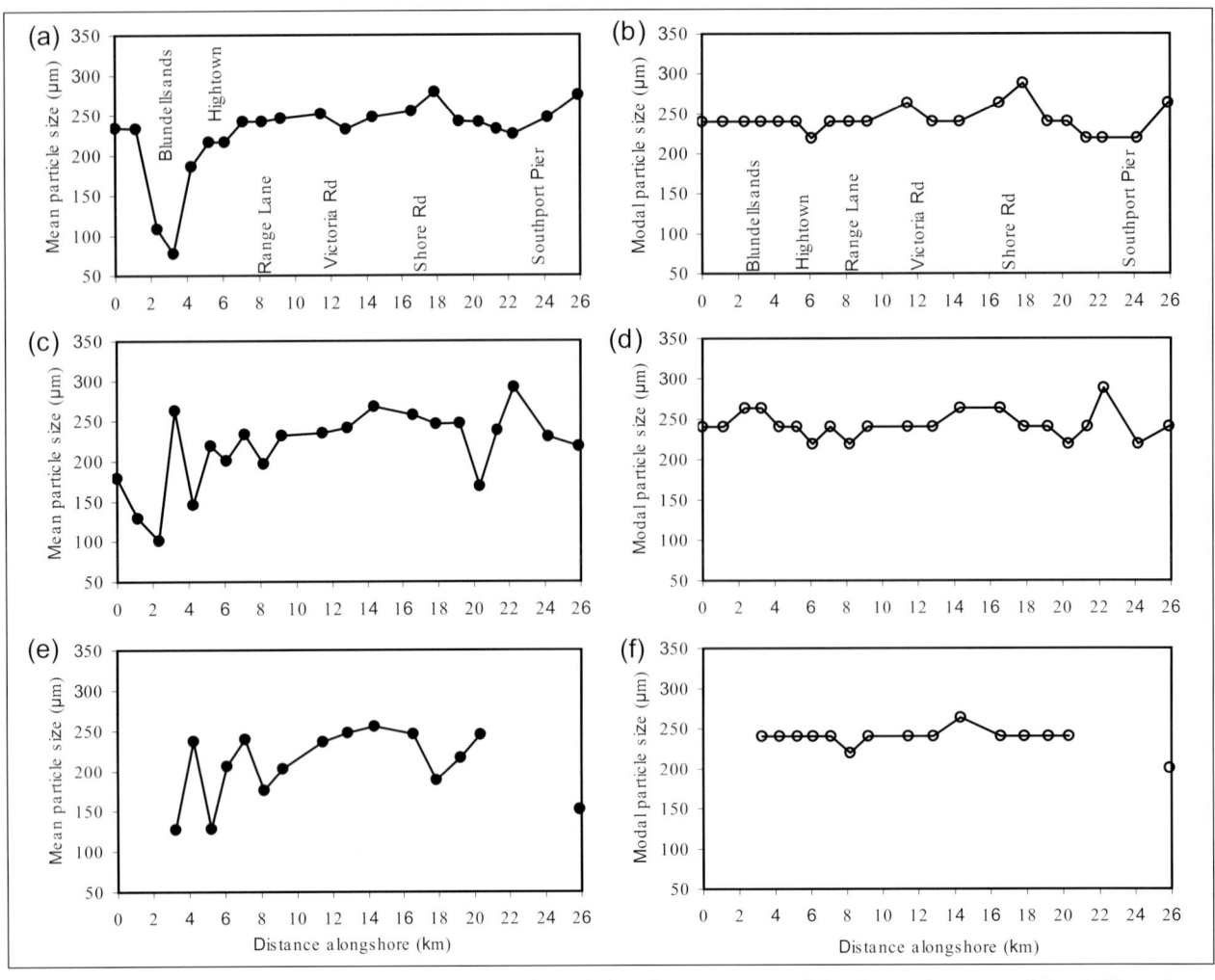

Figure 13. Longshore variation in the mean and modal sediment sizes of the dunes between Blundellsands and Southport, determined by laser granulometry: (a) mean size of frontal dunes, (b) modal size of the frontal dunes, (c) mean size of dunes 50 m inland from the shore, (d) modal size of dunes 50 m inland, (e) mean size of dunes 100m inland from the shore, and (f) modal size of dunes 100m inland from the shore. Data from Saye and Pye (2003).

AD 1720 there was major instability and sand invaded the mosslands behind Formby (Neal, 1993; Pye and Neal, 1993; Pye et al., 1995). There is documentary evidence that several settlements near the coast were lost at this time, including Argameols, near present day Birkdale, Old Aynesdale, and Ravenmeols (Ashton, 1909, 1920; Kelly, 1973). However, it is not clear whether these settlements were lost due to burial by blowing sand or to coastal erosion. Nor is it clear whether the whole coast was affected by simultaneous sand blowing and / or erosion.

Migration of dunes and sand sheets continued in parts of the area during the late 17th and early 18th centuries. Leases requiring the compulsory planting of starr grass (marram) were introduced as early as 1711 (Kelly,

1973), but sand-blowing continued and in 1739 the old Formby Chapel was buried. The problem may have been exacerbated by significant human use of the dune areas, notably for rabbit warrens, cattle grazing, and marram cutting for basket-making, but aeolian instability would naturally have been favoured at this time by high wind exposure and a general absence of trees. An Act of Parliament passed in 1740 made it an offence to cut marram, and planting became a common requirement of land leases. The first trees are thought to have been planted by the Formby Estate near the landward end of Lifeboat Road around 1795 but the majority of the conifer planting by the Formby and Ince Blundell Estates did not occur until after 1885 (Yorke and Yorke, 2008b).

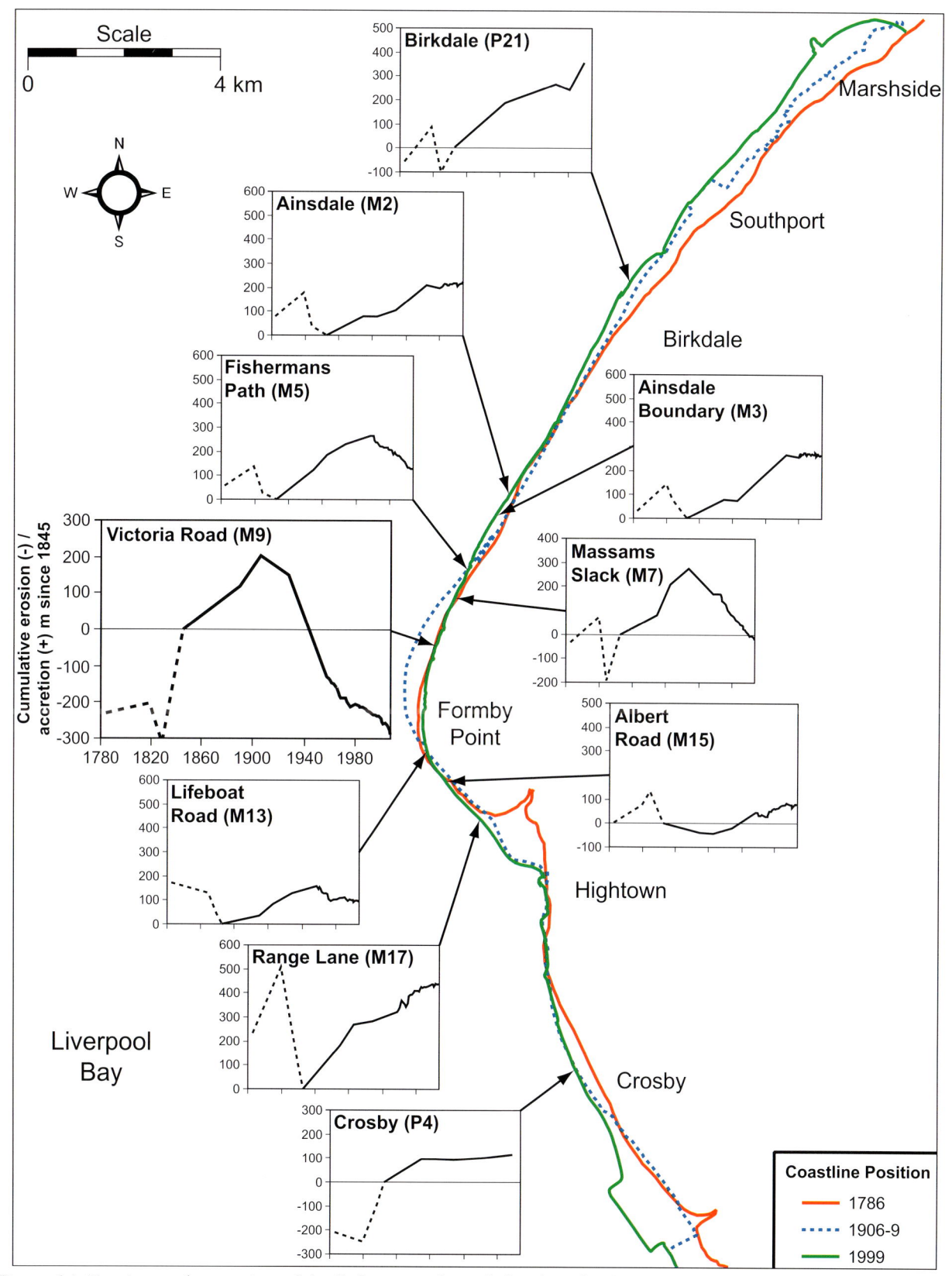

Figure 14. Erosion and accretion of the Sefton coastline, defined as the dune toe or sea defences, where present. Graphs show temporal trends at profile marker points digitized from historical maps (Yates 1786, Greenwood 1818, Hennet 1829, and the Ordnance Survey in 1845-48, 1889-93, 1906-09, 1924-26), aerial photography (flown in 1966, 1982 and 1999) and dune toe surveys undertaken by Sefton Council (and its predecessors) between 1958 and 2007, and Kenneth Pye Associates Ltd in February 2008. All graphs have the same scales as those indicated for Victoria Road. Changes between 1786 and 1845 are indicated by dashed lines due to the lower accuracy of the early maps. © Crown copyright. All rights reserved. Sefton Council Licence number 10001819 2010.

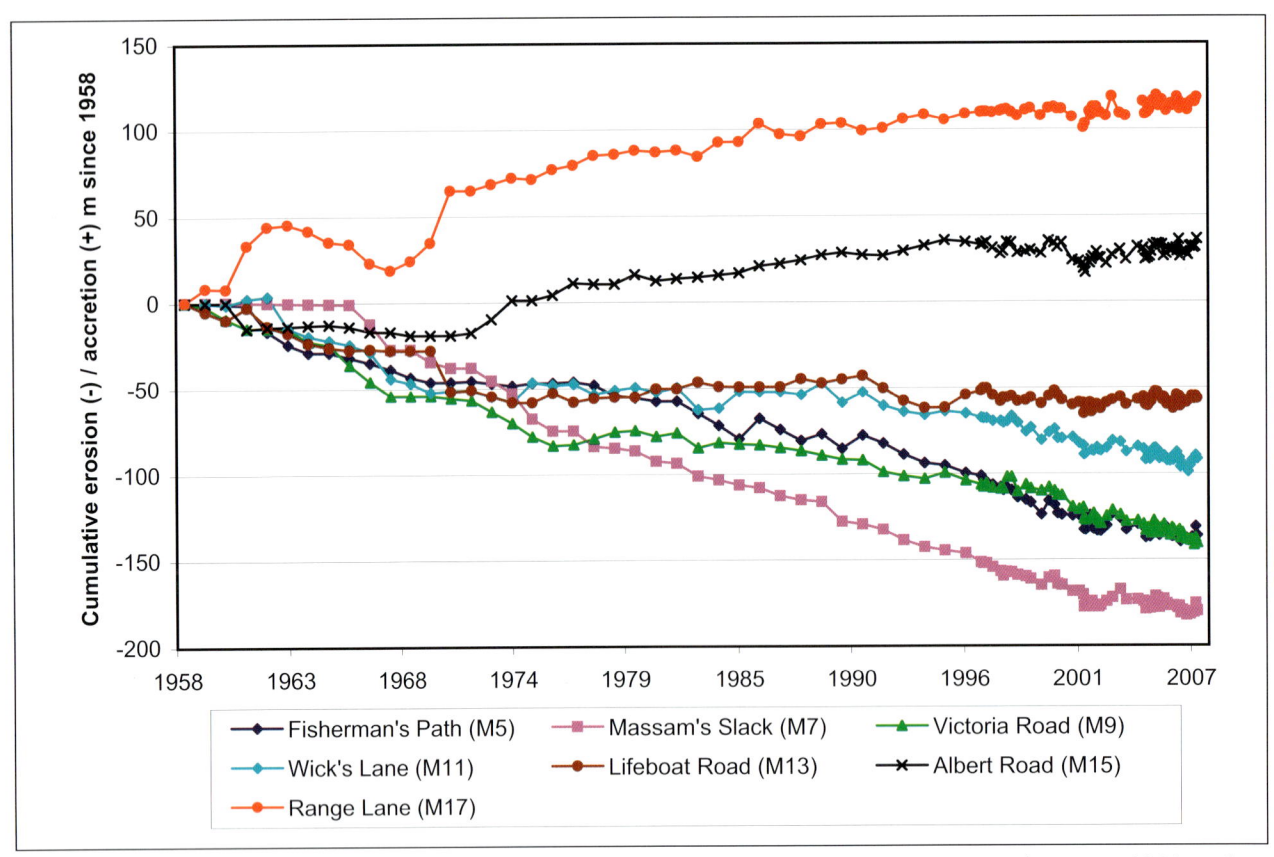

Figure 15. Changes in position of the dune toe at seven locations around Formby Point between 1958 and 2007, determined from ground measurements by Southport Corporation and Sefton Council.

Historical maps and charts provide qualitative evidence of changes to the coast and nearshore since the 16th century, although reliable surveys from which quantitative measurements can be made with reasonable confidence date only from the later 18th century. The first relatively reliable map is that by Yates dated 1786, which, based on comparison with later Ordnance Survey (OS) maps, has an accuracy of better than +/- 20 m in most places. Later maps by Greenwood (1818) and Hennet (1829) have a general accuracy of +/- 20 to 50 m, depending on location, compared with later OS maps (Pye and Blott, unpublished data). The First Edition Six Inch Ordnance Survey maps of the area were based on surveys between 1845 and 1848, with subsequent re-surveys in 1889-93, 1906-09, 1924-27. General OS guidance suggests that a survey error allowance of +/- 5m should be made for County Series Six Inch maps and +/- 3.5 m for post 1945 National Grid maps. However, additional errors can arise in comparing shoreline positions owing to differences in the way the 'coastline' was recorded and indicated on different map editions, and to digitization errors during the map analysis process.

Despite these sources of potential error, there is clear evidence that historical shoreline changes on the Sefton coast have been significant. Gresswell (1937, 1953c) first used OS Six Inch maps to quantify erosion and accretion at a number of locations between Ainsdale and Blundellsands. Gresswell's data have subsequently been updated by Pye (1977), Smith (1982), Pye and Smith (1988), Neal (1993), Pye and Neal (1994), Jay (1998) and Pye and Blott (this paper and unpublished data). The overall long-term changes at a number of selected points along the coast are summarised in Figure 15. Selected former shoreline positions are also shown superimposed on the lidar images in Figures 6, 7, 8 and 9.

Around much of Formby Point the position of the dune toe now lies well inland of its position at any previous time in the last 220 years. The most seaward position was recorded in 1906-09, since when the dune toe has retreated landwards by approximately 1 km. By contrast, to the north of the Ainsdale - Formby boundary, and to the south of a position approximately 200 m south of Lifeboat Road, the dune toe now lies further seaward than at any time in the previous two centuries.

Between Hightown and Blundellsands the position of the dune toe has not varied greatly over the past 200 years, although there has been a net erosional trend which has been influenced by migration of the River Alt across the foreshore.

The historical map evidence suggests that the shoreline of Formby Point north of Lifeboat Road advanced seawards between 1786 and 1900, after which time there was a sudden change to erosion which has continued to the present day (Figure 16). Net erosion is suggested at Lifeboat Road between 1786 and 1845; this resulted in destruction and re-location of the lifeboat station on a number of occasions in the decades after it was established in 1776 (Yorke and Yorke, 1982). Further south, at Albert Road and Alexandra Road, slow net erosion continued until 1900, after which time there was a change to progradation. At Range Lane, erosion between 1786 and 1845 was followed by rapid accretion. To the north of Formby Point, between Fisherman's Path and Shore Road, Ainsdale, apparent erosion of 20 - 200 m between 1786 and 1845 was followed by progradation of 220 - 300 m. Erosion began again at Fisherman's Path in the 1950's, since when approximately 150 metres of land has been lost. At the Formby - Ainsdale boundary, progradation effectively ceased in the 1950's and there has been little net movement of the shoreline since that time. Further north, the long-term progradation trend has continued.

Medium-term (years to decades)

Detailed monitoring of changes in the position of the dune toe around Formby Point was begun by Southport Borough Council in 1958, following advice from R.K. Gresswell, and was continued after the 1974 local government reorganization by Sefton Council. Initially there were seven monitoring points, but additional monitoring stations were established after 1980. Since 2000 surveys of the entire dune toe between Shore Road and Altcar have been made using global positioning equipment (Newton et al., 2008). Although the methods of survey have changed over time, and difficulties have arisen when some marker posts were lost during storms, the data provide a useful record of the annual, and sometimes shorter-term changes.

Figure 16 summarises the changes which have taken place at the original seven monitoring locations around Formby Point between 1958 and 2007. The figure shows a slight change in erosion / accretion rates at these stations after 1990, since when accretion at Range Lane and Albert Road has slowed, erosion has begun again at Lifeboat Road after a period of stability, and erosion rates between Wicks Lane and Fisherman's Path has shown a slight acceleration. However, only at Fisherman's Path has the average erosion rate since 1980 exceeded the longer term (1958-2007) average (Table 4). There is some evidence that an increase in the frequency and magnitude of storms and storm surges since the later 1980's may have contributed to this trend, although reduced management intervention since 1990 has probably also played a contributory role (Pye and Blott, 2008).

The impact of individual storms on the coast of Liverpool Bay is now well-documented (Pye, 1991; Jay, 1998; Pye and Blott, 2008). Deep Atlantic low pressure systems generate strong winds from SW, W and NW directions, and when winds from the direction of maximum fetch (WNW) are sustained for periods of a few hours or more predicted tidal levels may be raised by up to 2m (Lennon, 1963). High water levels and strong wave action may occur over a period of several successive tides, leading to several metres of dune erosion (Figure 16). Independent evidence (Allan et al., 2008) suggests an increase in the number of severe storms per decade across the UK and Ireland since the 1980's, following a minimum in the 1960's (Figure 17). However, there was a similar high incidence in the 1920's, and there is no convincing evidence that the recent increase is related to global warming.

There is also no evidence that changes in mean sea level have yet had any significant impact on dune erosion and accretion rates along the Sefton coast. Woodworth et al. (1999) calculated an overall average rate of relative rise

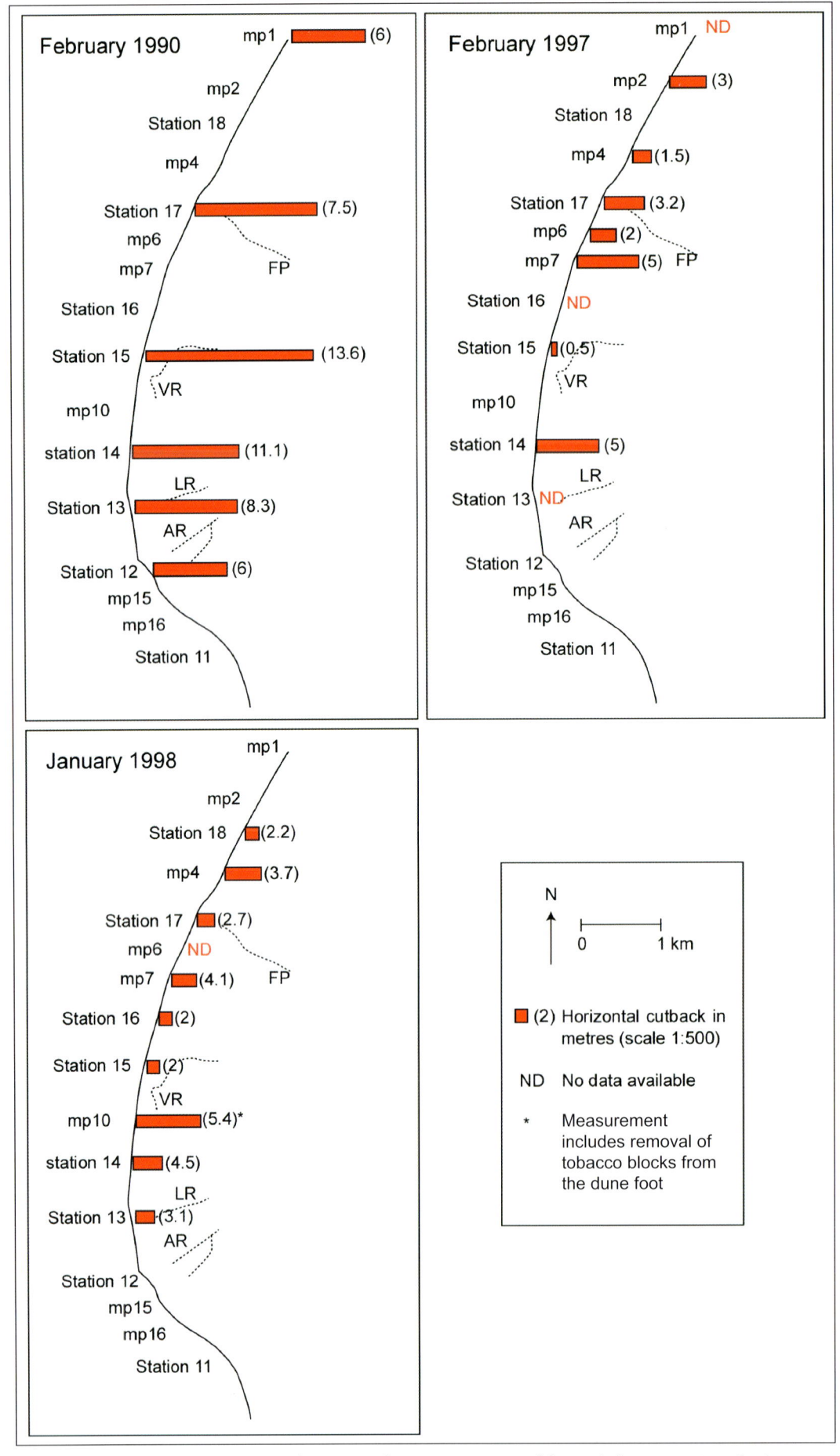

Figure 16. Summary diagram showing relative amounts of frontal dune erosion during three storms in February 1990, February 1997 and January 1998 (after Jay, 1998).

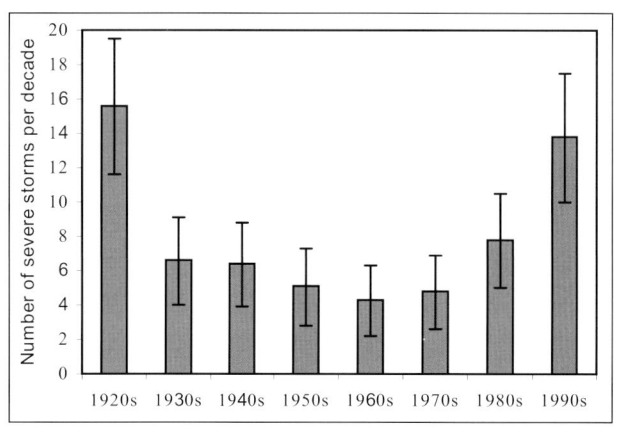

Figure 17. Number of severe winter storms (October to March) per decade over the UK and Ireland between 1920 and 1999 (data of Allan et al., 2008). Error bars shown are +/- one standard deviation.

in mean sea level at Liverpool of 1.23 +/- 0.12 mm yr for the period 1858 - 1996, based on composite tide gauge records for George's Pier, Prince's Pier and Gladstone Dock, and an average rate of 2.58 mm yr for the period 1959 - 1983 at Prince's Pier. Based on an assessment of geological evidence for the last 4000 years (subsequently presented in updated form by Shennan and Horton, 2002), Woodworth et al. concluded that the Mersey area is experiencing net land subsidence of 0.18 +/- 0.04 mm yr. More recent analysis of tide gauge records for Liverpool up to 2007 have provided little clear evidence of a significant acceleration in the rate of rise of mean seal level (Pye and Blott, 2008).

CONCLUDING COMMENTS

The Sefton coast possesses one of the largest and most geomorphologically significant dune systems in the United Kingdom. The dunes have developed over a period of more than 5000 years due to the onshore movement of sand from the eastern Irish Sea. Development of the dune system has apparently been episodic, with periods of widespread sand blowing and dune mobility separated by periods of stabilization by vegetation. The relative importance of changes in climate, shoreline movements and human activities as drivers of these changes remains uncertain and requires further research. Human activities have undoubtedly had a major influence in shaping the geomorphology of the dune system (Pye, 1990; Jones et al., 1993). Within the dune system itself, these activities have included management of rabbit warrens, cutting of marram, levelling of the dunes to form asparagus fields, sand mining, afforestation, landscape modification to create a large number of golf courses, military activities and more recent measures to control visitor pressure and to restore areas of 'degraded' dune landscape. Human activities in the adjoining estuaries, including construction of training walls, dredging and land reclamation, have also had a significant impact on regional sediment transport and coastal morphology (Smith, 1982; Pye and Neal, 1994). However, despite the strong human influences, the Sefton dunes retain many features of intrinsic geomorphological, as well as ecological, interest.

The prediction of future changes in shoreline position and dune mobility / stability is a difficult task owing to uncertainties regarding the nature of future sediment supply, nearshore bathymetry, magnitude and rate of sea level change, fluctuations in wind and wave conditions, storm surge frequency, and management interventions. Extrapolation of historical trends over the past 100years and application of simple models such as the Bruun Rule would suggest that there will be significant further shoreline retreat around Formby Point and further accretion at Altcar and Ainsdale (e.g. Lymbery et al., 2008; Esteves et al., 2009). However, a longer historical perspective suggests that it is quite possible that these trends could come to an end over the next 100 years if the regional sediment transport system changes back to a situation similar to that which existed before about 1890 (Pye, 1977; Pye and Blott, unpublished data).

Continued monitoring of the dunes, beaches and nearshore areas is essential in order to provide early warning of any future changes. Further research is also required to provide a better understanding of the potential sensitivity of the system to changes in environmental forcing factors.

ACKNOWLEDGEMENTS

We thank Lee Swift of the Environment Agency (Northwest Region) for providing access to the 2008 lidar data, and Paul Wisse and Graham Lymbery of

Sefton Council for provision of dune monitoring data and aerial photographs. Invaluable assistance with field surveys, sediment sampling and laboratory analysis was provided by Samantha Saye and Samantha Witton.

REFERENCES

Adams, M. and **Harthen, D.** (2007) *An Archaeological Assessment of the Sefton Coast, Merseyside.* Parts 1 & 2. National Museums Liverpool Field Archaeology Unit, Liverpool, 132p.

Allan, R., Tett, S. and **Alexander, L.** (2008) Fluctuations in Autumn-Winter sever storms over the British Isles: 1920 to present. *International Journal of Climatology* d29, 357-371.

Ashton, W. (1909) *The Battle of Land and Sea on the Lancashire, Cheshire and North Wales Coasts and the Origin of the Lancashire Sandhills.* Wm Ashton & Sons, Southport, 217p.

Ashton, W. (1920) *The Evolution of a Coastline. Barrow to Aberystwyth and the Isle of Man.* Edward Stanford Ltd., London, and WM. Ashton & Sons, Ltd, Southport, 302p.

Barron, J. (1938) *A History of the Ribble Navigation.* Preston Corporation and Guardian Press, Fishergate, Preston, 503.

Blott, S.J. and **Pye, K.** (2003) Sefton Coast Intertidal Sediment Survey (2003) *Results Summary and Interpretative Report.* External Investigation Report ER502, Kenneth Pye Associates Ltd., Crowthorne.

Blott, S.J., Pye, K., van der Wal, D. and **Neal, A.** (2006) Long-term morphological change and its causes in the Mersey estuary, NW England. *Geomorphology* 81, 185-206.

Brodrick, H. (1902) Martin Mere and its antiquities. *British Association Handbook*, Southport, pp. 198-204.

Brodrick, H. (1903) Martin Mere. *Annual Report of Southport Scientific Society* i, 5-18.

Cook, A.L.M. (1989) *Altcar - The Story of a Rifle Range.* Territorial, Auxilliary & Volunteer Reserve Association for the Northwest of England & The Isle of Man, Liverpool, 182.

Cowell, R.W. (2008) Coastal Sefton in the Prehistoric Period. In Lewis, J.M. and Stannistreet, J.E. (eds.) *Sand and Sea - Sefton's Coastal Heritage.* Sefton Council Leisure Services Department, Bootle, pp. 20-27.

Esteves, L.S., Williams, J.J., Nock, A. and **Lymbery, G.** (2009) Quantifying shoreline changes along the Sefton coast (UK) and the implications for research-informed coastal management. *Journal of Coastal Research Special Issue* 56, 602-606.

George, N. (1997) Liverpool Bay Natural Area. A Nature Conservation Area Profile. *English Nature*, Wigan, 65p.

Gonzalez, S. and **Huddart, D.** (2002) Formby Point and Hightown. In Huddart, D. and Glasser, N.F. (eds.) *The Quaternary of Northern England.* Geological Conservation Review Series 25, Joint Nature Conservation Committee, Peterborough, 569-588.

Gonzalez, S., Huddart, D. and **Roberts, G.** (1996) Holocene development of the Sefton coast: a multi-disciplinary approach to understanding the archaeology. In Sinclair, A., Slater, E. and Gowlett, J. (eds.) *Archaeological Science* 1995. Oxbow Books, Oxford, 289-299.

Gresswell, R.K. (1937) The geomorphology of the southwest Lancashire coastline. *The Geographical Journal* 90, 335-348.

Gresswell, R.K. (1953a) The physical landscape and landforms. In Smith, W., Monkhouse, F.J. and Wilkinson, H.R. (eds.) *A Scientific Study of Merseyside.* University of Liverpool Press, Liverpool, 37-48.

Gresswell, R.K. (1953b) The coast. In Smith, W., Monkhouse, F.J. and Wilkinson, H.R. (eds.) *A Scientific Study of Merseyside.* University of Liverpool Press, Liverpool, 49-52.

Gresswell, R.K. (1953c) *Sandy Shores in South Lancashire.* Liverpool University Press, Liverpool, 194p.

Gresswell, R.K. (1964) The origin of the Mersey and Dee estuaries. *Geological Journal* 4, 77-86.

Hale, W.G. (2005a) Out of the ice. In Hale, W.G. and Coney, A. (eds.) (2005) *Martin Mere. Lancashire's Lost Lake.* Liverpool University Press, Liverpool, 5-26.

Hale, W.G. (2005b) Mapping the Mere. In Hale, W.G. and Coney, A. (eds.) (2005) *Martin Mere. Lancashire's Lost Lake.* Liverpool University Press, Liverpool, 45-63.

Hale, W.G. (2005c) Draining the Mere. In Hale, W.G. and Coney, A. (eds.) (2005) *Martin Mere. Lancashire's Lost Lake.* Liverpool, 125-150.

Halliwell, A.R. (1973) Residual drift near the sea bed in Liverpool Bay: an observational study. *The Geophysical Journal of the Royal Astronomical Society* 32, 439-458.

Hopkins, J. and **Radley, G.** (2001) Sand dunes and The Habitats Directive: preparation of the UK National List. In Houston, J.A., Edmondson, S.E. and Rooney, P.J. (eds.) *Coastal Dune Management. Shared Experience of European Conservation Practice.* Liverpool University Press, Liverpool, 283-305.

Houston, J.A. and **Jones, C.R.** (1987) The Sefton Coast Management Scheme: project and process. *Coastal Management* 15, 267-297.

Howell, F.T. (1973) The sub-drift surface of the Mersey and Weaver catchments and adjacent areas. *Geological Journal* 8, 285-296.

HRS (1969a) *The Southwest Lancashire Coastline: A Report of the Sea Defences.* Report EX450, Hydraulics Research Station, Wallingford, 43p. + tables and figures.

HRS (1969b) *The Computation of Littoral Drift.* Report EX449, Hydraulics Research Station, Wallingford.

Huddart, D. (1992) Coastal environmental changes and morpho-stratigraphy in southwest Lancashire, England. *Proceedings of the Geologists Association* 103, 217-236.

Huddart, D., Gonzalez, S. and **Roberts, G.** (1999a) The archaeological record and mid-Holocene marginal coastal palaeoenvironments around Liverpool Bay. *Quaternary Proceedings* 7, 563-574.

Huddart, D., Gonzalez, S. and **Roberts, G.** (1999b) Holocene human and animal footprints and their relationships with coastal environmental change, Formby Point, NW England. *Quaternary International* 55, 29-41.

Innes, J.B. and **Tooley, M.J.** (1993) The age and vegetational history of the Sefton Coast dunes. In Atkinson, D. and Houston, J. (eds.) *The Sand Dunes of the Sefton Coast.* Museums and Galleries on Merseyside, Liverpool, 35-40.

Jackson, D.I., Jackson, A.A., Evans, D., Wingfield, R.T.R., Barnes, R.P. and **Arthur, M.J.** (1995) *The Geology of the Irish Sea.* British Geological Survey, United Kingdom Offshore Regional Report. HMSO, London, 123p.

Jay, H. (1998) *Beach - Dune Sediment Exchange and Morphodynamic Responses: Implications for Shoreline Management.* PhD Thesis, University of Reading.

Johnson, R.H. (1985) The imprint of glaciation of the West Pennine Uplands. In Johnson, R.H. (ed.) *The Geomorphology of North-west England.* Manchester University Press, Manchester, 237-262.

Johnson, R.H. and **Musk, L.F.** (1974) A comment on Dr. F.T. Howell's paper on the sub-drift surface of the Mersey and Weaver catchment and adjacent areas. *Geological Journal* 9, 209-210.

Jones, C.R., Houston, J.A. and **Bateman, D.** (1993) A history of human influence on the coastal landscape. In Atkinson, D. and Houston, J. (eds.) (1993) *The Sand Dunes of the Sefton Coast.* National Museums and Galleries on Merseyside, Liverpool, 18-20.

Kavanagh, K. (1997) *Significance of Spatial variations in sand Texture and Mineralogy Within the Sefton Coastal Dunefield, Northwest England.* MSc Dissertation, University of Reading.

Kelly, E. (1973) *Viking Village. The Story of Formby.* The Formby Society, Formby, 163.

Lennon, G. W. (1963) Identification of weather conditions associated with the generation of major storm surges along the west coast of the British Isles. *Quarterly Journal of the Royal Meteorological Society* 89, 381-394.

Lymbery, G. (2008) Working with the sea - changing attitudes to coastal defence. *Coastlines* Summer 2008, 4-5.

Lymbery, G., Wisse, P. and **Newton, M.** (2007) *Report on Coastal Erosion Predictions for Formby Point, Formby, Merseyside.* Unpublished Report, Sefton Council, Bootle, Merseyside, 33.

Milne, G. (2005) Maritime Liverpool. In Belchem, J. (ed.) Liverpool 800: Culture, Character and History. Liverpool University Press, Liverpool, 257-309.

Neal, A. (1993) *Sedimentology and Morphodynamics of a Holocene Coastal Dune Barrier Complex, Northwest England.* PhD Thesis, University of Reading.

Neal, A, and **Roberts, C.** (2001) Internal structure of a trough blowout, determined from migrated ground-penetrating radar profiles. *Sedimentology* 48, 791-810.

Newton, M., Lymbery, G. and **Wisse, P.** (2008) *A Description of Time Series Spatial Data Held and Used Within the Sefton Coast Database.* Sefton Council Technical services Department, Bootle, 21p.

Parker, W.R. (1971) *Aspects of the Marine Environment at Formby Point, Lancashire.* PhD Thesis, Liverpool University, 2 volumes.

Parker, W.R. (1975) Sediment mobility and erosion on a multi-barred foreshore (Southwest Lancashire, UK). In: Hails, J.R. and Carr, A.P. (eds.) *Nearshore Sediment Dynamics and Sedimentation.* John Wiley & Sons, London, 151-179.

Plater, A.J., Huddart, D., Innes, J.B., Pye, K., Smith, A.J. and **Tooley, M.J.** (1993) Coastal and sea level changes. In: Atkinson, D. and Houston, J. (eds.) *The Sand Dunes of the Sefton Coast.* National Museums and Galleries on Merseyside, Liverpool, 23-34.

Price, W.A. and **Kendrick, M.P.** (1963) Field and model investigations into the reasons for siltation in the Mersey estuary. *Proceedings of the Institution of Civil Engineers* 24, 473-518.

Pye, K. (1977) *An Analysis of Coastal Dynamics Between the Ribble and Mersey Estuaries With Particular Reference to Erosion at Formby Point.* BA Dissertation, University of Oxford, 137p.

Pye, K. (1990) Physical and human influences on coastal dune development between the Ribble and Mersey estuaries, northwest England. In Nordstrom, K.F., Psuty, N.P. and Carter, R.W.G. (eds.) *Coastal Dunes: Form and Process.* Wiley, Chichester, 339-359.

Pye, K. (1991) Beach deflation and backshore dune formation following erosion under storm surge conditions: an example from Northwest England. *Acta Mechanica Supplementum* 2, 171-181.

Pye, K. and **Blott, S.J.** (2008) Decadal-scale variation in dune erosion and accretion rates: an investigation of the significance of changing storm tide frequency and magnitude on the Sefton coast, UK. *Geomorphology* 102, 652-666.

Pye, K. and **Neal, A.** (1993) Late Holocene dune formation on the Sefton coast, northwest England. In Pye, K. (ed.) *The Dynamics and Environmental Context of Aeolian Sedimentary Systems.* Geological Society of London Special Publication 72, Geological Society Publishing House, Bath, 201-217.

Pye, K. and **Neal, A.** (1994) Coastal dune erosion at Formby Point, north Merseyside, England: causes and mechanisms. *Marine Geology* 119, 39-56.

Pye, K. and **Smith, A.J.** (1988) Beach and dune erosion and accretion on the Sefton coast, Northwest England. *Journal of Coastal Research Special Issue* 3, 33-36.

Pye, K. and **Tsoar, H.** (2009) *Aeolian Sand and Sand Dunes.* Springer-Verlag, Berlin, 458pp.

Pye, K., Stokes, S. and **Neal, A.** (1995) Optical dating of aeolian sediments from the Sefton coast, Northwest England. *Proceedings of the Geologists Association* 106, 281-292.

Pye, K., Blott, S.J., Short, B. and **Witton, S.J.** (2006) *Preliminary Investigation of Sea Bed Sediment Characteristics in Liverpool Bay.* External Research Report No. ER602, Kenneth Pye Associates Ltd., Crowthorne.

Reade, T.M. (1881) On a section of the Formby and Leasowe Marine Beds and superior Peat Beds, disclosed by the cuttings for the outlet sewer at Hightown. *Proceedings of the Liverpool Geological Society* 4, 269-277.

Roberts, G. and **Worsley, A.** (2004) Evidence for Human activity in mid-Holocene coastal palaeoenvironments of Formby, North West England. In Lewis, J.M. and Stannistreet, J.E. (eds.) *Sand and Sea - Sefton's Coastal Heritage.* Sefton Council Leisure Services Department, Bootle, 28- 43.

Roberts, G., Gonzalez, S. and **Huddart, D.** (1996) Inter-tidal Holocene footprints and their archaeological significance. *Antiquity* 70, 647-651.

Saye, S.E. (2003) *Morphology and Sedimentology of Coastal Sand Dune Systems in England and Wales.* PhD Thesis, University of London, 614p.

Saye, S.E. and **Pye, K.** (2003) *Particle Size Variation in Beach and Dune Sediments on the Sefton Coast, NW England: Nature and Significance.* Internal Research Report IR001, Kenneth Pye Associates Ltd., Crowthorne, 31p. + tables & figures.

Saye, S.E., van der Wal, D., Pye, K. and **Blott, S.J.** (2005) Beach-dune morphological relationships and erosion/accretion: an investigation at five sites in England and Wales using LIDAR data. *Geomorphology* 72, 128-155.

Shackleton, R.M. (1953) Geology. In Smith, W., Monkhouse, F.J. and Wilkinson, H.R. (eds.) *A Scientific Study of Merseyside.* University of Liverpool Press, Liverpool, 19-36.

Shennan, I. and **Horton, B.** (2002) Holocene land and sea level changes in Great Britain. *Journal of Quaternary Science* 17, 511-526.

Sly, P.G. (1966) *Marine Geological Studies in the Eastern Irish Sea and Adjacent Estuaries*. PhD Thesis, University of Liverpool, 2 vols.

Smith, A.J. (1982) *A Guide to the Sefton Coast Database*. Sefton Borough Council Engineer & Surveyor's Department, Bootle, 80p. + figures.

Smith, P.H. (1999) *The Sands of Time. An Introduction to the Sand Dunes of the Sefton Coast*. National Museums and Galleries on Merseyside, Liverpool, 196p.

Thomas, C.G., Spearman, J.R. and **Turnbull, M.J.** (2002) Historical morphological change in the Mersey estuary. *Continental Shelf Research* 22, 1775-1794.

Tooley, M.J. (1969) *Sea-Level Changes and the Development of Coastal Plant Communities During the Flandrian in Lancashire and Adjacent Areas*. PhD Thesis, University of Lancaster.

Tooley, M.J. (1970) The peat beds of the south-west Lancashire coast. *Nature in Lancashire* 1, 19-26.

Tooley, M.J. (1974) Sea level changes in the last 9000 years in northwest England. *The Geographical Journal* 140, 18-42.

Tooley, M.J. (1976) Flandrian sea level changes in West Lancashire and their implications for the 'Hillhouse Coastline'. *The Geological Journal* 11, 37-52.

Tooley, M.J. (1978) *Sea Level Changes in Northwest England During the Flandrian Stage*. Clarendon Press, Oxford.

Tooley, M.J. (1980) Theories of coastal change in North-West England. In Thompson, F.H. (ed.) *Archaeology and Coastal Change*. Society of Antiquaries, London, 74-86.

Tooley, M.J. (1985a) Sea level changes and coastal morphology in northwest England. In Johnson, R.H. (ed.) *The Geomorphology of Northwest England*. Manchester University Press, Manchester, 94-121.

Tooley, M.J. (1985b) Climate, sea-level and coastal changes. In Tooley, M.J. and Sheail, G.M. (eds.) *The Climatic Scene*. Allen and Unwin, London, 206-234.

Travis, C.B. (1926) The peat and forest bed of the south-west Lancashire coast. *Proceedings of the Liverpool Geological Society* 14, 263-277.

van der Wal, D., Pye, K. and **Neal, A.** (2002) Long-term morphological change in the Ribble estuary, northwest England. *Marine Geology* 189, 249-266.

Wheeler, D.J., Simpson, D.E. and **Houston, J.A.** (1993) Dune use and management. In Atkinson, D. and Houston, J. (eds.) *The Sand Dunes of the Sefton Coast*. National Museums and Galleries on Merseyside, Liverpool 129-150.

Wilson, P. (1985) The Mere Sands of Southwest Lancashire - A forgotten Flandrian deposit. *Quaternary Newsletter* 45, 23-26.

Wilson, P., Bateman, R.M. and **Catt, J.A.** (1981) Petrology, origin and environment of deposition of the Shirdley Hill Sand of Southwest Lancashire, England. *Proceedings of the Geologists Association* 92, 211-229.

Woodworth, P.L., Tsimplis, M.N., Flather, R.A. and **Shennan, I.** (1999) A review of the trends observed in British Isles mean sea level data measured by tide gauges. *Geophysical Journal International* 136, 651-670.

Wray, D.A., Wolverson Cope, F., Tonks, L.H. and **Jones, R.C.B.** (1948) *Geology of Southport and Formby*. Memoirs of the Geological Survey of Great Britain, England and Wales, HMSO, London, 54.

Wright, P. (1976) *The Morphology, Sedimentary Structures and Processes of the Foreshore at Ainsdale, Merseyside, England*. PhD Thesis, University of Reading.

Yorke, B. and **Yorke, R.** (1982) *Britain's First Lifeboat Station at Formby, Merseyside, 1776 - 1918*. The Alt Press, Formby, 73p.

Yorke, B. and **Yorke, R.** (2008a) Pine trees and asparagus: the development of a cultural landscape. In Lewis, J.M. and Stannistreet, J.E. (eds.) *Sand and Sea - Sefton's Coastal Heritage*. Sefton Council Leisure Services Department, Bootle, 96-114.

Yorke, B. and **Yorke, R.** (2008b) *A dangerous coast*. In Lewis, J.M. and Stannistreet, J.E. (eds.) *Sand and Sea - Sefton's Coastal Heritage*. Sefton Council Leisure Services Department, Bootle, 78-95.

Table 1. Sand volumes within (a) 200m and (b) 500 m of the dune toe (above 6.0 m OD) at profile locations A to L between Ainsdale and Hightown, derived from the 2008 lidar DTM.

Profile	Location	Volume width (m3) per metre	Maximum Height (m OD)	Minimum Height (m OD)	Average Height (m OD)	Stdev Height (m OD)
(a) Volumes within 200 m of dune toe						
A	South of Shore Road	417.8	10.4	6.1	8.2	1.3
B	Ainsdale Boundary Post	919.0	16.4	5.7	10.5	3.3
C	Fisherman's Path	302.8	9.8	5.7	7.5	1.0
D	Victoria Road	885.3	12.8	9.1	10.4	1.3
E	North of Wicks Lane	1968.9	24.2	12.6	15.8	4.8
F	North of Lifeboat Road	588.3	16.0	6.2	8.9	2.4
G	South of Lifeboat Road	1231.7	25.4	5.9	12.1	6.0
H	Devil's Hole	965.4	15.8	9.0	10.8	2.0
I	Range Lane	331.7	9.4	5.9	7.6	1.0
J	Altcar Range	533.7	14.8	5.1	8.5	3.2
K	Hightown	238.7	8.0	6.9	7.2	0.5
L	Fort Crosby	263.3	9.0	6.4	7.3	0.9
(b) Volumes within 500 m of dune toe						
A	South of Shore Road	1048.6	11.1	6.1	8.2	1.2
B	Ainsdale Boundary Post	1396.6	16.4	5.7	8.8	2.7
C	Fisherman's Path	1104.6	12.4	5.7	8.2	1.5
D	Victoria Road	1844.0	12.8	7.8	9.7	1.4
E	North of Wicks Lane	3080.2	24.2	8.0	12.1	4.6
F	North of Lifeboat Road	2859.5	21.4	6.2	11.7	3.9
G	South of Lifeboat Road	1716.3	25.4	5.8	9.4	4.5
H	Devil's Hole	1775.1	15.9	5.5	9.5	3.2
I	Range Lane	375.9	9.4	5.7	6.8	1.0
J	Altcar Range	533.7	14.8	3.8	6.2	2.8
K	Hightown	877.9	8.9	6.9	7.8	0.7
L	Fort Crosby	382.6	9.0	4.6	6.8	0.8

Table 2. Range and average values for grain size distribution parameters of samples from the beach (both untreated and carbonate / mud-free), frontal dunes and hind dunes on the Sefton coast (data from Jay, 1998; Kavanagh, 1997 and Saye, 2003).

Particle size parameter			Beach (Jay, 1998)	Beach (carbonate free) (Jay, 1998)	Frontal dunes (Kavanagh, 1997)	Hind (Kavanagh, 1997)	Frontal dunes (Saye, 2003)	Hind (Saye, 2003)
N			166	166	21	49	14	13
Mean	Min	μm	54.0	115.1	76.8	90.6	223.1	198.3
	Max	μm	282.5	279.0	279.1	291.9	325.1	256.9
	Mean	μm	212.1	218.2	225.3	201.0	251.3	233.0
	StDev	μm	38.2	30.1	48.5	49.0	25.7	17.2
	CV	%	18	14	22	24	10	7
Median	Min	μm	139.4	157.3	174.0	197.9	223.9	223.6
	Max	μm	283.1	279.9	279.6	289.4	280.3	257.4
	Mean	μm	217.8	219.5	237.8	233.6	245.5	240.7
	StDev	μm	30.2	28.5	21.4	16.8	15.1	9.8
	CV	%	14	13	9	7	6	4
Mode	Min	μm	152.5	152.5	219.4	200.3	223.4	223.4
	Max	μm	288.4	288.4	288.4	288.4	269.2	269.2
	Mean	μm	217.6	219.2	242.0	240.8	240.8	243.7
	StDev	μm	29.8	28.9	16.7	14.2	15.8	11.2
	CV	%	14	13	7	6	7	5
D90-D10	Min	μm	104.4	106.9	152.0	139.9	144.5	142.0
	Max	μm	279.2	303.3	421.9	469.4	705.9	204.7
	Mean	μm	167.0	163.0	192.9	211.7	216.3	168.0
	StDev	μm	24.1	20.5	66.2	73.0	154.1	15.7
	CV	%	14	13	34	34	71	9
Clay content	Min	%	0.0	0.0	0.0	0.0	0.0	0.0
	Max	%	5.7	3.7	5.1	4.0	0.0	0.3
	Mean	%	0.2	0.0	0.5	0.8	0.0	0.1
	StDev	%	0.6	0.3	1.4	1.2	0.0	0.1
	CV	%	299	830	260	140	0	130
Silt content	Min	%	0.0	0.0	0.0	0.0	0.0	0.0
	Max	%	33.0	10.5	30.6	25.8	0.0	5.6
	Mean	%	1.2	0.2	2.7	5.6	0.0	1.1
	StDev	%	3.4	1.0	7.6	6.8	0.0	1.7
	CV	%	280	515	278	122	0	152
Sand content	Min	%	61.3	85.8	64.3	70.2	100.0	94.1
	Max	%	100.0	100.0	100.0	100.0	100.0	100.0
	Mean	%	98.6	99.8	96.7	93.6	100.0	98.8
	StDev	%	4.0	1.3	8.9	7.9	0.0	1.8
	CV	%	4	1	9	8	0	2

Table 3. Range and average values for concentrations of major oxides and trace elements in 70 samples of dune sand taken along shore-normal transects at intervals between Birkdale and Hightown. Oxide data are in weight percent and trace element values in parts per million (µg/g). Data from Kavanagh (1997).

Major oxide / trace element	Minimum	Maximum	Mean	Standard deviation	Coefficient of variation
NaO	0.27	0.87	0.41	0.10	24
MgO	0.05	1.40	0.31	0.23	73
Al2O3	1.50	7.63	2.31	1.04	45
SiO2	59.43	100.00	91.29	7.50	8
P2O5	0.02	0.80	0.08	0.12	140
K2O	0.29	1.55	0.89	0.18	20
CaO	0.00	15.04	2.36	3.10	132
TiO2	0.08	1.29	0.17	0.15	93
MnO	0.01	0.11	0.03	0.01	54
Fe2O3	0.72	9.49	1.28	1.14	90
V	12	129	23	17	73
Cr	10	104	32	18	57
Co	5	20	7	3	39
Ni	6	41	10	5	53
Cu	1	132	10	21	206
Zn	1	645	41	89	218
Pb	7	1335	48	162	336
Rb	19	60	28	6	23
Sr	24	422	71	69	97
Y	0	177	5	21	441
Zr	18	1538	141	193	137

Table 4. Average erosion and accretion rates at selected stations between Ainsdale and Altcar, based on ground survey measurements between 1958 and 2007 (data recorded by Southport Corporation and Sefton Council).

Location	Marker Post	Accretion (+) / Erosion (-) Rate (m/yr) 1958 - 1980	1980 - 2007	1991 - 2007	1958 - 2007
South of Shore Road	3 / M1		+2.8	+2.7	
Ainsdale	2 / M2		+0.9	+0.6	
Ainsdale Boundary	1 / M3		+0.4	-0.4	
Ainsdale Nature Reserve	M4			-1.7	
Fishermans Path	A / M5	-2.5	-3.4	-3.7	-2.8
Formby Golf Course	M6			-2.7	
Massams Slack	B / M7	-3.9	-3.3	-3.2	-3.7
Freshfield	M8			-3.0	
Victoria Road	C / M9	-3.4	-2.3	-3.0	-2.9
Blundell Avenue	M10			-2.4	
Wicks Lane	D / M11	-2.3	-1.8	-2.6	-1.9
North Lifeboat Rd	M12			-1.4	
South Lifeboat Rd	E / M13	-2.5	-0.3	-0.3	-1.2
Alexandra Rd	M14			-0.7	
Albert Road	F / M15	+0.7	+1.0	+0.5	+0.7
Cabin Hill Reserve	M16			+1.0	
Range Lane	G / M17	+4.0	+1.1	+1.1	+2.4

SECTION B: BIOGEOGRAPHY

The birds of the Sefton Coast: A review

Steve White

Lancashire Wildlife Trust, Seaforth Nature Reserve, Port of Liverpool L21 1JD
swhite@lancswt.org.uk

ABSTRACT

A comprehensive list of all bird species recorded on the Sefton Coast in the past 100 years is published for the first time, together with a list of all species that have bred over the same period.

The significance of the area's birds for nature conservation is assessed, concluding that the Sefton Coast is of international and national importance for a total of 26 species of waders, wildfowl and seabirds during winter and spring and autumn migration periods. The coast is the most important site in Britain for two species of wader and in the top five for a further three.

Abundance data, based upon the British Trust for Ornithology's (BTO) Wetland Bird Survey, are given for the period 2001/02 to 2005/06 for these species of conservation importance and for waterfowl as a whole.

The relative importance of the main bird habitats, sand and mudflats, beaches, saltmarsh, grazing marsh, lagoons, sand dunes, scrub, woodland and the open sea, is described.

INTRODUCTION

The ornithological importance of the Sefton Coast is not always uppermost in the minds of scientists or nature conservationists, understandably distracted as they are by the plethora of rare plants, invertebrates, amphibians and reptiles, but, together with the Red Squirrel, it is its birds that are uppermost in the public eye.

A number of short papers and occasional articles (often unpublished) have dealt with the coast's avifauna over the years but none are widely available. Wood (1993) provided a useful review of this literature going back to the mid-19th century, together with an assessment of the current status of birds on the dunes – but he did not deal with other coastal habitats. McCarthy (2001) dealt in detail with the birds of Marshside and annual reports covering Seaforth Nature Reserve and Crosby Marine Park have been produced by the Lancashire Wildlife Trust since 1985.

Many birdwatchers are obsessed by rarities, which the Sefton Coast produces with some regularity. Representative recent examples include Pied Wheatear at Seaforth in 1997, Richard's Pipit on Crosby Marine Park in 2003, Pallas's Warbler in the Hightown Dunes in 2006, Fea's Petrel at sea off Formby Point in 1995 and Glossy Ibis at Marshside from 2006 to 2008. But there is much more to Sefton's birdlife than this.

The national and international importance of the coast's birds is recognised in the designation of its entire length as Special Protection Areas (the Ribble and Alt Estuaries from Crossens to Crosby, and the Mersey Narrows and North Wirral Shore at Seaforth). This statutory protection derives not from the presence of rare species but rather the sheer number and variety of birds that the coast supports.

The bald statistics are impressive. A total of 296 native and naturalised species (British Ornithologists Union Categories A, B and C) has been recorded at some time, accounting for 51% of all species ever recorded in Britain (Appendix 1). Around 200 of these are seen annually. It is impossible to put a precise figure on the total number of birds that make use of the coast each year, but some indication of this can be gained by adding up the largest annual counts for the well-monitored waterfowl species, which have fluctuated between 150,000 and 180,000 since the early 1990s. Taking into account the annual migrations in spring and autumn and the throughput of birds in winter, these figures can probably safely be doubled. Adding the unknown number of landbirds and the sometimes uncountable flocks of seabirds perhaps puts the numbers making use of the coast annually in the region of half a million.

The majority of these are waders feeding on the sand and mudflats the length of the shore while the tide is out and coming into favoured roost sites at high water. But the saltmarshes and grazing marshes at the northern end of coast, which are favoured by geese, swans and ducks, come a close second in importance. The lagoons at Seaforth/Crosby Marine Park (Plate 1), Southport Marine Lake and Marshside add considerably to the variety of species, while the mosaic of ornithological habitats is completed by the scrub, grassland and woods of the dunes, and the open sea.

Wintering and Passage Wildfowl and Waders

Assessing the contribution of the Sefton Coast to the estuarine complex of the Ribble, Alt, Mersey and Dee and Morecambe Bay, which together make up Britain's most important site for waterfowl, is confused by boundary and recording issues. The Alt Estuary (Crosby to Hightown) lies entirely within the boundaries of the Sefton Coast but the Ribble Estuary (Formby Point to Crossens) only partly so, extending northwards as far as Lytham St. Annes. Seaforth is part of the Alt Estuary for the purposes of the British Trust for Ornithology's Wetland Bird Survey (WeBS) but is joined with sites in both the Mersey and Dee Estuaries for its SPA designation. No statistics for the Sefton Coast are, therefore, published and the account that follows has been reconstructed from data for individual WeBS count sectors. They are published here with the permission of the BTO but with the caveat that some of the component counts may be unchecked.

The Sefton Coast is of international importance (supporting more than 1% of the European Union or East Atlantic Flyway populations) for 12 species and nationally important (supporting more than 1% of the British population) for a further 14 species, three of which are regarded as of special conservation concern in Europe and therefore also of international importance (Appendix 2).

Plate 1. The artificial lakes and lagoons of the Sefton Coast support a wide range of wintering wildfowl. These male Goldeneye, photographed at Seaforth, are amongst the most attractive. (Photograph by Steve Young).

Plate 2. Oystercatchers favour the northern sections of the Sefton Coast and are rarely seen in large numbers south of Formby Point. (Photograph by Steve Young).

10 of the 26 species mentioned previously are waders which, with the exception of Black-tailed Godwit, feed mainly on grazed saltmarshes and other grasslands especially at Marshside, exploit a variety of intertidal invertebrates on the extensive sand and mudflats that are revealed at low water from Seaforth to Marshside. Their distribution is determined largely by that of their prey and by freedom from human disturbance. Waders are more concentrated at high tide when roosts can form anywhere on the shore, but the largest numbers usually head towards traditional sites, the largest of which are at Seaforth, Formby Point (Taylor's Bank), Ainsdale-Birkdale Shore and Southport-Marshside. Here they are more susceptible to disturbance but access restrictions and wardening have in recent years gone some considerable way to alleviating this problem, especially at Ainsdale/Birkdale.

The average peak count for all wintering waders in recent years has been 130,000. The most numerous species are Knot, Dunlin, Bar-tailed Godwit and Oystercatcher (Plate 2); data for all nationally and internationally important species are shown in Table 1.

The Sefton Coast is the most important site in Britain for Bar-tailed Godwit, supporting almost a third of the national wintering population, and for both wintering and passage Sanderling. It is the second most important site for wintering and passage Grey Plover (after the Wash), the third most important for Knot (after the Wash and Morecambe Bay) and the fifth most important for Dunlin.

Wildfowl (swans, geese and ducks) numbers are significantly lower, averaging about 27,000 between 2001/02 and 2005/06. Seven species occur in internationally or nationally important numbers (Table 2) and, with two exceptions, Shelduck, which feeds on marine snails and other intertidal invertebrates, and Common Scoter, which is an entirely marine species in winter, are found predominantly on the coastal marshes. They divide into two groups: the grass-eaters, Pink-footed Goose and Wigeon that occur on the cattle-grazed saltmarshes of the Ribble Estuary, including

Table 1. Status of nationally and internationally important wader species on the Sefton Coast (2001/02 to 2005/06).

Species	5-year mean peak count	% international population	% British population
Knot	55220	12.3	19.7
Dunlin	24165	1.8	4.3
Bar-tailed Godwit	18126	15.1	29.2
Oystercatcher	15797	1.5	4.9
Grey Plover (passage)	6588	2.6	12.4
Grey Plover (winter)	4490	1.8	8.5
Sanderling (passage)	8294	6.9	27.6
Sanderling (winter)	2559	2.1	12.2
Redshank	1997	1.5	1.7
Curlew	1656	0.4	1.1
Black-tailed Godwit	1513	4.3	10.1
Ringed Plover (passage)	1330	1.8	4.4

Marshside; and the omnivorous but mainly seed- or invertebrate-eating dabbling ducks, Teal, Pintail and Shoveler, which mainly frequent the saltmarsh edge, lagoons and ditches. The management of the Ribble Estuary is focussed heavily upon the former group and they are by far the more numerous and the only species that occur in internationally important numbers.

A third group, the diving ducks, occur in regionally important numbers on the 'coastal lagoons': Pochard principally at Seaforth and Marshside, Tufted Duck mainly at Ainsdale Sands Lake and Seaforth, and Goldeneye and Scaup at Seaforth (although the latter species feeds predominantly on marine molluscs). Southport Marine Lake supports regionally important populations (>0.5% of the British populations) of two other waterfowl species, Mute Swan and Coot with average peak counts during 2001-2005 of 174 and 990 respectively.

Other wintering birds of the coastal saltmarshes and beaches

The saltmarshes, especially those that remain ungrazed at Marshside and Crossens, regularly support large numbers of seed-eating passerines (songbirds), in particular Skylark, Linnet and Reed Bunting. These, together with a large population of small mammals, attract small but regionally significant numbers of birds of prey, especially Hen Harrier, Merlin, Peregrine and Short-eared Owl.

Some of these same wintering passerines also frequent the dune grasslands and the strandline, where they are

Table 2. Status of nationally and internationally important wildfowl on the Sefton Coast.

Species	5-year mean peak count	% international population	% British population
Pink-footed Goose	c.5000	2	2
Shelduck	1623	0.5	2.1
Wigeon	15985	1.1	3.9
Teal	3744	0.9	2
Pintail	400	0.7	1.4
Shoveler	220	0.6	1.5

joined occasionally by more exotic Snow Buntings and Shorelarks.

Waders roosting at high tide on the beaches and sandbanks are frequently accompanied by large numbers of gulls, predominantly Black-headed, Herring, Common and Lesser Black-backed – birds which generally feed offshore or inland. Occasionally, after winter storms produce 'wrecks' of shellfish or starfish, truly vast numbers may gather to enjoy the spoils. Recent examples have included estimates of 100,000 Herring, 50,000 Black-headed and 25,000 Common Gulls between Ainsdale and Marshside in February 2002, and 3,700 Lesser Black-backed and 1,200 Great Black-backed Gulls between Ainsdale and Southport in February 1988 (White et al., 2008). These exceptional counts represented very large proportions of the national wintering populations for each species; winter numbers are usually much lower but there is little doubt that all five species are of at least national importance, while Common, Black-headed and Herring Gulls are almost certainly internationally important.

The open sea

Sefton's offshore waters form part of the proposed Liverpool Bay marine Special Protection Area, to be designated primarily for Common Scoter (a seaduck) and Red-throated Diver.

It is difficult to pin down exactly how many birds use 'Sefton's waters' as the winter flocks can be quite mobile. However, recent aerial surveys have regularly located up to 8,000 Common Scoter between Formby Point and the mouth of the Ribble (Cranswick et al., 2004), while counts of Red-throated Diver seen from Formby Point regularly exceed 100 in winter and especially in spring (Lancashire Bird Reports). Cormorant numbers regularly exceed 1,000 from late autumn onwards, feeding over a wide area but coming in to roost at Seaforth and less regularly to other parts of the coast.

Other wintering seabird species are quite scarce, although a regionally important flock of Scaup (another seaduck) occurs annually off Seaforth/Crosby Marine Park and Southport.

It is during migration periods, however – particularly in autumn – that the Sefton coast becomes most alive with seabirds, as species that breed in northern Britain and the Arctic move towards their southerly wintering grounds, often lingering for a while offshore, particularly between Formby Point and Seaforth. To a large extent, the occurrence of these species, amongst which skuas, shearwaters, auks and storm-petrels predominate, in Sefton waters is dependent upon weather conditions. The Sefton Coast is rather more important as a site for birdwatchers to enjoy the spectacle they provide than of special importance for the birds themselves.

More significant are the assemblies of Little Gulls in spring and Common Terns in autumn. About 500 Little Gulls arrive in April and remain until May, dividing their time between inshore waters between Crosby and the mouth of the Dee and feeding on chironomid midges over Seaforth and Crosby Marine Lake. This is the only significant spring assembly of Little Gulls in England and the largest in Britain (White, unpublished data).

Post-breeding Common Terns (Plate 3) begin to assemble in the Sefton section of Liverpool Bay from mid-July and are present in numbers until early September. Peak counts vary between 1,000 and 2,000 but probably 5,000 or so move through, feeding over inshore waters and roosting at Formby Point and Seaforth. This, the largest assembly of the species in Britain, attracts birds from all over Britain, Ireland and northern Europe (White, unpublished data).

Breeding birds

A total of 98 species is known to have nested on the Sefton Coast during the past 100 years or so but only 71 currently do so on an annual or near-annual basis (Appendix 3).

The greatest diversity of species breeds on the two nature reserves at Seaforth and Marshside, in part because of a lack of disturbance but more importantly because of the variety of their habitats, including permanent wetlands.

Plate 3. Common Terns once nested on the Formby Dunes but now only breed on rafts at Seaforth. (Photograph by Steve Young).

Ringed Plovers (Plate 4) and less frequently Little Ringed Plovers and Lapwings nest on the less disturbed areas of the beaches. The dunes support good populations of Skylark and Meadow Pipit, along with smaller numbers of several species of national conservation concern, including Grey Partridge, Cuckoo, Reed Bunting and Yellowhammer. Dune scrub supports a number of passerine species, importantly including migrant warblers such as Whitethroat, Lesser Whitethroat, Blackcap and Willow Warbler. The pine woodlands are dominated by very common species such as Woodpigeon, Coal, Blue and Great Tits and Chaffinch, but also support smaller numbers of Buzzard, Woodcock and Tawny Owl, and less regularly Siskin, Lesser Redpoll and Crossbill.

The demise in the 1940s of the dune colony of Common and some Arctic Terns, which numbered up to 900 pairs in the late 19th century, was attributed, despite the attentions of one of the RSPB's first wardening schemes, to a combination of shooting, disturbance and egg-harvesting. Common Terns became re-established in the late 1980s at Seaforth, where the breeding colony of up to 175 pairs is now of international importance, but Arctic Terns, which last attempted to breed on Ainsdale shore in 1962 (White *et al.*, 2008), remained extinct in Sefton until one pair nested unsuccessfully at Seaforth in 2008.

Whinchats were quite common breeders in the dunes until they were lost in 1980, only to be 'replaced' by the closely related Stonechat, which maintains a tenuous hold as a breeding species, mostly between Blundellsands and Ainsdale.

Other species which may have disappeared altogether in recent years are Long-eared Owl, Tree Sparrow and Corn Bunting, although all are still present on neighbouring farmland. Another species apparently heading rapidly towards local extinction is the Cuckoo, while Lesser Redpoll is becoming markedly scarce.

Set against these losses a number of new species have colonised during the past half-century, including

Plate 4: Ringed Plovers are almost the only species that nests on the beaches. As a consequence, their breeding numbers have declined dramatically in recent years due to increased disturbance. (Photograph by Steve Young).

Buzzard, Collared Dove, Green Woodpecker, Reed Warbler and Siskin. Nuthatches look set to become established, if they have not already done so. But most spectacular has been the arrival of the Avocet. Any sighting of Avocet in the north-west of England was a major event throughout the 20th century, but in 2002 two pairs nested at Marshside. This colony has shown steady growth to 27 pairs in 2007 and is now firmly established as a breeding species.

ACKNOWLEDGEMENTS

This paper could not have been produced without the efforts of the vast army of Sefton Coast birdwatchers who have submitted their records for publication in the Lancashire Bird Report over many years, the stalwart volunteers who carry out the monthly Wetland Birds Surveys and the BTO for allowing use of these data. Many thanks also to Steve Young for permission to use his photographs in this paper.

REFERENCES

Cranswick, P.A., Hall, C. and **Smith, L.** (2004) *All Wales Common Scoter survey: report on 2002/03 work programme.* Wetlands Advisory Service report to Countryside Council for Wales. Contract Science Report No. 615.

McCarthy, B. (2001) *Birds of Marshside.* Hobby Publications, Maghull.

White, S.J. (1995-2008) *Lancashire Bird Reports.* Lancashire and Cheshire Fauna Society.

White, S.J., McCarthy, B. and **Jones, M.** (eds.) (2008) *The Birds of Lancashire and North Merseyside.* Southport: Hobby Publications, Lancashire and Cheshire Fauna Society.

Wood, K.W. (1993) 'Birds'. In: Atkinson, D. and Houston, J. (eds.). *The Sand Dunes of the Sefton Coast.* Liverpool: National Museums and Galleries on Merseyside. 94-96.

Plate 5: Shelduck nest in rabbit burrows the length of the coast. Once the young hatch the parents take them to the shore where they feed on small marine snails. (Photograph by Steve Young).

APPENDIX 1
The Sefton Coast Species List
(Category A, B and C species)

Mute Swan *Cygnus olor*
Bewick's Swan *Cygnus columbianus*
Whooper Swan *Cygnus cygnus*
Bean Goose *Anser fabalis*
Pink-footed Goose *Anser brachyrhynchus*
White-fronted Goose *Anser albifrons*
Lesser White-fronted Goose *Anser erythropus*
Greylag Goose *Anser anser*
Snow Goose *Anser caerulescens*
Greater Canada Goose *Branta canadensis*
Barnacle Goose *Branta leucopsis*
Brent Goose *Branta bernicla*
Shelduck *Tadorna tadorna*
Mandarin Duck *Aix galericulata*
Wigeon *Anas penelope*
American Wigeon *Anas americana*
Gadwall *Anas strepera*
Teal *Anas crecca*
Green-winged Teal *Anas carolinensis*
Mallard *Anas platyrhynchos*
Pintail *Anas acuta*
Garganey *Anas querquedula*
Shoveler *Anas clypeata*
Pochard *Aythya ferina*

Ferruginous Duck *Aythya nyroca*
Tufted Duck *Aythya fuligula*
Scaup *Aythya marila*
Eider *Somateria mollissima*
Long-tailed Duck *Clangula hyemalis*
Common Scoter *Melanitta nigra*
Velvet Scoter *Melanitta fusca*
Goldeneye *Bucephala clangula*
Smew *Mergellus albellus*
Red-breasted Merganser *Mergus serrator*
Goosander *Mergus merganser*
Ruddy Duck *Oxyura jamaicensis*
Red-legged Partridge *Alectoris rufa*
Grey Partridge *Perdix perdix*
Pheasant *Phasianus colchicus*
Red-throated Diver *Gavia stellata*
Black-throated Diver *Gavia arctica*
Great Northern Diver *Gavia immer*
Little Grebe *Tachybaptus ruficollis*
Great Crested Grebe *Podiceps cristatus*
Red-necked Grebe *Podiceps grisegena*
Slavonian Grebe *Podiceps auritus*
Black-necked Grebe *Podiceps nigricollis*
Fulmar *Fulmarus glacialis*
Fea's Petrel *Pterodroma feae*
Cory's Shearwater *Calonectris diomedea*
Sooty Shearwater *Puffinus griseus*
Manx Shearwater *Puffinus puffinus*

Balearic Shearwater *Puffinus mauretanicus*
Little Shearwater *Puffinus assimilis*
European Storm-petrel *Hydrobates pelagicus*
Leach's Storm-petrel *Oceanodroma leucorhoa*
Gannet *Morus bassanus*
Cormorant *Phalacrocorax carbo*
Shag *Phalacrocorax aristotelis*
Bittern *Botaurus stellaris*
Night Heron *Nycticorax nycticorax*
Cattle Egret *Bubulcus ibis*
Little Egret *Egretta garzetta*
Great White Egret *Egretta alba*
Grey Heron *Ardea cinerea*
Black Stork *Ciconia nigra*
White Stork *Ciconia ciconia*
Glossy Ibis *Plegadis falcinellus*
Spoonbill *Platalea leucorodia*
Honey Buzzard *Pernis apivorus*
Black Kite *Milvus migrans*
Red Kite *Milvus milvus*
White-tailed Eagle *Haliaeetus albicilla*
Marsh Harrier *Circus aeruginosus*
Hen Harrier *Circus cyaneus*
Montagu's Harrier *Circus pygargus*
Goshawk *Accipiter gentilis*
Sparrowhawk *Accipiter nisus*
Buzzard *Buteo buteo*
Rough-legged Buzzard *Buteo lagopus*
Osprey *Pandion haliaetus*
Kestrel *Falco tinnunculus*
Red-footed Falcon *Falco vespertinus*
Merlin *Falco columbarius*
Hobby *Falco subbuteo*
Eleonora's Falcon *Falco eleonorae*
Peregrine *Falco peregrinus*
Water Rail *Rallus aquaticus*
Spotted Crake *Porzana porzana*
Corncrake *Crex crex*
Moorhen *Gallinula chloropus*
Coot *Fulica atra*
Common Crane *Grus grus*
Oystercatcher *Haematopus ostralegus*
Black-winged Stilt *Himantopus himantopus*
Avocet *Recurvirostra avosetta*
Stone Curlew *Burhinus oedicnemus*

Little Ringed Plover *Charadrius dubius*
Ringed Plover *Charadrius hiaicula*
Kentish Plover *Charadrius alexandrinus*
American Golden Plover *Pluvialis dominica*
Pacific Golden Plover *Pluvialis dominica*
Golden Plover *Pluvialis apricaria*
Grey Plover *Pluvialis squatarola*
Lapwing *Vanellus vanellus*
Knot *Calidris canutus*
Sanderling *Calidris alba*
Little Stint *Calidris minuta*
Temminck's Stint *Calidris temminckii*
White-rumped Sandpiper *Calidris fuscicollis*
Baird's Sandpiper *Calidris bairdii*
Pectoral Sandpiper *Calidris melanotos*
Curlew Sandpiper *Calidris ferruginea*
Purple Sandpiper *Calidris maritima*
Dunlin *Calidris alpina*
Broad-billed Sandpiper *Limicola falcinellus*
Buff-breasted Sandpiper *Tryngites subruficollis*
Ruff *Philomachus pugnax*
Jack Snipe *Lymnocryptes minimus*
Snipe *Gallinago gallinago*
Long-billed Dowitcher *Limnodromus scolopaceus*
Woodcock *Scolopax rusticola*
Black-tailed Godwit *Limosa limosa*
Bar-tailed Godwit *Limosa lapponica*
Whimbrel *Numenius phaeopus*
Curlew *Numenius arquata*
Terek Sandpiper *Xenus cinereus*
Common Sandpiper *Actitis hypoleucos*
Green Sandpiper *Tringa ochropus*
Spotted Redshank *Tringa erythropus*
Greenshank *Tringa nebularia*
Lesser Yellowlegs *Tringa flavipes*
Marsh Sandpiper *Tringa stagnatilis*
Wood Sandpiper *Tringa glareola*
Redshank *Tringa totanus*
Turnstone *Arenaria interpres*
Wilson's Phalarope *Phalaropus tricolor*
Red-necked Phalarope *Phalaropus lobatus*
Grey Phalarope *Phalaropus fulicarius*
Pomarine Skua *Stercorarius pomarinus*
Arctic Skua *Stercorarius parasiticus*
Long-tailed Skua *Stercorarius longicaudus*

Great Skua *Stercorarius skua*
Sabine's Gull *Xema sabini*
Kittiwake *Rissa tridactyla*
Bonaparte's Gull *Chroicocephalus philadelphia*
Black-headed Gull *Chroicocephalus ridibundus*
Little Gull *Hydrocoloeus minutus*
Ross's Gull *Rhodostethia rosea*
Mediterranean Gull *Larus melanocephalus*
Common Gull *Larus canus*
Ring-billed Gull *Larus delawerensis*
Lesser Black-backed Gull *Larus fuscus*
Herring Gull *Larus argentatus*
Yellow-legged Gull *Larus cachinnans*
American Herring Gull *Larus smithsonianus*
Iceland Gull *Larus glaucoides*
Glaucous Gull *Larus hyperboreus*
Great Black-backed Gull *Larus marinus*
Little Tern *Sternula albifrons*
Gull-billed Tern *Gelochelidon nilotica*
Black Tern *Chlidonias niger*
White-winged Black Tern *Chlidonias leucopterus*
Sandwich Tern *Sterna sandvicensis*
Forster's Tern *Sterna forsteri*
Common Tern *Sterna hirundo*
Roseate Tern *Sterna dougallii*
Arctic Tern *Sterna paradisaea*
Guillemot *Uria aalge*
Razorbill *Alca torda*
Black Guillemot *Cepphus grylle*
Little Auk *Alle alle*
Puffin *Fratercula arctica*
Feral Pigeon *Columba livia*
Stock Dove *Columba oenas*
Woodpigeon *Columba palumbus*
Collared Dove *Streptopelia decaocto*
Turtle Dove *Streptopelia turtur*
Ring-necked Parakeet *Psittacula krameri*
Cuckoo *Cuculus canorus*
Barn Owl *Tyto alba*
Little Owl *Athene noctua*
Tawny Owl *Strix aluco*
Long-eared Owl *Asio otus*
Short-eared Owl *Asio flammeus*
Nightjar *Caprimulgus europaeus*
Swift *Apus apus*

Little Swift *Apus affinis*
Kingfisher *Alcedo atthis*
Bee-eater *Merops apiaster*
Roller *Coracias garrulus*
Hoopoe *Upupa epops*
Wryneck *Jynx torquilla*
Green Woodpecker *Picus viridis*
Great Spotted Woodpecker *Dendrocopos major*
Lesser Spotted Woodpecker *Dendrocopos minor*
Skylark *Alauda arvensis*
Shorelark *Eremophila alpestris*
Sand Martin *Riparia riparia*
Swallow *Hirundo rustica*
House Martin *Delichon urbica*
Richard's Pipit *Anthus richardi*
Tree Pipit *Anthus trivialis*
Meadow Pipit *Anthus pratensis*
Red-throated Pipit *Anthus cervinus*
Rock Pipit *Anthus petrosus*
Water Pipit *Anthus spinoletta*
Yellow Wagtail *Motacilla flava*
Grey Wagtail *Motacilla cinerea*
Pied Wagtail *Motacilla alba*
Waxwing *Bombycilla garrulus*
Wren *Troglodytes troglodytes*
Dunnock *Prunella modularis*
Robin *Erithacus rubecula*
Nightingale *Luscinia megarhynchos*
Bluethroat *Luscinia svecica*
Black Redstart *Phoenicurus ochruros*
Redstart *Phoenicurus phoenicurus*
Whinchat *Saxicola rubetra*
Stonechat *Saxicola torquata*
Wheatear *Oenanthe oenanthe*
Pied Wheatear *Oenanthe pleschanka*
Black-eared Wheatear *Oenanthe hispanica*
Ring Ouzel *Turdus torquatus*
Blackbird *Turdus merula*
Fieldfare *Turdus pilaris*
Song Thrush *Turdus philomelos*
Redwing *Turdus iliacus*
Mistle Thrush *Turdus viscivorus*
Grasshopper Warbler *Locustella naevia*
Savi's Warbler *Locustella luscinioides*
Sedge Warbler *Acrocephalus schoenobaenus*

Reed Warbler *Acrocephalus scirpaceus*
Icterine Warbler *Hippolais icterina*
Blackcap *Sylvia atricapilla*
Garden Warbler *Sylvia borin*
Barred Warbler *Sylvia nisoria*
Lesser Whitethroat *Sylvia curruca*
Whitethroat *Sylvia communis*
Sardinian Warbler *Sylvia melanocephala*
Subalpine Warbler *Sylvia cantillans*
Pallas's Warbler *Phylloscopus proregulus*
Yellow-browed Warbler *Phylloscopus inornatus*
Wood Warbler *Phylloscopus sibilatrix*
Chiffchaff *Phylloscopus collybita*
Willow Warbler *Phylloscopus trochilus*
Goldcrest *Regulus regulus*
Firecrest *Regulus ignicapillus*
Spotted Flycatcher *Muscicapa striata*
Red-breasted Flycatcher *Ficedula parva*
Pied Flycatcher *Ficedula hypoleuca*
Bearded Tit *Panurus biarmicus*
Long-tailed Tit *Aegithalos caudatus*
Blue Tit *Cyanistes caeruleus*
Great Tit *Parus major*
Coal Tit *Periparus ater*
Willow Tit *Poecile montanus*
Nuthatch *Sitta europaea*
Treecreeper *Certhia familiaris*
Golden Oriole *Oriolus oriolus*
Red-backed Shrike *Lanius collurio*
Great Grey Shrike *Lanius excubitor*
Woodchat Shrike *Lanius senator*
Jay *Garrulus glandarius*

Magpie *Pica pica*
Chough *Pyrrhocorax pyrrhocorax*
Jackdaw *Corvus monedula*
Rook *Corvus frugilegus*
Carrion Crow *Corvus corone*
Hooded Crow *Corvus cornix*
Raven *Corvus corax*
Starling *Sturnus vulgaris*
Rose-coloured Starling *Sturnus roseus*
House Sparrow *Passer domesticus*
Tree Sparrow *Passer montanus*
Chaffinch *Fringilla coelebs*
Brambling *Fringilla montifringilla*
Serin *Serinus serinus*
Greenfinch *Carduelis chloris*
Goldfinch *Carduelis carduelis*
Siskin *Carduelis spinus*
Linnet *Carduelis cannabina*
Twite *Carduelis flavirostris*
Lesser Redpoll *Carduelis cabaret*
Common Redpoll *Carduelis flammea*
Crossbill *Loxia curvirostra*
Bullfinch *Pyrrhula pyrrhula*
Blackpoll Warbler *Dendroica striata*
Song Sparrow *Melospiza melodia*
White-crowned Sparrow *Zonotrichia leucophrys*
Lapland Bunting *Calcarius lipponicus*
Snow Bunting *Plectrophenax nivalis*
Yellowhammer *Emberiza citrinella*
Ortolan Bunting *Emberiza hortulana*
Reed Bunting *Emberiza schoeniclus*
Corn Bunting *Emberiza calandra*

APPENDIX 2

Species of national and international importance on the Sefton Coast (2001/02 to 2005/06)

Internationally Important	Nationally Important	Annex 1[1] (EU Birds Directive)
Pink-footed Goose	Shelduck (Plate 5)	Red-throated Diver[2]
Wigeon	Pintail	Little Gull
Teal	Shoveler	Common Tern
Oystercatcher	Common Scoter[2]	
Ringed Plover	Cormorant	
Grey Plover	Curlew	
Knot	Black-headed Gull[3]	
Sanderling	Common Gull[3]	
Dunlin	Herring Gull[3]	
Black-tailed Godwit	Lesser Black-backed Gull[3]	
Bar-tailed Godwit	Great Black-backed Gull[3]	
Redshank		

1. Species on Annex 1 of the Birds Directive are of special conservation concern in Europe and these nationally important populations are regarded as internationally important.

2. Species which occur entirely offshore.

3. Gulls are poorly monitored by WeBS and their numbers fluctuate widely between years with several species occurring in internationally important numbers in some winters; this assessment is based upon a 'guesstimate' of average numbers.

APPENDIX 3

Breeding species of the Sefton Coast in the 20th and 21st centuries

(Species in bold currently nest on a more or less annual basis; those in italics are now extinct on the Sefton Coast)

Mute Swan
Greylag Goose
Greater Canada Goose
Shelduck
Gadwall
Teal
Mallard
Garganey
Shoveler
Tufted Duck
Ruddy Duck
Grey Partridge
Red-legged Partridge
Pheasant
Little Grebe
Sparrowhawk
Buzzard
Kestrel
Moorhen
Coot
Oystercatcher
Avocet
Little Ringed Plover
Ringed Plover
Lapwing
Ruff
Snipe
Woodcock
Redshank
Black-headed Gull
Little Tern
Sandwich Tern
Common Tern

Arctic Tern
Feral Pigeon
Stock Dove
Woodpigeon
Collared Dove
Turtle Dove
Cuckoo
Barn Owl
Little Owl
Tawny Owl
Long-eared Owl
Nightjar
Green Woodpecker
Great Spotted Woodpecker
Skylark
Swallow
Meadow Pipit
Yellow Wagtail
Grey Wagtail
Pied Wagtail
Wren
Dunnock
Robin
Whinchat
Stonechat
Blackbird
Song Thrush
Mistle Thrush
Grasshopper Warbler
Sedge Warbler
Reed Warbler
Lesser Whitethroat
Whitethroat

Garden Warbler
Blackcap
Wood Warbler
Chiffchaff
Willow Warbler
Goldcrest
Spotted Flycatcher
Long-tailed Tit
Coal Tit
Blue Tit
Great Tit
Willow Tit
Treecreeper
Nuthatch
Jay
Magpie
Jackdaw
Carrion Crow
Starling
House Sparrow
Tree Sparrow
Chaffinch
Greenfinch
Goldfinch
Linnet
Lesser Redpoll
Siskin
Crossbill
Bullfinch
Yellowhammer
Reed Bunting
Corn Bunting

Dragonflies (Odonata)

Philip H. Smith

9 Hayward Court, Watchyard Lane, Formby L37 3QP
philsmith1941@tiscali.co.uk

Hall (1993) lists 14 species of Odonata for the Sefton Coast, 10 of which had bred. By 2008, this total had increased by 43% to 20 species (Table 1). The newcomers mainly fall into two categories: those with southern and eastern distributions in England which have moved north (*Aeshna mixta; Orthetrum cancellatum*) and long-distance migrants from continental Europe (*Anax parthenope; Sympetrum flaveolum; S. fonscolombii*) (Plate 1). These five seem to have been strongly influenced by warm summers, especially those of 1995, 1999, 2003 and 2006. The sixth addition, *Calopteryx splendens*, fits less easily into these groups. Although this insect has greatly extended its range in north-west England since the 1980s (Smith, 1998a), this may be as much to do with improving water-quality in the slow-moving rivers and streams in which it normally breeds, as to increasing temperatures. It is now abundant on Downholland Brook, just east of Formby, so the single duneland record at Sands Lake, Ainsdale, on 15-16th July 2006 was not entirely unexpected. More surprising is the regular occurrence of the peatland-breeding dragonfly *Sympetrum danae*, with major influxes in 1991, 2004 and 2008. However, this species is now known to disperse widely in late summer when it may appear in atypical habitats (Smith, 1998b).

Table 1. Check-list of dragonflies and damselflies recorded in the Sefton Coast sand-dune system up to 2008.

Species	English name	Status
Calopteryx splendens	Banded Demoiselle	R
Lestes sponsa	Emerald Damselfly	CB
Pyrrhosoma nymphula	Large Red Damselfly	R
Ischnura elegans	Blue-tailed Damselfly	CB
Enallagma cyathigerum	Common Blue Damselfly	CB
Coenagrion puella	Azure Damselfly	CB
Aeshna grandis	Brown Hawker	SB
Aeshna mixta	Migrant Hawker	CB
Aeshna juncea	Common Hawker	RB
Aeshna cyanea	Southern Hawker	RB
Anax imperator	Emperor	CB
Anax parthenope	Southern Emperor	V
Libellula depressa	Broad-bodied Chaser	RB
Libellula quadrimaculata	Four-spotted Chaser	CB
Orthetrum cancellatum	Black-tailed Skimmer	S
Sympetrum striolatum	Common Darter	CB
Sympetrum sanguineum	Ruddy Darter	SB
Sympetrum flaveolum	Yellow-winged Darter	V
Sympetrum fonscolombii	Red-veined Darter	V
Sympetrum danae	Black Darter	S

C = Common; S = Scarce; R = Rare; V = Vagrant; B = Breeder

Of the 20 dune-coast Odonata, 13 have probably bred, 11 of them regularly, *Aeshna mixta*, *Libellula depressa* and *Sympetrum sanguineum* being added to those cited as breeders by Hall (1993). The current Joint Nature Conservation Committee guidelines for north-west England state that any site supporting at least 11 regularly breeding dragonflies should be considered nationally important (Nature Conservancy Council 1989). French and Smallshire (2008) have recently proposed an updated methodology for determining "Key Odonata Sites" in Great Britain, taking into account nationally and regionally important taxa, species diversity, abundance and proof of breeding. Assessment based on these criteria suggests that the Sefton Coast dune system qualifies as a Key Site, *S. sanguineum* being our most notable regularly breeding species (Smith, 1997).

As mentioned by Hall (1993), the Sefton Coast Odonata rely for breeding almost entirely on artificial water-bodies, in particular scrapes and ponds dug in the dunes for conservation purposes during the 1970s and 1980s, mostly as breeding sites for Natterjack Toads (*Epidalea calamita*). There are about 50 of these ponds, scattered throughout the dune system, the best examples being in the central part of Ainsdale Sand Dunes National Nature Reserve (NNR) and within Birkdale Sandhills Local Nature Reserve (LNR). All have shelving margins and have become colonised by extensive stands of marginal, submerged and floating aquatic plants. During his detailed study of the Birkdale LNR ponds, Smith (2001) showed that the "best" sites for Odonata were characterised by permanent water, some form of shelter from prevailing winds, the presence of diverse aquatic vegetation, at least 50% open water and a relatively long margin.

Other important duneland sites for dragonflies are Sands Lake, Ainsdale-on-Sea, and Pinfold Pond at Ainsdale Sand Dunes NNR. The large Wicks Lake at Formby Point was excavated in the late 1970s as a recreational feature but later became a small nature reserve. By 1995, it was described as "prime dragonfly habitat" and was featured by Brooks and Lewington (1997). Unfortunately, this site has deteriorated in recent years due to the presence of large numbers of ducks that are fed by members of the public. These pollute the water and presumably

Plate 1. Sympetrum flaveolum, Birkdale. (Photograph by Philip Smith).

predate dragonfly nymphs. Few Odonata now occur there. However, the recent excavation of about 20, mostly small, ponds on Freshfield Dune Heath Nature Reserve provides further habitat for these insects, while several species have colonised the slack habitat on Birkdale Green Beach since 2000.

Although most Odonata on the Sefton Coast have a favourable conservation status at present, many of the older scrapes are now becoming too overgrown with emergent plants and may require management work, within the context of the Sefton Coast Nature Conservation Strategy, to retain a balance between open water and vegetation. Paradoxically, some of the Birkdale LNR ponds, which are visited daily by large numbers of dog-walkers, suffer from the opposite condition, having heavily trampled margins and a lack of aquatic vegetation. Birkdale Green Beach provides potentially extensive new habitat for Odonata but the wetlands here tend to dry up in mid-summer, unless rainfall is exceptional, thereby reducing breeding success.

REFERENCES

Brooks, S. and **Lewington, R.** (1997) *Field Guide to the Dragonflies and Damselflies of Great Britain and Ireland.* Hook, Hampshire: British Wildlife Publishing.

French, G. and **Smallshire, D.** (2008) Criteria for determining key Odonata sites in Great Britain. *Journal of the British Dragonfly Society.* 24(2) 54-61.

Hall, R.A. (1993) Dragonflies (Odonata). In: Atkinson, D. and Houston, J. (eds.). *The Sand Dunes of the Sefton Coast.* Liverpool: National Museums & Galleries on Merseyside. pp. 106-107.

Nature Conservancy Council (1989) *Guidelines for selection of biological SSSIs.* Peterborough: Nature Conservancy Council.

Smith, P.H. (1997) The Ruddy Darter *Sympetrum sanguineum* (Muller) in South Lancashire. *Journal of the British Dragonfly Society.* 13 (1) 27-29.

Smith, P.H. (1998a) The dragonflies of Lancashire and north Merseyside: an introduction to their distribution and status. *Lancashire Bird Report.* 1998 74-77.

Smith, P.H. (1998b) Dispersion or migration of *Sympetrum danae* (Sulzer) in South Lancashire. *Journal of the British Dragonfly Society.* 14 (1) 12-14.

Smith, P.H. (2001) Diversity of dragonflies in dune ponds at Birkdale Sandhills, Sefton Coast, Merseyside. *Journal of the British Dragonfly Society.* 17(1) 1-12.

Dune slacks on the Sefton Coast

Sally E. Edmondson

Department of Geography, Liverpool Hope University, Hope Park, Liverpool L16 9JD
edmonds@hope.ac.uk

ABSTRACT

The dune slacks of the Sefton Coast represent a nationally and internationally important biodiversity resource. In the light of current research elsewhere on dune slack dynamics, and the availability of a significant amount of data on the Sefton slacks, further research may significantly improve our understanding of dune slack vegetation dynamics. Current trends and predictions point to declining quality and quantity of the dune slack resource due to lack of dynamics, accelerated succession and climate change, with the notable exception of the accreting dune/dune slack system on the seaward edge of the dunes between Ainsdale and Birkdale. In the light of these trends, reflected nationally and internationally, it is suggested that large-scale dune re-mobilisation projects in a wider strategic framework represent the most viable option for sustainable management of both the dune slack resource and the dune system as a whole.

INTRODUCTION

This paper aims to review the dune slack resource in Sefton, identify trends important for nature conservation, and highlight some research and management options.

Description of the resource

The dune slacks of the Sefton coast are highly significant for biodiversity, being the habitat of a number of rare animals and plants. Humid dune slacks are an Annex 1 priority habitat in the EU Habitats and Species Directive. Sefton dune slacks have three Annex II species (Great Crested Newt *Triturus cristatus*, Natterjack Toad *Epidalea calamita* and Petalwort *Petalophyllum ralfsii*) and a number of nationally and regionally important species. Two hundred and fifty five vascular plant taxa are listed as occurring in dune slacks and scrapes in the Sefton Coast Partnership Area (Smith, 2005), representing 22% of the total list. Of these, there are 47 Species of Conservation Importance in North West England, 7 taxa that are Nationally Rare and 5 Nationally Scarce. The slacks also have a high diversity of lower plants including British Red Data Book listed mosses *Bryum warneum* and *B. neodamense* (Church *et al.*, 2004) and Lesser bearded stonewort *Chara curta* (Stewart and Church, 1992).

The Sefton Coast is a large hindshore dune system that, in past phases of geomorphological dynamism, has developed numerous, large, diverse dune slacks, representing 33% of the English dune slack resource (Radley, 1994; Gateley and Michell, 2004). The data collated by Edmondson *et al.* (2007) indicates a minimum of 114 ha of dune slacks, scrapes and open water on the dunes comprising 352 individual sites (Table 1), but this is an underestimate. Gateley and Michell (2004) define 152 hectares of National Vegetation Classification (NVC) slack vegetation types (see Table 2), all five types being present. The greatest number of slacks is in the wide blocks of open dune at the northern end of the system at Ainsdale and Birkdale. Cabin Hill National Nature Reserve (NNR), Altcar Rifle Range and Ravenmeols Local Nature Reserve (LNR) have an important number of slacks that are significant for nature conservation, but mostly result from human's modification of the dunescape, largely by sand-winning and coast defence works. The larger slacks are also north of Formby Point (Figure 1), mostly being secondary (formed by erosion of dunes down to the water table), aligned with the dominant westerly winds. The notable exception is the seaward edge of the dunes between Birkdale and Ainsdale (known locally as Birkdale Green Beach), where approximately 30ha of primary dune slack (formed by an area of beach plain being enclosed by a seaward new dune ridge) occurs parallel to the coast, transitional with the saltmarsh coast

Table 1. Number and size of dune wetlands (dune slacks, scrapes and open water) on the Sefton Coast. All figures are hectares. The wetlands included are those identified and located by the GIS layer in Edmondson et al. (2007). Note: the 14 management compartments vary significantly in area; they are arranged in geographical order from north to south.

Site	No. of slacks	Total slack area	Mean slack size	Standard deviation	Max.	Min.
Queens Jubilee Nature Trail	5	0.58	0.11	0.03	0.15	0.08
Birkdale frontal dunes	44	35.43	0.81	4.51	30	0.0001
Royal Birkdale Golf Course	16	2.06	0.13	0.33	30	0.0001
Hillside Golf Course	2	0.16	0.05	0.081	0.0001	0.26
Southport & Ainsdale Golf Course	2	0.03	0.02	0.01	0.02	0.01
Birkdale Sandhills LNR	25	20.87	0.83	0.70	2.25	0.03
Ainsdale Sandhills LNR	42	8.00	0.19	0.35	2.22	0.001
Ainsdale Sand Dunes NNR	140	35.68	0.25	0.39	2.47	0.01
Formby Golf Course	5	0.91	0.18	0.18	0.49	0.06
National Trust Formby Point	7	0.16	0.02	0.01	0.04	0.01
Lifeboat Road	9	0.16	0.22	0.01	0.04	0.01
Ravenmeols LNR	10	1.73	0.17	0.33	1.08	0.01
Altcar Rifle Range, & Cabin Hill NNR	29	5.68	0.20	0.40	0.49	0.06
Hightown, & West Lancs. Golf Course	16	2.04	0.13	0.16	0.52	0.01
TOTAL/OVERALL	**352**	**113.80**	**0.32**	**1.64**	**30.00**	**0.001**

to the north. These relatively natural landscapes of dune ridges and slacks, especially in the Birkdale frontal dunes where slack forming processes are currently operating, are also significant for geoconservation.

The numbers and areas of dune slacks shown in Table 1 are largely based on biological and hydrological factors. Many dune slack areas have developed by succession and infilling by blown sand to fixed dune habitats, these being geomorphologically slacks by origin.

Although this review focuses on dune slacks, they are an integral part of the dune system as a whole, geomorphologically, hydrologically and ecologically, and should not be considered in isolation.

Information about Sefton Coast dune slacks

The recent review of dune wetlands on the coast (Edmondson et al., 2007) demonstrates that a significant amount of information, spanning over fifty years, is available on Sefton dune slacks. There is also an important set of oblique aerial photographs taken in March 1995 that show the dune slacks flooded at the highest water table conditions in recent years.

Comparing the 1988 (Edmondson et al., 1988/89) and 2004 NVC surveys, Gateley and Michell (2004) highlighted some significant losses of slack habitat, suggesting that they result from processes such as scrub encroachment, nitrogen enrichment, slack infilling and also the lack of new slack formation. A negative trend seems clear, although the figures may be inaccurate because of differences in methodology between the two surveys (Edmondson et al., 2007). This comparison shows the importance of using available surveillance data to identify trends, and highlights the value of the intensive recording and reporting by P.H. Smith (ongoing) on the Sefton Coast, much in the form of

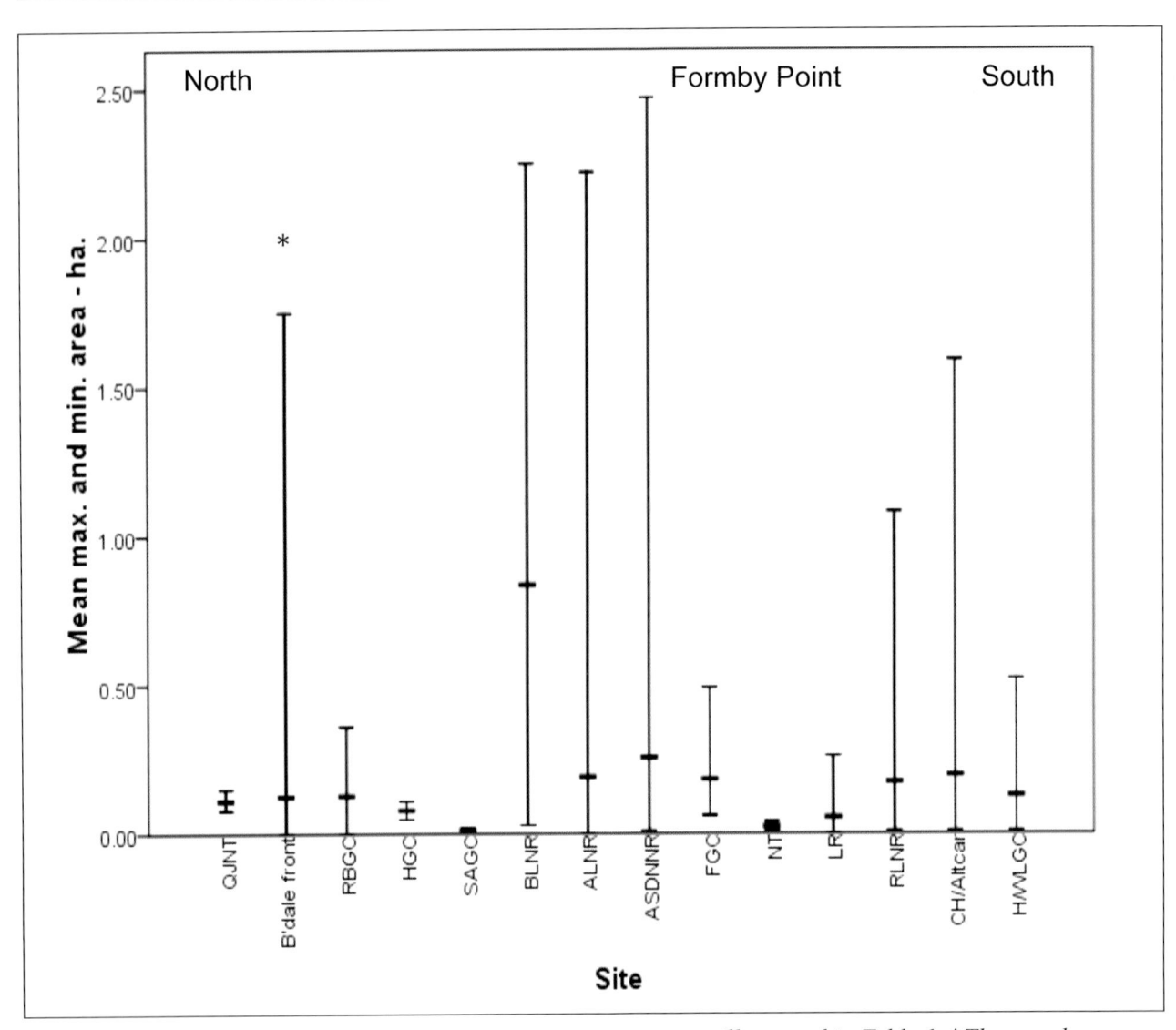

*Figure 1. Size of dune slacks in the 14 management compartments illustrated in Table 1. *The very large primary dune slack on the Birkdale frontal dunes (approximately 30 hectares), the largest slack on the coast from these data.*

unpublished reports. This includes a detailed study of the ecological value of dune wetlands in 1978 (Smith, 1978), a review of change over twenty years in the Birkdale frontal dune slacks (Smith, 2006), and ongoing surveillance of the development of the primary dune slack system between Ainsdale and Birkdale. Good records are maintained and reported on the populations and breeding success of Natterjack Toads (see Skelcher, 1996), but no significant systematic effort has recorded invertebrates from dune slacks.

The monthly record of water table levels in a series of 11 boreholes and a pond stage post, first established in February 1972 is one of the most valuable data-sets nationally for dune slacks (see Figure 2). Some additional boreholes have been added since the original eleven in 1972. The data-set has been referred to, and used, by several workers including current modelling of water-table response to future climate scenarios by Clarke and Sanitwong-Na-Ayutthaya at the University of Southampton. The boreholes were surveyed to Ordnance Datum by Edmondson in 1973, these levels being used later by Clarke (1980). Despite two GPS surveys in recent years, there is still concern about the accuracy of the levels of this important set of boreholes.

Ecological and hydrological issues for research

Davy *et al.'s* (2006) report on the eco-hydrology of dune habitats aimed to provide a foundation for further

Table 2. *Areas of NVC dune slack communities on the Sefton Coast relative to the total national resources. Sources: Radley (1994) and Gateley and Michell (2007) as reviewed in Edmondson et al. (2007).*

NVC dune-slack communities	Area, England	Area, Sefton Coast	% in Sefton
SD13 Sagina nodosa – Bryum pseudotriquetrum	11.3	8.7	77.0
SD14 Salix repens – Campylium stellatum	34.5	1.6	4.6
SD15 Salix repens – Calliergon cuspidatum	135.0	68.7	50.9
SD16 Salix repens – Holcus lanatus	227.4	68.6	30.2
SD17 Potentilla anserina – Carex nigra	56.7	4.0	6.9
Total	**465.8**	**151.6**	**32.5**

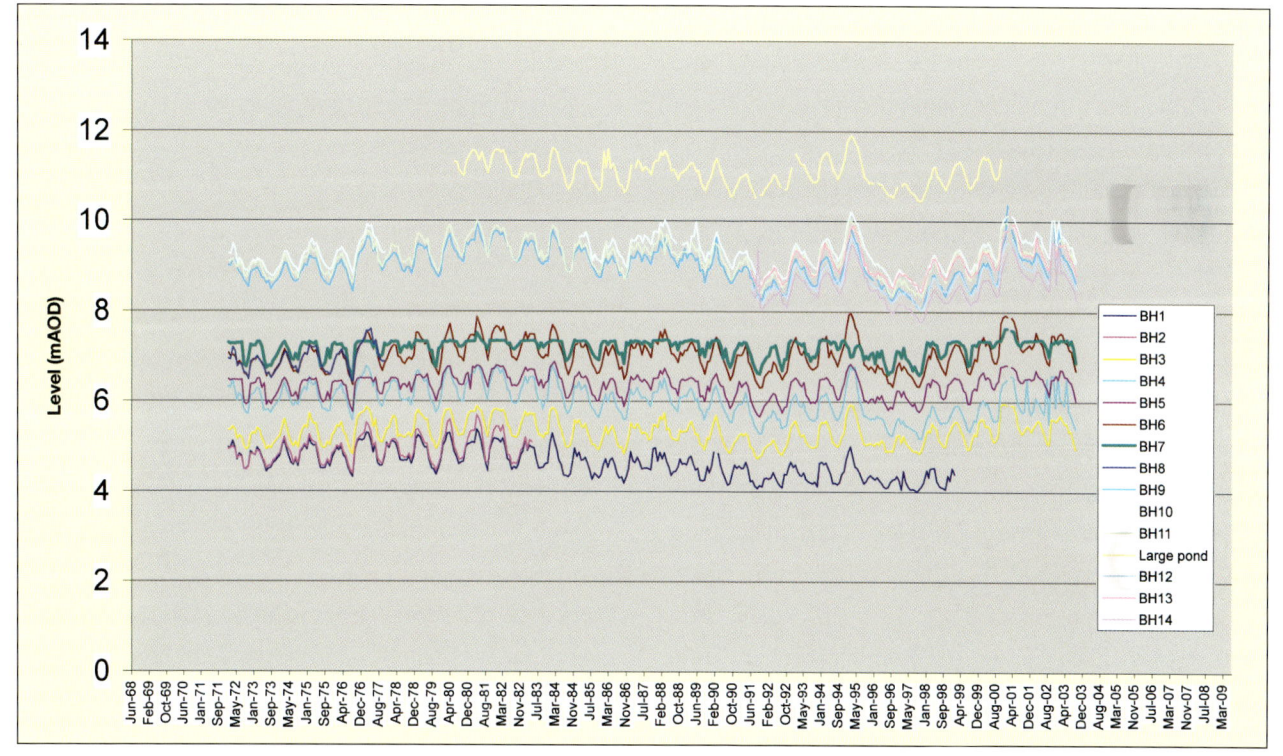

Figure 2. *Fluctuation of the water table in Ainsdale Sand Dunes National Nature Reserve Fluctuations are in metres above Ordnance Datum (y axis). Data: Natural England; diagram courtesy of Derek Clarke, University of Southampton.*

research on dune slacks in the UK. They propose a model for the classification of slack types (A – E) but there are some problems in applying this to the Sefton Coast, as the model assumes the dune aquifer to be in contact with an underlying, regional aquifer. Clarke (1980) concludes that this is not the case in Sefton as the sand is underlain by an impermeable layer. It is possible, however, to recognise attributes of slack types A, B, C and E on the Sefton dunes (Type D is at the boundary between the dune system and the inland area; there are no dune slacks in this zone in Sefton where the landward edge of the dune system is largely developed).

Type A is a young dune slack in the dynamic frontal dunes subject to mixing of fresh and saline groundwater. The large primary dune slack at Birkdale Green Beach is a possible example. The saline influence however, as far as is known, is largely from tidal inundation. Nothing is known of the dynamics of the dune water table and its interaction with the underlying saline water here. This area would appear to have some similarities in origin to Massam's Slack on Ainsdale Sand Dunes NNR, for which there is a long run of vegetation data, thus possibly providing the opportunity to build a model of vegetation development in large primary dune slacks.

Type B is a slack fed by precipitation input to the dunes, and where groundwater flow is directed towards the slack and water lost by evapotranspiration. It is assumed that most of the dune slacks in Sefton are of this type.

Type C is a slack where groundwater flows into the slack at the end with a higher water table, then flows through the slack (where some may be lost as transpiration) and exits at the down-gradient edge into the groundwater flow system. This is assumed to be the case in Slack 65 on Ainsdale Sand Dunes NNR (described in Edmondson et al., 2007 and illustrated by borehole 7 in Fig. 2), and possibly some other slacks. Payne's (1984) 'wet bryophyte slack' appears to be of this type. The case studies in Edmondson et al. (2007) identified a number of slacks where a landward – seaward vegetation zonation, or at least difference, is identifiable in the slacks, vegetation more typical of wetter conditions being found at the seaward end of the slacks. In addition to Slack 65 at Ainsdale, this can also be seen in Slack 11, and to certain extent Slack 18 on Birkdale Hills LNR. This pattern could result from the seaward flow of water as described above, held up at the seaward end by the physical barrier of the coastal road. Equally, the pattern could result from the seaward end of the slack being topographically lower as a result of the dynamics of sand movement at the time of slack formation.

Type E is a slack topographically high on the dune system kept damp by capillary water and only rarely flooding in very wet years. Clarke (1980) demonstrates the shape of the water table in Sefton using the water table data from Ainsdale Sand Dunes NNR. This shows the seaward half of a typical dome-shaped dune water table as described for example by Willis et al. (1959), thus it can be assumed that damp/dry dune slacks situated landward of the remaining dune areas will be of this type. An example would be Slack J on the Royal Birkdale Golf Course.

Davy et al. (2006) do not attempt to link the NVC slack vegetation communities to the differing hydrological conditions identified in slack types A – E. They suggest varying local hydrological dynamics will impact on the hydrochemistry and thus the vegetation. Work by Adema and others in The Netherlands (e.g. Adema and Grootjans, 2003) indicates links between the hydrology and successional sequences of dune slacks, due to factors such as Radial Oxygen Loss by early successional species of wet dune slacks affecting the availability of nitrate. There are also suggestions that these factors control stable states in some slack vegetation types. These findings have not been applied to understanding vegetation and vegetation change, and thus management decisions in Sefton. The detailed understanding of temporal and spatial variation in water chemistry and groundwater fluctuation provided by Jones et al. (2006) for Merthyr Mawr should also be taken in account in any future work in Sefton. The long run of high quality, monthly water-table data on Ainsdale Sand Dunes NNR provides a sound hydrological framework for any retrospective use of vegetation data in this context.

Davy et al. (2006) also refer to an 80 year cycle of Salix repens, and to the pivotal importance of this species in slack vegetation dynamics. This was also highlighted by the research of Edmondson (1991) on Ainsdale Sand Dunes NNR. Payne (1984) refers to the potential importance of disease in controlling the species and comments on the form of Salix repens in differing hydrological conditions. Better understanding of the role of this species, with respect to ecohydrology in particular, would be helpful. Davy et al. make only brief reference to the significance of inundation period, which has been thought by some workers on the Sefton Coast (e.g. Blanchard, 1952; Payne, 1984) to be significant for vegetation dynamics and the form of Salix repens. Based on work of Lammerts et al. (2001), Davy et al. suggest that local hydrological base level, seepage conductance and yield of water in low groundwater conditions may be more important.

A physically-based model of water-table balance in the dune system was produced by Clarke (1980). Using the Ainsdale borehole data, Clarke and Sanitwong-Na-Ayutthaya (2007) have now developed a model simulating daily water-balance that shows good agreement with measured fluctuations between 1972 and 2000. Running the model to 2080, using future weather data from the UK Climate Impacts Programme, suggests that there will be short sequences (3-4 years) of exceptionally wet years but also several long sequences (over 10 years) of very low water-table. Overall, average water-table is expected to fall by 1.1 – 1.3m by 2080. Without dune dynamics to excavate new slacks to this lower base level, this prediction forecasts significant losses of dune slack biodiversity. Erosion to the water table is restricted in Sefton to a few small patches in the frontal dunes, the only major deflation hollow being Devil's Hole on Ravenmeols LNR (Read 1995). This context highlights the significance of the dynamic, accreting situation at Birkdale, where primary slack formation is providing the only major new dune slack habitat on the coast, already a 'biodiversity hotspot' (Smith, 2007).

The persistence of some single-species dominated stands of vegetation is shown by the long data run in some Sefton Coast slacks. This is especially true of tall swamp/wet slack species, for example *Carex acuta,* and *Juncus* spp. in Massams Slack and *Carex acutiformis* in Slack 18 on Birkdale LNR. Blanchard (1952) maps these stands in Massams slack on Ainsdale Sand Dunes NNR, and the same patches are mapped by Robinson in 1971. Blanchard suggests that the stands were not present at the early stages of vegetation establishment in Massams, but does note their persistence once established. Further understanding of the ecological mechanisms and environmental controls controlling the establishment and persistence of these patches may give further insights into slack vegetation dynamics.

Increasing senescence of the dunes, with successional processes operating without dynamics to produce early seral stages, has been a concern of dune managers in Northwest Europe since the late 1980s. In Sefton, the grazing schemes introduced at Ainsdale in the 1990s (Simpson *et al.*, 2001) have controlled the biomass of competitive species in many dune slacks, but do not 'restart' succession. Although the general impacts of grazing in terms of vegetation structure and composition are well known, grazing exclosures (Edmondson, 1991) installed in four dune slacks on Ainsdale Sand Dunes NNR in 1974 would allow detailed study, for example of nutrient status, or the existence of alternating stable states, for sites whose detailed vegetation structure was known in 1974. In addition, these slacks are of variable hydrology with respect to Davy *et al.*'s (2006) classification.

The Sefton dunes experienced the enhanced successional trends observed widely after the 1950s with the arrival of myxomatosis (Edmondson *et al.*, 1993; Rhind *et al.,* 2001). There is a valuable record of change between the early 1950s and 1971 (Blanchard, 1971) that highlights this overall trend. Models of slack succession have been proposed by Blanchard (1952), Payne (1984) and Edmondson *et al.* (1993). It would be useful to review these models in the light of new research, including the eco-hydrological work cited in Davy *et al.* (2006). Blanchard's model proposes a successional series of bare wet sand through to acid heath. The origins of dune heath on the Sefton Coast have been discussed by Edmondson and Gateley (1996) who conclude that the important areas of dune heath on Southport and Ainsdale Golf Course (an unusual area in being one of the few remnant landward patches of dunes) are the only significant areas of primary dune heath that remain on the Sefton Coast, developed in old dune slacks.

In addition to the lack of grazing and dune dynamics, enhanced nitrogen deposition drives change towards more mesotrophic vegetation types and stability, a factor highlighted by Jones *et al.* (2004).

Payne (1984) notes accelerated succession as an edge effect around large slacks associated with species such as *Arrhenatherum elatius, Holcus lanatus,* and *Dactylis glomerata.* Edmondson (1991) also reports this effect, associated with tall *Salix repens* and *Rubus* spp. This zone of enhanced growth was first described in slacks near Southport by Salisbury (1952) who attributed it to the drift of *Salix repens* leaves to the edge of slacks during flooded periods

Whereas scrub clearance programmes have significantly improved the conservation status of many dune slacks, for example Ainsdale LNR and NNR (Simpson *et al.*, 2001), large and increasing areas of scrub persist in many of the slacks on Birkdale Sandhills LNR. Although the conservation value of dune scrub, and specifically wet dune scrub, has not been systematically reviewed either locally or nationally, the loss of biodiversity of earlier seral communities on a senescing dune system is cause for concern.

The Open Dune Restoration Project area of Ainsdale Sand Dunes NNR demonstrates the potential for recovery of dune slacks following Pine woodland clearance (see Edmondson, this volume). These slacks were species listed in 1960 and 1971 (Robinson 1971), Edmondson in 1974, and were mapped in both NVC surveys (1988 and 2003), thus data are available for further retrospective evaluation of this project.

The scrapes dug in dune slacks largely for Natterjack Toad breeding, in addition to hollows created by sand winning and coast defence works, have rapidly developed the specialised vegetation communities of early dune slack succession (see Edmondson *et al.*, 1993) that have been lost from many areas of the dunes. This small-scale, short-term, piecemeal approach to dune slack, or whole dune system, conservation is not, however, sustainable. A more long-term view and a wider geographical perspective is appropriate for management decisions, based on whole dune systems, not just dune slacks. It is suggested that large-scale re-mobilisation projects, set within this wider context, are the most viable option for sustainable management of dune biodiversity.

With appropriate caution due to the differing data types and resolutions, data available for the Sefton Coast could potentially reveal some important insights into vegetation change in the slacks, at the vegetation composition level, the distribution of stand type level, and also the vegetation structure as indicated by aerial and ground-based photographs. The quadrat data is suitable for analysis using Ellenberg values (Hill *et al.*, 1999) to indicate past environmental variables such as pH, wetness and fertility, and also for functional analysis of the species present based on Grime's (2001) triangular Competitor - Stress Tolerator - Ruderal (C-S-R) model. Some meaningful analysis may also be derived from slack species lists, using the slack itself as a 'quadrat'.

CONCLUSIONS

The Sefton Coast dune slacks are an important resource for nature conservation in both area and biodiversity, and in national and international contexts. There are important data-sets from the Sefton Coast dune slacks dating back over fifty years that could be used to gain a clearer understanding of their eco-hydrology for application to management. Davy *et al.* (2006) and other recent research provides an excellent context for such work.

Even with the amelioration in condition arising from grazing programmes on some sites, the dune slacks are increasingly successionally mature, a trend that will be exacerbated by enhanced atmospheric nitrogen deposition. Large areas of scrub occur on some sites, replacing the more diverse and specialized communities of young wet slacks. Without large scale dune dynamics naturally creating new dune slacks, and with the forecasted drop in water table levels due to climate change, the sustainability of the slack biodiversity is uncertain. In this context, the Green Beach at Birkdale is highly significant, being unique in its current and increasing biodiversity, and in being the only large, new incipient dune slack on the coast. Past excavation management projects have demonstrated the potential to create and/or ameliorate dune slack habitat, but are piecemeal in approach and not sustainable. A more appropriate approach to dune slack and whole dune system conservation would result from larger-scale re-mobilisation projects in a strategic national framework.

ACKNOWLEDGEMENTS

Phil Smith and Anna Cunningham were co-workers on the recent review of data on dune wetlands on the Sefton Coast from which this paper has arisen. Thank you also to Dr Derek Clarke for permission to use Figure 2.

REFERENCES

Adema, E.B. and **Grootjans, A.P.** (2003) 'Possible positive-feedback mechanisms: abiotic soil parameters in wet calcareous dune slacks.' *Plant Ecology* 167(1), 141-149.

Blanchard, B. (1952) *An ecological survey of the sand dune system of the south west Lancashire coast, with special reference to an associated marsh flora.* Unpublished PhD thesis, University of Liverpool.

Blanchard, B. (1971) *Unpublished report of a visit to Ainsdale*, 1971 ASDNNR.

Butcher, D., Gateley, P., Newton, M. and **Sinnott, D.** (1999-2002) *Royal Birkdale Golf Course Survey of Dune Slacks: Vegetation and Topography.* Unpublished report. Preston: University of Central Lancashire.

Church, J.M., Hodgetts, N.G., Preston, C.D. and **Stewart, N.F.** (2004) *British Red Data Books: mosses and liverworts.* Reprinted 2004. Peterborough: Joint Nature Conservation Committee.

Clarke, D. and **Sanitwong-Na-Ayutthaya, S.** (2007) *Water balance at Ainsdale National Nature Reserve 1972 – 2100.* Report of the Sefton Coast Partnership Research Meeting 14th July 2006, Edge Hill University.

Clarke, D. (1980) *Groundwater Balance of a Coastal Sand Dune System. A study of the water conditions in Ainsdale Sand Dunes National Nature Reserve, Merseyside.* PhD Thesis, University of Liverpool .

Davy, A.J., Grootjans, A.P., Hiscock, K. and **Petersen, J.** (2006) *Development of eco-hydrological guidelines for dune habitats – Phase 1.* English Nature Research Reports No. 696. Peterborough: English Nature.

Edmondson, S., Smith, P.H. and **Cunningham, A.** (2007) *Analysis of Current Information on the Dune Wetlands of the Sefton Coast. Natural England Project Reference MAR09-03-004.* Liverpool: Environment Agency, Liverpool Hope University, Natural England, Sefton Council.

Edmondson, S.E. (1974) *Vegetation Survey (Slacks) 1974 – repetition of N. Robinson's Vegetation Survey*: August 1974. (Unpublished Data, SE).

Edmondson, S.E. (1991) *Temporal and spatial variation in dune slack vegetation at Ainsdale, Merseyside.* M.Sc. (by research) thesis, University of Liverpool.

Edmondson, S.E. and **Gateley, P.S.** (1996) Dune Heath on the Sefton Coast Sand Dune System, Merseyside UK. In: P.S. Jones, M.G. Healy, and A.T. Williams (eds.) *Studies in European Coastal Management.* Cardigan: Samara Publishing Ltd / European Union for Coastal Conservation.

Edmondson, S.E. Gateley, P.S., Rooney, P.J. and **Sturgess, P.W.** (1993) Plant Communities and Succession. In: D. Atkinson and J. Houston (eds.) *The Sand Dunes of the Sefton Coast.* Liverpool: National Museums and Galleries on Merseyside. 65-84.

Edmondson, S.E., Gateley, P.S. and **Nissenbaum, D.A.** (1988/89) *National Sand Dune Vegetation Survey: Sefton Coast, Merseyside.* Report no. SC:NVC:88:00. Nature Conservancy Council.

Gateley, P.S. and **Michell, P.E.** (2004) *Sand Dune Survey of the Sefton Coast*, 2003/4. TEP.

Grime, J.P. (2001) *Plant strategies, vegetation processes and ecosystem properties.* Second edition. John Wiley and Sons, Chichester.

Hill, M.O., Mountford, J.O., Roy, D.B. and **Bunce, R.G.H.** (1999) Ellenberg's indicator values for British Plants. *ECOFACT 2a Technical Annex.* London: Detr/ CEH. HMSO.

Jones, M.L.M., Reynolds, B., Brittain, S.A., Norris, D.A., Rhind, P.M. and **Jones, R.E.** (2006) Complex hydrological controls on wet dune slacks. The importance of local variability. *Science of the Total Environment.* 372, 266-277.

Jones, M.L.M., Wallace, H.L. Norris, D.A. Brittain, S.A., Haria, S., Jones, R.E., Rhind, P.M., Reynolds, B and **Emmett, B.A.** (2004) Changes in Vegetation and Soil Characteristics in Coastal Sand Dunes along a Gradient of Atmospheric Nitrogen Deposition. *Plant Biology.* 6, 598-605.

Lammerts, E.J., Maas, C. and **Grootjans, A.P.** (2001) Ground water variables and vegetation in dune slacks *Ecological Engineering.* 17, 33-47.

Payne, K.R. (1984) *The Vegetation of Ainsdale Dunes.* Unpublished report to Nature Conservancy Council, held at Ainsdale Sand Dune NNR.

Radley, G.P. (1994) *Sand Dune Vegetation Survey of Great Britain; A National Inventory. Part 1: England.* Peterborough: Joint Nature Conservation Committee.

Read, S. (1995) *The development of Devil's Hole 1945 – 1995 Ravenmeols, Formby, Lancashire. A detailed topographic and ecological study.* B.Sc. dissertation held at Ainsdale Discovery Centre.

Rhind, P.M., Blackstock, Y.H., Hardy, H.S., Jones, R.E. and **Sandison, W.** (2001) The evolution of Newborough Warren with particular reference to the past four decades. In J.A. Houston, S.E. Edmondson & P.J. Rooney (eds.) *Coastal Dune Management: Shared Experience of European Conservation Practice.* pp. 262-270. Liverpool: Liverpool University Press.

Robinson, N.A. (1971) *Ainsdale Sand Dune National Nature Reserve Vegetation Survey: August 1971.* Unpublished report, Ainsdale NNR archive: ASD 6/4/8.

Salisbury, Sir E. (1952) *Downs & Dunes: Their plant life and its environment.* London: Bell.

Simpson, D.E., Houston, J.A. and **Rooney, P.J.** (2001) Towards Best Practice in the Sustainable Management of Sand Dune Habitats: 2. Management of the Ainsdale Dunes on the Sefton Coast. England. In J.A. Houston, S.E. Edmondson & P.J. Rooney (eds.) *Coastal Dune Management: Shared Experience of European Conservation Practice.* Liverpool: Liverpool University Press. pp. 262-270.

Skelcher (2006) *Sefton Coast Management Scheme/Partnership (1987-2006). Sefton Coast Natterjack Report.* Unpublished report.

Smith, P.H. (1978) *The ecological evaluation of wetland habitats in the north Merseyside sand-dune system.* Report to Merseyside County Council.

Smith, P.H. (2005) *An inventory of vascular plants identified on the Sefton Coast.* Unpublished report to Sefton Coast Partnership.

Smith, P.H. (2006) Changes in the floristic composition of sand-dune slacks over a twenty-year period. *Watsonia.* 26, 41-49.

Smith, P.H. (2007) The Birkdale Green Beach – a sand-dune biodiversity hot-spot. *British Wildlife.* 19, 11-16.

Stewart, N.F. and **Church, J.M.** (1992) *Red Data Books of Britain & Ireland: stoneworts.* Peterborough: Joint Nature Conservation Committee.

Willis, A.J., Folkes, B.F., Hope-Simpson, J.F. and **Yemm, E.M.** (1959) Braunton Burrows; The Dune System and its Vegetation Part I. *Journal of Ecology.* 47, 1-24.

Grasshoppers and crickets (Orthoptera)

Philip H. Smith

9 Hayward Court, Watchyard Lane, Formby, Liverpool L37 3QP
philsmith1941@tiscali.co.uk

Atkinson (1993) did not mention the House Cricket (*Acheta domesticus*) (Gryllidae) which was recorded on the Formby Point nicotine waste tip in the early 1970s, but not subsequently (personal observations). Three grasshoppers (Acrididae) are well-established and often abundant on the sand-dunes, namely the Field Grasshopper (*Chorthippus brunneus*), Mottled Grasshopper (*Myrmeleotettix maculatus*) and Common Green Grasshopper (*Omocestus viridulus*), the first two preferring more sparsely vegetated sites. The Common Groundhopper (*Tetrix undulata*) (Tetrigidae) seems to be rather scarce, though its small size and cryptic appearance make it less likely to be seen.

The only recent addition to the Sefton Coast's Orthoptera is the Short-winged Conehead (*Conocephalus dorsalis*) (Tettigoniidae) (Plate 1). It was first found by P.S. Gateley at Marshside in 2002 and by C. Felton at Birkdale Green Beach in 2005. Subsequent surveys revealed the insect is widespread and numerous at both Marshside and the Green Beach, with smaller populations in frontal dune slacks at Birkdale. The habitat is dense swamp vegetation, usually of Sea Club-rush (*Bolboschoenus maritimus*).

Plate 1. Conocephalus Green Beach, Birkdale. (Photograph by Philip Smith).

C. dorsalis has a mainly southern and eastern distribution in Britain but has recently spread northwards, reaching Morecambe Bay in 2005 and the Duddon Estuary by 2007. As the insect is generally flightless, it may be dispersed by eggs in sea-borne flotsam (Smith and Newton, 2007; Newton and Smith, 2008).

REFERENCES

Atkinson, D. (1993) Grasshoppers and crickets (Orthoptera). In: Atkinson, D. and Houston, J. (eds.) *The Sand Dunes of the Sefton Coast.* Liverpool: National Museums & Galleries on Merseyside. 107p.

Newton, J.M. and **Smith, P.H.** (2008) Short-winged Conehead *Conocephalus dorsalis* (Latreille) (Orthoptera: Tettigoniidae) in North-west England. *Journal of the Lancashire & Cheshire Entomological Society.* 130 28-30.

Smith, P.H. and **Newton, J.M.** (2007) *Conocephalus dorsalis* (Latreille) (Orthoptera: Tettigoniidae) in Merseyside and Lancashire. *British Journal of Entomology and Natural History.* 20 46-48.

Driftweed of the Sefton Coast: Its composition and origins

George Russell and Chris Felton

National Museums Liverpool, William Brown Street, Liverpool, L3 8EN

INTRODUCTION

Floristic analyses of driftweed are uncommon in modern algal literature. An investigation of the unattached algae in Port Erin Bay, Isle of Man, by Burrows (1958) included species lists but was based upon work carried out over a short period of time. Long-term studies tend to treat driftweed either in terms of biomass (Dion and Le Bozec, 1996; Fletcher, 1996) or as substrata for insects and other kinds of invertebrate life (see Andrews, 1991). An article by Hoek (1987) on the possible importance of driftweed in the recruitment of attached seaweed populations provides a very good introduction to the available literature but points mainly to the absence of information.

This study was the outcome of a recent survey of the marine and maritime algae of South Lancashire (Vice-county 59). Liverpool is particularly fortunate in having records of the local algal flora from over a century ago, with which the present flora may be compared for evidence of floristic change. These early records were made chiefly by Frederick Price Marrat (1820-1904) who was on the staff of Liverpool Museum for over 40 years and whose observations were eventually published by Gibson (1891). There is, however, a strong impression that the local records may have included species cast ashore from more distant regions. This practice was by no means unusual. Victorian marine biologists regularly inspected driftweed for the more uncommon species and a description of the procedure is given by Cocks (1853). Cocks specifies the types of shore and shore levels that he considers likely to be the most productive. He also instructs the collector to select only the rarer species and those least damaged and discoloured.

So, in order to draw a valid comparison with the Marrat records, an inspection of driftweed seemed necessary to supplement our knowledge of the attached algae but, after a few shore visits, it became evident that the driftweed was an interesting subject in its own right. Our early visits to Merseyside shores began in the 1990s, following no regular pattern, but systematic sampling from the Sefton coast began in the early 2000s. Since then, sampling has proceeded on an approximately monthly basis, chiefly from Ainsdale – Birkdale but also on occasion from Formby (Plate 1), Hightown and Hall Road. Seaweed was collected by hand from the shore line until no additional species were found and each sample was bagged up and returned to the laboratory for determination. Sampling stopped in December 2007. Notes were also kept on the reproductive states of the algae and their substrata. Permanent herbarium preparations and/or microscope slides were made when considered appropriate.

The driftweed flora

Seventy taxa were recorded comprising sixty-nine species and one hybrid. These are listed in Table 2 (page 192). The species composition of the flora gives a clear impression of an intertidal (eulittoral zone) origin for the bulk of the driftweed.

The number of occasions on which the species occurred was recorded and the results can be found in Table 1.

It is evident that the smallest class (1-5 records) contained the majority (78.6%) of the species and that the numbers of species decreased roughly by half in each succeeding class but showing a small increase in the largest. The distribution of numbers in Table 1 approximates to that expected in the 'Law of Frequencies' which is held to denote natural vegetation (Shimwell, 1971). This suggests that the driftweed flora is not seriously unrepresentative of the rocky-shore vegetation from which it came.

A total of 346 species records were obtained and these showed a marked seasonality in occurrence. 55 records

Plate 1. Drift weed at Formby. (Photograph by G. Russell).

were made during the spring months (March – May), 65 in summer (June – August), 125 in autumn (September – November) and 101 in winter (December – February). These seasonal differences are statistically significant (chi-squared = 36.37 $p<0.01$). The most productive month was November with 74 records and June the least with only 10.

Early in the investigation, the need for a simpler and non-taxonomic system became obvious. We therefore defined four types of driftweed according to their structural characteristics and assigned our records to these as follows: 1. VESICULATE SEAWEEDS (algae, usually large, with conspicuous air bladders); 2. INFLATED SEAWEEDS (algae, large or small, with irregular air-filled cavities in their tissues); 3. EPIPHYTES (algae, usually small, growing attached to other seaweeds, mostly from groups 1 and 2); 4. FELLOW TRAVELLERS (algae, large or small, which do not fit into the other categories and these may be entangled with other driftweed or with fishing line or perhaps attached to inanimate flotsam). Examples from groups 1 – 3 are shown in Figure 1.

Vesiculate seaweeds accounted for 28.6% of the species records, inflated seaweeds for 10.9%, epiphytes for 50.8% and fellow travellers for 9.5%. No measures of driftweed biomass were made during the survey but visual inspection suggested that the vesiculate forms accounted for at least 90% of the biomass.

Driftweed Origins

Knowledge of the Merseyside seaweed flora was helpful in determining whether a species was local in origin or had arrived from a more distant shore. An alga known to be distant could also point to the origins of others in the sample. For example, the brown vesiculate seaweed *Ascophyllum nodosum* (Figure 1 V2) is present locally (at Eastham) but the epiphytic red alga *Polysiphonia lanosa* (Figure 1 E1) no longer occurs in

Table 1. Driftweed species ranked according to numbers of records.

	\multicolumn{7}{c}{**Numbers of Records**}	
	1-5	6-10	11-15	16-20	21-25	26-30	>30	
Species	55	8	3	1	0	0	3	n=70

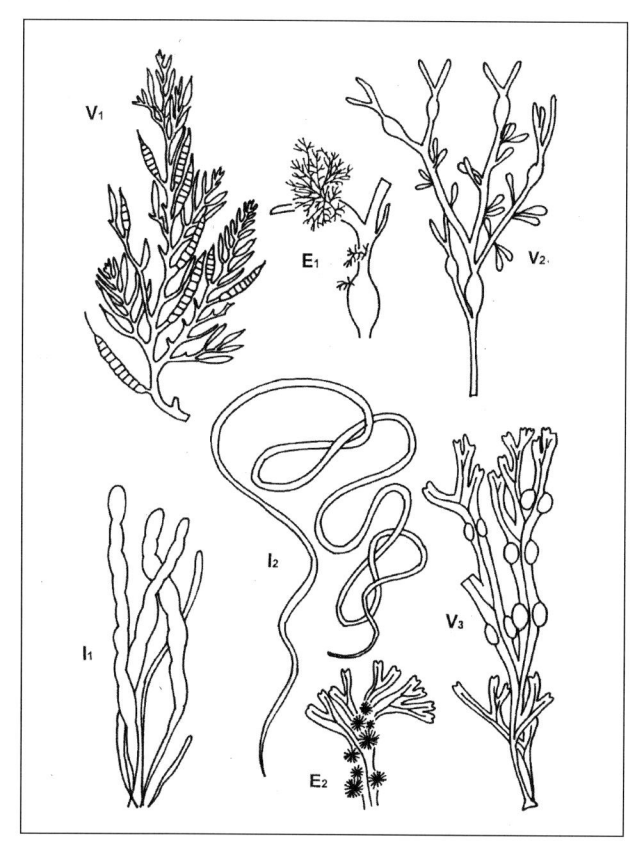

Figure 1. Guide to driftweed. (Redrawn and adapted from Newton (1931) and Taylor (1937)).

this region. So if drift *Ascophyllum* had *Polysiphonia* on its frond then the former must also have been foreign to Merseyside.

Other clues were morphological. The brown alga *Elachista fucicola* (Figure 2 E2) is plentiful locally but its filaments are usually less than 35μm in width, so drift *Elachista* with broader filaments was almost certainly distant in origins and hence also its host plant, usually *Fucus vesiculosus* (Figure 2 V3). Additional evidence was provided by the stones and rock fragments attached to the bases of some driftweed. Thus Bünter sandstone suggested a local source but other rock types (schist, slate, conglomerate, mudstone, siltstone, limestone) indicated more remote localities. Flotsam, less reliable than the rock, could sometimes give evidence of distant sources (labelled containers such as fish and milk boxes, coal sacks etc.).

On this evidence we concluded that only 20% of the records were possibly local in origin but the remaining 80% were most certainly from distant shores.

Identification of the exact locations of driftweed was more difficult. Some of the rock fragments could, with reasonable confidence, be identified with Anglesey. A milk box was also sourced to Anglesey but labelled fish boxes and coal sacks from S. Ireland have also come ashore along with more anonymous flotsam. All our evidence points therefore to N. Wales and S. Ireland as the most probable sources of Sefton driftweed.

The remaining problem was the period of time required for the passage of the driftweed to our coastline. The annual average speed of wind-driven surface water currents on the approaches to Liverpool seems to be about 1.76 cm/sec (Pingree, 2005). Extrapolating wildly from this figure, a floating seaweed would, on average, complete its passage from Amlych to Ainsdale in 60 days. Confirmation of this estimate was provided in the unlikely form of a plastic duck released from Dublin on 03-06-06 as part of a charity event and recovered at Ainsdale on 06-11-06, i.e. 156 days later. According to the above calculation it should have come ashore after 144 days. However, annual average current speeds necessarily include the relatively calm, summer months. During the autumn and winter with assistance from westerly gales, shorter transit times are to be expected and the generally excellent condition of the driftweed on arrival suggests that this may indeed be the case.

DISCUSSION

It is difficult to establish with any degree of certainty the role played by driftweed in the development of a local seaweed flora but in a few cases its importance is known. The best example from recent times is the invasive species *Sargassum muticum*. This vesiculate alga is native to coastal waters of Korea and Japan but it began to be recorded on the Pacific coast of N. America in the years following World War II. The exact date of its arrival in European coastal waters is not known but its range now extends from the Mediterranean to Sweden. Its presence in the British Isles was first reported in 1973 following its discovery on the Isle of Wight, since when it has spread to Wales, Ireland and Scotland.

The primary source of the European outbreak is thought

to be imported Pacific oyster *Crassostrea gigas* (Boaden, 1995). Subsequent spread, however, has been largely by passive transport of fertile plants or their branches in surface water. The specimens we recorded on the Sefton shore in September 2007 were certainly reproductive and therefore potentially capable of initiating a local population. To do so, however, would require the presence of suitable habitat and while some of the rock pools at New Brighton and Hilbre Island or the restored S. Docks might provide this, there is a general shortage of good habitat for it in the region.

The efficient spread of *S. muticum* owes much to its vesiculate structure and the dominance of vesiculate algae in the Sefton driftweed confirms this. A good example is provided by the 39 records of *Fucus vesiculosus* (vesiculate) and only 9 of *F. serratus* (evesiculate), six of which were as epiphytes of vesiculate algae. A single vesicle of *Ascophyllum nodosum* can exceed 25cm^3 in the volume of air contained and we have recorded an *Ascopyllum* plant which had carried ashore a stone weighing 890g. Vesiculate seaweeds with their cargoes of epiphytes and fellow travellers could well have played a part in the recruitment of the local seaweed flora but hard evidence is likely to remain elusive.

Driftweed has, however, another very minor part to play in Sefton coastal ecology. When it comes to rest at extreme high water mark it can form a nucleus for accretion of wind-blown sand and at the same time provide a source of organic material during the earliest phase of dune development. In similar fashion it can stabilise intertidal sand and mud as a precursor to saltmarsh development.

We conclude therefore that the Victorians had a point; driftweed really is a worthy subject for study.

REFERENCES

Andrews, J.W. (1991) *The Ecology of the Manx Strand Line*. Ph.D. Thesis, University of Liverpool, unpublished.

Boaden, P.J.S. (1995) The adventive seaweed *Sargassum muticum* (Yendo) Fensholt in Strangford Lough, Northern Ireland. *Irish Naturalists' Journal*, 25, 111-113.

Burrows, E.M. (1958) Sublittoral algal population in Port Erin Bay, Isle of Man. *Journal of the Marine Biological Association of the U.K.*, 37, 687-703.

Cocks, J. (1853) *The Seaweed Collector's Guide*. Van Voorst, London.

Dion, P. and **Le Bozec, S.** (1996) The French Atlantic Coasts. In: *Marine Benthic Vegetation. Recent Changes and the Effects of Eutrophication.* W. Schramm, P.H. Nienhuis (Editors), pp. 251-264. Springer-Verlag, Berlin.

Fletcher, R.L. (1996). The British Isles. In: *Marine Benthic Vegetation. Recent Changes and the Effects of Eutrophication.* W. Schramm, P.H. Nienhuis (Editors), pp. 223-250. Springer-Verlag, Berlin.

Gibson, R.J.H. (1891) A revised list of the marine algae of the L.M.B.C. district. *Proceedings and Transactions of Liverpool Biological Society*, 5, 83-143.

Hardy, F.G. and **Guiry, M.D.** (2003) *A Check-list and Atlas of the Seaweeds of Britain and Ireland*. British Phycological Society, London.

Hoek, C. van den (1987) The possible significance of long-range dispersal for the biogeography of seaweeds. *Helgoländer Meeresuntersuchungen*, 41, 261-272.

Newton, L. (1939) *A handbook of British seaweeds*. British Museum, London.

Pingree, R. (2005) North Atlantic and North Sea Climate Change: curl up, shut down, NAO and ocean colour. *Journal of the Marine Biological Association of the U.K.*, 85, 1301-1315.

Shimwell, D.W. (1971) *The Description and Classification of Vegetation.* Sidgwick & Jackson, London.

Taylor, W.R. (1937) *Marine algae of northwestern coast of North America.* University of Michigin Press, Michigan.

Table 2. List of marine algae collected from driftweed on the Sefton coast in the course of this investigation. Species are listed in alphabetical order of generic names in the major groups. All names follow Hardy and Guiry (2003) with more recent name changes given in parentheses.

CYANOPHYTA *Dermocarpa prasina*	
RHODOPHYTA *Acrochaetium secundatum* *Aglaothamion hookeri* *Ceramium deslongchampsii* *C. pallidum* *C. virgatum* *Choreocolax polysiphoniae* *Corallina officinalis* *Cryptopleura ramosa* *Erythrotrichia bertholdii* *E. carnea* *Hildenbrandia rubra* *Lomentaria articulata* *Mastocarpus stellatus*	*Palmaria palmata* *Phymatolithon lenormandii* *Pleonosporum borreri* *Plocamium cartilagineum* *Polyides rotundus* *Polysiphonia fibrata* *P. lanosa* *P. stricta* *Porphyra purpurea* *P. umbilicalis* *Pterothamnion plumula* *Phodochorton purpureum* *Titanoderma pustulatum*
PHAEOPHYTA *Ascophyllum nodosum* *Chorda filum* *Dictyota dichotoma* *Ectocarpus fasciculatus* *E. siliculosus* *Elachista fucicola* *E. scutulata* *Fucus ceranoides* *F. serratus* *F. spiralis* *F. vesiculosus* *F. spiralis* x *vesiculosus* *Halidrys siliquosa*	*Himanthalia elongata* *Hincksia granulosa* *Laminaria hyperborea* *L. saccharina* *Laminariocolax tomentosoides* *Litosiphon laminariae* *Pelvetia canaliculata* *Petalonia zosterifolia* *Pylaiella littoralis* *Sargassum muticum* *Sphacelaria cirrosa* *Spongonema tomentosum*
CHLOROPHYTA *Acrosiphonia arcta* *Blidingia marginata* *B. minima* *Cladophora hutchinsiae* *Enteromorpha compressa* (now *Ulva compressa*) *E. flexuosa* (now *Ulva flexuosa*) *E. intestinalis* (now *Ulva intestinalis*) *E. linza* (now *Ulva linza*) *E. prolifera* (now *Ulva prolifera*)	*Percursaria percursa* *Rhizoclonium riparium* *Spongomorpha aeruginosa* *Ulothrix flacca* *U. implexa* *Ulva lactuca* *U. rigida* *Ulvella lens*

Birkdale Green Beach

Philip H. Smith

9 Hayward Court, Watchyard Lane, Formby L37 3QP
philsmith1941@tiscali.co.uk

ABSTRACT

Birkdale Green Beach consists of a 4km-long strip of saltmarsh, sand-dune, dune-slack and swamp communities at least 200m wide that has developed since 1986 on the foreshore between Birkdale and Ainsdale. It originated as scattered patches of Common Saltmarsh-grass (*Puccinellia maritima*) on the open beach. These trapped silt and sand to form an intermittent embryo dune ridge which impeded drainage and created a series of seasonally flooded slacks, with both maritime and freshwater characteristics, backed by developing Alder (*Alnus glutinosa*) woodland. The vegetated area has grown exponentially with time, reaching about 62ha by 2005. Several examples of similar formations on this coast have been traced in the literature going back to the mid-19th century. Ongoing studies of flora and fauna show that the Green Beach is extremely biodiverse with large numbers of vascular plants, rare bryophytes, a large population of Natterjack Toads (*Epidalea calamita*), important nesting birds and several notable invertebrates. Because it contributes early successional stages to a largely senescent dune system, the Green Beach has a disproportionately high nature conservation value.

ORIGIN AND DEVELOPMENT

Birkdale Green Beach originated in 1986 as scattered patches of *P. maritima* on the open beach about 100m out from the dune edge on the foreshore between Birkdale (SD 321 163) and Ainsdale (SD 302 136) (Plate 1). These soon accumulated silt and blown sand to form low hummocks which became an intermittent line of embryo dunes (Plate 2), later colonised by more typical dune-forming grasses, such as *Elytrigia juncea* (Sand Couch), *Leymus arenarius* (Lyme-grass) and eventually *Ammophila arenaria* (Marram). The early years of this development are described by Edmondson *et al.* (2001). Behind the embryo dunes, salt-marsh vegetation was dominated at first by *P. maritima* but became much richer in species as time progressed. The new ridge impeded drainage of ground-water seawards from the dune aquifer and run-off from three land drains. The result was a series of seasonally-flooded, primary dune-slacks with both fresh-water and maritime characteristics, as described in the "Type A" slack of Davy *et al.* (2006). Over time, the slacks developed a complex mosaic of dune-slack, high-level salt-marsh and swamp plant communities. Alder germinating on old strand-lines from about 1998 has grown to produce parallel lines of incipient wet-woodland extending for over 2km and covering an area of 1.85ha in 2005 (Kristiansen, 2008) (Plate 3). Meanwhile, the Green Beach was still developing

Plate 1. Green Beach in 1987.
(Photograph by Philip Smith).

westwards, new embryo ridges cutting off sand-supply to the older ones which became low fixed-dunes. By 2005, the Green Beach was nearly 4km long and up to 200m wide; it had grown exponentially to an area of about 62ha from 2ha in 1989 (Figure 1). Currently, patches of *P. maritima* are still establishing on the foreshore seaward of the Green Beach, especially in the north, so its growth may continue for some time.

Reasons for Green Beach development

Birkdale is situated on the outer fringe of the Ribble Estuary which has been silting up for the last 150 years, the material being derived from reworked sandy glacial sediments in the south-east Irish Sea and from sand eroded from dunes at Formby Point, about 7km to the south-west (van der Wal *et al.*, 2002). Newton *et al.* (2007) show that siltation between 1913 and 2006 on the Marshside foreshore, about 5km north-east of Birkdale, resulted in a vertical height gain of 1.0-1.5m, while the Mean High Water Mark moved 400m seawards. In the same area, salt-marsh vegetation expanded south-westwards at a rate of 400m per annum from 2002 to 2005. It is inferred that similar trends have influenced the nearby Birkdale foreshore.

Usually, vegetation is prevented from growing on a sandy beach by the physical attrition of wave action, especially during winter gales. However, as the foreshore at Birkdale has become wider and higher in recent decades it has increasingly absorbed wave-energy. Now, even on windy days, at high spring-tides hardly any scouring takes place (personal observations).

Use of Ainsdale-Birkdale beach by cars since the 1930s, together with mechanised beach-cleaning, evidently inhibited embryo-dune formation beyond the existing dune frontage. Siltation eventually resulted in a wide expanse of foreshore sufficiently sheltered for Green Beach development. Although *Puccinellia* colonisation began in 1986, Sefton Council's decision in 1993 to restrict the driving and parking of motor vehicles on this

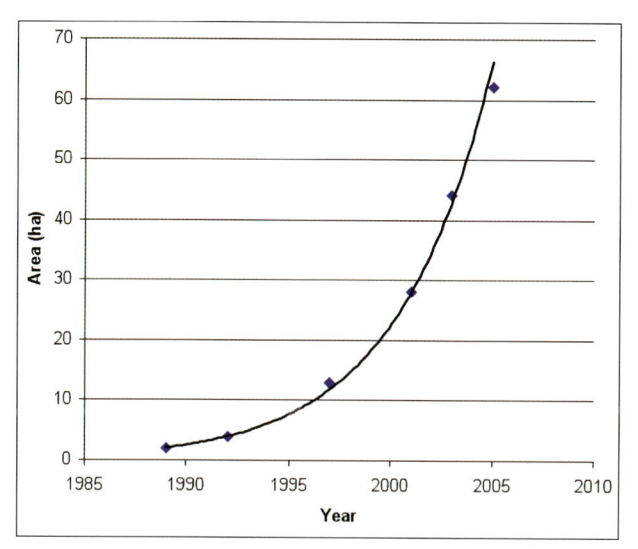

Figure 1. Changes in the area of Birkdale Green Beach, 1986 to 2005. Fitted line: exponential.

Plate 2. Embryo dunes formed by Common Saltmarsh-grass. (Photograph by Philip Smith).

Plate 3. Alder woodland on Birkdale Green Beach in 2008. (Photograph by Philip Smith).

section of shore, coupled with a change in the mechanical beach cleaning regime in 1997, probably allowed an increased rate of vegetation growth.

In spring 2005, a 400m stretch of beach at Ainsdale, contiguous with the Green Beach, was fenced to exclude vehicles in mitigation for a similar area of Special Protection Area foreshore at the nearby resort of Southport (which was treated with herbicide to remove vegetation). This resulted in the rapid formation of a "New" Green Beach, colonisation by 127 vascular plants being recorded by October 2008 (personal observations).

Other examples

Interestingly, this type of feature has developed before on the Sefton Coast. Thus, E.R. Beattie (cited in McNicholl, 1883: 140), writing about his memories of Southport in about 1862, states: *'From now onwards I watched the gradual growth of sandhills and marsh at the Birkdale end of the town, embryo sandhills in little hummocks forming about opposite the site of the Palace Hotel, and on the shore tufts of marsh grass began to appear, spreading and joining each other until what was a level expanse of tide-ribbed sand became a green expanse of marsh, and it was only on a spring tide that the sea reached the Promenade.'*

Allen (1932) gives a more detailed description of a 'sea-beach flora' that developed between Ainsdale and Freshfield in the early 1930s. Thirty-three vascular plants were identified, comprising a mixture of salt-marsh, freshwater and ruderal species very similar to those on the present-day Birkdale Green Beach. The Ainsdale Sand Dunes NNR archive has a photograph of this feature by R.K. Gresswell and others taken by D. Coult in the late 1940s of a similar but undescribed formation on Ainsdale beach. Both were soon washed away by winter gales.

Finally, at Birkdale from 1974 onwards, a 200m-long ridge formed in the same way, isolating a 50m-wide slack flooded by a surface-water drain. This feature persisted and became known as Tagg's Island (Edmondson *et al.*, 1993). The enclosed slack,

occasionally breached by the tide until 1983, after which this was prevented by the developing embryo dunes. At first, the slack was dominated by salt-marsh plants, soon replaced by freshwater species, including a large reed-bed (personal observations).

Elsewhere in Britain, this type of habitat seems to be rather rare, though similar examples have been described at Berrow Marsh, Somerset (Willis and Davis, 1960) and St Cyrus, Tayside (Gimingham, 1953; cited in Burnett 1964: 119). Very recently, Dutch scientists have begun to study "groen stranden" on the islands of Ameland, Rottumerplaat and Schiermonnikoog. Tidal Alder woodland seems to be particularly unusual, with possibly comparable habitat being found on the Fal Estuary, Cornwall (Packham and Willis, 1997), Loch Ridden, Argyll (H. McAllister, *in litt.* 2007), and Sandyhills Bay, Dumfries and Galloway (N.A. Robinson, *in litt.* 2006).

Vegetation classification

A National Vegetation Classification (NVC) survey (Gateley and Michell, 2002) deduced that the dominant NVC stand-type (Rodwell, 2000) in the northern sector of the Green Beach was S21c (*Bolboschoenus maritimus* swamp, *Agrostis stolonifera* sub-community) while, to the south, this was largely replaced by SM13 (*Puccinellia maritima* salt-marsh). In the central sector, a strip of S4d (*Phragmites australis* swamp, *Atriplex prostrata* sub-community) was mapped, while the embryo dunes were considered to be a mosaic of SD4 (*Elytrigia juncea* fore-dune) and SD6 (*Ammophila arenaria* mobile-dune).

Thomas (2005) thought that the developing Alder scrub habitat approximated to the NVC's W6 (*Alnus glutinosa – Urtica dioica* woodland), though Rodwell (1991) makes no mention of this community's occurrence in paramaritime conditions. Kristiansen (2008) found the Alder woodland was a very poor match to W7b: *Alnus glutinosa – Fraxinus excelsior – Lysimachia vulgaris* woodland, while the underlying vegetation, excluding Alder, was closest to MG11: *Festuca rubra – Agrostis stolonifera – Potentilla anserina* grassland, though again the statistical match is very poor.

Vascular plants

The number of vascular taxa found on the Green Beach increased progressively from two in 1986 to 288 in 2009 (personal observations). The rate of increase is now seen to be levelling off (Figure 2). Only 40 (14%) are non-native or introduced native species, probably because the site is remote from gardens. Many of the plants are associated with early successional stages of dune-slacks or brackish water and are scarce in other parts of the dune system.

Forty-six nationally or regionally notable taxa (16% of the flora) have been recorded: three Nationally Rare, six Nationally Scarce, one Vulnerable, two Near Threatened and 34 Species of Conservation Importance in North West England (SCIs) not included in any other category (Cheffings and Farrell, 2005; Regional Biodiversity Steering Group, 1999).

The Nationally Rare taxa comprise *Salix* x *friesiana* (a willow hybrid), *Limonium britannicum* ssp. *celticum* (Rock Sea Lavender), one plant of which was found in 2007, and *Schoenoplectus pungens* (Sharp Club-rush), the latter having spread naturally from a translocation site in the nearby Tagg's Island Marsh.

Among the Nationally Scarce taxa is Juncus balticus (Baltic Rush), which in England is known only from the Birkdale sand-dunes. Originally found on the Green Beach in 2000, this species was represented here by 15 patches in 2007.

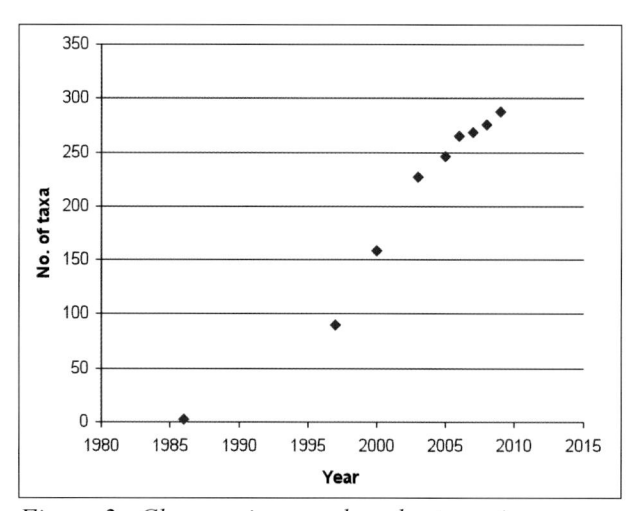

Figure 2. Changes in vascular plant species-richness, 1986 – 2009.

In June 2006, a Botanical Society of the British Isles field meeting resulted in several significant new finds, including *Blysmus rufus* (Saltmarsh Flat-sedge) (SCI), *Juncus compressus* (Round-fruited Rush) (Near Threatened) and *Dactylorhiza* x *wintoni* (a hybrid marsh-orchid), which had not been recorded on the Sefton Coast since 1986, 1933 and 1949 respectively. Also noted was *Carex remota* x *C. otrubae* (= *Carex* x *pseudoaxillaris*) (a hybrid sedge), not seen in South Lancashire since the 1890s.

Bryophytes

In a 2004 survey, David Holyoak (*in litt.*, 2006) recorded one liverwort and nine species of moss on the Green Beach. The mosses include two Nationally Rare taxa, *Bryum dyffrynense* and *B. warneum*, the latter being a UK Biodiversity Action Plan Priority Species. The population of *B. warneum* here is much larger than all its other British localities combined (Holyoak, 2002).

Vertebrates

Mammals recorded for the Green Beach include the Rabbit (*Oryctolagus cuniculus*), whose grazing has a marked effect on the vegetation, mainly in the northern third. Also noted are Stoat (*Mustela erminea*), Common Shrew (*Sorex araneus*) and Wood Mouse (*Apodemus sylvaticus*).

The site is important for breeding birds, including Ringed Plover (*Charadrius hiaticula*) (up to 15 pairs), Lapwing (*Vanellus vanellus*) (7 pairs) and Skylark (*Alauda arvensis*). A pair of Oystercatchers (*Haematopus ostralegus*) nested unsuccessfully in 2006 and 2007. In winter, the lagoons are used by Snipe (*Gallinago gallinago*) and Jack Snipe (*Lymnocryptes minima*).

Both Common Lizard (*Zootoca vivipara*) and Sand Lizard (*Lacerta agilis*) have been recorded under tide-line timber at the back of the Green Beach, together with Smooth Newts (*Triturus vulgaris*). Common Toads (*Bufo bufo*) breed in relatively small numbers in the drain outfalls and adjacent lagoons. The Natterjack Toad colonised in 2001 and now has a large breeding population, with 275 spawn strings counted in 2008 (personal observations).

Invertebrates

The Green Beach is rich in invertebrates. There are several uncommon solitary bees and wasps. Among beetles (Coleoptera), the nationally rare Northern Dune Tiger Beetle (*Cicindela hybrida*) has become well-established in recent years. Five species of Orthoptera have been recorded, including the Short-winged Conehead (*Conocephalus dorsalis*) which is a recent colonist in north-west England (Smith and Newton, 2007). In late August 2007, a colony of the Sandhill Rustic moth (*Luperina nickerlii* ssp. *gueneei*) was found on Green Beach embryo dunes (Burkmar and Jones, 2008), 315 adults being recorded the following year, (R. Burkmar *in litt.*, 2008). This rare sub-species is endemic to north-west England and North Wales. During August 2008, G. Jones (*in litt.*, 2008) and R. Burkmar recorded a further 69 species of moth, including *Agonopterix yeatiana* (new to both Lancashire vice-counties) and *Calamotropha paludella* (first for Lancashire since the 1950s). A study of Green Beach Alder invertebrates produced 104 taxa, mainly of beetles and spiders (two nationally scarce), 21 being potentially new records for the Sefton Coast (Kristiansen, 2008).

Assessment

The Birkdale Green Beach makes an outstanding contribution to the biodiversity of the Sefton Coast (Smith, 2007). In particular, it provides a significant area of pioneer habitat in a relatively mature dune system which has suffered from over-stabilisation in recent decades (Smith, 1999). This is reflected in a high species-richness, especially of vascular plants, the 288 taxa recorded (15% nationally or regionally notable) representing 25% of the entire Sefton Coast sand-dune flora (Smith, 2006). As a locality for rare mosses the Green Beach is internationally important (Holyoak, 2002). The recent development of Alder woodland is extremely unusual in such an exposed coastal location (Kristiansen, 2008) and is of at least regional significance, wet woodland being afforded a Habitat Action Plan in the North Merseyside Biodiversity Action Plan (Merseyside Biodiversity Group 2001).

The Green Beach is now considered to be one of the most important breeding sites for Natterjack Toads in Britain (J. Buckley *in litt.*, 2008). Notable breeding birds include Skylark, which is on the Red List of Species of Conservation Concern in the UK, while Lapwing and Ringed Plover are Amber listed (Brown and Grice, 2005). The Green Beach has the second largest concentration of breeding Ringed Plovers in Lancashire and North Merseyside (White *et al.*, 2008).

Recent discoveries of Northern Dune Tiger-beetle and Sandhill Rustic mean that the site is nationally important for invertebrates.

The Green Beach lies within the Sefton Coast SSSI and is proposed as an extension to the Ainsdale and Birkdale Sandhills Local Nature Reserve. It is also protected under the EU Birds and Habitats Directives as part of the Sefton Coast Special Protection Area and Special Area of Conservation within the Natura 2000 network.

Conservation management

Despite its recent origin, parts of the Green Beach are already beginning to show signs of increasing maturity with dense beds of Sea Club-rush and Common Reed dominating and replacing the more diverse vegetation in the northern and central sectors. This process may be slowed to some extent by Rabbit grazing. A potential problem with invasive Sea Buckthorn is being kept in check by active management but, as yet, no attempt has been made to control the Alder as this is a natural colonist. The continuing westwards growth of the Green Beach should maintain the representation of pioneer communities for at least the foreseeable future.

In an attempt to retain water longer for breeding Natterjacks, in spring from 2005 onwards, sand-dams were erected across the freshwater outflows near the southern end of the Green Beach. This has promoted successful metamorphosis in most years.

Large balks of timber and tree trunks are washed up from time to time. As they age and weather, these are increasingly used as nesting and basking sites by solitary bees and wasps, while such debris is also useful as hiding places for Natterjacks and lizards as well as invertebrates. It is therefore important that the logs are not removed during management operations.

Although visitor interest in the Green Beach has been encouraged in the north by the development by Sefton's Coast & Countryside Service of a way-marked footpath, the Velvet Trail, and the construction of a board-walk by Birkdale Civic Society, much of the site remains relatively quiet, this being crucial to its continuing importance for breeding birds.

A priority now is to continue and extend recording of the flora and fauna on this rapidly changing piece of coastline. In 2008, studies commenced on beetles of the strandline, on fungi and on the invertebrates of the Alder woodland, but there are opportunities for many more projects.

ACKNOWLEDGEMENTS

I am grateful to Inger Kristiansen and Richard Thomas for contributing unpublished data. M.P. Wilcox, E.F. Greenwood and Prof. R. Bateman assisted with the identification of some critical vascular plant taxa.

REFERENCES

Allen, M.J. (1932) Recent changes in the sea-beach flora at Ainsdale, Lancashire. *Northwestern Naturalist* 24 114-117.

Burkmar, R. and **Jones, G.** (2005) The Sandhill Rustic: a Conservation Flagship. *Atropos* 35 38-44.

Burnett, J.H. (ed.) (1964) *The Vegetation of Scotland.* Edinburgh and London: Oliver & Boyd.

Cheffings, C.M. and **Farrell, L.** (2005) *The Vascular Plant Red Data List for Great Britain. Species Status* 7: 1-116. Peterborough: Joint Nature Conservation Committee.

Davy, A.J., Grootjans, A.P., Hiscock, K. and **Petersen, J.** (2006) *Development of eco-hydrological guidelines for dune habitats – Phase 1.* Peterborough: English Nature Research Reports Number 696.

Edmondson, S.E., Gateley, P.S., Rooney, P. and **Sturgess, P.W.** (1993) Plant Communities and Succession. In: Atkinson, D and Houston, J. (eds.) *The Sand Dunes of the Sefton Coast.* Liverpool: National Museums & Galleries on Merseyside 65-84.

Edmondson, S.E., Traynor, H. and **McKinnell, S.** (2001) The development of a green beach on the Sefton Coast, Merseyside, UK. In: Houston, J.A., Edmondson, S.E. and Rooney, P.J. (eds.) *Coastal Dune Management. Shared Experience of European Conservation Practice.* Liverpool: Liverpool University Press. 48-58.

Gateley, P.S. and **Michell, P.A.** (2002) *NVC Survey of Salt-marsh, Ribble and Alt Estuaries, 2002.* Wigan: English Nature.

Holyoak, D.T. (2002) *Coastal mosses of the Genus Bryum. Report to Plantlife of work carried out in England during 2001.* Report no. 206. Peterborough: Plantlife/English Nature.

Kristiansen, I.L. (2008) *The Development and Conservation Value of Alder* (Alnus glutinosa) *Woodland on Birkdale Green Beach.* MSc Thesis. University of Edinburgh.

McNicholl, E.D. (ed.) (1883) *Handbook for Southport.* Southport: Robert Johnson & Co.

Merseyside Biodiversity Group (2001) *North Merseyside Biodiversity Action Plan 2001.* Maghull: Environmental Advisory Service.

Newton, M., Lymbery, G. and **Wisse, P.** (2007) *Report on the Evolution of Salt Marsh on the Sefton Coast.* Sefton Council Ainsdale-on-Sea: Costal Defence.

Packham, J.R. and **Willis, A.J.** (1997) *Ecology of Dunes, Salt Marsh and Shingle.* London: Chapman and Hall.

Regional Biodiversity Steering Group (1999) *A Biodiversity Audit of North West England.* 2 vols. Maghull: Environmental Advisory Service.

Rodwell, J.S. (ed.) (1991) *British Plant Communities Volume 1. Woodlands and scrub.* Cambridge: Cambridge University Press.

Rodwell, J.S. (ed.) (2000) *British Plant Communities Volume 5. Maritime communities and vegetation of open habitats.* Cambridge: Cambridge University Press.

Smith, P.H. (1999) *The Sands of Time. An introduction to the Sand Dunes of the Sefton Coast.* Liverpool: National Museums & Galleries on Merseyside.

Smith, P.H. (2006) An inventory of vascular plants for the Sefton Coast. *BSBI News* 102, 4-9.

Smith, P.H. (2007) The Birkdale Green Beach – a sand-dune biodiversity hotspot. *British Wildlife* 19 (1) 11-16.

Smith, P.H. and **Newton, J.M.** (2007) *Conocephalus dorsalis* (Orthoptera, Tettigoniidae) in Merseyside and Lancashire. *British Journal of Entomology & Natural History* 20, 46-48.

Thomas, R.L. (2005) *The origin of incipient alder (Alnus glutinosa (L) Gaertn.) woodland along the Sefton Coast, south of Southport.* MSc Thesis. University of Liverpool.

Van der Wal, D., Pye, K. and **Neal, A.** (2002) Long-term morphological change in the Ribble Estuary, northwest England. *Marine Geology* 189, 249-266.

White, S., McCarthy, B. and **Jones, M.** (eds.) (2008) *The Birds of Lancashire and North Merseyside.* Southport: Hobby Publications.

Willis, A.J. and **Davies, E.W.** (1960) *Juncus subulatus* in the British Isles. *Watsonia* 4, 211-217.

Further information is available on the Sefton Coast Partnership website: www.seftoncoast.org.uk

Non-native plants on the Sefton Coast sand-dunes

Sally E. Edmondson

Department of Geography, Liverpool Hope University, Hope Park, Liverpool L16 9JD
edmonds@hope.ac.uk

ABSTRACT

Introduction of non-native species is a major global threat to biodiversity. As the Sefton Coast sand dune system is an internationally important nature conservation site, there is concern about the increasing proportion of non-native plants. The most significant introductions on the Sefton Coast are attributed to deliberate planting (for example as part of landscaping and past stabilisation schemes), garden escapes, and propagules washed up in tidal litter. Recent surveys have demonstrated that the proportion of non-native flora on the Sefton Coast is growing, increasing the probability of invasive species threatening the conservation status of the dunes. Future introductions should be minimised and existing ones monitored with a view to taking action if a significant threat to native dune flora and fauna is identified.

INTRODUCTION

This paper aims to review the current status of non-native plants on the Sefton Coast sand dune system, and to consider their threat to sand-dune biodiversity.

The terminology associated with the study of non-native species is complex (see review in Pyšek *et al.*, 2002, p107). Here 'non-native' is used according to the definition of the Department of Environment, Food and Rural Affairs (Defra, 2003), 'a species introduced (*i.e.* by human action) outside its natural past or present distribution', and is synonymous with terms such as 'alien', 'non-indigenous' and 'exotic'. An 'introduction' is taken to mean the deliberate or accidental release of non-native species by people into the wild. Other terms are explained at the first time of their use in the text.

The introduction of non-native species is a major global threat to biodiversity, ranking only a little below direct habitat destruction in its significance (Cronk and Fuller, 2001). Introductions that become invasive have a severe detrimental effect on native species and ecosystems. In the UK, significant problems have arisen from some invasive non-natives (for example Grey Squirrel *Sciurus carolinensis*, Japanese Knotweed *Fallopia japonica*), though we have not suffered ecological disasters on the scale of those experienced in North America, southern Africa, Australasia and oceanic islands (Manchester and Bullock, 2000). Whilst most non-native introductions are benign (Williamson, 1996), the potential negative impact on biodiversity, and the billions of pounds incurred in direct economic costs of invasive species (Defra, 2003) demonstrates the importance of understanding, review and, if necessary, action to minimise future problems.

Non-native vascular plants of the Sefton Coast

Non-native plants, especially those that are 'naturalised' (*i.e.* form established, self maintaining populations in the wild), have long been recorded in the British flora, and in the Sefton region. Green's 'Flora of Liverpool' (1933) records non-natives, such as Snow-in-summer *Cerastium tomentosum,* Sea Buckthorn *Hippophae rhamnoides,* and Japanese Rose *Rosa rugosa* as naturalized on the Sefton Coast dunes and Savidge *et al.* (1963) presents an analysis of non-natives in the area. Traditionally, however, non-natives have been under-recorded (Hill *et al.*, 2005); only in recent years have records been systematically collected, particularly after the first edition of Stace (1991), which included many more non-natives than any previously published flora. The Sefton Coast higher plant species list has an increasing number of non-native species (Smith, 2005 with subsequent annual updates). The author has completed a systematic survey of housing/dune boundaries, recording the presence of introductions from adjacent gardens. All information in this paper, not otherwise referenced, is from these two sources.

There are some useful distinctions to be made amongst the range of non-native plants on the Sefton Coast dunes:

- Deliberate introductions; for example Corsican Pine, *Pinus nigra* ssp. *laricio*, which was widely planted in the late nineteenth and early twentieth century for dune stabilisation (Jones *et al.*, 1993).
- Introduced or 'escaped' by accident; for example the Evening Primroses *Oenothera* spp., that are now common on the dunes as garden escapes or possibly accidental imports through the docks. Common Evening-primrose *O. biennis* was first recorded on the dunes in 1801 at Crosby (Wheldon, 1913), and Large-flowered Evening-primrose *Oenothera glazioviana* was known from 1881 (Savidge *et al.*, 1963).
- Archaeophytes (plants introduced before 1500) (Preston *et al*, 2002) an example being Ground Elder *Aegopodium podagraria*, which occurs in dune grassland and on disturbed ground (Smith, 2005). This was an ancient introduction to gardens as a herb, and is now very common and naturalised throughout the British Isles (Clement and Foster, 1994).
- Neophytes (plants introduced after 1500) (Preston *et al.*, 2002), including the Evening-primroses mentioned above, or Hybrid Bluebell *Hyacinthoides non-scripta* X *H. hispanica,* that are spreading onto the dunes from housing boundary areas.
- 'Casuals'; i.e. plants that fail to persist longer than about five years (Preston *et al.*, 2002), as opposed to species that are established and reproducing with the likelihood of continued survival (Defra, 2003). Examples are Tomato *Lycopersicon esculentum*, and Geranium, *Pelargonium* X *hybridum,* frequently recorded near housing on the dunes.
- Another significant group consists of plants that persist, but may not reproduce. Thus, patches of Montbretia *Crocosmia* X *crocosmiiflora* and Bearded Iris *Iris germanica* occur regularly near housing boundaries but do not appear to spread, new patches apparently being the result of continued introductions.
- Some plants may become invasive. Typical examples are the woody species that were introduced for dune stabilization, such as Sea Buckthorn, White Poplar *Populus alba* and Corsican Pine. Another is New Zealand Pygmy weed, *Crassula helmsii*, an oxygenating aquatic plant introduced from ornamental ponds, first recorded as invasive in the UK in the 1980s, smothering existing vegetation (Dawson and Warman, 1987). It has been found invading a small number of slacks / scrapes on the Sefton Coast. These invasive plants are clearly a conservation threat. Williamson's (1993) 10:10 rule, which gives an adequate fit for British plant data, states that 10% of all introduced species become established and that 10% of established species become invasive. Thus, with the increase in non-native species, comes a greater probability of new invasives.

The most recent annual update to the Sefton Coast vascular plant species list (Smith, 2007) records fifteen new species, of which ten are non-native. There are currently 1092 taxa recorded on the 21 km^2 of the Sefton Coast sand-dune system, of which 381 (35%) are non-native. This compares with 50% non-native vascular taxa in vice-county 59 (South Lancashire), while all British vice-counties support an average of 40% (Stace and Ellis, 2004). The lower proportion on the Sefton Coast would be expected as the majority of the area is semi-natural habitat designated as Special Area of Conservation under the 1992 EU Habitats Directive (Directive 92/43/EEC on the Conservation of Natural Habitats and of Wild Fauna and Flora). In contrast, vice-county 59, and many other British vice-counties, contain large centres of population, thereby encouraging high proportions of non-native species (Gilbert, 1989) and the establishment of invasive species (Williamson and Fitter, 1996).

Resulting from consistent recording effort by one recorder on seven management compartments, Table 1 illustrates the occurrence of non-natives on the Sefton Coast. General predictions from these data (due to small sample number and variable size, shape and location)

Table 1. Number and proportion of introduced vascular plant species on seven sand-dune sites. All site species lists have been maintained by the effort of a single recorder (Smith, This volume).

Site	Area (ha)	Shore length (km)	Housing boundary length (km)	Total vascular plant species	No. of introduced plant species	% introduced plant species
Birkdale Green Beach	62	3.5	0	267	33	13
Southport Marine Lake dunes	6	0	0	176	30	17
Queen's Jubilee Nature Trail	9	0	0	229	45	20
Crosby Sand-dunes	8.5	0	2.0	141	31	22
Falklands Way	21	0	1.4	344	125	36
Kenilworth Road dunes	6	0	1.2	243	103	42
Cabin Hill NNR	26	0.4	0	354	53	15

are difficult although the high proportion on the Kenilworth Road dunes illustrates the impact of small patch size within a residential area. Direct field observations do, however, clearly indicate the source of most introductions.

Disturbed ground, often associated with residential areas, is the recorded habitat for 61% of the species listed in Smith's (2005) vascular plant species list. A large proportion of these are garden escapes. In decreasing order, other habitats recorded for non-natives are dune scrub (15%), fixed dunes (14%), woodland (11%), slacks ditches and scrapes (5%), dune grassland (4%), and dune heath, saltmarsh, strandline and mobile dunes, all less than 1%.

Major plant pests such as Japanese Knotweed, Giant Hogweed *Heracleum mantegazzianum* and Indian Balsam *Impatiens glandulifera* (Defra, 2003) are not frequent on the Sefton Coast, appearing occasionally in disturbed areas (Smith, 2005).

Sources, mechanisms and impacts

Site-based plant species lists (discussed above, for example Smith, 2005), published literature (e.g. Savidge et al, 1963; Jones et al., 1993 and reports in local naturalists journals), vegetation surveys (e.g. numerous unpublished reports by P. H. Smith lodged with Sefton Council; Edmondson et al, 1988/9), Edmondson's housing boundary survey (reported here), and field observations suggest three major mechanisms for the introduction of non-native plants to the Sefton Coast dunes:

1 Deliberate planting of woody species for dune stabilisation works.
2 Dumping of garden waste.
3 Delivery of propagules to the shore by the tides.

Examples of other mechanisms for introductions are given in Table 2.

Deliberate planting of woody species for dune stabilisation works

This is the most significant group of non-native species on the Sefton dunes, in both area and impact. Following the first experiments of Charles Weld Blundell in 1887, large areas were planted with conifers, mainly in the early 1900s through to 1925 (Jones et al., 1993), to stabilise mobile dunes, produce a timber crop and provide shelter for inland uses of the dunes, including agriculture and communication lines (Gateley and Michell 2004). The plantings were largely of Corsican Pine, but also included Scots Pine *Pinus sylvestris*, Lodgepole Pine *Pinus contorta*, Black Pine *Pinus nigra* ssp. *nigra*, and Maritime Pine *Pinus pinaster*.

Poplars *Populus* spp. and Sea Buckthorn (a native of the east coast of Britain, considered to have been introduced to the west coast) are the most significant of other woody non-native species, planted in the late nineteenth and early twentieth centuries, that have

Table 2. Some minor mechanisms for introduction of non-native plants on the Sefton Coast.

Mechanism of introduction	Examples
Anemochory (wind dispersal)	Canadian Goldenrod and Michaelmas-daisy both produce wind dispersed seeds and are frequent on disturbed areas of the dunes. The non-native moss *Campylopus introflexus* (Smith, 2004) is widespread on the grey dunes.
Zoochory (animals)	Cotoneasters *Cotoneaster* spp. are occasional on the open dunes at some distance from gardens or other sources. Sea Buckthorn can originate from seedlings. Both are spread by birds that feed on the fruits.
Past agricultural activity	Garden Asparagus is frequent on the fixed dunes originating from the formerly extensive asparagus growing on the dunes (Edmondson *et al.*, 1993).
Deliberate planting in landscape and amenity schemes	Pines are still being planted on the dunes. Japanese Rose has been planted in amenity areas, for example car parks as well as originating from gardens, and spreads onto open dunes (Gateley and Mitchell, 2004).
Deliberate planting by residents	Residents have reported planting on the dunes to the authors, e.g. Apple *Malus domestica* and Hybrid Bluebells. Gardening on the dunes has also been observed with a range of species planted, for example Shrub Ragwort *Brachyglottis* 'Sunshine' and *Berberis* X *ottawensis* 'Superba' (for the latter see Phillips and Rix, 1989).
Bird seed	Common Millet *Panicum miliaceum*, Canary-grass *Phalaris canariensis*, and Foxtail Bristle-grass *Setaria italica* have all been recorded as casuals on the Hightown dunes in the proximity of a garden aviary.
Picnics / litter	Grape-vine *Vitis vinifera* has been recorded on Birkdale LNR at some distance from gardens; assumed to originate from a seedling.

become invasive. White Poplar, Balm of Gilead *Populus* X *jackii* and Grey Poplar *Populus* X *canescens* were planted in the mid twentieth century as windbreaks around residential areas and for sand stabilisation on golf courses (Centre for Marine and Coastal Studies, 2000).

Dunes planted with, or invaded by, these woody species are 'fossilised' in a condition where sand movement can no longer occur. They can no longer act as dynamic dunes, providing early successional communities for specialist sand dune biota, the characteristic landscapes of a dune coast, or the mechanism for flexibility in the face of environmental change.

Few plant species persist beneath the canopy of mature pine plantations, apart from patches of the moss *Hypnum cupressiforme* and Broad Buckler-fern *Dryopteris dilatata*, and dense swards of Sand Sedge *Carex arenaria* around the margins and in woodland glades (Edmondson *et al.*, 1988/9a; Sturgess and Atkinson 1993). **These plantations** cover about 300 ha of the Sefton dunes, i.e. 15% of the whole system (Gateley and Mitchell, 2004). Thus, this large area of the dune system retains little of the native dune vegetation, with the notable exceptions of the rare Dune

Helleborine *Epipactis dunensis* and Green-flowered Helleborine *E. phyllanthes*.

Mature pines seed freely onto the open dunes becoming a component of the mixed scrub (Edmondson *et al.*, 1988/9a) that has required clearance to promote favourable conservation status of open habitats. They also create shelter that has accelerated the development of scrub on the landward side of the plantations, causing further habitat deterioration. In addition, the plantations lower the dune water table through enhanced interception and evapotranspiration (Clarke, 1980; Betson *et al.*, 2002) thus negatively affecting the base level for geomorphological activity and the ecohydrology of the dune slacks. The landward pine woodlands have provided a refuge for a population of Red Squirrel *Sciurus vulgaris*.

The Open Dune Restoration Project on Ainsdale Sand Dune National Nature Reserve has successfully cleared 27ha of poor quality frontal pine woodland and associated scrub (Phase 1 in 1992, 11ha, Phase 2 in 1995, 16ha). Following clearance, the availability of nutrients released from the pine litter layer, initially encouraged the invasion of nitrophiles such as Rosebay Willowherb *Chamerion angustifolium* and Creeping Thistle *Cirsium arvense* on the dunes, as predicted by Sturgess and Atkinson (1993). However a TEP (2000) survey indicated that 31 out of 34 quadrats on the grey dunes of the area were of a recognizable (albeit poorly matched) sand dune or dune slack community. The 2003 National Vegetation Classification (NVC) Survey (Gateley and Michell, 2004) mapped these areas as mosaics of SD10 *Carex arenaria* dune community, SD19 *Phleum arenarium-Arenaria serpyllifolia* dunes annual community and SD6 *Ammophila arenaria* mobile dune, the presence of SD6 mobile dune indicating the initiation of dune dynamics, a major aim of the pine plantation removal (Centre for Marine and Coastal Studies, 2000). TEP (2000) reported a more rapid recovery of dune slack communities, positive impacts including the successful return of Natterjack Toad breeding and the appearance of large populations of the endangered Yellow Bartsia *Parentucellia viscosa*. This project demonstrates the restoration potential following clearance of pines, but has met with some public concern, resulting in significant financial and human resources being allocated to investigating the appropriateness of the project (Centre for Marine and Coastal Studies, 2000; Edmondson and Velmans, 2001).

In some Poplar patches, remnants of the dune vegetation may persist for some time but scrub encroachment mostly results in significantly enhanced nutrient status with the associated arrival of nitrophiles such as Common Nettle *Urtica dioica* and Bramble *Rubus fruticosus* and the replacement of native dune flora and fauna.

Sea buckthorn has nitrogen fixing nodules associated with its roots (Stewart and Pearson, 1967) thus significant increases in soil nitrogen are associated with the development of SD18 *Hippophae rhamnoides* dune scrub (Rodwell, 2000). Much has been removed, along with stands of other shrub species by ongoing operational and 'emergency habitat restoration' efforts (Rooney, 2001). Most of the remaining stands on the Sefton coast are well established and are of the SD18b *Urtica dioica-Arrhenatherum elatius* sub-community type. Under the dense canopy of mature stands the only typical associates are sparsely distributed nitrophilous species, frequently Bittersweet *Solanum dulcamara* and Common Nettle (Gateley and Michell, 2004). In continental northwest Europe, where Sea Buckthorn is native, mature stands can be more species rich than the stands on the Sefton Coast (Isermann, 2008a), although declines in the typical, light demanding dune species resulting from the shrub cover are similar (Isermann *et al.*, 2007).

Explosive spread of these invasive, woody species, together with native shrubs such as birch *Betula* spp., hawthorn *Crataegus monogyna*, and willows *Salix* spp., was a result of release from rabbit-grazing pressure following the introduction of myxomatosis in the 1950s. Significant conservation effort has been required to maintain favourable conservation status of the dune system (Simpson *et al.*, 2001) although extensive stands of scrub of both native and non-native species persist, for example on the Birkdale dunes.

In summary, these introduced woody species have been very invasive on the Sefton dunes, have been the cause of a significant loss of dune biodiversity and habitat quality, and have necessitated expenditure of significant resources in habitat restoration.

Dumping of garden rubbish

The high incidence of garden escapes along the dune/housing boundaries was first formally recorded in the 1988 NVC survey (Edmondson et al., 1988/9). Survey has shown that the source of these garden escapes is the dumping of garden rubbish, very few introductions resulting from unassisted spread from the gardens to the dunes. The housing boundary survey recorded a garden waste tip every 19m, and a plant introduction every 12m (mean figures: counts divided by boundary length).

Sycamore *Acer pseudoplatanus* is now abundant on the dunes. Listed as a garden escape in Clement and Foster (1994) this plant has almost certainly been planted in various parts of the dune system but is widely naturalised throughout the British Isles and would have arrived independently by anemochory.

The concentration of non-native species in domestic gardens probably constitutes the biggest source of potentially invasive, non-native plants (Smith *et al.*, 2006). The impact of garden escapes on the Sefton Coast dunes, however, is spatially much more limited than the woody species described above. Some anemochorous species such as Canadian Goldenrod *Solidago canadensis* and Confused Michaelmas-daisy *Aster novi-belgii* occur frequently in patches throughout the dunes but most are restricted to a clearly detectable zone of modified dune vegetation, averaging 4.3m wide, along housing boundaries. This vegetation is characterized by an enhanced concentration of non-native species (garden escapes) and also plants characteristic of mesotrophic conditions, contrasting with the nutrient-stressed substrates required by dune vegetation. The input of organic matter by dumping garden rubbish releases this nutrient stress, allowing nitrophilous plants, such as Bramble, Cock's-foot *Dactylis glomerata*, Common Nettle, Creeping Thistle and Mugwort *Artemisia vulgaris,* to become dominant. Propagules and throw-outs in garden rubbish become established on the dunes, aided by the disturbance caused by the dumping.

Some garden escapes are casuals, but a number are established along the housing boundary, the ten most frequent being Bearded Iris, Bluebells (Hybrid and Spanish), Butterfly Stonecrop *Sedum spectabile*, Greater Periwinkle *Vinca major*, Meadow Crane's-bill *Geranium pratense* (native in the UK but not on the Sefton dunes), Montbretia, Red Hot Poker *Kniphofia uvaria*, Russian Vine *Fallopia baldschuanica,* Snow-in-summer and Soapwort *Saponaria officinalis*. Most of these species are persistent perennials that spread in gardens, thus plants with roots and soil are dumped in garden waste. No long-term data on the behaviour of these plants is available, but the patches of some species, for example Bearded Iris and Red Hot Poker, are relatively stable in the dune sward and do not spread significantly. Their occurrence may however, be increasing due to persistent inoculation.

Some introductions spread within the boundary zone. Thus, Bluebells increase quite rapidly by seeding, as does Snow-in-summer, which can also spread vegetatively and patches of these species occur at some distance from housing boundaries. Russian-vine and Broad-leaved Everlasting-pea *Lathyrus latifolius* (both fast-growing climbing/scrambling plants), spread vegetatively to form large patches on the dunes that can outcompete and replace the native dune vegetation. Such species are therefore potentially invasive, although factors that might encourage their spread are unknown.

Less robust species, such as Snow-in-summer and Bluebells coexist with native dune species thus must modify the vegetation. These plants have the potential to spread more widely onto the dunes, as have the Evening-primroses in the past. The cumulative effect on nature conservation value of the dunes of such species continuing to spread is not known. The small

Australasian native moss *Campylopus introflexus*, first recorded in the UK in 1941 (Smith, 2004), is widespread on the grey dunes and its occurrence is known to reduce the plant diversity, particularly in respect of other bryophytes and lichens (van der Meulen *et al.*, 1987). This suggests that there is a danger in assuming these species to be benign, their very presence being indicative of some impact on dune habitats.

Japanese Rose has been introduced both as a garden escape and through landscape planting around car parks and amenity areas. It is further spread by zoochory (Isermann, 2008). Gateley and Michell (2004) recorded 13 patches of vegetation dominated by Japanese Rose, as well as frequent occurrences of the species in stands of mixed scrub. Japanese Rose dominated patches retain few typical dune species, are sometimes monospecific and show evidence of spread onto open dunes. This species occurs widely on British dunes and is listed as an indicator of unfavourable dune condition (Joint Nature Conservation Committee, 2004). It is one of the most problematic invasive species in coastal areas of Germany, where control or clearance are extremely difficult (Isermann, 2008).

Delivery of propagules by tides

The presence of non-native species on high strand lines at the Green Beach, Birkdale, (e.g. Garden Angelica *Angelica archangelica*) is evidence for this mechanism. Other garden species recorded on the Green Beach (which is isolated from housing) include Annual Sunflower *Helianthus annuus*, Garden Lobelia *Lobelia erinus*, Garden pea *Pisum sativum*, Rose Campion *Lychnis coronaria*, Snapdragon *Antirrhinum majus*, Tomato and Yellow-eyed grass *Sisyrinchium californicum* (Smith, 2003).

This, and other sources of introductions, collectively have less impact numerically and spatially than the woody species and garden escapes discussed above.

DISCUSSION

Non-native vascular plants are widespread and increasing on the Sefton Coast and have significant, negative impacts, including loss of biodiversity and the need for expenditure on control and habitat restoration. These negative impacts support Isermann's (2008; 303) advice that 'the prohibition of planting alien species on dunes' should be encouraged. The potential danger is illustrated by the recent invasion of Rum Cherry *Prunus serotina* on the Amsterdam Watersupply Dunes. This species, planted to promote soil improvement in association with pine plantations, was only thought to be a problem in the 1980s, for native species regeneration in coastal woodlands. Since 1989, it has undergone an explosive spread onto the open, grey dunes replacing the native vegetation over wide areas (Ehrenberg, 2008). The change in dynamics of this species illustrates the importance of vigilance in surveillance, and a precautionary approach in nature conservation management policy and practice on the dunes. Perhaps fortunately, only a few specimens of Rum Cherry are known from the Sefton Coast.

Several factors are important when considering the invasion potential arising from the continued presence and increased numbers of non-natives. Despite some criticism, Williamson's (1993) 10:10 rule remains a useful generalisation (Richardson and Pyšek, 2006); the more species that become established, the higher the probability that one or more will become invasive.

The size of the inoculation, time, propagule pressure and chance events are all important in determining whether a species will invade (Reymanek *et al.*, 2005a).

The large inoculation, numerically and spatially, of pines and associated woody species, clearly has been a factor in their consequent invasion onto the dunes. The frequent and consistent occurrence of some garden escapes along housing boundaries also constitutes a significant inoculation size. Residence time in the receiving plant community also increases the probability of invasions, particularly in its effects on propagule pressure, time encouraging increased numbers of propagules thus a greater chance of dispersal (Reymanek *et al.*, 2005b). Residence time can be more important than plant traits in explaining the distributions of non-natives, and is associated with a lag phase before exponential spread (Pyšek and Jaosik,

2005). The continued presence of established garden escapes on the dunes, together with continued inoculation by garden waste dumping, can therefore be taken as a warning of future potential invasion.

Most introduced species will not become established and invasive. The number of non-native species in a region is the product of the number of species introduced and their survival rate in the receiving plant community due to chance factors, competition, herbivory, pathogens and inappropriate adaptation for the new environment (Lonsdale, 1999). The invasibility of dune plant communities thus must also be considered.

The roles of disturbance, competitive release, resource availability and propagule pressure are considered to be key factors in community invasibility (Richardson and Pyšek, 2006). The importance of disturbance for introductions on the Sefton Coast is shown by the high proportion occurring on disturbed ground (Smith, 2005). Starting from the premise that an invasive species must have access to essential resources (light, water, nutrients), the fluctuating resource theory of invasibility (Davis et al, 2000) proposes that increases in availability of these resources may facilitate invasion. Although light would not be limiting, the natural dune substrates are nutrient poor and (on the dune ridges) dry thus an increase of water or nutrient availability would increase invasibility. At the housing boundaries, the dumping of garden waste on the dunes simultaneously disturbs and adds nutrients and propagules, thus establishment of garden escapes in this zone is not surprising. Although residence time will increase invasion probability, nutrient and water limitation may be a factor in restricting these species to the modified boundary vegetation zone. Increased nutrient status resulting from succession, or atmospheric nitrogen deposition (Jones et al., 2004) may release invasion opportunities on the open dunes. 'Invasional meltdown' (Simberloff and Von Holle, 1999) could then ensue, with non-native species enhancing the susceptibility of the dunes to more invading species.

Climate change may also alter the invasibility of the dune vegetation. Although likely to decrease water availability at least in the summer, increased temperatures, altered rainfall patterns and increased winter storminess forecasted by the UK Climate Impacts Programme may impact on dune vegetation in such a way as to increase invasion opportunities. Pampas-grass Cortaderia selloana, for example, may become invasive with higher summer temperatures. It occurs rarely on disturbed ground in the Sefton dunes, but is invasive on dunes in Spain. Dukes and Mooney (1999) concluded that the most important elements of global change (including nitrogen deposition and habitat fragmentation) are likely to increase the spread of invasive species. Using horizon-scanning, Sutherland et al. (2008) list 'facilitation of non-native invasive species through climate change' and 'invasional meltdown'as a medium likelihood, high-threat, issue for UK biodiversity.

A final question surrounds the extent to which non-native species can become integrated with the native flora without negative impacts on biodiversity. Manchester and Bullock (2000) identify a group of non-native species that they consider to be benign or even beneficial, including some charismatic species that are now considered part of the UK biodiversity. The Evening-primroses, two of which are now abundant on fixed dunes and disturbed ground, could be described as charismatic in their striking visual impact. They have thus 'invaded' the dunes, but are considered as 'naturalised' rather than 'invasive' because they do not dominate and exclude other species of the native dune flora. Several such species are now widely distributed on the dune system; Garden Asparagus Asparagus officinalis ssp. officinalis and Garden Tulip Tulipa gesneriana are frequent on disturbed ground and fixed dunes. These species are not considered invasive or viewed as a significant conservation threat. Spring Beauty Claytonia perfoliata is abundant in dune scrub and disturbed ground, sometimes in dense (although short-lived) stands. Snow-in-summer is common along housing boundaries and appears to be spreading. What is the combined net impact of these species on native dune biodiversity? Is it appropriate to call them benign? Further detailed research may provide site- and species-specific answers to this question for the Sefton Coast but it is important to consider these questions in the wider context of global change. Biological invasions are recognized as an important element of global change (Dukes and Mooney, 1999), some of which is natural, but most is accelerated by human-induced factors such as climate change and nitrogen deposition. Models of future changes in native species distributions resulting from climate change (e.g. Berry *et al.*, 2002) indicate that the baseline of current distributions on the dunes may not continue to be valid for future assessment of impacts of non-native species. Evidence presented here does, however, suggest that combining the threat of climate change, with continued introduction and acceptance of non-native species, is risky for sustainability of the dune biodiversity.

This paper focuses on the distribution and impacts of non-native species on the Sefton Dunes. The attributes of invasive species are not unique; many are shared with native species (Thompson *et al.*, 1995). Birch, Willows and Hawthorn have all invaded dune and slack habitats and been cleared in habitat restoration projects. Indicators of unfavourable condition include native invasive species such as Bramble and Creeping Thistle (Joint Nature Conservation Committee, 2004). Wider issues, such as enhanced nitrogen deposition (Jones et al., 2004), or climate change, and local issues such as grazing pressure will affect the invasive potential of native and non-native species alike.

CONCLUSIONS

Non-native species are an established and increasing element of the Sefton Coast dune flora. Woody species planted for sand stabilization, shelter and landscaping are the most significant in terms of area and impact. They have become invasive and have had severe negative impacts on the dune biodiversity. Garden escapes are frequent on the dunes and inoculation continues, resulting in a zone of modified boundary vegetation around housing areas. The actual and potential impacts of these and other non-natives on dune biodiversity are not known. Increased residence time, disturbance and/or environmental change (including climate change) may reduce the resistance of the dune vegetation to invasion by species that are currently restricted. We should minimize any future introductions and be vigilant of existing established species, with a view to taking action if a significant threat to native dune flora and fauna is identified.

ACKNOWLEDGEMENTS

I am indebted to Phil Smith for constantly collecting data on plants on the Sefton Coast, for Table 1 and for his comments.

REFERENCES

Berry, P.M., Dawson, T.P., Harrison, P.A. and **Pearson, R.G.** (2002) Modelling potential impacts of climate change on the bioclimatic envelope of species in Britain and Ireland. *Global Ecology and Biogeography*, 11, (6), 453-462.

Betson, M., Connell, L. and **Bristow, C.** (2002*)* *The Impact of Forestry on Coastal Geomorphology at Newborough Warren/Yn ys Llanddwyn NNR, SSSI, pSAC. Hydrogeology Volume 5 Groundwater Modeling of Newborough Warren.* Contract number: FC 73-05-18 Unpublished report commissioned by the Countryside Council for Wales. Birkbeck College.

Centre for Marine and Coastal Studies (2000) *Review of Dune Restoration – Ainsdale Sand Dunes National Nature Reserve, Final Report.* Centre for Marine and Coastal Studies, University of Liverpool. Report for English Nature Refs GL155 and GL173.

Clarke, D. (1980) *The Groundwater Balance of a Coastal Dune System: A study of the water table conditions in Ainsdale Sand Dunes National Nature Reserve, Merseyside.* Unpublished PhD Thesis, University of Liverpool.

Clement, E.J. and **Foster, M.C.** (1994) *Alien Plants of the British Isles* Botanical Society of the British Isles, London.

Crawley, M.J. (1987) What makes a community invasible? In S. J. Gray, M. J. Crawley, and P. J. Edwards (eds.) *Colonization, succession, and stability.* Blackwell Scientific, Oxford. pp. 429-453.

Cronk, Q.C.B. and **Fuller, J.L.** (2001) *Plant Invaders: The threat to Natural Ecosystems.* Earthscan, London.

Davis, M.A., Grime, J.P. and **Thompson, K.** (2000) Fluctuating Resources in Plant Communities: A General Theory of Invasibility. *Journal of Ecology,* 88(3), 528-534.

Dawson, F. H. and **Warman, E. A.** (1987) *Crassula helmsii* (T. Kirk) cockayne: Is it an aggressive alien aquatic plant in Britain? *Biological Conservation, 42, 4,* 247-272

Department for Environment, Food and Rural Affairs (2003) *Review of non-native species policy.* Department for Environment, Food and Rural Affairs, London.

Department for Environment, Food and Rural Affairs (2008) News Release: non-native species to be tackled http://www.defra.gov.uk/news/2008/080528a.htm accessed May 2008.

Dukes, J.S. and **Mooney, H.A.** (1999) Does global change increase the success of biological invaders? *Trends in Ecology and Evolution,* 14 (4), 135-140.

Edmondson, S.E., Gateley, P.S. and **Nissenbaum, D.A.** (1988/9a) *National Sand Dune Vegetation Survey: The Sefton Coast.* Division 6 Ainsdale Sand Dunes National Nature Reserve Nature Conservancy Council Report 917.

Edmondson, S.E., Gateley, P.S., Rooney, P. and **Sturgess, P.W.** (1993) Plant Communities and Succession. In D. Atkinson and J. Houston, (eds.) *The Sand Dunes of the Sefton Coast,* National Museums and Galleries on Merseyside. 65-84.

Edmondson, S.E. and **Velmans, C.** (2001) Public perception of nature management on a sand dune system. In J.A. Houston, S.E. Edmondson and P.J. Rooney (eds.) *Coastal Dune Management: Shared Experience of European Conservation Practice* Liverpool University Press, Liverpool. pp. 206-218.

Ehrenberg, A. (2008) *Invasion and management of* Prunus serotina *in the Amsterdam Watersupply Dunes at the Dutch coast: can it be stopped? Presentation to the International Dune Conference: Changing Perspectives in Coastal Dune Management* 31 March - 3 April 2008, Liverpool, UK.

Gateley, P.S. and **Michell, P.E.** (2004) *Sand Dune Survey of the Sefton Coast, 2003/4.* TEP (Consultancy Report for Sefton Council).

Gilbert, O.L. (1989) *The Ecology of Urban Habitats.* Chapman and Hall, London.

Green, C.T. (ed.) (1933) *The Flora of the Liverpool District.* T. Buncle and Co., Arbroath.

Hill, M., Baker, R., Broad, G., Chandler, P.J., Copp, G.H., Ellis, J., Jones, D., Hoyland, C., Laing, I., Longshaw, M., Moore, N., Parrott, D., Pearman, D., Preston, C., Smith, M. and **Waters, R.** (2005) *Audit of non-native species in England.* English Nature, Peterborough.

Isermann, M. (2008) Effects of *Rosa rugosa* invasion in different coastal dune vegetation types. In B. Tokarska-Guzik, J.H. Brock, G. Brundu, L. Child, C.C. Daehler and P. Pyšek (eds) *Plant Invasions: Human perception, ecological impacts and management* Backhuys Publishers, Leiden. pp. 289-306.

Isermann, M. (2008a) Expansion of *Rosa rugosa* and *Hippophae rhamnoides* in coastal grey dunes; effects at different spatial scales. *Flora.* (on-line).

Isermann, M., Diekman, M. and **Heemann, S.** (2007) Effects of the expansion by *Hippophae rhamnoides* on plant species richness in coastal dunes. *Applied Vegetation Science.* 10, 33-42.

Joint Nature Conservation Committee (2004) *Common Standards Monitoring Guidance For Sand Dune Habitats.* Joint Nature Conservation Committee.

Jones, C.R., Houston, J. A. and **Bateman, D.** (1993) A History of Human Influence on the Coastal Landscape. In D. Atkinson and J. Houston (eds.) *The Sand Dunes of the Sefton Coast* National Museums and Galleries on Merseyside, pp. 3-17.

Jones, M.L.M., Wallace, H.L., Norris, D., Brittain, S.A., Haria, S., Jones, R.E., Rhind, P.M., Reynolds, B.R. and **Emmett, B.A.** (2004) Changes in vegetation and soil characteristics in coastal sand dunes along a gradient of atmospheric nitrogen deposition. *Plant Biology,* 6(5), 598-605.

Lonsdale, W.M. (1999) Global patterns of plant invasions and the concept of invisibility. *Ecology,* 80, 1522-1536.

Manchester, S.J. and **Bullock, J.M.** (2000) The impacts of non-native species on UK biodiversity and the effectiveness of control. *Journal of Applied Ecology,* 37: 845-864.

Phillips, R. and **Rix, M.** (1989) *The Garden Plant Series: Shrubs.* Macmillan, London.

Preston, C.D., Pearman, D.A. and **Dines, T.D.** (2002) *The New Atlas of the British and Irish Flora: An Atlas of the Vascular Plants of Britain, Ireland, The Isle of Man and the Channel Islands.* Oxford University Press, Oxford.

Pyšek, P. and **Jaosik, V.** (2005) Residence time determines the distribution of alien plants. In S. Inderjit (ed.) *Invasive Plants: ecological and agricultural aspects.* Birkhauser Verlag, Basel. pp. 77-96.

Pyšek, P., Sádlo, J. and **Bohumil, M.** (2002) Catalogue of alien plants of the Czech Republic. *Preslia, Praha,* 74, 97–186.

Reymanek, M., Richardson, D.M. and **Pyšek, P.** (2005a) Plant invasions and invasibility of plant communities. In E. van der Maarel (ed) *Vegetation Ecology.* Blackwell, Oxford. 332-55.

Reymanek, M., Richardson, D.M., Higgins, S.I., Pitcarin, M.J. and **Grokopp, E.** (2005b) Ecology of invasive plants: state of the art. In H.A. Mooney, R.M. Mack, J.A. McNeely, L. Neville, P. Schei, amd J. Waage (eds) *Invasive alien species: searching for solutions.* Island Press, Washington. 104-161.

Richardson, D.M. and **Pyšek, P.** (2006) Plant invasions: merging the concepts of species invasiveness and community invisibility. *Progress in Physical Geography,* 30(3), 409-431.

Rodwell, J.S. (2000) *British plant communities. Volume 5: Maritime communities and vegetation of open habitats.* Cambridge University Press, Cambridge.

Rooney, P.J. (2001) The Sefton Coast Life Project: A Conservation Strategy for the Sand Dunes of the Sefton Coast, Northwest England. In J.A. Houston, S.E. Edmondson and P.J. Rooney (eds.) *Coastal Dune Management: Shared Experience of European Conservation Practice.* Liverpool University Press, Liverpool. pp. 243-254.

Savidge, J.P., Heywood, V.H. and **Gordon, V.** (eds.) (1963) *Travis's Flora of South Lancashire.* Liverpool Botanic Society.

Simberloff, D. and **Von Holle, B.** (1999) Positive interaction of non-indigenous species: invasional meltdown? *Biological Invasions.* 1. 21-32.

Simpson, D.E., Rooney, P. and **Houston, J.A.** (2001) *Towards best practice in the sustainable management of sand dune habitats:* 2. Management of the Ainsdale dunes on the Sefton Coast. In J.A. Houston, S.E. Edmondson and P.J. Rooney (eds) *Coastal Dune Management: Shared Experience of European Conservation Practice.* Liverpool University Press, Liverpool, 262-271.

Smith, A.J.E. (2004) *The Moss Flora of Britain and Ireland.* Second edition. Cambridge University Press, Cambridge.

Smith, P.H. (2003) *Further studies on the vascular plants of the Birkdale Green Beach.* Unpublished report to Sefton Council.

Smith, P.H. (2005) *An inventory of vascular plants identified on the Sefton Coast.* Unpublished report to Sefton Coast Partnership.

Smith, P.H. (2007) *Further additions to the Inventory of Vascular Plants for the Sefton Coast.* Unpublished report to Sefton Coast Partnership.

Stace, C. (1991) *New Flora of the British Isles.* Cambridge University Press, Cambridge.

Smith, R.M., Thompson, K., Hodgson, J.G., Warren, P.H. and **Gaston, K.J.** (2006) Urban domestic gardens (IX): Composition and richness of the vascular plant flora, and implications for native biodiversity. *Biological Conservation,* 129, 3. 312-322.

Stace, C.A. and **Ellis, R.G.** (2004) *Area and species richness of British vice counties. Botanical Society of the British Isles News,* 97, 15-19.

Stewart, W.D.P. and **Pearson, M.C.** (1967) Nodulation and nitrogen-fixation by *Hippophaë rhamnoides* L. in the field. *Plant and Soil,* 26(2), 348-360

Sturgess, P.W. (1993) Clear-felling Dune Plantations: Studies in Vegetation Recovery on the Sefton Coast. In D. Atkinson and J. Houston, (eds.) *The Sand Dunes of the Sefton Coast,* National Museums & Galleries on Merseyside. pp. 85-89.

Sturgess, P.W. and **Atkinson, D.** (1993) The clear-felling of sand-dune plantations: soil and vegetational processes in habitat restoration. *Biological Conservation,* 66, 171-183.

TEP (2000) *Ainsdale Sand Dunes National Nature Reserve. NVC survey of Dune Restoration Area.* Unpublished report for English Nature, report ref. 421.001.

Sutherland, W.J., Bailey, M.J., Bainbridge, I.P., Brereton, T., Dick, J.T.A., Drewitt, J., Dulvy, N.K., Dusic, N.R., Freckleton, R.P., Gaston, K.J., Gilder, P., Green, R.E., Heathwaite, A.L., Johnson, S.M., Mcdonald, D.W., Mitchell, R., Osborn, D., Owen, R.P., Pretty, J., Prior, S.V., Prosser, H., Pullin, A.S., Rose, P., Stott, A., Tew, T., Thomas, C.D., Thompson, D.B.A., Vickery, J.A., Walker, M., Walmsley, C., Warrington, S., Watkinson, A.R., Williams, R.J., Woodroffe, R. and **Woodroof, H.J.** (2008) Future novel threats and opportunities facing UK biodiversity identified by horizon scanning. *Journal of Applied Ecology,* 45, 821-833.

Thompson, K., Hodgson, J.G. and **Rich, T.C.G.** (1995) Native and alien invasive plants: more of the same? *Ecography,* 18, 390-402.

Van der Meulen, F., Van der Hagen, H. and **Kruijsen, B.** (1987) *Campylopus introflexus. Invasions of a moss in Dutch Coastal dunes.* Proceedings of the Koninklijke Nederlandse Akademie van Wetenschappen Series C 90. 73-80.

Wheldon, J.A. (1913) The *Oenothera* of the South Lancashire Coast Lancashire. *Naturalist,* 6, 205-210.

Williamson, M. (1993) Invaders, weeds and the risk from genetically manipulated organisms. *Experientia,* 49, 3, 219-224.

Williamson, M. (1996) *Biological Invasions* Chapman and Hall, London.

Williamson, M. and **Fitter, A.** (1996) The Characters of Successful Invaders. *Biological Conservation*, 78, 163-170.

The status of the Sand Lizard *Lacerta agilis* on the Sefton Coast: A recent study

Dave Hardaker

23 Forefield Lane, Crosby, Merseyside L23 9TG
davehardaker@yahoo.co.uk

Plate 1. Male and Female Sand Lizards in typical habitat. (Photograph by Dave Hardaker).

ABSTRACT

The Sand Lizard (*Lacerta agilis*) is a rare and endangered species, the Sefton Coast being one of its last remaining British strongholds. In the past fifty years it has suffered from both habitat loss and changes resulting in a substantial population decline, numbers falling from an estimated high point of 8-10,000 to a low of 250 in the 1970s (Corbett, 1974.) The accuracy of these figures is difficult to establish and no recent systematic surveys have taken place. Currently, there is little known about the status and distribution of Sand Lizards on the Sefton coast. To address this gap in information, members of the North Merseyside Amphibian and Reptile Group have embarked upon a long term population study. It was decided to conduct intensive, saturation monitoring at four known breeding colonies. There is evidence of an increase in coastwide numbers to perhaps 1000-1500 adults.

INTRODUCTION

The Sand Lizard ranges across northern and central Europe but is declining over much of its range (Berglind, 2005). In the UK, natural populations have become extinct in many counties including Cornwall, Devon, Wiltshire, Kent, Sussex, Hampshire, Berkshire and the whole of north and central Wales. It is now mainly restricted to its former strongholds of Dorset, Surrey and North Merseyside, where its local history and past population are unclear. In his *Handbook for Southport,* McNicholl (1883) states of reptiles: '*…none of the family of snakes but an abundance of other kinds.*' It is likely that, during this period, a distinction between

Sand and Common Lizards was not made but the word 'abundance' is significant.

The Merseyside population reaches the northern limit of its native British range and is separated by over 300km from southern colonies. As a result of geographical isolation over thousands of years, this population has developed into a distinct colour form, with the breeding males a rich lime green as distinct from the emerald or bottle green of southern animals.

The Sand Lizard is found in fragmented areas of dune habitat in loose colonies, its decline having been documented from perhaps 8-10,000 post war to as few as 250 in the 1970s (Corbett, 1974) or 'a few hundred' (Jackson, 1979). Simms (1966) estimated numerical strength on the basis of recapture data, the upper limit being 700 individuals. However, by 1971 Shorrock (1973) reports the species in a critical state and near to total extinction, citing destruction of major breeding sites by holiday camp and housing development at Ainsdale. A survey between April 1986 and August 1989 yielded a total of 228 sightings (Lunn and Wheeler, 1989). In an attempt to bolster the population, a joint HCT/NCC captive breeding programme released 227 young lizards at three sites at Ravenmeols in the 1980s (Atkinson and Houston, 1993).

This species' presence is a reason for the designation of both the Sefton Coast SSSI and SAC. It is listed under the Wildlife and Countryside Act 1981 and the EU Habitats Directive as a species of European significance. The Sand Lizard is also included in the UK Biodiversity Steering Group Report as a priority species for conservation action and is the subject of a Species Action Plan in the North Merseyside Biodiversity Action Plan (NMBAP). The purpose of the latter is to promote Sand Lizard conservation issues, including regular monitoring of populations.

Previous studies have been sporadic and population estimates uncertain. This provided the rationale for North Merseyside Amphibian and Reptile Group (NMARG) to embark upon a long term study, concentrating on a limited number of sites in an attempt to collect more precise data. The work began in 2005, initially monitoring hatchlings in autumn to identify suitable study areas. In 2006, a total of 121 adults and juveniles, with 53 hatchlings, was recorded over 35 recording days. It became clear that cumulative totals of 'new' animals continued to rise up to hibernation, suggesting that not all individuals were being seen. In 2007, saturation monitoring was intensified, forming the basis for the current study.

Methods

The main aims of the study were to establish a clear relationship between 'effort' and data collected and to determine an accurate estimate of population size in the study area. It was quickly established that cumulative totals improved relative to effort, therefore saturation monitoring techniques were used with repeated visits to defined areas, when recorders were available, in 'suitable' weather conditions. Ideally, these were a calm, dry day which began sunny, clouded over mid morning and remained warm, perhaps hazy and humid throughout.

The study area falls within the Sefton Coast SSSI and was divided into four 'sites,' with 'foci' populations, three of which are in the Ainsdale area (Hectad SD3010) and were loosely geographically connected. The fourth site lies 10km to the north within the Southport boundary (Hectad SD3519). The term 'foci' is defined by Herpetological Conservation Trust (HCT) as Sand Lizard population centres, usually associated with favourable topographic features and/or high habitat quality. Foci hold a higher density of Sand Lizards than the surrounding areas, perhaps 30+ animals per hectare. Each of the study sites contained a range of habitats, for example sandy banks with grass and scrub, which were monitored up to five times per visit.

The technique was to approach each potential basking spot quietly with a minimum of disturbance, visually searching from a standing position. A circular route was taken and then usually repeated. During 2007, between March 31st and October 21st over sixty days in total were spent in the field, ranging from a minimum of two hours to up to eight hours per visit. For the purposes of this paper, only records of 'new' animals are examined. Where possible, new lizards were photographed, their unique body patterns being used to separate individuals. We did not record 'rustles,' i.e. the sound of a lizard as it disappears into the undergrowth before a sighting could take place. Observations were also made on the feeding behaviour of individuals. A ten figure grid-reference was taken for each lizard using a Garmin Etrex GPS device. No attempt was made to catch or handle any animals and recorders were fully licensed.

Habitat

The study area did not include the coastal frontal dunes, focusing more on well established, fixed dune and other, less traditional, habitat where the dominant plants were largely Creeping Willow (*Salix repens*), Dewberry (*Rubus caesius*), Marram (*Ammophila arenaria*) and other grasses such as Red Fescue (*Festuca rubra*). As indicated in the literature (e.g. Cooke, 1993), Sand Lizards often preferred south or south-east facing banks, especially those affording good ground cover with some loose sand nearby, such as Rabbit (*Oryctolagus cuniculus*) burrow spoil. Gravid females particularly favoured sites with suitable refuges, basking areas and egg-laying habitat within a short distance (Plate 2). Some females were remarkably sedentary, several never being found more than a metre or so away from a favourite basking spot. Lizards rarely basked in a totally exposed position, preferring to remain at least part hidden in vegetation, but did occasionally use discarded garden material, fence posts or wood dumps where they were easier to see (Plate 3).

Food

Food items recorded consisted mainly of abundant invertebrate species such as grasshoppers (Orthoptera, Acrididae) and harvestmen (Opiliones) which either moved within catching distance or were actively pursued. One female was seen to catch a social wasp (Vespidae) and then, within a few seconds, leap back violently, spitting out the wasp which presumably had stung it inside the mouth.

Plate 2. Female basking on sand. (Photograph by Dave Hardaker).

Population

Between late March and the end of October 2007, a total of 348 Sand Lizards were recorded in the study area: 106 males, 89 females, 75 juveniles and 80 hatchlings (Table 1). The sex ratio is 1.2 males per female. 'Juveniles' are defined as sexually immature individuals, not hatched in the year of recording; i.e. sub-adult animals between approximately eight months to slightly less than two years old.

Males emerged from hibernation first and were much more in evidence up to the middle of May, after which sex ratios began to even out and were equal by June. In July, we recorded more new females than males. Most new lizards were recorded in April (84), with a surprising reduction in May (22), perhaps due to poor weather, then peaking again at 63 in July before falling away rapidly by late September, with the onset of hibernation, when a sharp increase in hatchlings took place. By October, no new adults were recorded.

Figures 1 and 2 show that 'new' individuals were found at a fairly constant rate during the season, with some

Plate 3. Male basking on sand. (Photograph by Dave Hardaker).

indication of a slowing in the finding rate later in the summer for males and juveniles. The data suggest that not all individuals were located during the recording period which was terminated by hibernation. The number of hatchlings was probably under-recorded due to the short recording period, as they enter hibernation six to eight weeks after emerging. Their small size (Plate 4) also makes them harder to find in thick vegetation. In the event, the cool, wet summer of 2007 may well have been poor for Sand Lizard hatchlings, thereby contributing to the low numbers found.

Apart from a peak in April, as the rate of adults found increased over the season the proportion of juveniles (Plate 5) fell away and then rose again slightly as adults began to hibernate, suggesting that increased adult activity possibly drove them away from the best basking and feeding areas or forced them to remain hidden. Adult males become aggressive during the summer and I have noted females attempting to escape their attention. Paul Hudson (pers. com.) has also noted a decrease in juveniles in relation to increased adult activity.

Excluding hatchlings, the estimated population of all four sites in total was 269 adults and juveniles within a total area of 7.5ha giving an overall mean density of 36/ha. Most workers monitoring Sand Lizards agree that, even in optimum conditions, only about one in three animals are on view at any one time and the figure may be as low as one in five (Frazer, 1983; Beebee, 2000). Although Figure 1 shows that not all new animals were found, the saturation monitoring technique is thought to be more efficient than other methods and probably resulted in about 50% of animals being recorded. This gives a total estimated population of 540 adults and juveniles at a density of 72/ha in prime habitat across the study area.

Table 1. Sightings of 'new' lizards per month (2007).

	March	April	May	June	July	August	September	October
Males	2	43	10	16	21	11	3	
Females		15	7	16	29	10	12	
Juveniles	3	27	5	7	13	6	12	2
Hatchlings						1	27	52

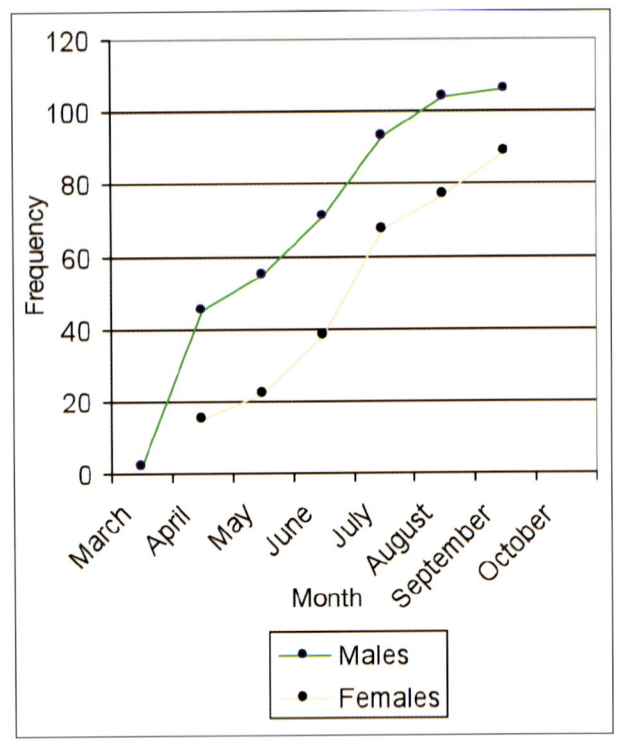

Figure 1. Cumulative sightings of 'new' lizards per month - males and females.

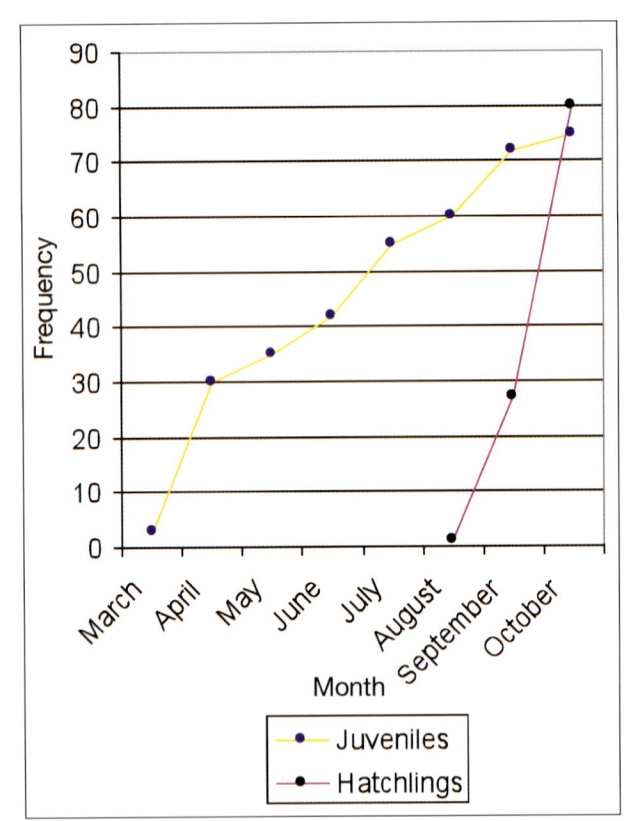

Figure 2. Cumulative sightings of 'new' lizards per month - juveniles and hatchlings.

Other sites from Birkdale to Altcar outside our main study area are known to hold Sand Lizard colonies. Monitoring of some of these areas by members of NMARG has reported viable colonies in the last three years (P. Hudson and J. Newton pers. com.). Using a 'Map-info' program containing the HCT data-base of previous records, eight other main Sand Lizard foci were identified with an estimated area of 11ha. From the density figure from our study, the population of the additional sites is estimated at 792, giving an overall total for adults and juveniles of about 1330.

More systematic research is needed in these outlying areas but it may now be possible to provide a better estimate of the overall Sefton Coast population size. This is hampered by the patchy and aggregated distribution making Sand Lizard detectability very variable, however the 2007 data for breeding adults suggests a minimum total of 900-1000 and, if sub-optimal habitat is included, perhaps as many as 1200-1500.

Conclusions

The aim of this study was to devise a meaningful and repeatable method for assessing the status of the Sand Lizard on the Sefton coast. It has been demonstrated that saturation monitoring is useful for gathering accurate population data. Compared with previous estimates, our 2007 data suggest an increase in numbers; however this is difficult to quantify against the backdrop of an animal known to be declining across its entire range.

Although this study demonstrates that the status of the Sand Lizard on the Sefton coast is in a stronger position than recently thought, the species is still extremely vulnerable, occurring in numbers representing only 10-15% of historical estimates.

Future prospects

The NMBAP proposes a series of actions which are designed to ensure that any strategic management plans take account of the Sand Lizard's requirements (Merseyside Biodiversity Group, 2001). NMARG has recently recommended amendments to the BAP and has drawn attention to a range of significant issues. These include the desirability of controls over the recreational use of the dunes and the resultant ecological impact as Sand Lizards generally inhabit the 'wilder' sections of

Plate 4. Hatchling found in Ainsdale. (Photograph by Dave Hardaker).

Plate 5. Juvenile Sand Lizard. (Photograph by Dave Hardaker).

dune habitat including untrampled sandy banks for basking and egg-laying. That considered, our study has also shown that Sand Lizards can at times demonstrate a remarkable degree of tolerance to human activity and during the active season, may bask within a few metres of visitors, trains or back gardens. It is a delicate balance however and important to draw a clear distinction between human 'activity' and actual 'disturbance'. Maintenance and way-marking of paths and management of habitat requirements of known colonies, as described by Cooke 1993, is essential in areas of greatest public pressure to minimize impacts of pollution, erosion and disturbance by, for example, dogs.

As the Sand Lizard is on the edge of its range in Britain, our cool summers mean that it can only survive in suitable habitats. Therefore, the species is particularly vulnerable to loss, fragmentation, change or excessive recreational use. Near built development, domestic cats (*Felis catus*) are known to take lizards (Larsen and Henshaw, 2001) and recent surveys by the Mammal Society indicate that cats are the main predators of British wildlife (Woods *et al.*, 2003). We have observed them stalking in basking spots, particularly on railway verges and we know of at least one female lizard taken by a cat in 2007. Fragmented habitat is likely to be more susceptible to such predation as cats roam what is left, leaving little or no opportunity for Sand Lizard recolonization as adjoining populations are effectively isolated.

Litter also is hazardous, discarded bottles proving to be death-traps for lizards. Fires cause obvious destruction to vegetation and have been known to destroy colonies. Collection for the pet trade was an issue until the late sixties and, although difficult to quantify, probably contributed to the Sand Lizard's decline. Currently however, this does not seem to pose a serious threat as the species is fully protected under the Wildlife and Countryside Act 1981 (as amended).

HCT points out that genetic 'bottle necking' may become a problem in fragmented sites. Captive breeding and re-introduction programs, as have taken place on the Sefton Coast, may help to improve the genetic mix.

Scrub and tree encroachment, if left unchecked, is perhaps the most immediate threat to the Sefton Sand

Lizard populations, causing shading of basking sites and loss of sand patches for egg laying. From a management standpoint, this is a particularly pressing issue. Effective conservation of this charismatic creature will require a long term management strategy as recommended in the NMBAP.

Addendum 2008 - Additional Population Information

Monitoring and collection of Sand Lizard data continued into 2008, largely focusing on four population foci in the frontal dunes between Formby and Birkdale not covered in the 2007 study. The data below therefore represent new animals.

The same saturation monitoring techniques were employed, resulting in a total of 144 adults and juveniles being recorded: 69 males, 49 females, 26 juveniles and 8 hatchlings (Table 2). The sex ratio is 1.4 males to females compared to 1.2 in the inland areas covered by the 2007 study. Because they emerge earlier and are more conspicuous in the frontal dunes due to colour and behaviour, especially early in the season, males are more likely to be recorded than females, thereby possibly biasing the sex ratio.

Figures 3 and 4 show that the rate at which new animals were found slowed down during July and August, a similar finding to that in 2007. As in previous years, males emerged from hibernation first, with females being recorded in greater numbers as the season progressed. It was noticeable that frontal-dune individuals often emerged earlier in the day due to rapid heating of open sand. Then, in hot, sunny conditions, they were rarely observed after mid-morning until reappearing again in the late afternoon. On a few, exceptionally hot days in midsummer, no lizards were evident in the frontals after 0900 hours.

May was by far the most productive month with 85 individuals being recorded compared with only 22 across the entire study area during that month in 2007. This was partly the result of one exceptional day (3rd May) in the Ainsdale area when group monitoring resulted in 41 individuals being found. Also relevant were favourable weather conditions during that month, enabling eight successful recording days. Poor weather and lack of coverage later in the season account for lower hatchling numbers and fewer new adults and juveniles recorded.

2009

A continuation of the research in 2009 resulted in three additional foci being monitored. Table 3 shows a total of 59 new animals recorded: 22 males, 27 females, 10 juveniles and 77 hatchlings.

These foci were monitored later in the season, this might account for the unusual female to male sex ratio as active males in April would probably have been missed. Also, on one 600m^2 south-facing, fixed-dune bank, we recorded 13 adult females and one male in late July and August at a density of 208/ha. Such unusual concentrations of females suggests some frontal-dune animals may move after egg laying to nearby richer food sources, increased shelter and also to avoid predators such as Stoats (*Mustela erminea*).

A particular highlight of 2009 was the rediscovery of a colony of Sand Lizards at Hightown thought extinct. Common Lizards are abundant in this area and past records for Sand Lizards may be unreliable. However at least one adult male was seen by the author around 1980; none has since been confirmed. Saturation monitoring in recent years resulted in a single, small female and an adult male being located in late May 2009 in the Hightown Dunes. By the end of the recording season we had identified five males, five females and two juveniles. In October, six hatchlings were found whose distribution suggests two, or possibly three females bred. This appears to be single foci representing a tiny, fragmented and vulnerable population threatened with extinction as it is cut off from the larger colonies to the north by the River Alt Estuary. It is hoped that hatchlings from the Natural England captive breeding program will be introduced to supplement this colony to ensure its future.

Also noteworthy was the high number of 77 hatchlings, with 32 of these located in the Formby frontals.

Table 2. Sightings of new Sand Lizards per month in the frontal dunes (2008).

	March	April	May	June	July	August	September	October
Males		11	44	8	3	2	1	
Females		2	26	10	5	2	3	1
Juveniles		1	15	2	4	2	1	1
Hatchlings							3	5

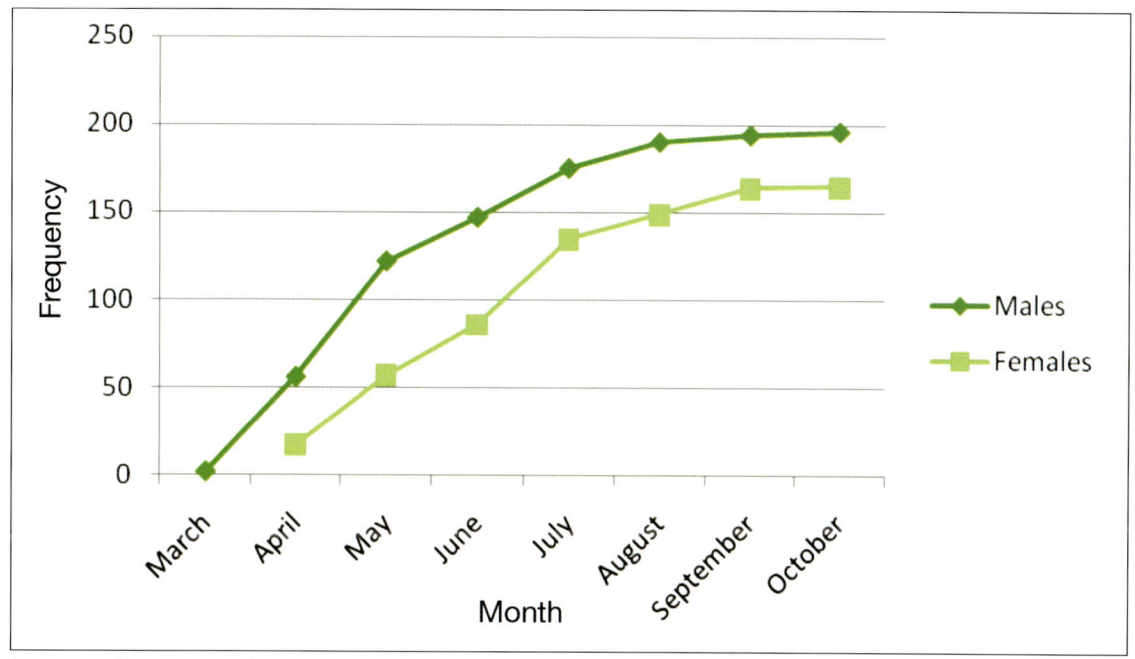

Figure 3. Total cumulative sightings for new males and females per month, 2007 to 2009.

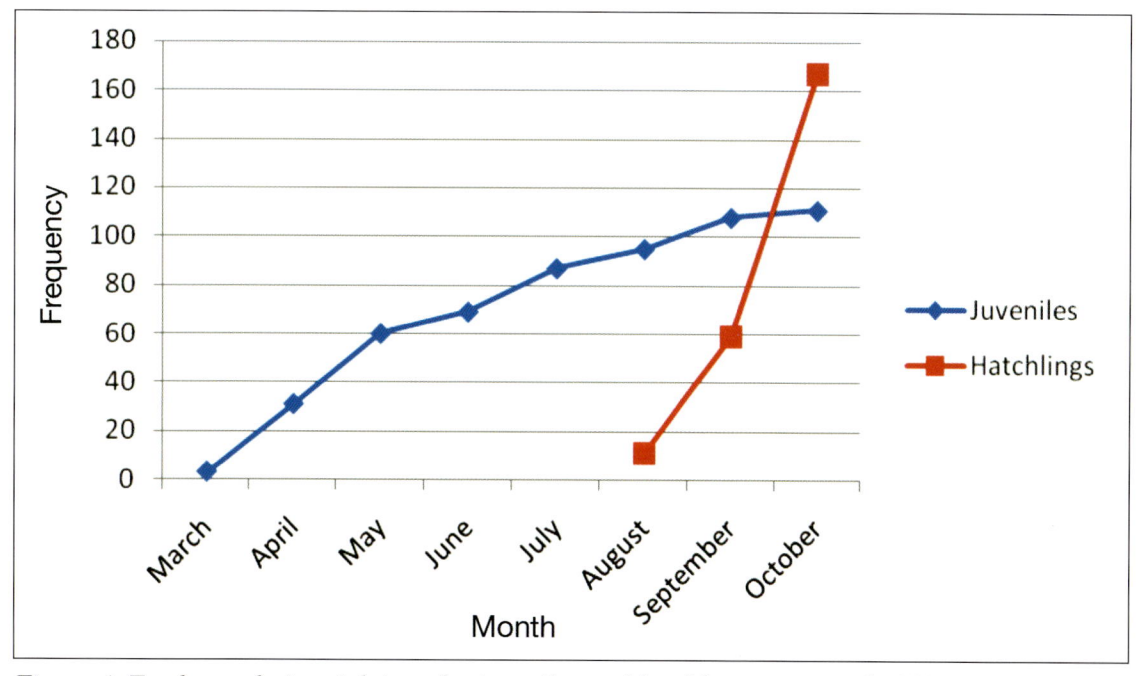

Figure 4. Total cumulative sightings for juveniles and hatchlings per month, 2007 to 2009.

Common Lizards were abundant at some sites, particularly in the frontal dunes at Ravenmeols and Hightown, often in sparse Marram where large females may be confused with female Sand Lizards by casual observers.

Egg Laying

In June, gravid females were seen digging egg burrows in the early evening. This rarely observed behaviour provided valuable information on breeding site preferences, timing and weather conditions. The day had been very hot becoming cooler in the early evening. The females chose a relatively flat, open area of sunny, dry, root-free sand on a south-facing, gentle incline on mobile dunes immediately inland of an embryo-dune ridge dominated by Sand Couch (*Elytrigia juncea*) but where Marram Grass (*Ammophila arenaria*) was established but not too dense. The technique was to dig in frantic bursts with the front legs sending fine sand flying out behind. The animal disappeared headfirst into the burrow, sand appearing to collapse behind and on top of it as digging progressed. Finally, only the tip of the tail was visible before the entire body was completely covered, the whole exercise taking approximately 30 minutes. The lizards appeared highly vulnerable to predation at this time. Shortly afterwards, there was hardly any visible evidence of a burrow in the soft, dry sand, the animals presumably digging themselves out after completing egg-laying. The absence of dune-slack vegetation indicated that this area was well drained. At least three burrowing females were seen within a distance of 10m, suggesting that some animals may travel from surrounding, fixed vegetation to optimum egg laying sites. These burrows, (nests) appear quite distinct from burrows dug for hibernation or for cover (see Plate 6). On three occasions, P. Hudson (*in litt*. 2009) has found Sand Lizard nests with eggs buried in sand on inland sites with about a 30° incline. As a member of the Natural England captive breeding programme, he notes that females in enclosures appear to prefer sand slopes, although they also use level sites. Hudson also notes that the eggs, usually 5 to 9, are laid approximately 100mm below the surface at a depth where enough moisture prevents desiccation. Interestingly, Simms (1970) states: 'Neither in the field nor in the vivarium were Sand Lizards ever seen digging in flat ground.'

The total number of new lizards to date over the three year study period is 472 adults and juveniles: 196 males, 165 females and 111 juveniles giving an average sex ratio of 1.2 males to females (Table 4). Using the figures and the same calculation adopted in the 2007 study, the total of 472 adults and juveniles recorded gives an estimated grand total of 944 across our study area.

Figures 3 and 4 show a consistent pattern over three years of new males, females and juveniles being found, initially, at a constant rate, slowing down as the season progressed, again suggesting that not all new animals were found during the recording period which was terminated by hibernation. The sharp upward surge in hatchling records coincides with newly emerged animals between August and October, the graph emphasizing the likelihood of significant under-recording due largely to the short recording season and difficulty in finding hatchlings.

Concluding remarks

The large extent of the frontal-dunes hampered data collection in 2008/9 as finding lizards is more difficult and time consuming in this habitat. This was not helped by yet another wet summer, with a limited number of

Table 3. Sightings of new Sand Lizards per month across three new foci (2009).

	March	April	May	June	July	August	September	October
Males			12	1	5	2		2
Females			7	3	15	2		
Juveniles			9		1			
Hatchlings						8	18	51

Table 4. Total sightings of new lizards per month across the study area, 2007 to 2009.

	March	April	May	June	July	August	September	October
Males	2	54	66	25	28	15	4	2
Females		17	40	29	49	14	15	1
Juveniles	3	28	29	9	18	8	13	3
Hatchlings						11	48	108

recording days possible. The distribution of Sand Lizards in these areas is uneven and patchy, particularly in large areas of bare, open sand. However our work in 2008/9 has shown that there are additional Sand Lizard foci along the frontal-dunes not yet covered by our study, the 2008/9 data reinforcing the earlier conclusion that the total population of Sand Lizards on the Sefton Coast lies between 1200 and 1500 adults and juveniles. It is hoped to continue this research.

ACKNOWLEDGMENTS

Thanks are again due to members of the NMARG, especially Mike Brown for his tireless enthusiasm for monitoring and as group recorder; also to Phil Smith for his support, useful comments and editing.

REFERENCES

Atkinson, D. and **Houston, J.** (eds.) (1993) *The Sand Dunes of the Sefton Coast.* Liverpool: National Museums and Galleries on Merseyside.

Berglind, S. (2005) *Population Dynamics and Conservation of the Sand Lizard on the Edge of its Range.* Digital Comprehensive Summaries of Uppsala Dissertations from the Faculty of Science and Technology no. 41. University of Uppsala.

Cooke, A.S. (1993) The Habitat of Sand Lizards *Lacerta agilis* L. on the Sefton Coast. In: Atkinson, D. and Houston, J.(eds.) *The Sand Dunes of the Sefton Coast.* Liverpool: National Museums and Galleries on Merseyside. pp. 123-126.

Corbett, K. (1974) *Field Survey to Determine the Status of the Sand Lizard in SW Lancashire.* Unpub. report. Peterborough: Nature Conservancy Council.

Plate 6. Sand Lizard egg burrow. (Photograph by Dave Hardaker).

Larsen, C.T. and **Henshaw, R.E.** (2001) Predation of the Sand Lizard (*Lacerta agilis*) by the domestic cat (*Felis catus*) on the Sefton Coast. In: Houston, J.A., Edmondson, S.E. and Rooney, P.J. (eds.) *Coastal Dune Management. Shared Experience of European Conservation Practice.* Liverpool: Liverpool University Press. pp. 140-154.

Lunn, J. and **Wheeler, D.** (1989) *A Sand Lizard Conservation Strategy for the Merseyside Coast.* Unpublished NCC Report. ASD 6/5/3/3/2.

Merseyside Biodiversity Group (2001) *North Merseyside Biodiversity Action Plan.* Unpublished report. Maghull Environmental Advisory Service.

McNicholl, E.D. (ed.) (1883) *Handbook for Southport.* Southport: Robert Johnson and co.

Jackson, H. (1979) The decline of the sand lizard population on the sand dunes of the Merseyside Coast, England. *Biological Conservation.* 16. 177-193.

Shorrock, H. (1973) Report on Mammals, Reptiles and Amphibians 1969-1971. *Lancashire and Cheshire Fauna Society Publication no. 63.* 9-18.

Simms, C. (1966) The status of amphibians and reptiles in the dunes of southwest Lancashire, 1961-64. *Lancashire and Cheshire Fauna Society 36th report.* 7-9.

Simms, C. (1970) *The Lives of British Lizards.* Norwich: Goose and Son.

Woods, M. et al. (2003) Predation of Wildlife by Domestic Cats (Felis catis) in Great Britain. *Mammal Review* 33: 174-88.

Sefton Coast rare plants

Philip H. Smith

9 Hayward Court, Watchyard Lane, Formby L37 3QP
philsmith1941@tiscali.co.uk

ABSTRACT

The Sefton Coast sand-dune system supports 179 nationally and regionally notable vascular plants, twelve being included in the North Merseyside Biodiversity Action Plan either in their own right or under the category "Sefton Coast Rare Plants". Ten species or groups have been the subject of recent intensive monitoring and are included in this paper as follows, their conservation status being described by Cheffings and Farrell (2005) and Regional Biodiversity Steering Group (1999):

- Grey Hair-grass *Corynephorus canescens*: Nationally Rare.
- Baltic Rush *Juncus balticus*: Nationally Scarce; the only English locality.
- Baltic Rush hybrids *J. balticus* x *J. inflexus; J. balticus* x *J. effusus* (= *J.* x *obotritorum*): Nationally Rare, the former endemic to Britain.
- Sharp Club-rush *Schoenoplectus pungens*: Nationally Rare; the only British locality.
- Hybrid willows *Salix* x *angusensis, S.* x *doniana, S.* x *friesiana*: Nationally Rare, the first endemic to Britain.
- Dune Wormwood *Artemisia campestris* ssp. *maritima*: Nationally Rare.
- Common Wintergreen *Pyrola minor*: Species of Conservation Importance in North West England; endangered in Lancashire, Greater Manchester and North Merseyside.
- Smooth Cat's-ear *Hypochaeris glabra*: Nationally Vulnerable.
- Early Sand-grass *Mibora minima*: Nationally Rare; the only native English locality.
- Isle of Man Cabbage *Coincya monensis* ssp. *monensis*: Nationally Scarce; endemic to Britain.

Grey Hair-grass (*Corynephorus canescens*)

Corynephorus canescens (Grey Hair-grass) is a Nationally Rare, perennial grass of distinctive and attractive appearance that occurs on consolidated sand-dunes, sandy shingle and open sand in both coastal locations and inland. In Britain, it is largely confined to the coasts of East Anglia and Jersey but is also found rarely in Breckland. Other scattered sites are generally thought to be introductions, with the exception of the Sefton Coast (Smith, 2008a). Although the *New Atlas of the British and Irish Flora* maps it as native here (Preston *et al.*, 2002), there is still debate about its origin (Smith, 2008a).

Edmondson *et al.* (1993) incorrectly ascribe its discovery on the Sefton Coast to J.A. Wheldon. The finder, in 1919, was actually F.W. Holder, though he did not identify it at the time (Travis, 1929). In the 1920s and 1930s, two populations were known on outer high dunes between Formby and Freshfield but, although Holder photographed one site in 1937 (Roberts and Smith, 2007), the descriptions given are so vague that precise locations are now impossible to determine. These colonies eventually disappeared but the plant was rediscovered in the early 1980s on inner dunes at Southport and Ainsdale Golf Course (Smith, 2008a). D.A. Nissenbaum also found it in abundance on railway ballast near Ainsdale (Edmondson *et al.*, 1993).

The Sefton Coast LIFE-Nature Project mapped the golf course populations in 1996-98, finding 15 colonies at Southport and Ainsdale (Figure 1) and two small patches on the eastern edge of Hillside Golf Course. That survey was repeated in 2007 by Smith (2008a). He could not refind six colonies on Southport and Ainsdale, probably due to coarsening of the sward, but added nine new ones and discovered a small population (110 plants) on the adjacent Ainsdale Recreation Ground (Figure 2). Some of the mapped sites on the golf course were now much larger than in the earlier survey and the

Figure 1. Locations of Corynephorous canescens *colonies 1996-1998 Southport and Ainsdale golf course.*
© Crown copyright. All rights reserved. Sefton Council Licence number 100018192 2010.

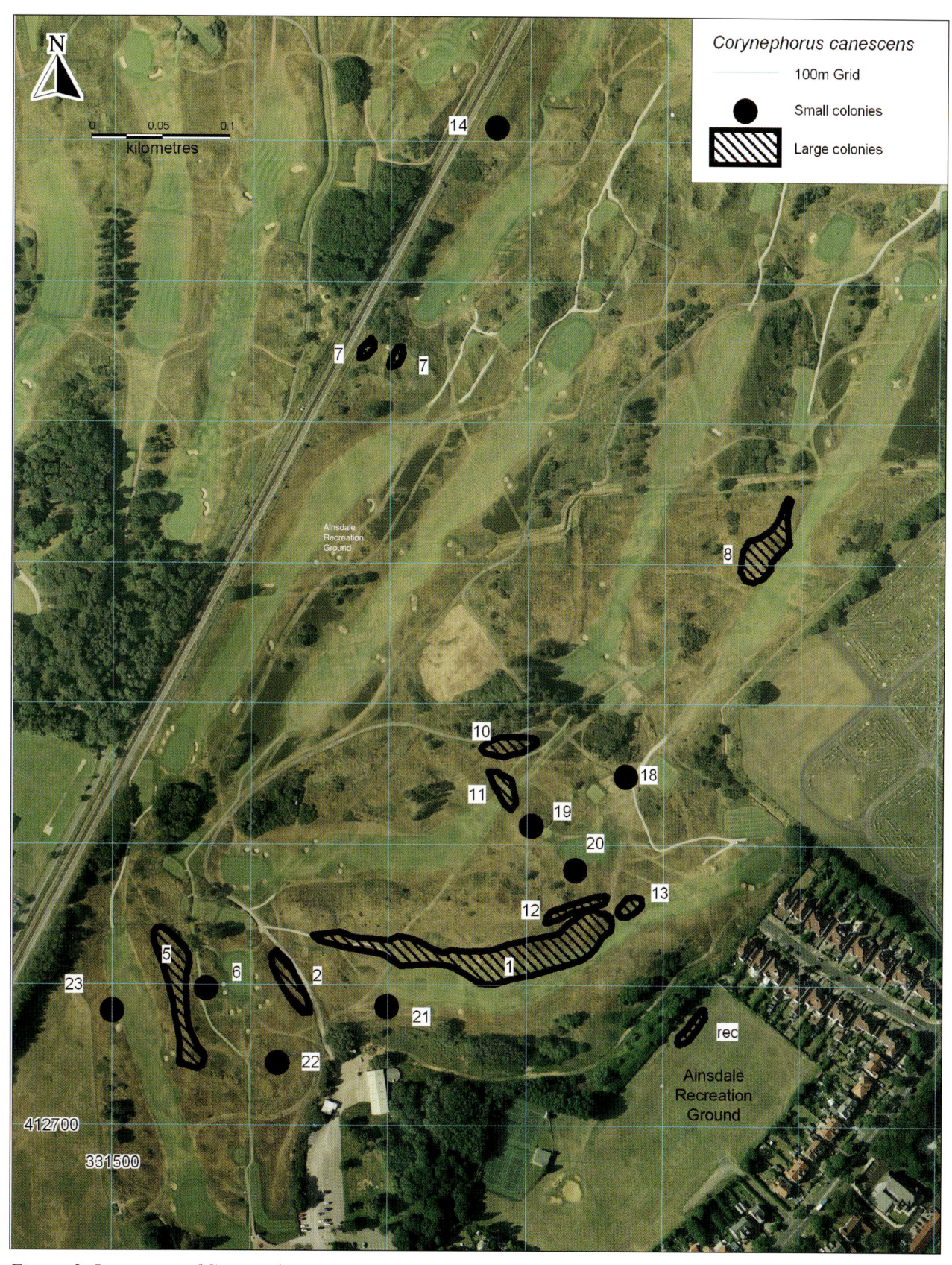

Figure 2. Locations of Corynephorous canescens *colonies 2007 Southport and Ainsdale golf course.*
© *Crown copyright. All rights reserved. Sefton Council Licence number 100018192 2010.*

total population of *C. canescens* was estimated at nearly 10,000 plants, occupying about 0.86ha.

The preferred habitat of *C. canescens* at Southport & Ainsdale Golf Course is described in both National Vegetation Classification (NVC) surveys of the Sefton Coast. Edmondson *et al.* (1988/89) found extensive stands of the grass in "a possible variant of SD13b", later reclassified as SD11a (*Carex arenaria – Cornicularia aculeata* dune, *Ammophila arenaria* sub-community) (Rodwell, 2000), with much Common Bent (*Agrostis capillaris*) and mosses, or as a co-dominant with Sheep's-fescue (*Festuca ovina*). In contrast, Gateley and Michell (2004) attribute this vegetation to U1 (*Festuca ovina – Agrostis capillaris – Rumex acetosella* grassland) (Rodwell, 1992), which is characteristic of grazed swards on acid substrates in the pastoral upland fringe. The U1 community has similarities to SD12 (*Carex arenaria – Festuca ovina – Agrostis capillaris* dune grassland), and grades into it, but generally lacks typical coastal dune plants. However, the fixed-dune communities on Southport & Ainsdale Golf Course were poor statistical fits to known NVC communities (P.S. Gateley, pers. com.).

In his earlier survey of Sefton Coast dune-heath, Gateley (1995) recorded *C. canescens* in two golf course quadrats, describing the habitat as "*Polytrichum*-dominated low dunes". The soils were generally acid with pHs of 3.43-4.44 at the surface and 3.82-4.62 at 10cm depth. Smith (2008a) did not take NVC samples but recorded all vascular plants associated with *C. canescens* in 2007. In his opinion, most of the grassland was probably closer to SD12 than to U1 but, in a few places it had similarities to SD11.

Smith (2008a) often found *C. canescens* on fixed-dune crests or on quite steep south or west-facing slopes. Such sites will be well-drained and susceptible to summer dryness, which presumably gives a competitive advantage to this relatively drought-tolerant species. However, Marshall's (1967) findings that it requires sand accretion for vigorous growth and a long life-span do not explain the grass's success at the golf course where there appears to be little sand movement. The Ainsdale recreation ground colony is probably a relict of that first described in the early 1980s. It has persisted for at least 25 years, apparently assisted by regular mowing which creates bare ground for seedling establishment. Although small, this population is of particular interest, being the only one readily accessible without special permission.

In Merseyside, *C. canescens* is near the northern limit of its European distribution, which closely matches the 15°C July isotherm. The climate in northern Britain is cold enough to have a cumulative effect on flowering date and seed germination, sufficient to postpone the emergence date of seedlings beyond the critical time for survival (Marshall, 1968). Therefore, recent increases in summer temperatures may well have benefited reproduction and survival of Sefton Coast populations. Most of the existing populations on the golf course appear healthy with plenty of young plants. There appear to be extensive areas of potentially suitable habitat in the roughs to allow for further spread. Although no evidence of Rabbit (*Oryctolagus cuniculus*) grazing was found and the grassy roughs are not mowed (M. Mercer, pers. com.), the extreme infertility of the old, heavily leached dune soils here, coupled with small-scale disturbance from course activity, seem capable of maintaining *C. canescens* populations. Since about 1999, the Southport & Ainsdale Club, with the support of Countryside Stewardship funding, has pursued a programme of "rough" management designed to restore and conserve the dune and dune heath character of the course (Gill, 2004). However, the long-term future of the plant may be threatened by nitrogen deposition from air pollution which is reaching critical levels in North West England, perhaps sufficient to cause vegetation change (Gateley and Michell, 2004).

As this is a nationally rare plant, the abundance of *C. canescens* at Southport & Ainsdale Golf Course has great conservation significance. Indeed, based on Trist's (1998) data, this may be the largest population in Britain outside North Norfolk and the Channel Is. The presence of this species is also listed as a reason for the notification of the Sefton Coast Site of Special

Scientific Interest, which was extended to include the Southport & Ainsdale Golf Course in 2000.

ACKNOWLEDGEMENTS

I am grateful to the Southport & Ainsdale Golf Club for giving permission for the 2007 survey and to the Course Manager, Mike Mercer, for information and guidance. Patricia Lockwood very kindly assisted with much of the field work. Steve Cross was extremely helpful in searching the Liverpool Museum herbarium and library for records and early literature sources. Dave Earl kindly provided records from the *New Flora of South Lancashire* Project. Michelle Newton and Graham Lymbery assisted with the preparation of distribution maps.

Baltic Rush (*Juncus balticus*) and its hybrids

Juncus balticus (Baltic Rush) is a Nationally Scarce plant of sand-dune slacks and other damp areas in maritime sand, mud or peat. In Britain, it is largely confined to the north and north-east coasts of Scotland and the Hebrides (Preston *et al.*, 2002). Its only English locality is a relatively small area of the Birkdale Sandhills where it was discovered in 1913 (Smith, 1984). Two Baltic Rush hybrids: *J. balticus* x *J. inflexus* and *J. balticus* x *J. effusus* (= *J.* x *obotritorum*) also occur as largely infertile clones in the sand-dunes of Merseyside and Lancashire, the former being endemic to the region. In the UK, *J.* x *obotritorum* is otherwise known only from Orkney.

The history and known status of *J. balticus* and its hybrids on the Sefton Coast up to the early 1990s were reviewed by Edmondson *et al.* (1993), based largely on Smith (1984). The latter major study was repeated in 2003/04 (Smith, 2006). He found 86 patches of *J. balticus* with a total area of 185m^2 at 10 Birkdale sites. It had disappeared from six northern Birkdale slacks, including the Royal Birkdale Golf Course, but had consolidated and extended its range southwards, colonising three new slacks in the Birkdale frontal dunes and the Birkdale Green Beach. There had been a major decline in old slacks east of the coastal road at Birkdale. Overall, the total area of patches had increased 34% since 1982.

Analysis of quadrat samples taken using NVC methodology shows that *J. balticus* can grow in a wide range of dune-slack, swamp, salt-marsh and damp grassland communities under varying water-table and base-status conditions. The plant appears to be a good coloniser of young, sparsely vegetated wet-slacks and then may persist for many years before declining as the habitat becomes dryer and more heavily vegetated. Losses from the northernmost slacks at Birkdale are attributed to their colonisation by dense clumps of *Hippophae rhamnoides* (Sea Buckthorn), later removed. Currently, the largest area of the rush is in a slack that was churned up by illegal motorcycle scrambling in the early 1980s. The plant has colonised the open ground here and also on Birkdale Green Beach where 15 patches have been found since 2000.

The two *J. balticus* hybrids have both been recorded three times in the wild state in the region but all the clones of *J.* x *obotritorum* have become extinct. However, two clones of this hybrid were taken into cultivation and have been successfully translocated back to Sefton Coast dune slacks and ponds (Plate 1). Smith (2006) gives precise details of the locations and areas of the four extant sites. One of the wild clones of the *inflexus* hybrid was lost to sand-blow at Ainsdale Sand Dunes National Nature Reserve (NNR) in the late 1980s; the others at Lytham St Annes Local Nature Reserve (LNR), Lancashire and at Birkdale Sandhills LNR are thriving. The two Sefton Coast clones have been cultivated and translocated, Smith (2006) describing five extant sites in dune slacks where the plants are spreading vegetatively. Further translocations, especially of *J.* x *obotritorum*, may be justified to ensure their future conservation.

The future survival of *J. balticus* and its hybrids on the Sefton Coast will depend on effective site protection and positive habitat management, especially to maintain early successional vegetation and a high water-table. Ideally, this should be achieved through implementation of the Sefton Coast Nature Conservation Strategy. Birkdale Green Beach provides an extensive area of

Plate 1. Mike Wilcox examining hybrid Baltic Rush at Hightown sand dunes. (Photograph by Philip Smith).

ostensibly suitable habitat which should assist this process.

Sharp Club-rush (*Schoenoplectus pungens*)

Edmondson *et al.* (1993) describe the discovery of *Schoenoplectus pungens* (Sharp Club-rush) at Massams Slack, Ainsdale by W.G. Travis in 1909 and its subsequent history to about 1990. Travis concluded that the plant was native but this is still a matter for debate, Preston *et al.* (2002) mapping it as "alien" in the *New Atlas of the British and Irish Flora*. Ainsdale material cultivated at Liverpool University was transplanted by D.E. Simpson to four sites at Birkdale Sandhills LNR in 1990 (Simpson,1990) and the plant has flourished there. In 1999, D. Wrench found a patch of *S. pungens* on Birkdale Green Beach. It is presumed that it had spread naturally from one of the nearby translocation sites (Plate 2).

Smith (2005a) conducted a detailed monitoring exercise in June 2004, finding five discrete colonies at Birkdale with a total area of 173m^2, the largest patch (90m^2) being that on the Green Beach. The plant is continuing to grow well in two types of vegetation: the NVC's S21c (*Bolboschoenus maritimus* swamp, *Agrostis stolonifera* sub-community) (Rodwell, 1995) and SD16 (*Salix repens – Holcus lanatus* dune-slack) (Rodwell, 2000). In 2006/07, large Grey Willow (*Salix cinerea*) bushes that were competing with *S. pungens* at two of the translocation sites were removed by Sefton Coast and Countryside Service.

Plate 2. Sharp Club-rush at Birkdale. (Photograph by Philip Smith).

Rare hybrid willows

The Sefton Coast supports about 29 *Salix* (willow) taxa, 14 of them hybrids. Three of the latter, *Salix* x *friesiana*, *S.* x *doniana* and *S.* x *angusensis*, are nationally rare, the last named being endemic to Britain. All involve the abundant *S. repens* (Creeping Willow) as one of the parents (Table 1).

S. x *friesiana* and *S.* x *doniana* were first noted here in the 1940s (Savidge *et al.*, 1963) but the first record of *S.* x *angusensis* was as recent as 1993, when an unknown willow, found by N.A. Robinson and M. Gee at Ainsdale Sand Dunes NNR, was subsequently identified by R. D. Meikle (Meikle and Robinson, 2000). Further studies by Michell (2001) and Wilcox (2005) have sought to clarify identification features of this very difficult taxon.

The BSBI *Vascular Plant Atlas Update Project* (http://www.bsbimaps.org.uk/atlas/) gives a better picture of the occurrence of these rare plants in the British Isles and shows that the Sefton Coast is particularly well represented in their national distributions (Table 1). Often these hybrids turn up in low numbers, perhaps only one or two individuals at a particular location. On the Sefton Coast, however, two of them are relatively numerous.

In 2007, a survey began with the aim of locating (with GPS technology) and mapping all individuals of the three rare willow hybrids in the dune system. Only three individuals of *Salix* x *angusensis* are currently fully authenticated but, due to the difficulty of distinguishing this hybrid from *S.* x *friesiana*, others may well await discovery. About 10 bushes of *S.* x *doniana* were already known but the survey located four new ones at Lifeboat Road, singles at both Cabin Hill NNR and Falklands Way, Ainsdale, and nine at Hightown, raising the number of extant *S.* x *doniana* bushes to 25 (Plate 3). Visits to eight locations along the coast realised the remarkable total of about 250 bushes of *S.* x *friesiana*, the largest concentration being 111 at Cabin Hill NNR.

As well as being of inherent scientific interest, the information collected will have direct relevance to the planning of dune scrub management on the coast.

Dune Wormwood (*Artemisia campestris* ssp. *maritima*)

While recording plants at Crosby Sand-dunes on 17th April 2004, M.P. Wilcox and P.H. Smith found a patch of an *Artemisia* that they did not recognise. Material sent to E.J. Clement was identified as *Artemisia campestris* ssp. *maritima*, a sub-species thought new to Britain (Smith and Wilcox, 2006). Initially, it was

Plate 3. Rare hybrid willow Salix x doniana at Lifeboat Road, Formby. (Photograph by Philip Smith).

Table 1. British Isles and Sefton Coast distribution of rare hybrid willows according to BSBI Vascular Plant Atlas Update Project (October 2008).

Taxon	Parentage	No. of post-1986 hectads in Br. Isles	No. of Sefton Coast hectads
S. x *friesiana*	*S. repens* x *S. viminalis*	10	3
S. x *doniana*	*S. repens* x *S. purpurea*	4	3
S. x *angusensis*	*S. repens* x *S. viminalis* x *S. cinerea*	7	4

thought to be an introduced plant but Clement (2006) and Twibell (2007) argued that the sub-species should be considered native to the British Isles; in particular, J. Twibell drew attention to a small colony which had been known for some years at Crymlyn Warren sand-dunes in South Wales. Soon afterwards, the Sefton Coast locality was mapped as native under the new name "Dune Wormwood" in the BSBI *Vascular Plant Atlas Update Project* (http://www.bsbimaps.org.uk/atlas/).

On 6th September 2007, a visit with the Liverpool Botanical Group found that the original plant at Crosby had been joined by three smaller, younger individuals. The parent plant has grown in size from 200 x 170cm in 2005 to 300 x 240cm currently. All the plants lie within 49m of each-other on a fixed dune dominated by *Festuca rubra* (Red Fescue) with occasional *Leymus arenarius* (Lyme-grass), *Lotus corniculatus* (Bird's-foot-trefoil), *Medicago lupulina* (Black Medick) and *Trifolium* spp. (clovers) (Smith, 2008b). A full list of associated vascular plants is given by Smith and Wilcox (2006).

In August 2008, the site was visited by Dr Twibell and Joan Valles, the latter being Professor of Botany at the University of Barcelona with a particular interest in the evolution of *Artemisia*. With permission, samples were taken for genetic studies which may throw further light on the native status of these plants. Prof. Valles (*in litt.*, 2008) emphasises the importance of the Crosby dunes for botanical conservation (Plate 4).

The Crosby locality is within a Site of Local Biological Interest designated under the Sefton Unitary Development Plan. Although the plants are within about 5m of a well-used footpath, the area is fenced and little disturbed, the habitat appearing suitable for further spread.

ACKNOWLEDGEMENTS
I am grateful to Patricia Lockwood for finding two of the new individuals.

Common Wintergreen (*Pyrola minor*)
Pyrola minor (Common Wintergreen) was recently placed on the Lancashire Wildlife Trust's list of endangered vascular plants in north Merseyside,

Plate 4. Professor Joan Valles studying Dune Wormwood at Crosby Marina in 2008. (Photograph by Philip Smith).

Greater Manchester and Lancashire (Lancashire Wildlife Trust, 2007). This plant has declined throughout its scattered British range due to changes in land-use and management, often as sites become too dry (Preston *et al.*, 2002). In modern times, the only South Lancashire (vice-county 59) occurrences have been in the River Darwen valley west of Blackburn in 2000 and 2002 and on the Sefton Coast sand-dunes, Merseyside (D.P. Earl *in litt.*, 2008).

The first Sefton Coast record was in 1957 at "Ainsdale sands". Twelve sites are shown on a map of rare plants on Ainsdale Sand Dunes NNR drawn up in 1976 by the then warden, K.R. Payne (Ainsdale Sand Dunes NNR archive), but no systematic monitoring seems to have taken place until a comprehensive survey was organised in June/July 2007 (Smith, 2008c). Ten colonies were found on the NNR, covering an area of 811m^2 and supporting over 2000 plants (Plate 5). Unexpectedly, two new colonies were discovered by P.A. Lockwood on the National Trust estate at Formby Point. These have low hundreds of plants in 102m^2. Two additional colonies were located at Formby Point in 2008. All the sites are clearings or tracks in pine woodland and have a similar habitat type – an acidic type of dry slack with a high frequency of mosses and lichens. The vascular plant community has low diversity, being represented mainly by species that tolerate base-poor conditions. At most sites, regeneration of young trees – *Pinus* (pine), *Quercus* (oak), *Betula* (birch) and/or *Acer*

Plate 5. Common Wintergreen at Ainsdale NNR (Copyright Maria Knowles).

pseudoplatanus (Sycamore) – is taking place. Therefore, management may be required to prevent over-shading if current colonies are to be conserved. Several of the 1976 NNR sites no longer support *P. minor*, having become dense birch woodland (Smith, 2008c).

ACKNOWLEDGEMENTS

Patricia Lockwood and the Liverpool Botanical Group played major roles in the 2007 survey. Alice Kimpton kindly provided transport to remote parts of Ainsdale Sand Dunes NNR.

Smooth Cat's-ear (*Hypochaeris glabra*)

Hypochaeris glabra (Smooth Cat's-ear) is a native annual of open, summer-parched grasslands and heathy pastures, on usually acidic, nutrient-poor, sandy or gravely soils, also occurring in dune grassland or on sandy shingle. It is scattered throughout Britain from the Channel Islands to Scotland, there being only four recent records in the latter country. The plant is mainly concentrated in the East Anglian Breckland and Sandlings, the Surrey heaths and the Welsh Borders (Wilson and King, 2003). *H. glabra* is declining in semi-natural habitats, having disappeared in many areas as a result of agricultural improvement or loss of grazing (Preston *et al.*, 2002). Its conservation status is given as "vulnerable" and it is a UK Species of Conservation Concern (Cheffings and Farrell, 2005). In North-west England, *H. glabra* is notified as a Species of Conservation Importance (Regional Biodiversity Steering Group, 1999), being poorly represented in most vice-counties in this region (Smith, 2008d).

The first record of *H. glabra* on the Sefton Coast was by Hon. J.L. Warren (Lord de Tabley) who found it in 1866 on sandhills half-a-mile north of Crosby (Brown, 1875). A handful of South Lancashire vice-county records followed in the late 19th and early 20th centuries but, prior to 2007, only three Sefton Coast sightings have been traced in the previous 50 years (Smith, 2008d).

On 19th June 2007, two colonies of *H. glabra* were found by chance on heavily Rabbit grazed acid grassland at the National Trust's Formby Point property. More systematic searches of similar habitat resulted in 28 colonies being recorded at Lifeboat Road, Formby Golf Course and Ainsdale Sand Dunes NNR, as well as at Formby Point. These cover a total area of about 2.4ha and contain over 5200 plants (Smith, 2008d).

The associated plant communities are relatively species-poor, dominated by *Agrostis capillaris* (Common Bent) and *Carex arenaria* (Sand Sedge). High rates of occurrence were also found for *Arenaria serpyllifolia* (Thyme-leaved Sandwort), *Centaurium erythraea* (Common Centaury), *Crepis capillaris* (Smooth Hawk's-beard), *Erodium lebelii* (Sticky Stork's-bill), *Lotus corniculatus*, *Sedum acre* (Biting Stonecrop) and *Senecio jacobaea* (Common Ragwort). This suggests a neutral to somewhat acidic substrate at most sites. The habitat is mostly fixed-dune with a short, open sward, and a substantial cover of mosses and lichens. Except on Formby Golf Course, there is invariably evidence of heavy rabbit-grazing, while Ainsdale Sand Dunes NNR sites are also winter-grazed by sheep. Moderate human trampling is a factor at some localities, but heavy recreational pressure seems to result in an absence of *H. glabra*.

Although samples were not taken to determine NVC communities, the associated species suggest that most of the sites accord with the SD12 *Carex arenaria-Festuca ovina-Agrostis capillaris* dune grassland, though some stands may be closer to, or include patches of, SD19 *Phleum arenarium-Arenaria serpyllifolia* dune annual community (Rodwell, 2000).

In view of the paucity of previous records for the Sefton

Coast and the nationally declining status of the species, the abundance of *H. glabra* on the Sefton Coast sand-dunes in 2007 is extraordinary. As the plant is small and rather shy of flowering, it could have been overlooked in the past. However, the intensity of botanical recording over many years in this area, and the large size of the populations found, make this unlikely. Being an annual, this species may be susceptible to adverse environmental conditions, especially at time of germination and establishment of seedlings. The exceptional weather conditions of 2007 may have provided ideal conditions for a population explosion. Interestingly, although it was not possible to repeat the intensive survey, in 2008 several additional colonies were found at Lifeboat Road and on Altcar Rifle Range (personal observations).

Although two of the largest sites contain recently planted pine trees, most of the habitat of *H. glabra* on the Sefton Coast dunes appears to be self-sustaining, requiring little active management. At most sites, rabbit-grazing and human trampling seem sufficient to maintain the open habitat favoured by the plant. However, the recently noted trend towards denser swards, perhaps influenced by aerial nitrogen deposition (Gateley and Michell, 2004), could represent a long-term threat to this species.

ACKNOWLEDGEMENTS

I am extremely grateful to Patricia Lockwood for assistance in the field and to Alice Kimpton for drawing attention to colonies at Formby Golf Course and Ainsdale NNR and for transport to the Dune Restoration Area. Steven Cross kindly looked up old records and references in the World Museum Liverpool library and herbarium, while Dave Earl provided records from the *New Flora of South Lancashire* Database. Michelle Newton assisted with the preparation of distribution maps.

Early Sand-grass (*Mibora minima*)

Mibora minima (Early Sand-grass) is a nationally rare annual, said to be the smallest grass in the world (Rich, 1997). Its only native locality in England is on dunes to the west of Southport Marine Lake where it was found by Dave Earl and Joyce Buckley-Earl in April 1996. The plant may have been overlooked previously because of its diminutive size and early flowering season. Its habitat is semi-fixed dune with rather sparse vegetation and a high proportion of bare sand, often on the edges of informal footpaths (Smith 2005b) (Plate 6).

Two detailed surveys of *M. minima* distribution at Southport have been carried out, mapped areas occupied by the plant increasing by 47% from 1465m^2 in 1999 to 2158m^2 in 2004 (Smith, 2005b). The plant requires a very open vegetation structure in which to seed, this being maintained by a slow input of blown sand from the adjacent foreshore, together with locally intense human trampling and moderate Rabbit grazing. However, in 2007, part of the site was fenced off by a private land-owner, thereby controlling public access and reducing trampling effects. It remains to be seen whether this will adversely affect the plant in the long term.

Isle of Man Cabbage (*Coincya monensis* ssp. *monensis*)

Coincya monensis ssp. *monensis* (Isle of Man Cabbage) is a nationally scarce annual or short-lived perennial, mainly found by the sea on open sand-dunes in north-west England, south-west Scotland and south Wales, and is one of few British endemics (Preston *et al.*, 2002) (Plate 7).

Edmondson *et al.* (1993) included this species in a list of plants having very localised distributions on the Sefton Coast, mentioning its occurrence in disturbed areas around Southport Marine Lake, at a site in the Birkdale frontal dunes and on a sandy verge at Blundellsands.

The Marine Lake colony, discovered by R.A. Hall and D.E. Nissenbaum in 1989 had 347 plants, rising to 874 in 1997 (Brummage, 1997), but then declining to 281 in 2004 (Smith, 2004). At Birkdale frontal dunes, 55 plants were found in July 1983 in a sandy dune hollow. This population increased to 168 plants in 1986, followed by a steady decline to extinction in about 1993 as the habitat changed to a closed plant community with much *Hippophae rhamnoides* (Smith and Hall, 1991).

The long-established Blundellsands colony was

progressively destroyed by housing development until, by 1989, it was restricted to a sandy footpath on the north side of Park Drive. In 1992, this habitat was destroyed when the footpath was top-soiled and turfed over but, before this happened, a rescue operation organised by the Lancashire Wildlife Trust and Sefton Ranger Service translocated 385 young plants to ostensible suitable dune sites nearby at Hall Road and Crosby Marine Park (Rooney, 1992). The introduction sites were monitored at one or two year intervals. Only 30 plants were found in 1994 but, thereafter, the populations usually increased. By 2007, there were 701 plants at Hall Road and 622 at Crosby, a total of 1323 (Figure 4). The area of duneland occupied by *C. m. monensis* plants increased fourteen-fold between 1998 and 2007, reaching a total of over 3,600m^2 (Figure 5). Further details of this successful conservation exercise, including lists of associated vascular plants, are given in Smith (2007).

The future of *C. monensis* on the Sefton Coast will depend on habitat protection (all three extant localities lie within Sites of Local Biological Interest designated under Sefton Council's Unitary Development Plan) and appropriate management through the Sefton Coast Nature Conservation Strategy. The plant appears to require rather thinly vegetated duneland with plenty of bare sand for colonisation. At present, this is maintained by moderate levels of pedestrian activity and patchy grazing by Rabbits. The site at Southport Marine Lake has recently been fenced by a private land-owner and there is concern that reduced levels of trampling may adversely affect the plant's habitat.

Plate 7. Isle of Man Cabbage at Hall Road Dunes. (Photograph by Philip Smith).

Plate 6. Early Sand-grass at Southport Marine Lake. (Photograph by Philip Smith).

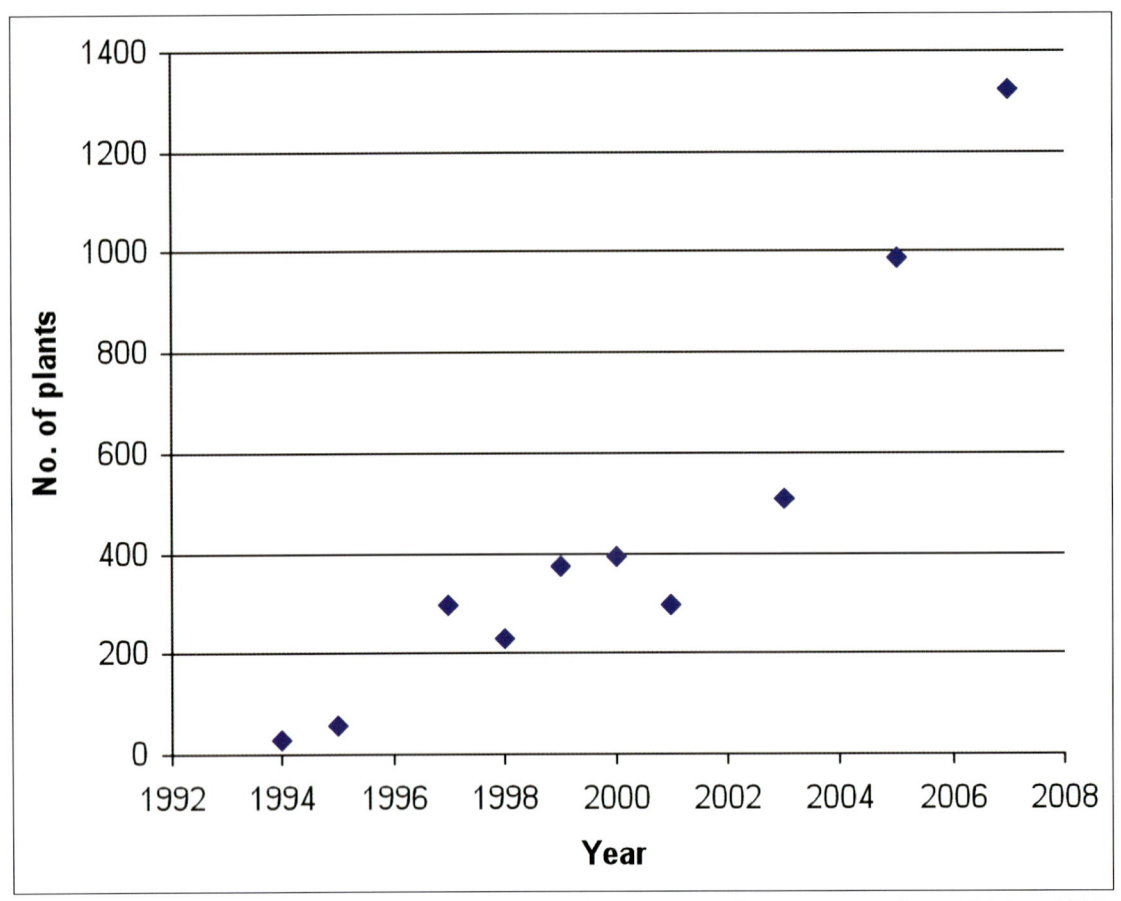

Figure 4. Total number of Isle of Man Cabbage plants at translocation sites from 1994 to 2007.

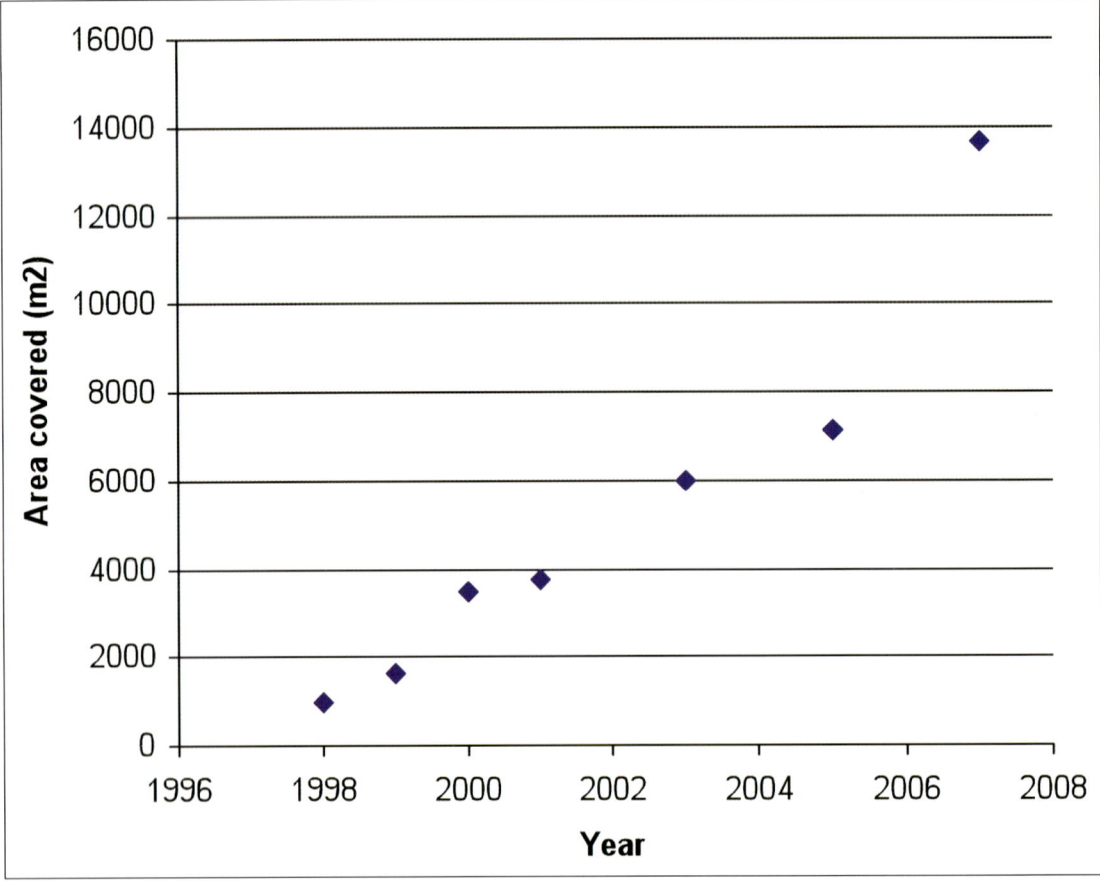

Figure 5. Area occupied by Isle of Man Cabbage plants at both translocation sites, 1998 to 2007.

REFERENCES

Brown, R. (1875) *Second Appendix to the Flora of Liverpool.* Liverpool: Liverpool Naturalists Field Club.

Brummage, M.K. (1997) *Marine Lake, Southport. Wildlife Survey.* Unpub. report. Bootle: Sefton Coast Life Project.

Cheffings, C.M. and **Farrell, L.** (eds.) (2005) *The Vascular Plant Red Data List for Great Britain.* Species Status 7, 1-116. Peterborough: Joint Nature Conservation Committee.

Clement, E.J. (2006) Could Artemisia campestris subsp. maritima be native? *BSBI News.* 103, 4.

Edmondson, S.E., Edmondson, M.R., Gateley, P.S., Rooney P.J. and **Smith, P.H.** (1993) Flowering plants (Angiospermae). In Atkinson, D. and Houston, J. (eds.) *The Sand Dunes of the Sefton Coast.* Liverpool: National Museums & Galleries on Merseyside. pp. 109-118.

Edmondson, S.E., Gateley, P.S. and **Nissenbaum, D.A.** (1988/89) *National Sand Dune Vegetation Survey: the Sefton Coast.* Report no. 917. Peterborough: Nature Conservancy Council.

Gateley, P.S. (1995) *Sefton Coast Dune Heath Survey 1993-1994.* Bootle: Sefton Metropolitan Borough Council.

Gateley, P.S. and **Michell, P.E.** (2004) *Sand Dune Survey of the Sefton Coast.* TEP, Warrington. Bootle: Sefton Metropolitan Borough Council.

Gill, B. (2004) Conservation management at Southport and Ainsdale Golf Club. *Coastlines.* Summer 2004 17.

Lancashire Wildlife Trust (2007) *Endangered Plant Species Working Group.* Minutes of the meeting held 03/04/07. Unpublished Preston: Lancashire Wildlife Trust.

Marshall, J.K. (1967) Biological Flora of the British Isles, no. 105. *Corynephorus canescens (L.) Beauv. Journal of Ecology.* 55 207-220.

Marshall, J.K. (1968) Factors limiting the survival of *Corynephorus canescens* (L.) Beauv. in Great Britain at the northern edge of its distribution. *Oikos.* 19 206-216.

Meikle, R.D. and **Robinson, N.A.** (2000) A new record for *Salix x angusensis (Salicaceae) Rech.* f. from the Ainsdale Sand Dunes National Nature Reserve, S. Lancs. v.c. 59. *Watsonia.* 23 327-330.

Merseyside Biodiversity Group (2001) *North Merseyside Biodiversity Action Plan.* Unpub. report. Maghull: Environmental Advisory Service.

Michell, P.E. (2001) *A morphological characterisation of the rare hybrid willows* Salix x friesiana Anderss. *and* Salix x angusensis Rech. f. *on the Sefton Coast.* B.Sc. Dissertation. Edge Hill University College.

Preston, C.D., Pearman, D.A. and **Dines, T.D.** (eds.) (2002) *New Atlas of the British and Irish Flora.* Oxford: Oxford University Press.

Regional Biodiversity Steering Group (1999) *A Biodiversity Audit of North West England.* 2 vols. Maghull: Environmental Advisory Service.

Rich, T. (1997) Wildlife reports: Flowering plants – England. *British Wildlife.* 8 (5) 328.

Roberts, F.J. and **Smith, M.E.** (2007) Joseph Norman Frankland (1904-1995): a tribute. *BSBI News.* 106 13-15.

Rodwell, J.S. (ed.) (1992) *British Plant Communities volume 3. Grasslands and montane communities.* Cambridge: Cambridge University Press.

Rodwell, J.S. (ed.) (1995) *British Plant Communities volume 4. Aquatic communities, swamps and tall-herb fens.* Cambridge: Cambridge University Press.

Rodwell, J.S. (ed.) (2000) *British Plant Communities volume 5. Maritime communities and vegetation of open habitats.* Cambridge: Cambridge University Press.

Rooney, P.J. (1992) *Isle of Man Cabbage, Park Drive, Blundellsands.* Unpub. report. Bootle: Sefton Ranger Service.

Savidge, J. P., Heywood, V.H. and **Gordon, V.** (eds.) (1963) *Travis's Flora of South Lancashire.* Liverpool: Liverpool Botanical Society.

Simpson, D.E. (1990) *The conservation of Scirpus americanus on the Sefton Coast: the introduction of this species to Tag's Island.* Unpub. project plan. Ainsdale Sand Dunes NNR archive.

Smith, P.H. (1984) The distribution, status and conservation of *Juncus balticus Willd.* in England. *Watsonia.* 15 15-26.

Smith, P.H. (2004) *Isle of Man Cabbage at Southport Marine Lake, July 2004.* Unpub. report. Bootle: Sefton Coast Partnership.

Smith, P.H. (2005a) *Schoenoplectus pungens* on the Sefton Coast. *BSBI News.* 98 30-33.

Smith, P.H. (2005b) *Mibora minima* on the Sefton Coast, Merseyside. *BSBI News.* 99 33-34.

Smith, P.H. (2006) Revisiting *Juncus balticus Willd.* in England. *Watsonia.* 26 57-65.

Smith, P.H. (2007) Successful translocation of *Coincya monensis ssp. monensis* on the Sefton Coast, Merseyside. *BSBI News.* 106 16-19.

Smith, P.H. (2008a) *Corynephorus canescens (L.) P. Beauv.* on the Sefton Coast, Merseyside. *Watsonia.* 27 149-157.

Smith, P.H. (2008b) Increase in *Artemisia campestris ssp. maritima* at Crosby sand dunes, Merseyside. *BSBI News.* 107 28-29.

Smith, P.H. (2008c) *Pyrola minor* on the Sefton Coast, Merseyside. *BSBI News.* 108 3-6.

Smith, P.H. (2008d) Population explosion of *Hypochaeris glabra L.* on the Sefton Coast, Merseyside in 2007. *Watsonia.* 27 159-166.

Smith, P.H. and **Hall, R.A.** (1991) The Isle of Man Cabbage in South Lancashire. *Lancashire Wildlife Journal.* 1 31-34.

Smith, P.H. and **Wilcox, M.P.** (2006) *Artemisia campestris subsp.maritima,* new to Britain, on the Sefton Coast, Merseyside. *BSBI News.* 103 3.

Travis, W.G. (1929) *Scirpus americanus Pers.* and *Weingaertneria canescens Bernh.* in Lancashire. *North Western Naturalist.* 4 175-177.

Trist, P.J.O. (1998) The distribution of *Corynephorus canescens (L.) P. Beauv. (Poaceae)* in Britain the Channel Islands with particular reference to its Conservation. *Watsonia.* 22 41-47.

Twibell, J.D. (2007) On the status of *Artemisia campestris ssp. maritima* as a native. *BSBI News.* 104 21-23.

Wilcox, M.P. (2005) *Further investigations into the parentage of* Salix x friesiana *Andersson and putative* S. x angusensis *Rechinger f. (Salicaceae) in the British Isles.* B.Sc. dissertation. Edge Hill University College.

Wilson, P. and **King, M.** (2003) *Arable Plants – a field guide.* Old Basing, Hants: Wild Guides.

A potential role for Habitat Suitability Mapping in monitoring rare animals on the Sefton Coast

Richard Burkmar

40 Old Vicarage Road, Horwich, Bolton. BL6 6QT
rich.burkmar@tesco.net

ABSTRACT

The Sefton Coast is home to a number of animals that are especially well adapted to exploit the dunes and associated habitats. Rarities include vertebrates such as Sand Lizard (*Lacerta agilis*) and invertebrates like Sandhill Rustic Moth (*Luperina nickerlii gueneei*). A common feature of many rare species on the coast is that they are difficult to survey and, as a consequence, population monitoring typically involves time-consuming surveys by specialists. The result is that monitoring effort is patchy at best. This paper describes a technique called *habitat suitability mapping* and explores its potential for supplementing direct monitoring of taxa to keep track of the condition of the coast as a habitat for its rarer animals.

Multiple Logistic Regression is a well-established technique for developing models that describe distributions of plants and animals as a function of measurable habitat features. Such models can be combined with data collected on habitat features alone to produce habitat suitability maps for organisms. Geographical Information Systems (GIS) provide us with the tools to combine these models and data. Parsimonious models may require data to be collected on relatively few features by non-specialists, thus providing a cost-effective technique for monitoring habitat condition in relation to rare animals on the Sefton Coast.

INTRODUCTION

The Sefton Coast's rare animal taxa are often difficult to survey and monitor. The very rarity of vertebrates like Sand Lizard (*Lacerta agilis*) and invertebrates such as Sandhill Rustic Moth (*Luperina nickerlii gueneei*), Northern Dune Tiger Beetle (*Cicindela hybrida*) and Vernal Mining-bee (*Colletes cunicularius celticus*) can be a barrier to effective monitoring since patchy distributions make it difficult to locate all populations with confidence. In general, effective monitoring of different animal taxa requires markedly different techniques and – very often – specialist skills.

With resources and specialist skills at a premium, monitoring of animal taxa tends to play second fiddle to monitoring of habitat quality – normally by using plant community quality as a proxy. Rightly so: as a general rule of thumb, conservationists know that if they get the habitat 'right' then the animals and plants that are specialists of those habitats will tend to prosper. But this *is* only a rule of thumb. Different 'dune specialists' often have different habitat requirements when examined in detail, and changing the habitat in a certain direction may well benefit one taxon at the expense of another.

Notwithstanding the lack of resources, land managers have a duty – sometimes a legal duty – to ensure that their operations on the coast have regard for the welfare of the rare animals found there. So given the restrictions on resources and expertise, how can land managers best fulfil this duty? A growing amount of research is pointing to a suite of techniques, which together can be called 'habitat suitability mapping', as a pragmatic way forward. These techniques hold out the promise that the distribution of an animal can be predicted across a habitat by relatively simple measurement of habitat and environmental variables that are functionally important to that animal. So it might be possible to use habitat suitability as a proxy for measuring the distribution of a rare animal for a fraction of the cost and effort of monitoring the animal directly.

Habitat Suitability Mapping

The goal of habitat suitability mapping is to monitor a

relatively small number of easily measured habitat variables at many locations across a site and translate these into a measure of how suitable the habitat is for the target species[1] at each of those locations. But in order to translate the habitat variables into a measure of suitability for the target species – and indeed to establish *which* habitat variables should be measured in the first place – we need first to develop a distribution model for the species. In this relatively new field, species distribution models are not available 'off the shelf' and a considerable investment of effort will normally be required up front in order to establish a model. Even in cases where distribution models for a target species have been established elsewhere, the model may not be transferable if it was designed to operate:

- at a different geographical scale;
- in another geographical region where different abiotic conditions (e.g. temperature) predominate; or
- in another habitat where different biotic conditions predominate.

Vanreusel *et al.* (2007) and Guisan *et al.* (2002) discuss the problems of model transferability in more detail. The upshot is that it will not normally be possible to design a habitat suitability monitoring programme for a species without first carrying out an intensive initial study within the area of interest to:

- identify which are the functionally important habitat variables for the target species;
- combine these in a quantitative model that will produce a measure of the suitability of the habitat for the target species at any given location.

Development of a species distribution model will require an initial study to develop a data model with reference to both the ecological theory driving the distribution of the target species and the statistical model that will be used for the habitat suitability mapping. This process is discussed in more detail below and elucidated by reference to a specific Sefton Coast example: that of the Sandhill Rustic Moth.

Ecological, data and statistical models

An initial study must draw upon previous work into the ecology of the target species in order to determine a list of candidate habitat variables that may influence its distribution across a habitat. The normal assumption is that ecological niche theory – or some variant of it – can be invoked to explain the distribution of the target species (Austin, 2007). In a review of the ecological theory of species distribution models Austin (2007) notes that too many accounts make no explicit reference to the underlying ecological theory and lack any discussion of the relationship between ecological theory and the data and statistical models employed.

The initial study must be carried out with, at the very least, *some* idea of the ecology driving the distribution of the target species but, just as importantly, it must also be designed with a *clear* idea of the data and statistical models that will be used to construct the distribution model. An ideal species distribution model would be capable of predicting the *population* of the target species at a given location: in other words it would account for both distribution and abundance. But to build a model of this kind, the initial study would have to reliably measure abundance. In reality, abundance data are nearly always too difficult to obtain and our data models normally record only presence (or absence) at a particular location. As a result the distribution models that we build from these data are only capable of predicting the distribution of the target species; they cannot predict abundance. This single important restriction plays a major part in determining the kind of statistical model that can be employed in habitat suitability mapping.

Typically, data models will consist of a number of presence and absence records for the target species and, for each of these records, measures of several habitat variables deemed possibly influential in its distribution. An increasingly well established statistical technique that can operate on data models such as these is Multiple Logistic Regression (e.g. Menard, 2001; Pampel, 2000) which is a class of Generalised Linear Models (GLM). Most modern statistical packages have an implementation of GLM.

Linear models for predicting presence and absence of animals and plants from habitat features based on logistic regression now "*form the backbone of the modelling approaches*" (Rushton *et al.*, 2004) and are widely reported in the literature both for plants (e.g. Engler *et al.*, 2004, Zaniewski *et al.*, 2002) and animals (e.g. Dennis and Eales, 1997; Oostermeijer and van Swaay, 1998; Peeters and Gardeniers, 1998; Cowley *et al.*, 2000; Binzenhofer *et al.*, 2005; Strauss and Biedermann, 2005; Vanreusel *et al.*, 2007).

Sandhill Rustic: a potential case study

The Sandhill Rustic Moth (*Luperina nickerlii*) is represented in the British Isles by four separate sub-species, all with relatively restricted distributions: ssp. *knilli* is only known from the Dingle Peninsula in Ireland; ssp. *leechi* is found on a single beach in Cornwall; ssp. *demuthi* occurs on coastal saltmarshes in North Kent, Essex and Suffolk; and ssp. *gueneei* occurs on dune systems along the North Wales, Cheshire and Merseyside coasts (Waring *et al.*, 2003).

Ssp. *gueneei* (subsequently referred to here as the 'Sandhill Rustic') is listed as 'vulnerable' in the Red Data Book (Shirt, 1987) and confined to sand dune systems in North Wales, Cheshire and Lancashire. Although not a UK BAP priority species, it is the subject of Species Action Plans in the following local BAPs: North Merseyside, Cheshire, Flintshire, Denbighshire and Conwy. There are a number of scattered populations along the North Wales, Cheshire and Merseyside coasts (Wallace and Judd, 2003).

Historically, in Lancashire, the Sandhill Rustic was found both on the Sefton Coast and at St Annes but has never been regularly recorded. The last record at St. Annes was made in 1987 and searches made in 2007 proved fruitless (personal communications, Adrian Spalding, 2007 and Steve Palmer, 2007). On the Sefton Coast, the moth was regularly recorded at Formby/Altcar from the 1980s onwards but was last seen at Altcar in 2003. Searches for the moth at Altcar failed to locate it in 2004 and in 2007 (personal communication, Graham Jones, 2007).

Plate 1. Sandhill Rustic Moth on the Green Beach, 2007. (Photograph by Richard Burkmar).

In 2007, a significant new colony of Sandhill Rustic was discovered at Birkdale Green Beach (Plate 1), a relatively new and rapidly changing feature of the Sefton Coast (Smith, 2007). The colony appears to be extensive and there is abundant 'suitable' habitat within the mosaic of vegetation communities that comprise the Green Beach. However natural geomorphological and ecological processes mean that the Green Beach is changing rapidly and to what extent it will remain suitable for the Sandhill Rustic in the future is not known. Wallace and Judd (2003) found that storms at Gronant-Talacre could destroy a high proportion of the suitable habitat over the course of a single winter period. In the face of such a rapidly changing habitat, a habitat suitability model for Sandhill Rustic could help us to keep track of how the moth is likely to be faring year on year. The author plans to undertake an initial study in order to develop a habitat suitability model for Sandhill Rustic on the Sefton Coast during the summer and early autumn of 2008.

Sandhill Rustic: ecology

We are in a strong starting position with Sandhill Rustic because the moth's habitat requirements have been studied and described qualitatively by Wallace and Judd (2003) and Wallace (2006) for the colony at Gronant-Talacre Dunes, Point of Ayr, Flintshire. Their work has identified many of the likely functional habitat features for Sandhill Rustic and points to several biotic and abiotic features that have an important influence on the ecology of the moth.

Not surprisingly, the moth shows a very strong association with the food-plant of the caterpillar – Sand Couch (*Elytrigia juncea*) – but at Point of Ayr there were often considerable areas of Sand Couch that were not utilized by the moth. Many other habitat features are likely to influence the distribution of the moth and some potential candidates identified by Wallace and Judd are:

- the amount of bare sand;
- the likelihood of immersion by high tides;
- the amount of Red Fescue (*Festuca rubra*); and
- the age structure of the Sand Couch.

The Sandhill Rustic's ecology has evolved so that it occupies a particular niche represented by combinations of habitat features, such as these, within certain favourable ranges. In advance of the fieldwork to quantify the association between the distribution of the moth and these features, it will be important to identify all possible candidate features and determine how these will be measured and represented in the data model (see next section).

The field study involves systematically searching a pre-defined study area on the Green Beach to locate as many moths as possible during its flight season (second half of August and the first half of September). At every location where the moth is found, each of the variables representing the functional habitat features will be measured. Areas that are systematically searched but which do not produce moths will be used to generate negative records (see the next section).

Since we are looking for associations between the moth and its habitat, attracting moths to light in order to 'locate' them is not a suitable technique. Instead, moths must be actively searched for in their natural resting positions using the beam of a powerful torch (Spalding, 2002). Females of the Sandhill Rustic are much more sedentary than males and therefore more likely to be found in association with the 'ideal' habitat. Therefore if enough females are found, the subsequent analysis of habitat suitability will be confined to them, but males will also be used if larger sample sizes are required for statistical reasons.

Sandhill Rustic: data and statistical models

After identifying a range of candidate functional habitat features such as those listed in the previous section, it will be important to select one or more variables that can represent each of them. In some cases this is straightforward; for example the amount of bare sand can be represented by a direct estimate of the amount of sand visible from above within a specified area. In other cases it is much less straightforward; for example how does one measure the likelihood of immersion by high tides? A variable based on the absolute elevation of a location might be most accurate, but neither instrumentation nor expertise will be available to measure this. Instead a proxy variable is required and in this case we may use something based on the evidence of strandlines at a location; for example the distance of the location from the nearest evidence of strandline or the amount of strandline material within a certain distance of the location.

During the field study, the variables representing functional habitat features will be measured wherever a moth is found. But a data model based only on presence records like these is not amenable to statistical analysis with Multiple Logistic Regression. Therefore, following a method similar to that used by Vanreusel *et al.* (2007), steps will be taken to produce an equal number of absence records directly after surveying for moths has been completed. This method can be summarised thus:

- using GIS, delineate all parts of the study area further than some distance (to be decided) from presence records;
- within these areas, randomly generate points to represent absence records (the number of points generated will be equal to the number of presence records obtained for the moth); and
- return to the field and, at each of the locations generated above, record the functional habitat variables.

The result will be a data model consisting of an equal number of presence and absence records. Note that it is important that areas used to generate absence records must have been searched as thoroughly for moths as those where the moths were found, otherwise the two sets are not comparable. A hypothetical data model is illustrated in Table 1.

The dependent variable is the presence/absence of the moth and the variables representing the functional habitat features are the independent variables. We can attempt to use Multiple Logistic Regression on such a data model to build a distribution model that will predict the probability of the moth occurring at a given location based on combinations of the independent variables. In effect we aim to produce a model that, hypothetically, looks something like that shown in Box 1.

It is not just a case of plugging all the data in and – *hey presto* – a model is produced at the end of it! There is a certain amount of art, as well as science, involved in determining a meaningful set of independent variables to explain a significant amount of the distribution of the target species. The inclusion of independent variables in the final model must never be unquestioningly accepted without consideration of the ecological meaning of their inclusion.

The ideal model is a *parsimonious* one: that is one in which a large amount of the distribution of the target species is explained with a relatively small set of functional habitat variables. It may be that including 20 independent variables in the final model explains 70% of the target species distribution compared to 65% explained by 5 of the variables. But including 20 independent variables would be considered 'over-fitting' the model (Menard, 2001). A model with five variables would, by comparison, be considered

Table 1. Hypothetical data model representing the habitat preferences of the Sandhill Rustic.

Moth present/ absent	% cover *Elytrigia juncea*	% cover *Festuca rubra*	% cover bare sand	Age of stand (years)	Height of grass (cm)	Dune height (cm)	Dune aspect	Distance from top of foreshore (m)	Distance to closest strandline (m)
P	50	0	10	2	40	30	SW	30	5
P	0	0	20	1	30	50	S	3	0
P	50	0	40	3	20	80	S	56	0
P	30	10	60	2	15	30	W	23	10
P	20	0	30	3	25	40	S	67	15
A	40	0	20	4	35	40	N	23	25
A	30	15	30	3	60	70	NW	120	45
A	10	20	10	2	20	80	W	45	10
A	5	0	10	3	60	30	SW	66	0
A	15	20	10	4	20	60	NW	45	50

Box 1. Hypothetical distribution model for the Sandhill Rustic.

Probability of moth occurring at location X = 0.12

+ 0.005 * % cover of Elytrigia at location X

+ 0.13 * age of Elytrigia at location X

- 0.0015 * % cover of Festuca rubra at location X

- 0.03 * distance to nearest strandline

Logistic regression does not work directly with probabilities but with the natural logarithm of the odds – or 'logit' – which must then be converted to probabilities, but the principle is the same.

parsimonious. An over-fitted model is likely to include variables that improve the performance of the model a small amount by chance due to idiosyncrasies in the data used to derive the model. When the model is used on data other than those from which it was derived, the inclusion of such variables becomes meaningless. A parsimonious model includes only those variables that are likely to be truly significant and meaningful when the model is applied to data other than those from which it was derived.

Most modern statistical packages available for personal computers include a GLM implementation which is capable of performing Multiple Logistic Regression. For the study of the Sandhill Rustic, a package known simply as 'R' will be used (R Development Core Team, 2007). This is a free program available for download on the internet and is widely used within the scientific community.

Sandhill Rustic: the habitat suitability model

A distribution model, such as that illustrated in Box 1, can help us to understand the ecology of the target species and is a useful way of describing its habitat requirements quantitatively. But a further practical benefit is that the model can be taken forward and used as the basis for a habitat suitability model to assess habitat condition for the target species without directly surveying or monitoring the target species itself.

It is hoped that a parsimonious distribution model for the Sandhill Rustic will identify a small number of functional habitat variables that can be used to predict the likelihood of the moth occurring at any given location. In subsequent years, habitat suitability studies for the Sandhill Rustic could proceed thus:

- identify a study area of interest within which the habitat suitability for Sandhill Rustic is to be assessed;
- generate a range of sample locations within that study area – either randomly or on a regular grid (this would most conveniently be done using GIS – see Figure 1a);
- input the grid references for the sample locations into handheld GPS;
- send field surveyors to measure the functional habitat variables (identified by the distribution model) at each of the sample points in the GPS;
- input the results into a spreadsheet – each record will include an easting & northing (from the grid reference) and a measure of each functional habitat variable;
- implement the distribution model as a formula in a column of the spreadsheet – this will result, for each record, in a predicted probability of the moth occurring at the associated location;
- use a GIS to display the resulting probabilities in the spreadsheet spatially (see Figure 1b); and
- use the GIS to interpolate a probability surface over the entire study area representing the probability of Sandhill Rustic occurring (see Figure 1c).

The GIS can be used to illustrate regions of potentially suitable habitat and areas of less suitable habitat as well as quantify their areas. Year on year it would be possible, using this technique, for land managers to

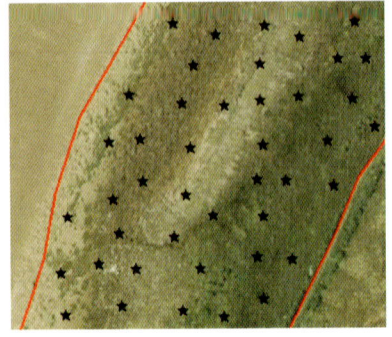

a. sampling   b. applying the model   c. interpolation

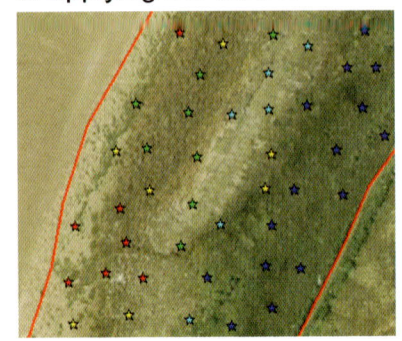

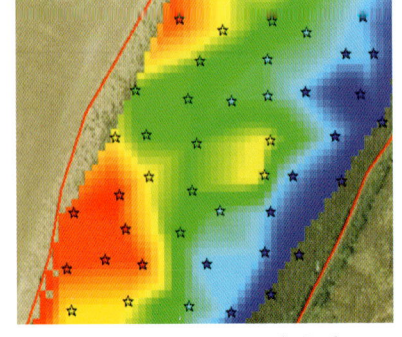

Figure 1.Stages of implementing a habitat suitability model. © Crown copyright. All rights reserved. Sefton Council Licence number 100018192 2010.

objectively assess the quality of their habitat for particular target species such as Sandhill Rustic without undertaking expensive and time-consuming direct monitoring.

CONCLUSIONS

Coastal dune systems are often subject to a great amount of change over short periods of time and the distribution and abundance of some of the animal taxa that depend on them are consequently affected. It is often not feasible to undertake direct regular monitoring of many animal taxa of dune systems because of limited resources. Habitat suitability mapping using distribution models based on Multiple Logistic Regression may provide a cost-effective proxy method for monitoring such species. An examination of these techniques, with reference to the Sandhill Rustic moth as a case study, has illuminated the process of constructing a habitat suitability model.

Clearly the major overhead in this process is the 'initial study' required to develop the species distribution model. However such studies should not require much, if any, additional effort than those already periodically carried out to survey some of the Sefton Coast's rare animal taxa; for example those for Northern Dune Tiger Beetle (Judd, 2003), Vernal Mining Bee (Taylor, 2004) and Sand Lizard (Hardaker, This volume). It would only require a small change in focus to ensure that future studies of this sort include objectives to identify functionally important habitat variables and produce quantitative distribution models for the target species.

Habitat suitability models cannot *replace* direct monitoring of our target species and we will never be able to accept their predictions of habitat suitability unquestioningly. From time to time it will be necessary to undertake direct monitoring to confirm the distribution and/or abundance of the target species; and to test and refine the habitat suitability models. Habitat suitability models do not account for many other factors that can affect the abundance and distribution of species – for example pathology. But whilst pathogens and other factors undoubtedly play an important role in the population dynamics of our target species, there is usually little that land managers can do about them. Habitat suitability models have the tremendous practical appeal that they concentrate on measuring the very thing that land managers can affect – habitat quality.

REFERENCES

Austin, M. (2007) Species distribution models and ecological theory: A critical assessment and some possible new approaches. *Ecological Modelling.* 200(1-2) 1-19.

Binzenhofer, B., Schroder, B., Straus, B., Biedermann, R. and **Settele, J.** (2005) Habitat models and habitat connectivity analysis for butterflies and burnet moths - The example of *Zygaena carniolica* and *Coenonympha arcania*. *Biological Conservation.* 126(2) 247-259.

Cowley, M.J.R., Wilson, R.J., Leon-Cortes, J.L, Gutierrez, D., Bulman, C.R. and **Thomas, C.D.** (2000) Habitat-based statistical models for predicting the spatial distribution of butterflies and day-flying moths in a fragmented landscape. *Journal of Applied Ecology.* 37(1) 60-72.

Dennis, R.L.H. and **Eales, H.T.** (1997) Patch occupancy in *Coenonympha tullia* (Muller, 1764) (Lepidoptera: Satyrinae): habitat quality matters as much as patch size and isolation. *Journal of Insect Conservation.* 1(3) 167-176.

Engler, R., Guisan, A. and **Rechsteiner, L.** (2004) An improved approach for predicting the distribution of rare and endangered species from occurrence and pseudo-absence data. *Journal of Applied Ecology.* 41(2) 263-274.

Guisan, A., Edwards, T.C. and **Hastie, T.** (2002) Generalized linear and generalized additive models in studies of species distributions: setting the scene. *Ecological Modelling.* 157(2-3) 89-100.

Hardaker, D. (This volume) The Status of the Sand Lizard, *Lacerta agilis*, on the Sefton Coast: A Recent Study. In: Proceedings of Sefton's Dynamic Coast Conference. Merseyside: Sefton Council Technical Services - Coastal Defence.

Judd, S. (2003) *Status and Biology of the Northern Dune Tiger Beetle* Cicindela hybrida (L.) *in Britain, with special reference to the Sefton Coast, Merseyside*. Peterborough: English Nature.

Menard, S.W. (2001) *Applied Logistic Regression Analysis Second Edition*. London: Sage Publications, Inc.

Oostermeijer, J.G.B. and **van Swaay, C.A.M.** (1998) The relationship between butterflies and environmental indicator values: a tool for conservation in a changing landscape. *Biological Conservation*. 86(3) 271-280.

Pampel, F.C. (2000) *Logistic Regression: A Primer*. London: Sage Publications Inc.

Peeters, E. and **Gardeniers, J.J.P.** (1998) Logistic regression as a tool for defining habitat requirements of two common gammarids. *Freshwater Biology*. 39(4) 605-615.

R Development Core Team (2007) R: A language and environment for statistical computing, Vienna, Austria: R Foundation for Statistical Computing. Available at: http://www.R-project.org.

Rushton, S.P., Ormerod, S.J. and **Kerby, G.** (2004) New paradigms for modelling species distributions? *Journal of Applied Ecology*. 41(2) 193-200.

Shirt, D.B. (1987) *British Red Data Books: 2. Insects*. Peterborough: Nature Conservancy Council.

Smith, P.H. (2007) The Birkdale Green Beach-a sand-dune biodiversity hotspot. *British Wildlife*. 19(1) 11-16.

Spalding, A. (2002) In the field: Searching for Sandhill Rustic *Luperina nickerlii* (Freyer). *Atropos*. 17 50-52.

Strauss, B. and **Biedermann, R.** (2005) The Use of Habitat Models in Conservation of Rare and Endangered Leafhopper Species (Hemiptera, Auchenorrhyncha). *Journal of Insect Conservation*. 9(4) 245-259.

Taylor, S. (2004) *Colletes cunicularis L - the vernal mining bee. Survey of Nesting Burrows at Ainsdale Sand Dunes National Nature Reserve, 2004. Ainsdale NNR, Sefton*: English Nature.

Vanreusel, W., Maes, D. and **Van Dyck, H.** (2007) Transferability of Species Distribution Models: a Functional Habitat Approach for Two Regionally Threatened Butterflies. *Conservation Biology*. 21(1) 201-212.

Wallace, I.D. (2006) *The Sandhill Rustic Moth Luperina nickerlii gueneei on Gronant Dunes: Establishing a baseline for Common Standards Monitoring*. Countryside Council for Wales.

Wallace, I.D. and **Judd, S.** (2003) *The Sandhill Rustic Moth Luperina nickerlii gueneei Doubleday at Gronant – Talacre Dunes, Point of Ayr, Flintshire*. BHP Billiton Petroleum Ltd.

Waring, P., Townsend, M. and **Lewington, R.** (2003) *Field Guide to the Moths of Great Britain and Ireland*. Hook, Hampshire: British Wildlife Publishing.

Zaniewski, A.E., Lehmann, A. and **Overton, J.M.** (2002) Predicting species spatial distributions using presence-only data: a case study of native New Zealand ferns. *Ecological Modelling*. 157(2-3) 261-280.

END NOTES

1. The word 'species' has been used in this paper in preference to 'taxon' and 'taxa' for the sake of readability, but the reader should bear in mind that the techniques described here can, of course, be applied to taxa at levels other than species (for example sub-species).

An inventory of vascular plants for the Sefton Coast

Philip H. Smith

9 Hayward Court, Watchyard Lane, Formby, Liverpool L37 3QP
philsmith1941@tiscali.co.uk

ABSTRACT

An inventory of vascular plants (species, sub-species and hybrids) identified on the Sefton Coast was completed in 2005, listing 1177 taxa for the coastal zone as a whole and 1055 for the sand-dune system. Further field observations from 2006 to 2008, together with the availability of a draft *New Flora of South Lancashire* has enabled further additions. The 2008 revision includes 1282 taxa on the coast and 1143 for the dunes. Of these, 38.4% of the coastal taxa and 35.4% of the sand-dune plants are introduced. Only about 50 taxa are thought to have become extinct in the last 150 years. A total of 188 taxa (179 in the dunes) is nationally or regionally notable. Habitat analysis shows that by far the largest number and proportion of plants (33.4%) is dependent on disturbed ground, followed by freshwater wetland (18.8%) and fixed-dunes (14.2%). The lowest numbers are associated with mobile and embryo-dunes (1.1%) and the strandline (1.1%). The Sefton duneland probably has the greatest botanical diversity of any British dune system but comparisons are difficult because of the paucity of data from other sites.

The Sefton Coast vascular plant inventory

In 1999, an attempt was made to draw up a provisional inventory of all the vascular plants (species, sub-species and hybrids), reliably identified on the Sefton Coast. This used a wide range of sources, beginning with *Travis's Flora of South Lancashire* (Savidge *et al.*, 1963), to create lists for both the entire coastal zone, which extends from Bootle Docks to Crossens, and the sand-dune system, consisting of the 2100ha area of blown sand between Crosby and Southport. A total of 971 taxa was recorded, 881 of which occurred in the dunes (Smith, 1999a). Subsequently, intensive survey work was undertaken for the *New Flora of South Lancashire Project*, resulting in a large number of new records, an updated inventory being produced in 2005. This lists 206 additional taxa, revised totals being 1177 for the coast and 1055 in the dunes. It specifies non-native and introduced native taxa (both archaeophytes and neophytes), the total of "aliens" for the coast being 435 (37.0% of the total) and for the dunes 348 (33.0%).

Much of the increase in numbers of taxa between 1999 and 2005 was attributable to better recording of non-native plants (including garden-escapes) and hybrids, the totals of which increased by 145 and 45 respectively.

Using the criteria of Cheffings and Farrell (2005), Hill *et al.* (2004), Lang (2004), Preston *et al.* (2002), and the Regional Biodiversity Steering Group (1999), nationally and regionally notable taxa are indicated under the following headings: Nationally Rare; Nationally Scarce; Endangered; Vulnerable; Near Threatened; Species of Conservation Importance in North West England. A total of 177 notable taxa is listed (15% of the total). Brief details are given on the status and main habitats of all the plants. This study is summarised by Smith (2006).

Although formal recording for *The New Flora* finished in 2005, botanical field work along the coast continued and new discoveries were made, adding 25 taxa in 2005/06, 18 in 2007 and 17 in 2008. Drafts of *The New Flora of South Lancashire* also became available, including a number of mainly historical records previously missed. Adding these raised the 2008 inventory total to 1282 for the coast and 1143 for the dune system. The new totals of nationally and regionally notable plants are 188 for the coast and 179 for the dunes, while the revised proportions of non-native taxa are 38.4% and 35.4% for the coastal zone and dunes respectively (Table 1).

A surprisingly small number of vascular plants is considered to have become extinct on the coast since recording began about 150 years ago, as many as 16 taxa (10 native) thought lost in 1999 having been rediscovered since that date. The revised total of plants now believed extinct is 52 (36 native). However, a number of species, especially in critical genera such as *Hieracium* and *Taraxacum*, which can only be determined by experts, has not been seen for some years.

Several of the larger genera are particularly well-represented on the coast. Particular efforts have been made to record *Cotoneaster* (16 taxa), *Epilobium* (17), *Rubus* (35) and *Salix* (29), the diversity of the latter genus being quite remarkable for a lowland area. Also noteworthy are *Carex* (24), *Juncus* (18) and *Veronica* (18).

Table 2 gives an analysis of the main habitat types in which the plants are found. "Disturbed ground", which includes sites affected by human activities and grazing animals such as Rabbits (*Oryctolagus cuniculus*), is by far the most important habitat, with 33.4% of occurrences. The second ranked habitat is freshwater wetland (18.8%), mainly dune-slacks which support many duneland specialities. Fixed-dune is also important (14.2%), while dune-scrub (9.1%), dune-grassland (8.1%) and woodland (both deciduous and conifer) support fewer species. The lowest numbers of plants are found in mobile/embryo dunes and the strand-line (both 1.1%), presumably because of the extreme environmental conditions associated with these habitats.

Vascular plant species richness on the Sefton Coast is evidently extremely high, being about 56% of the total vice-county flora according to Stace and Ellis (2004), though their figures are now somewhat out-of-date. This diversity may be attributed, in part, to the large study area, the wide range of habitats present, the abundance of calcareous substrates and also the geographical position of the coast which supports species with both northern and southern British distributions (Smith, 1999b). Of particular interest is the total of 188 nationally and regionally notable taxa, this being 39% of all the notable plants listed for Cumbria, Lancashire, Merseyside, Greater Manchester and Cheshire (Regional Biodiversity Group, 1999).

While just over a third of the coast's flora is non-native, this is not a particularly high figure in the regional or national context. Thus, Stace and Ellis (2004) show that the average proportion of alien taxa in British vice-

Table 1. Summary of the Vascular Plant Inventory data (2008).

	Coastal zone	Sand-dune system
No. of taxa	1282	1143
No. of species	1070	958
No. of sub-species	71	64
No. of hybrids	125	107
Native taxa	783	733
Introduced taxa	492	405
% introduced	38.4	35.4
Extinct	52	48
Nationally Rare	14	14
Nationally Scarce	15	14
Endangered	3	3
Vulnerable	17	16
Near Threatened	17	15
Species of Conservation Importance	146	138
Total nationally & regionally notable	188	179

Table 2. Main habitats occupied by vascular taxa in the inventory.

Habitat	No. of occurrences	%
Disturbed ground	506	33.4
Slacks, scrapes & ditches	283	18.8
Fixed-dunes	214	14.2
Dune-scrub	138	9.1
Dune-grassland	122	8.1
Woodland	118	7.8
Salt-marsh	54	3.6
Dune-heath	44	2.9
Mobile & embryo dunes	17	1.1
Strand-line	17	1.1

Note: Many plants occur in more than one habitat.

counties is 40%, while in South Lancashire it is 50%. However, the number of neophytes becoming established on the coast is undoubtedly increasing, largely due to the prevalence of garden waste dumping. Although most are low-impact neophytes, a small number of invasive aliens is causing actual or potential ecological problems (Smith, 1999b).

Although comparative data are hard to find, it is likely that Sefton has the most plant species-rich coastal dune system in Britain. For example, the similarly-sized Newborough Warren, Anglesey, is thought to have about 600 vascular taxa, while Braunton Burrows in Devon supports around 500, these figures being only 53% and 44% respectively of the Sefton dunes total.

These data confirm the outstanding botanical importance of the Sefton Coast, reflected in the many national and international conservation designations that apply to the coastal zone. Most recently, in 2007, *Plantlife International* included the Sefton Coast in its list of 155 *Important Plant Areas* (IPAs) in the UK, part of a Europe-wide network of sites that are of exceptional botanical richness (Plate 1).

Plate 1. Grass-of-Parnassus at Cabin Hill NNR. (Photograph by Philip Smith).

ACKNOWLEDGEMENTS

A great many people contributed records to this study. I am particularly grateful to D.P. Earl, S.E. Edmondson, P.S. Gateley, the late V. Gordon, P.A. Lockwood and M.P. Wilcox.

REFERENCES

Cheffings, C.M. and **Farrell, L.** (eds.) (2005) *The Vascular Plant Red Data List for Great Britain. Species Status* 7 1-116. Peterborough: Joint Nature Conservation Committee.

Hill, M.O., Preston, C.D. and **Roy, D.B.** (2004) *PLANTATT Attributes of British And Irish Plants: Status, Size, Life History, Geography and Habitats.* Huntingdon: Centre for Ecology & Hydrology.

Lang, L. (2004) *Britain's Orchids. A guide to the identification and ecology of the wild orchids of Britain and Ireland.* Old Basing, Hampshire: Wild Guides.

Preston, C.D., Pearman, D.A. and **Dines, T.D.** (eds.) (2002) *New Atlas of the British & Irish Flora.* Oxford: Oxford University Press.

Regional Biodiversity Steering Group (1999) *A Biodiversity Audit of North West England.* 2 vols. Maghull: Environmental Advisory Service.

Savidge, J.P., Heywood, V.H. and **Gordon, V.** (eds.) (1963) *Travis's Flora of South Lancashire.* Liverpool: Liverpool Botanical Society.

Smith, P.H. (1999a) A provisional inventory of vascular plants for the Sefton Coast. Unpublished report to the Sefton Coast Life Project.

Smith, P.H. (1999b) *The Sands of Time. An introduction to the sand-dunes of the Sefton Coast.* Liverpool: National Museums & Galleries on Merseyside.

Smith, P.H. (2006) An inventory of vascular plants for the Sefton Coast. *BSBI News* 102 4-9.

Stace, C.A. and **Ellis, R.G.** (2004) Area and species-richness of British vice-counties. *BSBI News* 97 15-19.

Common Lizard *Zootoca vivipara*

Dave Hardaker

23 Forefield Lane, Crosby, Merseyside L23 9TG
davehardaker@yahoo.co.uk

ABSTRACT

The status and distribution of the Common Lizard on the Sefton Coast have never been systematically studied. It has recently become a target species in the Biodiversity Action Plan, monitoring by the North Merseyside Amphibian and Reptile Group (NMARG) suggesting a widespread, low density population, locally common in some areas. This species uses a wider range of habitats than the Sand Lizard (Lacerta agilis); however, where their ranges overlap, they will coexist. A coastwide population of 2-3000 Common Lizards is tentatively estimated.

INTRODUCTION

The status and distribution of the Common Lizard *Zootoca vivipara* on the Sefton coast are unclear (Atkinson and Houston, 1993). There is evidence that this species is quite well represented overall and in some areas can be described as locally 'common,' though never in very high densities. It is more widespread and adaptable than the Sand Lizard (*Lacerta agilis*) and has been recorded as far south as 'The Key Park', Blundellsands, occurring in variable densities northwards at least as far as Birkdale.

Plate 1. Male Common Lizard. Note the slimmer build, smaller head and lack of brilliant green of the male Sand Lizard. (Photograph by Dave Hardaker).

Individuals can frequently be seen basking on grassy railway verges, walls, fence posts, dune banks and even garden rockeries. They are particularly well represented in the Hightown, Freshfield and Ainsdale areas, occurring mainly on the fixed, more densely vegetated dune areas. They may also use, or even share, the same habitat as Sand Lizards and often coexist where their ranges overlap on the dunes.

Very little is known about their hibernation, however during a conservation task in February 2008, two adult males in good condition were found buried just below the surface on a loosely vegetated, south facing sandy bank. They revived within minutes and were released, apparently none the worse.

'The Common Lizard is a more northerly animal than the Sand Lizard and its temperature requirements are lower.' (Frazer 1983). This is certainly the case on the Sefton coast with Common Lizards (Plates 1 and 2) generally emerging around mid-March, even in modest sunshine. At this time they appear less secretive, are more temperature tolerant and even on cool days will bask in full view on south facing aspects where any heat can be trapped. They also have a longer season and can be seen on the coast to at least late October. Like the Sand Lizard however, they do not like very hot days, particularly in mid-summer when basking is limited to short excursions in and out of the sun with most time being spent in vegetation.

On the Sefton coast, Common Lizards show a variety of colour forms, ranging from light to very dark chocolate brown as well as the less frequent, pale green.

Nationally, the Common Lizard is thought to be in decline and, as a result, has become one of the target species in the Biodiversity Action Plan. It is also a useful indicator species and therefore important in

Plate 2. Juvenile Common Lizards. (Photograph by Dave Hardaker).

assessing the health and condition of habitats such as dunes and heathlands. However, despite being our most widespread reptile, the Common Lizard has rarely been systematically surveyed. For example, virtually nothing is known about the status and distribution of this species in the Merseyside area away from the Sefton coast.

Population

Simms (1966; 1970) suggests that, at least since 1940, and particularly since 1960, the Common Lizard has been more widespread and abundant on the dunes than the Sand Lizard. However, a 2007 survey by the NMARG found similar numbers of both species within their study area. More widely across the Sefton coast however, populations show a patchy distribution. Here, the Common Lizard does not seem to occur at the high densities occasionally found on some Cumbrian moorland areas (Hudson 2007, pers comm.) or in Kent, where 600-800/ha have occasionally been recorded (Bradley 2007, pers. comm.). Members of the NMARG report seeing them frequently, often in double figures, in areas such as Hightown, where they replace the Sand Lizard and for which they can be mistaken as they will bask in sandy habitat. They can even be found among tide line debris where they may hunt sand-hoppers (Amphipoda: Talitridae).

A recent intensive study into the status of the Sand Lizard on the Sefton coast by NMARG recorders (Hardaker, This volume) found evidence that Sand and Common Lizards share similar habitat, food and breeding requirements and will bask in mixed groups, often huddled together. Although there could be competition for limited resources, no aggressive encounters between these two species were recorded. An overlap in food and habitat requirements was also noted by Simms (1966).

A 2007 survey of Common Lizards on the Freshfield Dune Heath Nature Reserve recorded 34 individuals in six visits in 2ha (a density of approximately 17/ha) in a Heather (*Calluna vulgaris*) community. This probably represents only about half of the actual population, as new animals were recorded at a constant rate per visit. The 35ha site contains about 17ha of dune heath, some of which is shaded by Gorse (*Ulex europaeus*) and Silver Birch, (*Betula pendula*), much of the remainder being suitable Common Lizard habitat. From an estimated density of approximately 30-40/ha, a likely population for the reserve is 250-300. Recent scrub management in line with the North Merseyside Biodiversity Action Plan and creation of log-pile hibernacula by conservation volunteers have improved the dune heath habitat and should help to ensure continuity of the population. In the absence of systematic saturation monitoring, a tentative estimate for the Sefton Coast Common Lizard population is 2000-3000 individuals.

REFERENCES

Atkinson, D. and **Houston, J.** (eds.) (1993) *The Sand Dunes of the Sefton Coast*. Liverpool: National Museums and Galleries on Merseyside.

Frazer, D. (1983) *Reptiles and Amphibians in Britain*. Collins.

Hardaker, D. (This volume) The status of the Sand Lizard on the Sefton Coast. In: Worsley, A.T., Lymbery, G., Holden, V.J.C. and Newton, M. (eds) *Sefton's Dynamic Coast: proceedings of the conference on coastal geomorphology, biogeography and management 2008.* Coastal Defence. Sefton MBC Technical Services.

Simms, C. (1966) The status of amphibians and reptiles in the dunes of southwest Lancashire, 1961-64. *Lancashire and Cheshire Fauna Society 36th report.* 7-9.

Simms, C. (1970) *The Lives of British Lizards*. Goose and Son. Norwich.

Mosses and liverworts of the Sefton Coast

David T. Holyoak

8 Edward Street, Tuckingmill, Camborne, Cornwall TR14 8PA
david@holyoak9187.fsnet.co.uk

ABSTRACT

The Sefton Coast has a long and distinguished history as an important area for rare coastal bryophytes. It became famous in 1855 when the *Bryologia Britannica* announced discovery on the "flat sandy shore at Southport" of the moss *Bryum marratii* new to science, along with the first British record of *B. calophyllum* and the second of *B. warneum*. Later years of the nineteenth century produced records of several other rare mosses, including *Amblyodon dealbatus*, *Bryum knowltonii*, *B. neodamense*, *Catoscopium nigritum* and *Meesia uliginosa*, along with the liverworts *Moerckia hibernica* and *Petalophyllum ralfsii*. However, loss of early-successional habitats during the late nineteenth and twentieth centuries has led to local extinction of *Bryum calophyllum*, *B. knowltonii*, *B. marratii*, *Catoscopium* and others. Although the Ainsdale dunes had become the best known British locality for *Bryum neodamense*, research published in 2006 showed this moss to be no more than an inconstant phenotype of the common *B. pseudotriquetrum*, so it has since lost its special legal protection.

Nevertheless, several important bryophytes still occur. Through the early years of the twenty-first century the Birkdale Green Beach has provided a habitat for the largest populations of *Bryum warneum* ever reported in the British Isles, along with large amounts of the newly described and nationally scarce *Bryum dyffrynense*. Strong populations of *Petalophyllum* also persist in some dune-slacks, although these have struggled to maintain their numbers while populations in south-western England and south Wales have shown dramatic increases. The main concern for the future is to maintain sufficient open, moist, early-successional habitat for the rarer bryophytes.

INTRODUCTION

The dune system on the northwest coast of England between Liverpool and Southport (the Sefton Coast) is one of the most important in Europe. It is the largest dune system in England, some 28 km in length, and of outstanding interest for wildlife (Gateley and Michell, 2004), including numerous rare bryophytes.

The populations of two bryophyte "Priority Species" within the UK Biodiversity Action Plans remain of national importance, the liverwort *Petalophyllum ralfsii* (Plate 1) (Wilson) Nees and Gottsche ex Lehm. and the moss *Bryum warneum* (Röhl.) Blandow ex Brid. (Plate 2), the latter having larger populations on the Birkdale Green Beach than are known at any other site in western Europe.

This short account gives a historical review of the rare coastal bryophytes, from their initial discovery and excessive specimen collecting by the Victorian bryologists to the modern concerns with habitat loss and species conservation.

The nineteenth century

The Lancashire coast became famous for its rare mosses when two pages of Addenda at the front of William

Plate 1. The liverwort Petalophyllum ralfsii with young capsules. (Photograph by D.T. Holyoak).

Plate 2. The moss Bryum warneum with capsules nearly mature, on Birkdale Green Beach. (Photograph by D.T. Holyoak).

Wilson's *Bryologia Britannica* (1855) announced discovery on the "flat sandy shore at Southport" of the moss *Bryum marratii* Hook.f. and Wilson new to science, along with the first British record of *Bryum calophyllum* R.Br. and the second of *B. warneum*. All three were discovered by Mr. W.M. Marrat of Liverpool in 1854. Initial identification of the *B. calophyllum* as a British species was particularly challenging because it was previously known only from specimens collected in 1823 in the Arctic at Igloolik (Melville Island, Canada), during Capt. Parry's voyage.

Later years of the nineteenth century produced the first records from the region of several other rare mosses, including *Amblyodon dealbatus* (Hedw.) Bruch and Schimp., *Bryum knowltonii* Barnes (at Birkdale, 1855), *B. neodamense* Itzigs. ex Müll.Hal. (at Southport, 1857), *Catoscopium nigritum* (Hedw.) Brid. (abundant, at Ainsdale, 1864) and *Meesia uliginosa* Hedw. along with the liverworts *Moerckia hibernica* (Hook.) Gottsche (at Southport, 1863) and *Petalophyllum ralfsii* (at Ainsdale, 1861) (Braithwaite 1888-1895; Pearson 1902; Savidge *et al.* 1963). During the Victorian era, large numbers of specimens of the rarer species were collected from the region and exchanged by bryologists (especially G.E. Hunt, D.A. Jones, J.A. Wheldon, W. Wilson and J.B. Wood), so that they are represented in many herbaria (Holyoak, 2001).

Bryum marratii was recorded from 1854 until 1888 (possibly 1898), but not subsequently. In 1854 it was "plentiful on the shore from Southport to Churchtown" (note by F.P. Marrat in Herb. Palgrave) and in 1860 it was also discovered near Ainsdale (Savidge *et al.*, 1963; Holyoak, 2001).

The twentieth century

Energetic collecting of specimens of most of the rarer bryophytes for exchange continued well into the 1920s (e.g. by J.A. Wheldon and W.G. Travis), but there were few new discoveries. *Bryum mamillatum* Lindb. was newly recorded from "edge of sand-dune, Freshfield" in 1908 by J.A. Wheldon (Savidge *et al.*, 1963), but this taxon is now regarded as a rare and probably somewhat aberrant form of *B. warneum* (Holyoak, 2004).

Bryum knowltonii was last recorded in 1920 (specimen in Herb. HERBARIUM OF NATIONAL MUSEUMS AND GALLERIES OF WALES, CARDIFF: Holyoak, 2001). *B. calophyllum* persisted much longer, with confirmed specimen records dating from 1904 (Freshfield, collected by S.J. Owen and D.A. Jones, HERBARIUM OF NATIONAL MUSEUMS AND GALLERIES OF WALES, CARDIFF) to 1973 (Ainsdale dunes in wet sand, leg. J. Appleyard, HERBARIUM OF NATIONAL MUSEUMS AND GALLERIES OF WALES, CARDIFF) and unconfirmed reports as late as 1983 (M.J. Wigginton), but no records since then (Holyoak, 2001).

Although there are post-1950 records of *Amblyodon dealbatus* and *Meesia uliginosa* from the Sefton Coast dune-slacks, *Catoscopium nigritum* had apparently disappeared before 1950 (Hill *et al.*, 1994).

Vegetation succession in the dune-slacks leading to reduction in the extent of open habitats probably led to local extinction of *Bryum calophyllum*, *B. knowltonii*, *B. marratii* and *Catoscopium nigritum*. By the late 1990s it was becoming apparent that some of the rarer bryophyte species that still survived in the dune-slacks were also at risk from loss and degradation of habitat, particularly *Bryum neodamense*, *B. warneum* and *Petalophyllum ralfsii*. Critchley (1999) identified encroachment of scrub as a threat to *B. neodamense* and advocated grazing for the removal of this vegetation as a part of conservation work that was also by then recognised as necessary to protect *P. ralfsii* and Natterjack Toads *Epidalea calamita*.

Energetic efforts by conservation managers to remove Sea Buckthorn *Hippophae rhamnoides* scrub and other invasive plants were begun. Deepening and scraping out of dune-slack pools ("re-profiling") during the 1990s to benefit Natterjack Toads was having an added advantage in creating small patches of open, sandy, early-successional habitat suitable for *B. neodamense*, *B. warneum* and *P. ralfsii*.

The twenty-first century

The UK Biodiversity Action Plans did much to encourage field surveys and other conservation work on rare coastal bryophytes in Britain. The plant-conservation charity Plantlife organised much of this work in England, with funding from English Nature, from 1996 onwards for *Petalophyllum ralfsii* and 2000 onwards for several of the rare coastal *Bryum* species.

On the Sefton Coast the first really detailed surveys of the status of *P. ralfsii* were carried out by the Sefton Coast Life Project with fieldwork mainly by Dan Wrench (anon. 1998) and an analysis of factors affecting the local distribution of this species was described by Hughes (1977). In 1997 its total population on the Sefton Coast was reported as 7887 thalli in 47 sub-populations (anon. 1998), whereas another survey in 2001 revealed large declines, with only 3147 thalli counted (Holyoak, 2002a). Over the same period *P. ralfsii* had increased at its sites in S. Wales and SW England. The studies on the Sefton Coast and elsewhere have emphasised that local habitat conditions are important for survival of *P. ralfsii* populations, which need short open vegetation in dune-slacks, especially in trampled habitat adjacent to pathways; they decline as natural succession or reduced grazing pressure lead to taller vegetation or when pathways fall into disuse (see the expanded Species Action Plan: anon., 2001; a detailed "Species Dossier" for *P. ralfsii* with conservation recommendations can be downloaded from: http://www.plantlife.org.uk/uk/plantlife-saving-species-under-our-care.html#bryophytes).

Surveys of rare coastal *Bryum* species were carried out around the coasts of England and Wales during 2001 (Holyoak, 2002b; 2002c). These failed to refind *B. calophyllum*, *B. knowltonii* or *B. marratii* on the Sefton Coast, where they are assumed to have become locally extinct. However, the Nationally Rare and declining *B. warneum* was refound not only as a few small patches in dune-slacks but also in much larger amounts on the floristically varied Birkdale Green Beach. On the new foredune habitat of the Green Beach (Smith, 2007) more than a hundred patches of *B. warneum* were recorded along a strip of coast 1.4 km in length, the largest patch covering an area of 170 x 150 cm. The total amount of *B. warneum* on the Green Beach thus exceeded the combined total population then known from all other localities in England and Wales by several orders of magnitude and the largest patch measured there contained far more plants than were present in aggregate at any other single site (Holyoak, 2002c). In 2002 it was correctly predicted that this great abundance would be a short-lived phenomenon, since vegetation succession was proceeding rapidly and leading to a reduction in extent of unshaded habitats with partly bare damp sand. The main concern for the future will therefore be to maintain sufficient open, moist, early-successional habitats on the Green Beach for *B. warneum* if continued coastal sedimentation ceases to provide them. A detailed "Species Dossier" for *B. warneum* with conservation recommendations can be downloaded from:

http://www.plantlife.org.uk/uk/plantlife-saving-species-under-our-care.html#bryophytes

During the studies carried out by Plantlife, an unidentified species of *Bryum* was found repeatedly on damp coastal sands, initially at Dungeness (Kent), then on the Welsh coast (e.g. at Morfa Dyffryn) and also in large quantities on the Green Beach. This previously undescribed taxon allied to *B. bicolor* Dicks. and *B. dichotomum* Hedw. was eventually named as *Bryum dyffrynense* Holyoak (Holyoak, 2003). It now appears to be Nationally Scarce rather than really rare in Britain, with records from coastal sites in England and Wales from E. Kent and W. Cornwall northwards to Westmorland. Its wider range includes scattered localities on the N. and W. coasts of Ireland and coastal sites in NW. France (Bray dunes, Dept. Nord) and Belgium (Zeebrugge).

For many years the Ainsdale dune-slacks had been the best known British locality for *Bryum neodamense* (cf. Critchley, 1999; Church *et al.*, 2001; Holyoak, 2002c). However, field research there and in Ireland revealed occurrence of many plants which were difficult to identify because they show characters intermediate between those of *B. neodamense* and the much commoner *B. pseudotriquetrum* (Hedw.) P.Gaertn. *et al.* Studies of herbarium material showed such intermediates to be widespread in Europe, leading to molecular studies which confirmed that this moss is no more than a distinctive looking but inconstant phenotype of *B. pseudotriquetrum*, which develops in some basic habitats that are periodically flooded (Holyoak and Hedenäs, 2006). *B. neodamense* has subsequently been removed from the list of plants afforded special protection under the Wildlife and Countryside Act.

Phytosociological studies of sand-dune vegetation became easier when the last volume of the series describing the National Vegetation Classification (NVC) was published (Rodwell, 2000). However, coverage of bryophytes in the NVC was somewhat incomplete because it largely ignored vegetation types dominated by cryptogams. This also applies to the recent surveys of the Sefton Coast sand-dune vegetation using NVC techniques (Gateley and Michell, 2004), but a more detailed analysis of the role played by bryophytes there was given by Callaghan and Ashton, (2007).

For some years, fieldwork by several members of the British Bryological Society has been underway to prepare a new bryophyte flora of vice-county 59 (South Lancashire), which will include comprehensive species lists and distribution maps for all bryophyte species occurring on the Sefton Coast. Their continuing surveillance of the bryophytes is likely to be necessary to safeguard the rare species currently known and also to watch for new arrivals. Events over the past decade emphasise that dynamic natural coastal processes and vegetation succession on the Sefton Coast have led to both rapid increases and decreases among the rarer bryophytes, during which the conservationists have largely been spectators. Rising sea-levels and climatic warming will doubtless lead to further habitat changes, and challenges for bryophyte conservation.

REFERENCES

anon. (1998) *The Sefton Coast dune system: survey of Petalophyllum ralfsii (Petalwort).* Unpub. report. Formby: Sefton Coast Life Project.

anon. (2001) *Species action plans for plants, Petalwort.* London: English Nature and Plantlife.

Braithwaite, R. (1888-1895) *The British Moss Flora.* Vol. II. London: L. Reeve.

Callaghan, D.A. and **Ashton, P.A.** (2007) Bryophyte clusters and sand dune vegetation communities. *Journal of Bryology.* 29 213-221.

Church, J.M., Hodgetts, N.G., Preston, C.D. and **Stewart, N.F.** (2001) *British Red Data Books, mosses and liverworts.* Peterborough: Joint Nature Conservation Committee.

Critchley, R.E. (1999) *Factors affecting the occurrence of Bryum neodamense, a nationally rare moss of the Sefton Coast, Merseyside.* B.Sc. Dissertation. University of Nottingham.

Gateley, P. and **Michell, P.** (2004) *Sand dune survey of the Sefton Coast, 2003/2004.* Bootle: Sefton Council.

Hill, M.O., Preston, C.D. and **Smith, A.J.E.** (1994) *Atlas of the Bryophytes of Britain and Ireland.* Vol. 3. Mosses (Diplolepideae). Colchester: Harley Books.

Holyoak, D.T. (2001) *Coastal mosses of the Genus Bryum.* Report to Plantlife on work carried out during 2000. Report no. 178. Plantlife.

Holyoak, D.T. (2002a) *Petalwort Petalophyllum ralfsii*. Report to Plantlife on work carried out in England and Wales during 2001 and 2002. Report no. 202. Plantlife.

Holyoak, D.T. (2002b) *Coastal mosses of the Genus Bryum.* Report to Plantlife on work carried out in Wales during 2001. Report no. 203. Plantlife.

Holyoak, D.T. (2002c) *Coastal mosses of the Genus Bryum.* Report to Plantlife on work carried out in England during 2001. Report no. 206. Plantlife.

Holyoak, D.T. (2003) A taxonomic review of some British coastal species of the Bryum bicolor complex, with a description of *Bryum dyffrynense* sp. nov. *Journal of Bryology.* 25 107-113.

Holyoak, D.T. (2004) Taxonomic notes on some European species of Bryum (Bryopsida: Bryaceae). *Journal of Bryology.* 26 247-264.

Holyoak, D.T. and **Hedenäs, L.** (2006) 'Morphological, ecological and molecular studies of the intergrading taxa *Bryum neodamense* and *B. pseudotriquetrum* (Bryopsida: Bryaceae)'. *Journal of Bryology.* 28 299-311.

Hughes, K. (1977) *The factors which affect the distribution of Petalophyllum ralfsii on the Birkdale Coast.* B.Sc. Dissertation. Geographical Studies. Liverpool Hope University College.

Pearson, W.H. (1902) *The hepaticae of the British Isles.* vol. I. London: Lovell Reeve and Co.

Rodwell, J.S. (ed.) (2000) *British plant communities.* vol. 5. Maritime communities and vegetation of open habitats. Cambridge: Cambridge University Press.

Savidge, J.P., Heywood, V.H. and **Gordon, V.** (eds.) (1963) *Travis's flora of south Lancashire.* Liverpool: The Liverpool Botanical Society.

Smith, P.H. (2007) The Birkdale Green Beach – a sand-dune biodiversity hotspot. *British Wildlife.* 19 11-16.

Wilson, W. (1855) *Bryologia Britannica; containing the mosses of Great Britain and Ireland.* London: Longman, Brown, Green, and Longmans.

The Hall Road - Hightown shingle beach

Philip H. Smith

9 Hayward Court, Watchyard Lane, Formby, Liverpool L37 3QP.
philsmith1941@tiscali.co.uk

ABSTRACT

The only shingle (or more strictly, cobble) beach formation in South Lancashire (vice-county 59) occurs on a 2km stretch of foreshore between Hightown and Hall Road. It is formed from water-worn house bricks and other rubble eroding from a coast-protection embankment tipped here from about 1942 to the 1970s. Four sections, totalling 915m, have become partly vegetated since the first appearance of characteristic shingle plants in the mid-1970s. The oldest section (0.21ha) near Hightown includes vegetation that accords with the National Vegetation Classification's (NVC) SD1 (*Rumex crispus – Glaucium flavum* shingle community). "Classic" shingle plants include Yellow-horned Poppy (*Glaucium flavum*) (125 plants in 2007), Sea-kale (*Crambe maritima*) (11 plants) and Rock Samphire (*Crithmum maritimum*) (53 plants). Three much younger sections (0.18ha) nearer to Hall Road are poor statistical fits to NVC communities and resemble sea-cliff or strand-line types. Although the shingle is of artificial origin, the vegetation has developed naturally and has a high nature conservation value (Plate 1).

INTRODUCTION

The only shingle (or more strictly speaking "cobble") beach in Sefton, and indeed the sole example of this habitat in South Lancashire (vice-county 59), is found on a 2km section of shoreline between Hall Road, Blundellsands and Hightown. It is formed entirely from water-worn house bricks and other rubble eroding from a coast-protection embankment. The rubble embankment was tipped from about 1942, initially using Liverpool bomb-damage debris. Tipping continued at intervals until the early 1970s, when spoil

Plate 1. Sea-kale on the Hall Road shingle. (Photograph by Philip Smith).

from the construction of the second Mersey Tunnel was used (Edmondson *et al.*, 1988).

For most of its length the embankment has been cliffed by wave-action and there is little or no shingle vegetation. Elsewhere, erosion is less severe and plants have colonised the brick rubble to produce, in places, a characteristic vegetation type. This is the SD1 *Rumex crispus – Glaucium flavum* community defined and described by Rodwell (2000). Nationally, the community is characterised by the near constant presence of Curled Dock (*Rumex crispus* ssp. *littoreus*), Yellow-horned Poppy, Sea-kale (Plate 1) and Sea Beet (*Beta vulgaris* ssp. *maritima*), among others. Where conditions are sandier, Sea Sandwort (*Honckenya peploides*) occurs, but Rock Samphire is only very occasional (Rodwell, 2000).

Vegetated shingle is a relatively rare habitat in Britain, largely confined to the coasts of the warmer south and east of England, though fragments occur as far north as the Clyde. Sea-kale, Yellow-horned Poppy and Rock Samphire are considered to be thermophilous and largely limited by the July 17.5° maximum (Rodwell, 2000).

In the Northwest of England, vegetated shingle is particularly rare, most of it being in Cumbria (60ha). The Regional Biodiversity Group (1999) incorrectly states that it is absent south of north Lancashire.

Vegetated shingle is listed in Annex 1 of the EU Habitats Directive and is a UK Key Habitat in the Biodiversity Action Plan (BAP) (UK Biodiversity Group, 1999). However, perhaps because of its limited extent and artificial origin, shingle does not appear in the north Merseyside BAP (Merseyside Biodiversity Group, 2001).

Origin and development of the shingle vegetation

Shingle vegetation seems not to have occurred at Hightown much before 1975 when Yellow-horned Poppy was first found (personal observations). By 1984, this species had disappeared (Smith, 1984) but the first NVC survey of the Sefton Coast in 1988 found an SD1-type community at the northern end of the embankment, described as "Brick rubble foreshore with scattered plants of *Glaucium flavum* and some *Crambe maritima*." The mapped SD1 is about 115m long and 10m wide at National Grid Reference SD296023 (Edmondson *et al.*, 1988). Sea-kale was first recorded in 1988 (Smith, 1999), while Rock Samphire appeared in about 1993 (SGS Environment, 1995).

Floristic studies in 1999 and 2000 listed 63 vascular taxa on the northern 250m of the Hightown shingle, including 122 plants of Yellow-horned Poppy, five of Sea-kale and five of Rock Samphire (Smith, 2000).

By the second NVC survey of the Sefton Coast, the Hightown SD1 community had increased in area from 0.12ha to 0.19ha but, according to Gateley and Michell (2004), was still restricted to the northern site, although they did map a narrow zone of SD2 (*Honckenya peploides-Cakile maritima* strandline), covering 0.22ha, extending north from the SD1 community.

Another study in 2003 noted that the southernmost 400m section of shingle at Hall Road had become partly consolidated, with much blown sand and was, in places, becoming vegetated. There were two large individuals of Sea-kale, much Sea Beet and occasional large patches of Sea Sandwort, but no Yellow-horned Poppy was found (Smith, 2003).

The 2007 survey

A further survey was conducted in June 2007. Four areas of vegetated shingle were identified, descriptive data being summarised in Table 1.

Site 1 is the old-established beach at Hightown, occupying the northern section of weathered brick-rubble and now extending for about 425m. Since the previous studies, rubble has been moved northwards by long-shore drift. The new extension is more sparsely vegetated and sandier; it merges with fore-dunes dominated by Sand Couch (*Elytrigia juncea*) with much Sea Sandwort and was mapped as SD2 by Gateley and Michell (2004). The southern section is partly protected

Table 1. Data from four shingle sites surveyed in 2007.

Site	Grid reference	Area (m²)	Length (m)	Maximum width (m)	No. of vascular plants
1	29584 02306 – 29570 02730	2100	425	8	64
2	29850 00640 – 29811 00799	600	200	4	32
3	29713 01130 – 29701 01171	125	50	3	22
4	29674 01271 – 29587 01501	1100	240	5	27

to seaward by a narrow storm-beach over 200m long and 1-1.5m high, first described by SGS Environment (1995). There were 125 individuals of Yellow-horned Poppy, 53 of Rock Samphire and 11 of Sea-kale. All counts are higher than previous figures.

The other sites extend northwards from the Hall Road carpark. Curled Dock is present throughout. At Site 2, the gently shelving shingle beach merges with embryo dunes dominated by Sand Couch. There is no Rock Samphire or Yellow-horned Poppy but three plants of Sea-kale are present just north of the car park and Sea Sandwort is locally abundant. The plantlife is much less diverse than that at Hightown. Site 3 is a short, narrow section which merges with coarse grassland on the rubble embankment. Of the "classic" shingle plants, only Sea Beet and Curled Dock are present. The final stretch, Site 4, finishes just south of the Alt training bank and has an indistinct inland boundary to coarse grassland. Sea Beet becomes progressively more abundant northwards, while Sea Sandwort is locally frequent.

In total, 75 vascular taxa were identified on the four shingle sites, six (8.2%) being non-native. One of these, Japanese Rose (*Rosa rugosa*), has increased in frequency since the 2003 survey. There are one nationally and seven regionally notable taxa (Table 2). The total area of vegetated shingle is now about 0.39ha, extending for a linear distance of 915m.

NVC communities

As the Hightown shingle community was described as recently as 2004 (Gateley and Michell, 2004), it was decided to concentrate efforts on recording the three vegetated sections nearer to Hall Road. Using NVC methodology (Rodwell, 2000), five 2 x 2m samples were recorded on Sites 2 and 3 and seven on the larger Site 4. These were subject to TABLEFIT analysis (Hill, 1996) to determine the degree of fit to known communities. The results are summarised in Table 3.

The vegetation of Site 2 accords with SD1a *Rumex crispus – Glaucium flavum* shingle, typical sub-

Table 2. Nationally and regionally notable vascular plants recorded on the Hall Road - Hightown shingle in 1999 and 2007.

Taxon	English name	Freq. 1999	Freq. 2007	Status
Crambe maritima	Sea-kale	r	o	SCI
Crithmum maritimum	Rock Samphire	r	lf	SCI
Eryngium maritimum	Sea Holly	o	r	SCI
Euphorbia paralias	Sea Spurge	r	la	SCI
Glaucium flavum	Yellow-horned Poppy	o	o	SCI
Phleum arenarium	Sand Cat's-tail		r	SCI
Polygonum oxyspermum raii	Ray's Knotgrass		r	SCI
Vulpia fasciculata	Dune Fescue		r	NS

NS = nationally scarce; SCI = Species of Conservation Importance in North West England.
a = abundant; f = frequent; o = occasional; r = rare; l = locally.

Table 3. TABLEFIT analysis of Hall Road samples in 2007.

Site	NVC code	Community	Goodness of fit (%)
2	SD1a	*Rumex crispus – Glaucium flavum* shingle; typical sub-community	28
3	MC6	*Atriplex hastata – Beta vulgaris* sea-bird cliff	29
4	SD3	*Matricaria maritima – Galium aparine* strandline	30

community, but the level of fit (28%) is poor.

Site 3 is best described by MC6 *Atriplex hastata* (now *A. prostrata*) – *Beta vulgaris* sea-bird cliff community, but again the confidence is poor (29%). This community is most characteristic of rocky coastal sites where there is a combination of high maritime influence and intense disturbance (both nutrient deposition and physical damage) by sea-birds. The latter factor does not apply at Hall Road but essentially similar mixtures of *Atriplex* spp. and *Beta* can be found on strandline debris in sandy and shingle foreshores, as here (Rodwell, 2000).

Site 4 shows a poor fit (30%) to SD3 *Matricaria* (now *Tripleurospermum*) *maritima – Galium aparine* strandline. This community is typical of sandy shingle strandlines with drift detritus around more sheltered shores in the cooler, wetter north of Britain (Rodwell, 2000).

Discussion

Since 2004, three new areas of vegetated shingle have developed north of Hall Road with a total area of about 0.18ha, adding to the long-established section at Hightown (0.21ha). However only the southernmost section (Site 2) accords with SD1 and the statistical fit is poor. The other two sections support communities more typical of sand and shingle strandlines. The poor degree of fit to NVC communities suggests that these areas of vegetation are of recent origin and have not yet matured enough to accord with a known community.

The 75 vascular taxa recorded in 2007 are a mixture of maritime, shingle, sand-dune and ruderal specialists. One Nationally Scarce and seven Species of Conservation Importance in North West England (Cheffings and Farrell, 2005; Regional Biodiversity Group, 1999) were found. As the habitat provides harsh conditions for plant growth, the number of species found is relatively low, ranging from 22 to 64 per site. As might be expected, the oldest site at Hightown supports the largest number of taxa, as there has been time for more plants to become established. This is also the largest site and there appears to be a predictably positive relationship between species-richness and site area (Figure 1).

Counts of the number of Sea-kale, Yellow-horned Poppy and Rock Samphire plants show that their populations are increasing with time. As these "classic" shingle plants are thermophilous, they may be responding to warmer summers. The three species are rare in the vice-county (D.P. Earl *in litt.* 2008).

Although the shingle beach is of artificial origin, the vegetation has developed naturally, perhaps aided by arrival of propagules from the North Wales coast (Gateley and Michell, 2004). This is the only known example of vegetated shingle in South Lancashire, the habitat being rare in North West England. It therefore has a high nature conservation value and could be considered for inclusion in the North Merseyside Biodiversity Action Plan. Although there are few alien plants, the invasive Japanese Rose may eventually justify control.

A level of statutory protection is afforded, as the northern section lies within the Sefton Coast SSSI/SAC, while the three southern sections are adjacent to the SSSI/SAC boundary and are part of a Site of Local Biological Interest designated under the Sefton Unitary Development Plan.

It will be interesting to follow the future development of the artificial shingle and its plantlife, particularly in the

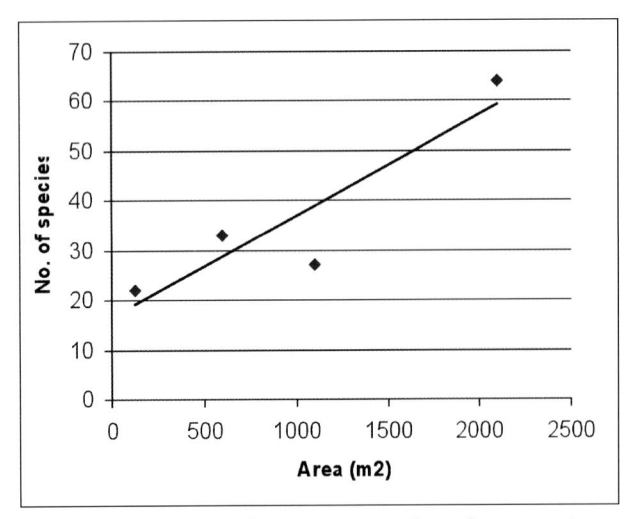

Figure 1. Relation between vascular plant species-richness and site area, Hall Road – Hightown shingle.

context of long-standing proposals for coast protection works in this area.

A more detailed account of the shingle beaches, with species lists, is available on the Sefton Coast Partnership website: www.seftoncoast.org.uk

ACKNOWLEDGEMENTS

I am grateful to David Earl for information from the New South Lancashire Flora Project. Pauline Michell provided much appreciated assistance on the analysis of NVC plant communities.

REFERENCES

Cheffings, C.M. and **Farrell, L.** (2005) *The Vascular Plant Red Data List for Great Britain. Species Status* 7 pp. 1-116. Peterborough: Joint Nature Conservation Committee.

Edmondson, M.R., Edmondson, S.E. and **Gateley, P.S.** (1988) *National Sand Dune Vegetation Survey, Sefton Coast. Division 1, Seaforth-Hightown*. Report no. SC:NVC:88:00. Peterborough: Nature Conservancy Council.

Gateley, P.S. and **Michell, P.E.** (2004) *Sand Dune Survey of the Sefton Coast*. Warrington: TEP.

Hill, M.O. (1996) *TABLEFIT V1.0, for identification of vegetation types*. Huntingdon: Institute of Terrestrial Ecology.

Merseyside Biodiversity Group (2001) *North Merseyside Biodiversity Action Plan 2001*. Maghull: Environmental Advisory Service.

Regional Biodiversity Group (1999) *A Biodiversity Audit of the North West*. 2 vols. Maghull: Environmental Advisory Service.

Rodwell, J.S. (ed.) (2000) *British Plant Communities volume 5. Maritime communities and vegetation of open habitats*. Cambridge: Cambridge University Press.

SGS Environment (1995) *Hightown Coastal Works. Ecology, landscape & peat bed impact assessment*. Sefton Council Ainsdale: Report no. HO2751/V2/02-95.

Smith, P.H. (1984) *The special plants of Hightown sand-dunes and salt-marsh*. Unpublished report. Lancashire Trust for Nature Conservation.

Smith, P.H. (1999) *A reappraisal of the special plants of Hightown Dunes and Meadows*. Unpublished report. Ainsdale: Sefton Coast Life Project.

Smith, P.H. (2000) *Additions to the vascular plants of the Hightown shingle bank, Sefton Coast*. Unpublished report. Ainsdale: Sefton Coast Life Project.

Smith, P.H. (2003) *A note on the Hall Road shingle beach, Sefton Coast*. Unpub. report. Ainsdale: Sefton Coast Partnership.

UK Biodiversity Group (1999) *Tranche 2 Action Plans – Volume V: Maritime species and habitats*. Peterborough: English Nature.

Northern Dune Tiger Beetle *Cicindela hybrida* L. a 'flagship species' of high conservation value on the Sefton Coast sand dunes

Dr Stephen Judd

World Museum Liverpool, National Museums Liverpool, William Brown St, Liverpool L3 8EN
steve.judd@liverpoolmuseums.org.uk

Plate 1. Adult C. hybrida. *(Photograph by Paul Wisse).*

ABSTRACT

Northern Dune Tiger Beetle *Cicindela hybrida* L. (Plate 1) is at the north-westerly extreme of its range in Britain with isolated populations on the Drigg Coast, Cumbria and Sefton Coast, Merseyside. It is one of five tiger beetle species recorded in Britain and is classified as 'vulnerable' in the UK *Red Data Book* and as a 'priority' species in the UK *Biodiversity Action Plan*.

This study clarifies the status and distribution of *C. hybrida* in Britain, which has been the subject of much confusion. A summary of known autecology is also provided to inform the discussion about future conservation management requirements for the species.

INTRODUCTION

C. hybrida is the subject of extensive experimental and field-based published research. The desynchronised life cycle and population structure was investigated by Simon-Reising *et al.* (1996), building on earlier biological work by Faasch (1968). Field populations have been used by Dresig (1980, 1981, 1983 and 1984) to determine water loss, thermoregulation and predation rates. Evans (1965) described the feeding method, and the role of sight and memory in food capture is discussed by Swiecimski (1957). Further insight is provided by unpublished studies of the species in North West England. An autecological investigation by Aldridge (1974) at Ainsdale Sand Dunes National Nature Reserve, contained many original observations. Distribution patterns were studied by Gore (1997), and Starkings (1999) investigated adult and larval habitat requirements at Formby Point and Ravenmeols dunes, also on the Sefton Coast. Copestake (1999), evaluated Cumbrian habitat requirements.

Widely distributed in mainland Europe and Siberia, except for the extreme north, *C. hybrida* is a very thermophilous, relatively narrowly adapted species, occurring in habitats that experience extreme insolation, heat and desication. On the continent, its habitats are characterised by open sand with little or no vegetation, high day-time temperatures and low humidity (Swiecimski, 1957; Lehmann, 1978; Dresig, 1980; Simon-Reising *et al.*, 1996; Van Dooren, 1999). The species is also found on fine gravel (Lindroth, 1985). In Britain, the occurrence of *C. hybrida* on the Sefton Coast has been extensively recorded (e.g. Ellis, 1889; Chaster and Burgess Sopp, 1903; Sharp, 1908; Britten, 1943; Flint, 1964; Thomas, 1967; Eccles, 1991). National distribution was most recently discussed and mapped by Luff (1998).

British Distribution
Survey effort

A total of 16 person days was devoted to searching for *C. hybrida* between 1999 and 2003 in order to determine the species' range and distribution on the Sefton Coast and to search potentially suitable sites within its current and historic range in western Britain (Judd, 2003a). To assist this, a desktop survey of historic and contemporary records was undertaken. Records were mapped using Mapinfo. In addition, land ownership and management details for sites supporting *C. hybrida* were investigated. The results are summarised below.

National Status

Cicindela hybrida is an exclusively coastal species in Britain and is restricted in distribution to two sites in North West England – covering three adjacent ten kilometre squares on the Sefton Coast and one on the Drigg Dunes, Cumbria (Figure 1). It was formerly thought to be more widely distributed in the UK, but records in Luff (1998) for North Lancashire, South Cumbria, Wirral, North West Wales (two sites), Cornwall and Norfolk, are incorrect.

Historically, the closely related Dune Tiger Beetle *C. maritima* Latreille and Dejean was regarded as a subspecies, or form, of *C. hybrida* and confusion between the two species has persisted to the present day. In Britain, the exclusively coastal distributions of both species do not overlap. However, in mainland Europe and Siberia, where *C. hybrida* is widely distributed, except for the extreme north, they occur together in riparian biotopes (Lindroth, 1985 and 1992; Trautner and Geigenmuller, 1987).

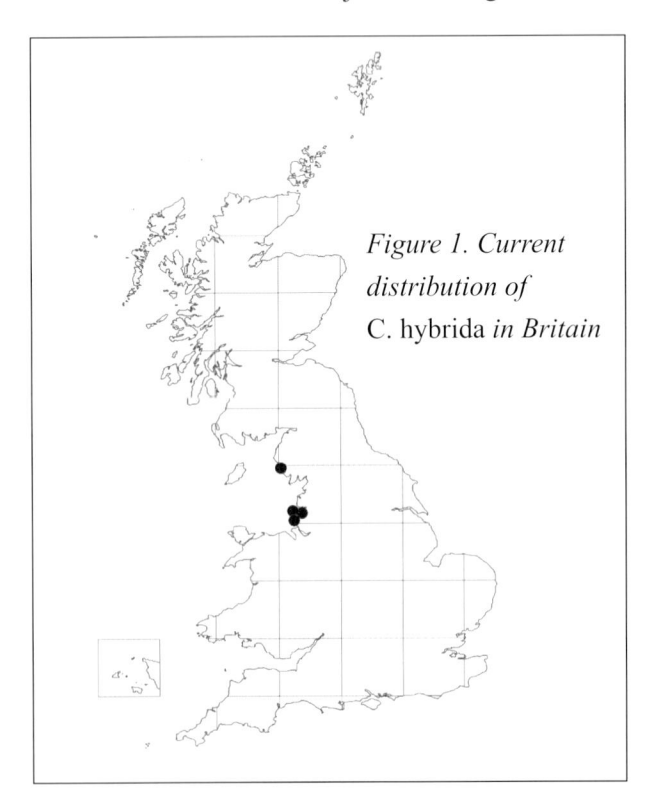

C. hybrida was not recorded during visits to the following dune systems in West and North Wales - Ynyslas Dunes, Dyfi NNR Wales (Mike Bailey and Carl Clee pers. comm.); Morfa Dinlle, Morfa Harlech, Newborough Warren and other Anglesey sites (Adrian Fowles and Mike Howe pers. comm.); Gronant and Talacre (Judd, 2003b). In England, the species was not recorded from Wirral or Fylde sand dune systems, nor did it occur on remnant dunes south of Crosby (SD300990) or north of Southport (SD331180) on the Sefton Coast. It was also absent from all modified, heavily vegetated and stable parts of the Sefton Coast Dune system. In Cumbria, the species was not recorded from Eskmeals Nature Reserve, or Eskmeals MOD site, to the south of the Drigg population, or from dune habitat to the north, from Allonby to Silloth (Copestake, 1999). It is also no longer recorded from three sites where it was historically known - Carnforth (SD47-70), undated Hancock Museum specimen; Isle of Walney Sandhills 1872; Wallasey various records, last recorded date by Rev. CE Tottenham in 1948.

Only three locations on Drigg Dunes support *C. hybrida* (Copestake, 1999); Alongside the estuary of the River Irt (SD071951 – SD079957); Alongside the estuary of the River Esk Ravenglass Nature Reserve (SD064970); and on foreshore dunes near the access road (SD048984 – SD050980).

Sefton Coast Status

C. hybrida has been regularly recorded in varying numbers on the Sefton Coast since the 1880s. During this survey, it was extensively found along a 15 kilometre stretch (Figure 2) and occurred on five of the eight major land units, all within the Sefton Coast SSSI - managed by Sefton Council, National Trust, English

Figure 1. Current distribution of C. hybrida *in Britain*

Plate 2. Typical C. hybrida *habitat. (Photograph by Lynne Collins).*

Nature, MOD and the Duchy of Lancaster. It was recorded from 105, separate 100 metre squares and there are paper records for an additional 44, post-1970, 100 metre squares. The species was recorded almost continuously between Birkdale in the north and Hightown in the south. The most southerly record was from Hall Road Crosby – a slight extension to the previous known range on the Sefton Coast. Most records are from mobile fore dunes (Plate 2). However, *C. hybrida* also occurs along sandy tracks and on eroded fixed dunes, particularly at Ainsdale NNR and Formby, in 'blow-outs' and wherever there is extensive open sand. There are even records from sandy clearings in the pine plantations. The association between *C. hybrida* and bare sand is clearly demonstrated when distribution is mapped in relation to aerial photographs (Figure 3).

Large areas of totally bare sand, which lack suitable food, densely vegetated and too stable areas, north-facing and tree-covered dunes, are all unsuitable habitat and explain the species' absence from some coastal dune areas. However, adjacent habitat, which is ignored by the species, is potentially important for foraging. Adult burrows are never located in fixed dune areas or vegetation margins. North-facing dune systems are unsuitable habitat, and this is one reason why *C. hybrida* does not occur in some coastal dune areas of Cumbria such as Silloth, Eskmeals MOD site and Eskmeals Nature Reserve (Copestake, 1999).

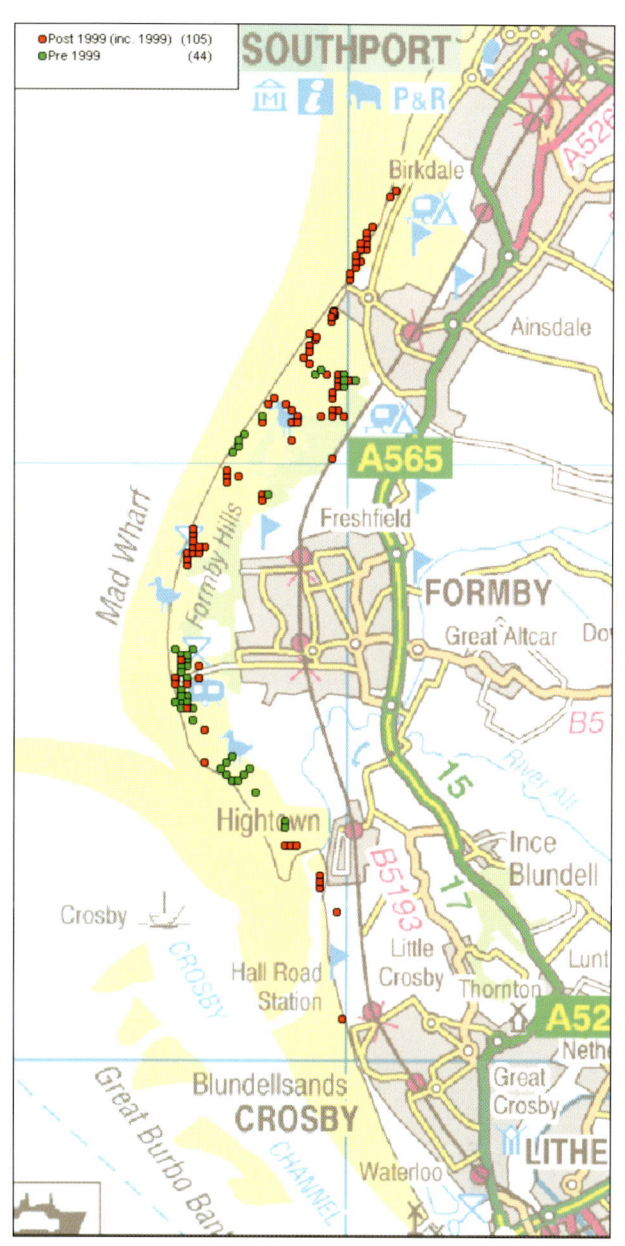

Figure 2. General distribution of C. hybrida *on the Sefton Coast. © Crown copyright. All rights reserved. Sefton Council Licence number 100018192 2010.*

Autecology

Tiger beetles have a fascinating life history and are wonderfully adapted to their environment. A detailed understanding of tiger beetle biology is an essential requirement to inform future conservation management of *C. hybrida* in Britain.

Activity and Thermoregulation

Adult *C. hybrida* are thermal specialists with high-energy requirements, and a high resting metabolic rate. The start and finish of daily activity is determined by

Figure 3. Aerial photographs of the Sefton Coast (north to south) showing the distribution of C. hybrida *in relation to open sand. © Crown copyright. All rights reserved. Sefton Council Licence number 100018192 2010.*

○ 1978-1998
○ 1999-2003

specific temperature thresholds, below which hunting is not profitable (Dresig, 1980). The average duration of the daily activity period on the Sefton Coast is 5.5 hours (Starkings, 1999). Light levels or a specific phase relationship with sunrise does not influence activity and the species is adapted to function optimally at 36°C, when prey is most abundant (Dresig, 1980). Adults are induced to emerge onto the sand surface, between 07.00hrs and 10.00hrs when a burrow temperature threshold of ca. 19°C is reached; mean emergence occurs when surface temperature is 28°C and maximum numbers are recorded when surface temperature is 34°C - 42°C, between 10.00hrs and 11.00 hrs (Dresig, 1980).

Three successive phases are recorded for adult *C. hybrida* during the daily increase in ambient temperature (Dresig, 1984) : (i) basking and foraging; (ii) proportional control (stilting) during which behaviour patterns are used to maintain an optimal body temperature in spite of a rise in ambient temperatures; (iii) shuttling between sun and shade.

Initially, about 20% of the adult *C. hybrida*'s time is used for movement; the remainder is spent sun-basking on the sand (Dresig, 1980, 1981). The body touches the ground in order to gain heat by conduction and is orientated at a right angle to the sun, often by the animal selecting a slope facing the sun to gain heat by radiation. As the temperature rises during the day, more time is used searching for prey.

C. hybrida has a maxithermal strategy to cope with sand surface temperatures (Dresig, 1980). On the Sefton Coast this varies between 19.5°C and 47°C, when air temperature is 18°C (Aldridge, 1974). Body temperature stabilises at 35°C, and further increase is prevented by stilting - raising the body away from the hot surface. This generally occurs in response to a rise in environmental temperature above 18°C, and is invariably found at temperatures above 28°C (Dresig, 1980). The time spent searching then remains at a constant high level. Convective heat loss is also increased by flight. Dresig (1980) found that *C. hybrida* is able to fly at all temperatures at which normal activity occurs, but the number of spontaneous flights, increases when surface temperatures are above 40°C.

Adults disappear by digging into the sand when temperatures become intolerably high and when the surface cools to 28°C. The number of animals begins to decline shortly before noon and continues to fall during the afternoon. This decline occurs in two phases – a rapid decrease following the peak, and another during the last part of the activity period. The last animals disappear in late afternoon. Typically, Sefton Coast adults retired underground between 14.00hrs and 15.00hrs (Starkings, 1999).

Adults are inactive at night and on rainy or cold days, when they remain buried in the sand, although it is believed that some spend the resting period beneath vegetation (Dresig, 1980). On the Sefton Coast, Starkings (1999) found that adult overnight burrows (Plate 3) were on average 3.6cm long and occurred on south-facing dune flanks that receive maximum insolation.

Daily activity patterns of tiger beetle larvae are influenced by a combination of burrow temperatures

Plate 3. Burrow of adult C. hybrida. *(Photograph by Guy Knight).*

and soil desiccation (Pearson, 1988). Larvae plug their burrows prior to periods of inactivity and before moulting. Tiger beetle larvae are active over a wide range of temperature and are usually inactive between 35°C - 40°C (Knisley, 1986).

Predators and Parasites

Cicindela hybrida is effectively adapted to avoid predation. It has cryptic/ disruptive colouration which makes it extremely difficult to see when standing still and burrows in sand. It escapes by flight, moves unpredictably and 'startle flashes' when the brilliant metallic abdomen is exposed (hence the old name of 'sparklers'). Tiger beetles also have defence (pygidial) glands that exude volatile chemicals such as benzolcyanide and benzaldehyde (Forsyth, 1970; Blum *et al.* 1981; Moore and Brown, 1971).

There are no predation or parasitism records for *C. hybrida* in Britain, although Kestrels are suspected to be possible predators on Ainsdale NNR (Aldridge, 1974). This is surprising, because adult tiger beetles generally, despite their keen eyesight, are known to be under extreme predation pressures and are eaten by badgers, foxes, shrews, lizards, toads and various birds (Hori, 1982). Insect predators are asilid flies, but only in flight, dragonflies, histerid beetles and ants (Knisley and Pearson, 1984).

The most important natural enemies of tiger beetle larvae, world-wide, are parasitic bombyliid bee flies in the genus *Anthrax* (Knisley, 1986). Parasitism rates from bee flies may be from 10% to over 70% for some larval populations of some species (Knisley and Pearson, 1981; Palmer, 1982; Bram and Knisley, 1982; Mury Meyer, 1987). The tiphiid wasp *Methocha articulata* Lat, a parasite of tiger beetle larvae has not yet been associated with *C. hybrida*. The closest record to British populations of *C. hybrida* is for the Lleyn Peninsula in Wales. Close observation of *C. hybrida* larval burrows, for evidence of parasitic attack, is required.

Foraging and Attack

Food is a limiting resource in the life cycle of tiger beetles - adult females at low feeding levels produce significantly fewer eggs and larvae than females at high feeding levels (Pearson and Knisley, 1985).

Adults

Most daily activity time is spent searching for prey and the rate of predation in *C. hybrida* is strongly dependent on ambient temperature (Dresig, 1981). Prey includes small ants, small beetles, spider-hunting wasps, small spiders and flies, nymphs of true bugs, springtails and moth larvae, together with dead insects lying on the sand (Dresig, 1981). The expected prey capture rate is three per hour. Speed of locomotion, handling time and success rate are not particularly temperature dependent because of the species' thermoregulatory abilities (Dresig, 1981).

The eye of the adult *C. hybrida* is optimised for visual hunting in a flat environment and the bright habitat has resulted in the evolution of apposition eyes (Brannstrom, 1999). Adults can sight prey from a distance of 20-30cm (Faasch, 1968) and, in experiments, attack creeping maggots 8–12mm in length, from a distance of 10cm (Friederichs, 1931). They assume an alert, raised stance when stalking prey (Swiecimski, 1957), often remain stationary for up to four minutes, before relocating a few metres away and fly infrequently (Copestake, 1999). The body is aligned to the long axis of the prey, thus placing the prey directly in their limited stereoscopic visual field (Swiecimski, 1957). Prey movement is the most important stimulus and adults make intermittent fast, short sprints, darting towards prey (Copestake, 1999).

Plate 4. Adult C. hybrida *with captured wolf spider. (Photograph by Lynne Collins).*

This stop-and-go movement enables them to continually relocalise moving prey within their field of vision (Pineda, 1999).

Cicindela hybrida engages in aggressive behaviour towards insects it cannot overcome (Dresig, 1981). If the prey is big, it will bite once, retreat a few centimetres, and lunge forwards and bite again. This is repeated until the prey is immobilised, after which it is torn and eaten (Swiecimski, 1957). A considerable number of prey items cannot be overcome but the tiger beetle still approaches and attacks indiscriminately (Dresig, 1981). If unsuccessful it releases it at once and runs away. If the prey is small it will start to chew it immediately (Dresig, 1981).

Larvae

The ecology and natural history of tiger beetle larvae are not well known. The three larval instars are sedentary ambush predators, lying in wait at the entrance to their burrows (Shelford, 1908; Willis, 1967). Their tunnel walls are sealed with saliva to reduce the possibility of collapse and the pit may serve to trap prey (Shelford, 1908). Larvae have characteristic zig-zag shaped bodies, which grip the sides of the burrow (Sutton and Browne, 2001) (Figure 4). The enlarged 8th segment in the middle of the abdomen has two hooked spines on the dorsal surface that fit into two small grooves in the tunnel wall and anchor the larva. The fifth segment of the abdomen is also equipped with a pair of these anchoring spines. Their large flattened heads (which are used to excavate the burrows) lie flush with the soil surface, and perfectly fit the burrow entrance (Shelford, 1908).

Larvae are rarely seen feeding in the field. Insects and spiders are usually 'snatched' as they pass the burrow entrance. A dark object against a light background releases the prey-catching behaviour (Faasch, 1968). Larvae can rear backwards up to half their length out of the tunnel and grasp the prey with their mandibles, and will even race out of the burrow with surprising agility to catch prey further afield (Linssen, 1959). This is dragged down and is consumed at the bottom of the burrow (Sutton and Browne, 2001).

First instar tiger beetle larvae require only one meal of sufficient size to moult (Willis, 1967). They will attack virtually any moving object below a maximum size, apart from some noxious species (Friedrichs, 1931; Willis, 1967; Larochelle, 1977; Mury Meyer, 1987). As a larva progresses through each successive stadium, its prey range expands upward, while the lower limit of prey size remains constant (Mury Meyer, 1987).

Larval food availability affects survival, developmental rate, adult size and emergence time (Hori, 1982). Larvae at low feeding levels take significantly longer to pass through all three larval stages. Their pupae and emergent adults are significantly smaller than those individuals raised at higher feeding levels (Pearson and Knisley, 1985).

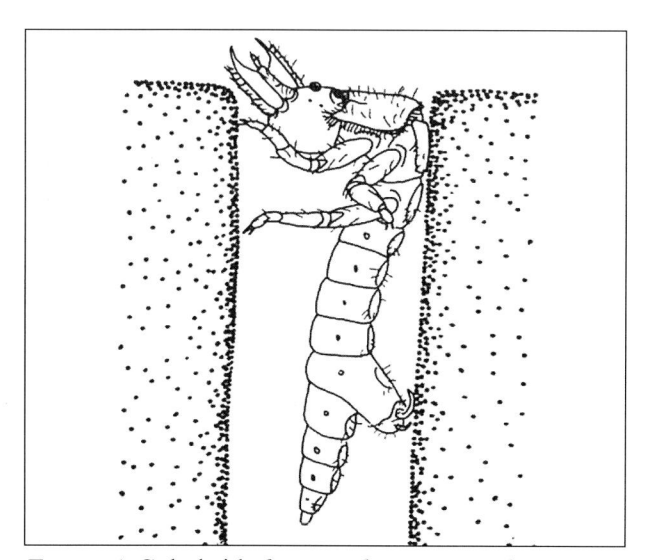

Figure 4. C. hybrida *larva in burrow. Re-drawn from Faasch (1968).*

Population Structure

Tiger beetle population dynamics and community structure are influenced by abiotic factors, resource limitation and competition, predation, parasitism and mutualism (Kinsley and Juliano, 1988).

Cicindela hybrida is a mobile species that can colonize suitable habitat quickly and efficiently. However, the ability to colonize is limited. Small-scale migration between adjacent habitat patches in a study on a 1.5 hectare area of relict Bavarian inland dune was rare although it did increase at specific periods of the year (Simon-Reising *et al*, 1996). Mark-recapture experiments at this location showed that the beetle tended to stay within a single habitat patch even though other suitable patches were located just a few metres away. The home range of males was significantly larger than that of females and they were recaptured at a distance of more than 600m across arable land.

On the Sefton Coast, Aldridge (1974) observed that adults did not have separate territories and roamed freely in a random fashion throughout suitable habitat. They kept spaced-out by moving away from other adults, unless mating. In a dispersal experiment involving 28 beetles, with equal numbers of each sex, the average distance travelled by males was 22.4 metres and 43 metres for females.

The estimated population size within a ca.0.7 hectare area at Ainsdale NNR during August 1973 was 55, increasing to 132 in May 1974 with equal numbers of males and females. The estimated population size of hibernating adults in April/ May was 280, after which population size decreased rapidly. The second peak of 310 in the second half of July, originated from adults emerging after mid-June (hibernating instars). A larval population of four tunnels per square metre of optimal habitat was estimated by Starkings (1999) for the Sefton Coast.

Desynchronised Life Cycle

Some observers have described a rather simple life cycle for *C. hybrida* with only one generation per year. However, in reality, two vastly desynchronised populations co-exist (Simon-Reising *et al.*, 1996).

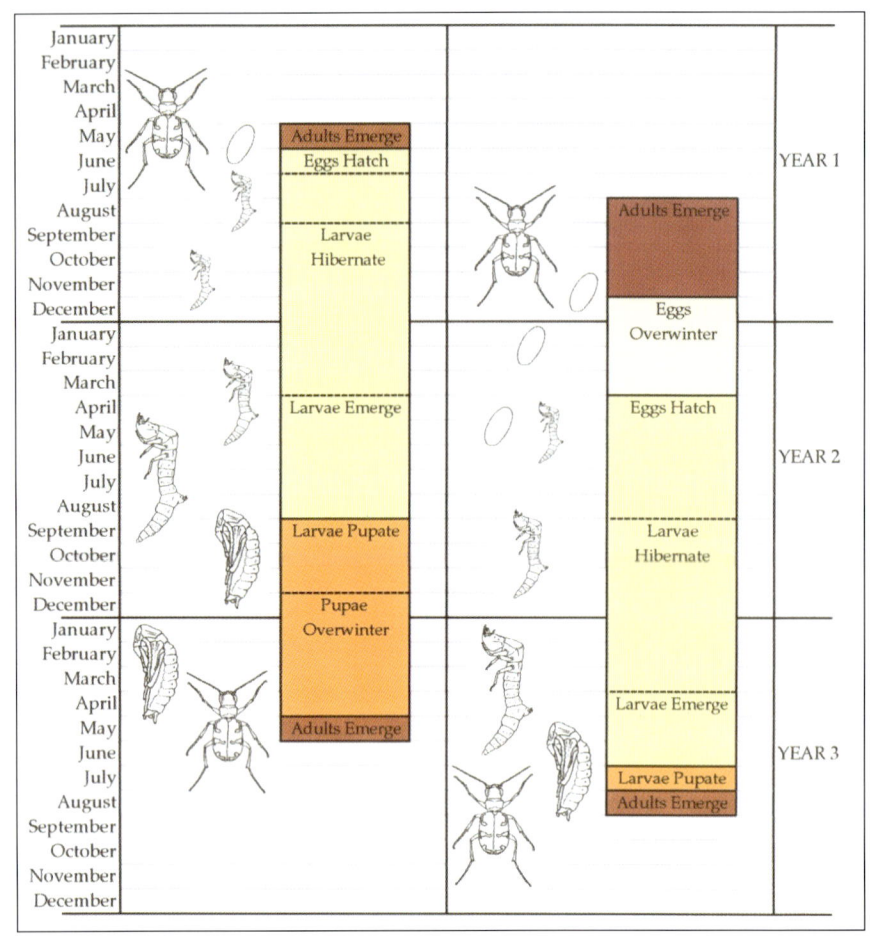

Figure 5. Desynchronised life cycle of C. hybrida. (Interpreted by Guy Knight).

Adult *C. hybrida* are encountered between April and September, with distinct peaks at the end of April and in early August (Aldridge, 1974). The spring population consists entirely of individuals which have over-wintered as adults (Luff, 1993). The second peak originates from adults emerging after mid-June from over-wintered larvae (Figure 5).

The normal life cycle for *C. hybrida* is two years. The May and August peaks in adult numbers represent two separate, co-existing populations (the three months time lapse prevents competition between identical stages in the development of the two populations) (Simon-Reising *et al.*, 1996). This means that different developmental stages hibernate simultaneously and are active during spring and summer. The different generations contribute to the adults of a given year in the following way:

- Adults resulting from eggs in the given year (hibernating as second or third instar larvae).

- Adults resulting from eggs laid the year before (hibernating as third instar, imago or first instar larvae).

A desynchronised life cycle lowers the risk of *C. hybrida* becoming extinct at a site. Differences in development speed and summer quiescence lead some specimens to 'switch' from one generation to another (Simon-Reising *et al.*, 1996).

Mating occurs in May and August. The male *C. hybrida* (Plate 5) seek females using similar aggressive behaviour to prey searching (Faasch, 1968). He rushes at the female from a distance of about 15cm, or less, in a series of quick movements and grips the base of the female's thorax with his mandibles with his front legs raised off the ground (Pearson, 1988). Copulation starts within one or two minutes, and continues for between 10–20 minutes. Contact guarding – where the male continues to grip the female until forcibly brushed away may be a characteristic of the family and improves the likelihood of paternity for the riding male (Willis, 1967; Pearson 1988).

Female tiger beetles use their ovipositor to carefully select for soil type and moisture prior to egg laying

Plate 5. Pair of C. hybrida. (Photograph by Lynne Collins).

(Pearson, 1988). Oviposition is usually initiated by the female touching the substrate with her antennae and biting the soil with her mandibles (Pearson, 1988). The ovipositor is extended and the sclerotised gonopophyses, together with thrusts of the abdomen, are used to excavate a hole (Pearson, 1988). The female *C. hybrida* lays eggs singly and at a depth of about 5mm. These take about 30 days to hatch (Faasch, 1968).

There are three larval instars. Larvae inhabit vertical cylindrical burrows (Plate 6), which they dig at the site where they hatch. The larva loosens the soil with its mandibles and uses its head and pronotum as a shovel to carry the soil (Pearson, 1988). At the surface it flips the soil off backwards. It repeats this procedure to enlarge the tunnel after each larval moult (Shelford, 1908; Willis, 1967). The diameter of the burrow entrance corresponds to the width of the larval head and pronotum and can be used to distinguish the instar of the larva. An increase in burrow diameter after a period of plugging indicates moulting has occurred (Knisley and Pearson, 1984).

Soil moisture and associated temperatures are critical for larval activity and survival. Under desiccating moisture levels, the larvae plug the burrow, retreat to the bottom 15 – 30 cm below the surface and become inactive until soil moisture levels rise sufficiently. In some instances they will leave their burrow and dig a new one in a more suitable spot.

First instar *C. hybrida* larvae occur in groups of eight to ten individuals, which have hatched from eggs laid by one female (Simon-Reising *et al.*, 1996). After moulting, the second instar is morphologically identical as the first instar but increases in length from 6.5 mm to 11 mm. As a result of this increase the diameter of the burrow mouth increases to an average of 2.73mm. About two thirds of larvae found in April are second instar and mortality rates are ca. 15% at this stage (Simon-Reising *et al.*, 1996). Third instar larvae are up to 14mm long and the burrow mouth width is ca. 3.81mm (Simon-Reising *et al.*, 1996). The burrow is closed for pupation and the pupal cell is made in the soil (Plate 7).

Conservation Management

Cicindela hybrida is a very rare insect of high conservation value in Britain. It is the subject of Species Action Plans in both the UK and North Merseyside Biodiversity Action Plans, being a "priority" species in the former, and classified as "vulnerable" in Britain (UK Biodiversity Group, 1999; Merseyside Biodiversity Group, 2008). It has become extinct at four of its six known sites in North West England and was not, as previously thought, ever recorded from Cornwall, Norfolk or Wales. An estimate by English Nature that up to 75% of the UK population is found on the Sefton Coast, one of only two remaining sites, is probably too low. The species recently became extinct at Eskmeals Dunes, Cumbria and suitable habitat at nearby Drigg is disappearing (Copestake, 1999). It is quite possible, that the Sefton Coast might soon become the only known location for *C. hybrida* in Britain.

Plate 6. Larval burrows of C. hybrida. *(Photograph by Guy Knight).*

Plate 7. Third instar C. hybrida *larva excavated from its burrow. (Photograph by Guy Knight).*

The two remaining populations of *C. hybrida* in Britain are protected. Both Drigg and the Sefton Coast are SSSI's, although most surprisingly, *C. hybrida* is not a named *feature* for either site. Drigg Dunes is a Local Nature Reserve covering 383 hectares and the Sefton Coast, which appears to have a well established and stable population along a 15 kilometre stretch of dunes, is part of the Liverpool Bay Natural Area and includes 4,605 hectares of land, of which 2,100 hectares are sand dune. Despite loss to development, it is still the largest area of open dune landscape in England (Doody, 1991) and comprises four nature reserves and land managed by the National Trust and MOD.

The two most important limiting resources for tiger beetles are availability of oviposition sites and food (Knisley, 1986; Pearson, 1988). British populations of *C. hybrida* may be subject to greater environmental stress than populations on mainland Europe and, as a consequence, habitat requirements may be more precise (Starkings, 1999). Occurrence is related to a number of factors, including vegetation cover and type, soil type, availability of bare ground, gradient of slope, temperature and direction of slope. The desynchronised two-year life cycle lowers the risk of dramatic declines in population size due to unfavourable periods.

Sandy ground which is bare or sparsely vegetated and open to sunlight is important for burrow construction, hunting, behavioural thermoregulation and the maintenance of the high body temperature necessary for prey capture. On the Sefton Coast, Starkings (1999) calculated that suitable bare sandy habitat frequented by adults is less than ten percent of total open dune area. Vegetation cover on the open dunes varies between 0 – 15% but is typically 6%. The average gradient of slope is 15°. Depending on the time of day, south, south-west or south-east facing slopes are often very sheltered and receive a lot of sun.

When an area of larval habitat becomes endangered and disappears, so does the species it supports (Dunn, 1999). Future conservation work for *C. hybrida* must address larval habitat requirements. Larvae occur with adults in the same habitat and, because they live in permanent burrows, their distribution and abundance can be accurately determined (Knisley, 1986). The female is the key determiner for tiger beetle distribution, and by selecting oviposition sites she provides optimum conditions for survival and larval development (Shelford, 1908). Because the larvae are relatively immobile and the habitat requirements are more circumscribed than that of adults, the availability of larval habitat is the main limiting factor in the population levels of the beetles.

Cicindela hybrida is reliant on bare exposed, south-facing profiles of sand in very early successional habitat and becomes locally extinct as habitat matures. As such, localised declines in the abundance and extent of suitable breeding habitat due to natural processes are to be anticipated and can be tolerated, but the losses should not be chronic and remedial measures should be employed if thresholds are breached. Simon-Reising *et al*'s (1996) studies on *C. hybrida* led them to conclude that the quality of the matrix between neighbouring populations turns out to be as important as habitat quality itself. This concept corresponds to the landscape level and stresses the significance of natural processes in the dune system.

Dune stabilisation and a low rate of new dune and slack formation are the main conservation threats for *C. hybrida* on the Sefton Coast. Among several reasons for this are: a four to five-fold increase in dune scrub area between 1950 and 1980; conifer-planting and the introduction of broadleaved tree and shrub species in the late nineteenth century and a crash in the rabbit population (Smith, 2000). Dunes at Birkdale, in particular, and inland at Ainsdale LNR, have very little open sand and do not support the insect. It is worth noting that, the apparent extinction of the tiger beetle *Cicindela abdominalis* at a southeastern Virginia pines barrens habitat was believed to be the result of increased vegetation encroachment, which eliminated open areas needed by this species (Knisley and Hill, 1992).

General management should aim to retain a reasonable transition of dune vegetation, especially semi-fixed yellow dune. The current restoration of open dune habitat from pine woodland on the NNR will benefit the insect by creating more available bare sand habitat. In Bavaria, the species was not recorded from two areas of remnant inland sand dune that had been planted with pine woods (Simon-Reising *et al.*, 1996). However, within a few weeks of a dune restoration programme being completed, three males were recaptured after crossing more than 600m of arable land.

The grazing of livestock, since 1990, by English Nature on the Sefton Coast is intended to help maintain open dune landscape at Ainsdale NNR and LNR and provide valuable habitat niches in which many rare invertebrates such as *C. hybrida* find homes (Brown, 2000). This in part replaces the role historically provided by rabbits, which had been warrened in previous centuries and which, in addition disturbed the ground with their burrowing. However, grazing sheep might have been a contributory cause of larval tiger beetle mortality at a site studied by Simon-Reising *et al.* (1996).

Human impact on habitat, particularly off road vehicle activity, was reported to be the factor responsible for the decline of the tiger beetle *Cicindela oregona* along an Arizona stream edge (Schultz, 1988) and *Cicindela dorsalis* on coastal beaches in East Florida California. However, other common species may benefit from the activities of man, which create edges and cleared, open areas (Knisley and Hill, 1992). Physical damage to the dunes at Drigg has resulted from vehicles, heavy foot traffic and grazing animals (Copestake, 1999). However, on the Sefton Coast these activities can be beneficial and *C. hybrida* is frequently recorded along paths and landrover tracks and appears to be thriving, in areas immediately adjacent to the popular beach at Ainsdale LNR.

Recreational pressure is potentially damaging for larval habitat and burrows are particularly vulnerable at areas such as Formby Point, where the National Trust have recorded visitor numbers of ca. 90,000 per year. This environmental stress is a possible reason why Gore (1997) found that, although there was more suitable habitat at Formby Point, the species was recorded almost twice as frequently at less pressurised Raven Meols Dunes. Alternatively, dune recontouring management to stabilize the dunes at Formby may have influenced these findings.

Copestake (1999) observed that *C. hybrida* could potentially be seriously damaged, at Drigg, by over-collecting and should be further protected. Collecting is not a potential threat on the Sefton Coast, but should be discouraged, unless it relates to a particular scientific objective.

In the long-term, suitable open dune habitat for *C. hybrida* may disappear on the Sefton Coast. The potential for dune accretion between Ainsdale and Southport has been largely prevented by heavy recreational pressures, car parking and mechanical beach-cleaning which destroys the embryo-dune vegetation (Smith, 2000). By 2050, the coastline along a five kilometre frontage at Formby Point is predicted to erode by between 150 and 270 metres. Ultimately, if this 'coastal squeeze' continues, the dunes will disappear (Centre for Marine and Coastal Studies, 2003).

CONCLUSION

Tiger beetle species have their major economic impact as indicators of habitat degradation and serve as a focal point for habitat preservation and protection (Pearson, 1988).

An assessment of past and present distribution patterns for *C. hybrida* in the UK, coupled with detailed understanding of the species' autecology, will inform and enhance conservation management on the Sefton Coast SSSI.

The Sefton Coast Site of Special Scientific Interest (SSSI) supports a nationally significant invertebrate assemblage, including 33 British *Red Data Book* and 264 Nationally Scarce species and in North West England is the 'hotspot' for invertebrate biodiversity. A small number of invertebrate species such as *C. hybrida*, known from only two localities in Britain, are of very high conservation value and should be evaluated as potential named *features* for the SSSI – either individually or as an assemblage. *Cicindela hybrida* is an indicator species of site quality and can be used as a target species to monitor the future 'health and integrity' of the SSSI and in particular mobile dune and open sand habitat. The specialised invertebrate assemblage associated with mobile dunes and bare sand will benefit from any positive habitat management for this species.

The identification of key invertebrate indicator species of habitat quality for all major habitats on the SSSI, which represent a suite of other significant invertebrate species in a particular habitat assemblage, will assist conservation management. Invertebrates need to be effectively integrated within conservation management systems on the SSSI in order to be fully protected, appreciated, understood and enjoyed. *Cicindela hybrida* is a photogenic 'flagship species' for the Sefton Coast and should be used to present positive messages regarding invertebrate conservation.

ACKNOWLEDGEMENTS

I would like to thank the following: - Dave Boyce (Scarce Ground Beetle Project Coordinator) and Roger Key (English Nature) for commissioning the *C. hybrida* survey and for their advice and support. Mike Bailey, Adrian Fowles and Mike Howe (CCW); Martin Luff, Adrian Spalding and Stella Turk for providing distribution information. David Copestake for information on the Cumbrian population of *C. hybrida*. Mike Bigmore, Carl Clee, Chris Felton and Guy Knight (National Museums Liverpool) for assisting with field survey. Rob Wolstenholme and Lynne Collins (Natural England) for providing detailed distribution information on Ainsdale NNR, access to Reserve files and photographs of *C. hybrida*. Paul Wisse (Sefton Borough Council) for kindly providing the lead photograph and producing Sefton Coast distribution maps for *C. hybrida*. Phil Smith for comments on an earlier draft of this paper.

REFERENCES

Aldridge, A.C. (1974) *The distribution and autecology of the tiger beetle* Cicindela maritima *on the Ainsdale Sand Dunes National Nature Reserve.* Unpublished Report. 91.

Blum, M.S., Jones, T.H., House, G.J., and **Tschinkel, W.R.** (1981) Defensive secretions of tiger beetles: Cyanogenetic basis. *Comparative Biochemistry and Physiology.* 69B 903-904.

Bram, A.L. and **Knisley, C.B.** (1982) Studies of the bee fly, *Anthrax analis* (Bombyliidae) parasitic on tiger beetle larvae (Cicindelidae). *Virginia Journal of Science.* 33 99.

Brannstrom, A. (1999) *Visual ecology of insect superposition eyes.* Unpublished Dissertation. Lund University.

Britten, H. (1943) *The Coleoptera of the Isle of Man.* Arbroath: T. Buncle & Co. Ltd.

Brown, S. (2000) *Ten years on – sheep grazing on the Sefton Coast. English Nature North West Team.* http:/www.seftoncoast.org.uk/articles/00summergrazing.html

Chaster, G.W. and **Burgess Sopp, E.J.** (1903) *Coleoptera of the Southport District. Reprinted from the British Association Handbook.*

Centre for Marine and Coastal Studies (2003) *Environmental impact assessment Ainsdale Sand Dunes National Nature Reserve.* Unpublished Draft Scoping Report for English Nature.

Copestake, D.R. (1999) *An investigation into the habitat requirements of the BAP tiger beetle,* Cicindela hybrida *L. at Cumbrian coast sites.* Unpublished Report for English Nature. 18.

Doody, J.P. (ed.) (1991) *Sand dune inventory of Europe*. Peterborough: Joint Nature Conservation Committee/European Union for Coastal Conservation.

Van Dooren, T. (1999) Spatial and temporal aspects of life history strategies affecting population structure. http://bio-www.uia.ac.be/bio/deco/cici.html.

Dresig, H. (1980) Daily activity, thermoregulation and water loss in the tiger beetle *Cicindela hybrida*. *Oecologia*. 44 376-389.

Dresig, H. (1981) The rate of predation and its temperature dependence in a tiger beetle, *Cicindela hybrida*. *Oikos*. 36 196-202.

Dresig, H. (1983) A time budget model of thermoregulation in shuttling ectotherms. *Journal of Arid Environments*. 8 191-205.

Dresig, H. (1984) Control of the body temperature in shuttling ectotherms. *Journal of Thermal Biology*. 9 229-233.

Dunn, G.A. (1999) Biology of tiger beetles. http://members.aol.com/YESedu/biology.html.

Eccles, T.M. (1991) Beetles (Coleoptera). In: Atkinson, D. and Houston, J. (eds.) *The sand dunes of the Sefton Coast*. Liverpool: National Museums and Galleries on Merseyside. pp. 99-103.

Ellis, J.W. (1889) *The coleopterous fauna of the Liverpool district*. Liverpool: Turner, Routledge & Co.

Evans, M.E.G. (1965) The feeding method of *Cicindela hybrida* L. (Coleoptera: Cicindelidae). *Proceedings of the Royal Entomological Society*. London. (A) 40 61-67.

Friederichs, H. (1931) Breiträge zur morphologie und physiologie der sehorgane der cicindelinen (Col.). *Zeitschrift fuer Morphologie und Oekologie der Tiere*. 21 (1) 1-172.

Faasch, H. (1968) Beobactungen zur biologie und zum verhalten von *Cicindela hybrida* L. und *Cicindela campestris* L. und experimentelle analyse ihres beutefangverhaltens. *Zoologische Jahrbuecher Abteilung fuer Systematik Oekologie und Geographie der Tiere*. 95 472-522.

Flint, J.H. (1959-1964) *Entomological survey of Freshfield Dunes*. Unpublished Report.

Forsyth, D.J. (1970) The structure of the defence glands of the Cicindelidae, Amphizoidae and Hygrobiidare (Insecta: Coleoptera). *Journal of Zoology*. 160 51-69.

Gore, D. (1997) *The distribution of tiger beetles along the Sefton Coast*. Undergraduate Dissertation. Liverpool Hope University College, 54.

Hori, M. (1982) The biology and population dynamics of the tiger beetle *Cicindela japonica* (Thunberg). *Physiology and Ecology Japan*. 19 77-212.

Judd, S. (2003a) *Status & biology of the Northern Dune Tiger Beetle* Cicindela hybrida *(L.) in Britain with special reference to the Sefton Coast*. English Nature Contract No. SGBP01/04.

Judd, S. (2003b) *Invertebrate biodiversity and conservation management at Talacre Dunes and The Warren SSSI, Point of Ayr, Flintshire*. Commissioned Report for BHP Billiton Petroleum.

Knisley, C.B. (1986). Habitats, food resources and natural enemies of a community of larval *Cicindela* in southeastern Arizona (Coleoptera: Cicindelidae). *Canadian Journal of Zoology*. 65 1191-1200.

Knisley, C.B. and **Hill, J.M.** (1992) Effects of habitat change from ecological succession and human impact on tiger beetles. *Virginia Journal of Science*. 43 (1B) 133-132.

Knisley, C.B. and **Juliano, S.A.** (1988) Survival, development, and size of larval tiger beetles: effects of food and water. *Ecology.* 69 (6) 1983-1992.

Knisley, C.B. and **Pearson, D.L.** (1981) The function of turret building behaviour in the larval tiger beetle, *Cicindela willistoni* (Coleoptera: Cicindelidae). *Ecological Entomology.* 6. 401-410.

Knisley, C.B. and **Pearson, D.L.** (1984) Biosystematics of larval tiger beetles in the Sulphur Springs Valley, Arizona. Descriptions of species and a review of larval characters for Cicindela (Coleoptera: Cicindelidae). *Transactions American Entomological Society.* 110 465-551.

Larochelle, A. (1977) Notes on the food of tiger beetle larvae. *Cicindela.* 9 13-14.

Lehmann, F. (1978) Notes on the hibernation of *Cicindela hybrida* L. (Coleoptera: Carabidae). *Cordulia.* 4 (2) 78.

Lindroth, C.H. (1985) *The Carabidae (Coleoptera) of Fennoscandia and Denmark. Fauna Entomologica Scandinavica* 15 (1). Leiden: E.J. Brill/Scandinavian Science Press Ltd., 225.

Lindroth, C.H. (1992) *Ground beetles (Carabidae) of Fennoscandia, a zoogeographic study, Part I.* Andover: Intercept.

Linssen, E.F. (1959) *Beetles of the British Isles.* London: The Wayside and Woodland Series, Warne.

Luff, M.L. (1993) *The Carabidae (Coleoptera) larvae of Fennoscandia and Denmark. Fauna Entomologica Scandinavica,* 27. Leiden: E.J. Brill, 187p.

Luff, M.L. (1998) *Provisional atlas of the ground beetles (Coleoptera: Carabidae) of Britain.* Huntingdon: Institute of Terrestrial Ecology, 194p.

Merseyside Biodiversity Group (2008) *North Merseyside Biodiversity Action Plans 2008.* Maghull: Environmental Advisory Service.

Moore, B.P. and **Brown, W.V.** (1971) Benzaldehyde in the defensive secretion of a tiger beetle (Coleoptera: Carabidae). *Journal of the Australian Entomological Society.* 10 142-143.

Mury Meyer, E.J. (1987) Asymmetric resource use in two syntopic species of larval tiger beetles (Cicindelidae). *Oikos.* 50 167-175.

Palmer, M.K. (1982) Biology and behaviour of two species of *Anthrax* (Diptera: Bombyliidae), parasitoids of the larvae of tiger beetles (Coleoptera: Cicindelidae). *Annals of the Entomological Society of America.* 75 61-70.

Pearson, D.L. (1988) Biology of tiger beetles. *Annual Reviews of Entomology.* 33 123-147.

Pearson, D.L. and **Knisley, C.B.** (1985) Evidence for food as a limiting resource in the life cycle of tiger beetles (Coleoptera: Cicindelidae). *Oikos.* 45 161-168.

Pineda, P.M. (1999) Foraging, pursuit and attack mechanisms of adult tiger beetles (Coleoptera: Cicindelidae). http:/www.colostate.edu/Depts/Entomology/courses/en507/papers_1999/pineda.htm.

Schultz, T. D. (1988) Destructive effects of off-road vehicles on tiger beetle habitat in central Arizona. *Cicindela.* 20 25-29.

Sharp, W.E. (1908) *The Coleoptera of Lancashire and Cheshire.* St Albans: Gibbs & Bamforth Ltd.

Shelford, V.E. (1908) Life-histories and larval habits of the tiger beetles (Cicindelidae). *Journal of the Linnaean Society, Zoology.* 30 157-184.

Simon-Reising, E.M., Heidt, E. and **Plachter, H.** (1996) Life cycle and population structure of the tiger beetle *Cicindela hybrida* L. (Coleoptera: Cicindelidae). *Deutsche Entomologische Zeitschrift.* 43 (2) 251-264.

Smith, P.H. (2000) Classic wildlife sites. The Sefton Coast sand dunes, Merseyside. *British Wildlife.* October 28-37.

Starkings, M.J. (1999) *Habitat requirements of adult and larval* Cicindela hybrida *(L.) (Coleoptera: Cicindelidae) on the Sefton Coast sand dune system, Merseyside.* MSc Thesis. Liverpool Hope University.

Sutton, P.G. and **Browne, D.E.** (2001) British tiger beetles. *Bulletin of the Amateur Entomologists Society.* 60 21-35.

Swiecimski, J. (1957) The role of sight and memory in food capture by predatory beetles of the species *Cicindela hybrida* L. (Coleoptera, Cicindelidae). *Polskie Pismo Entomologiczne.* 26 205-232.

Thomas, J. (1967) *The beetles of the Ainsdale Nature Reserve.* Undergraduate Dissertation. London University.

Trautner, J. and **Geigenmuller, K.** (1987) *Illustrated key to the Cicindelidae and Carabidae of Europe.* Aichtal, Germany: Josef Margraf.

UK Biodiversity Group (1999) *Tranche 2 Action Plans - Volume IV: Invertebrates.* pp. 43-45. Peterborough: English Nature.

Willis, H.L. (1967) Bionomics and zoogeography of tiger beetles of saline habitats in the Central United States (Coleoptera: Cicindelidae). *The University of Kansas Science Bulletin.* 47 (5) 145-313.

SECTION C: MANAGEMENT

Marine Drive, Southport – Sea Defence Improvements

Graham Lymbery

The Ainsdale Discovery Centre, The Promenade, Shore Road, Ainsdale-on-Sea, Southport, PR8 2QB
Graham.lymbery@technical.sefton.gov.uk

The author was involved with the design and construction of the sea defence improvements scheme and has made use of material written at the time of construction in this review paper.

INTRODUCTION

The proposed construction of new sea defences along Marine Drive, Southport became a frequent topic in the local press from 1993 when public discussions commenced on the form the defences should take, through to the completion of phase 1 in 1998. At the time concerns were raised by the public about costs, benefits, visual impacts and scale of the structure. This paper sets out to briefly describe the evolution of Southport which led to the need for the seawall, the design and construction of the seawall, along with the concerns raised and benefits predicted, followed by a reflection 10 years on from completion of the first phase.

Early Reclamations

The town of Southport, on the southern fringes of the Ribble Estuary has, since its establishment, experienced rising beach levels, pushing the high-water contour away from the town. The coastline (or High Water Mark) in 1736 lay along what is now Lord Street (Figure 1), which was then sand dunes and saltmarsh (Ashton 1920); the 1845 coastline was established at the Promenade with the construction of a seawall. In 1860, the Pier was constructed from the Promenade across the shore to the South Channel, where vessels trading with Preston and local pleasure craft could berth in a pool of permanently deep water known as the Bog Hole (Newton et al., 2007). As the High Water Mark receded more land was reclaimed with the construction of the first section of the Marine Lake south of the Pier in 1887. In 1894 Marine Drive was constructed in the form of a causeway, covered only by the highest tides.

In the early 20th Century, the South Channel was sustained by the fresh-water input of the Crossens Channel and the scouring effect of water entering and leaving the new Marine Lake. However, the Channel progressively silted up, partly due to the training of the River Ribble and partly by land reclamations in the inner estuary (Smith, 1982).

There has been extensive reclamation in Southport (Figure 2) and the southern reaches of the Ribble Estuary since the early 1800s. This is evidenced not only by comparison of historic maps and photographs, but is also apparent as you walk along the pier that would once have been over the beach for its entire length, with fishing boats and excursion steamers moored at its seaward end (Plate 1).

Plate 1. Photograph of the view from the end of Southport pier around 1901, showing vessels lightering the Bog Hole. (Source: Barron 1938, page 172).

An important area of reclamation for Southport was the construction of Marine Drive and its subsequent linking to the Coastal Road in 1974. In 1959 a new embankment was built on the shore, extending Marine Drive 1,500 metres north east to Fairway. A level of 24 feet above Ordnance Datum (7.3m AOD) was selected for the crest of this embankment and this level was

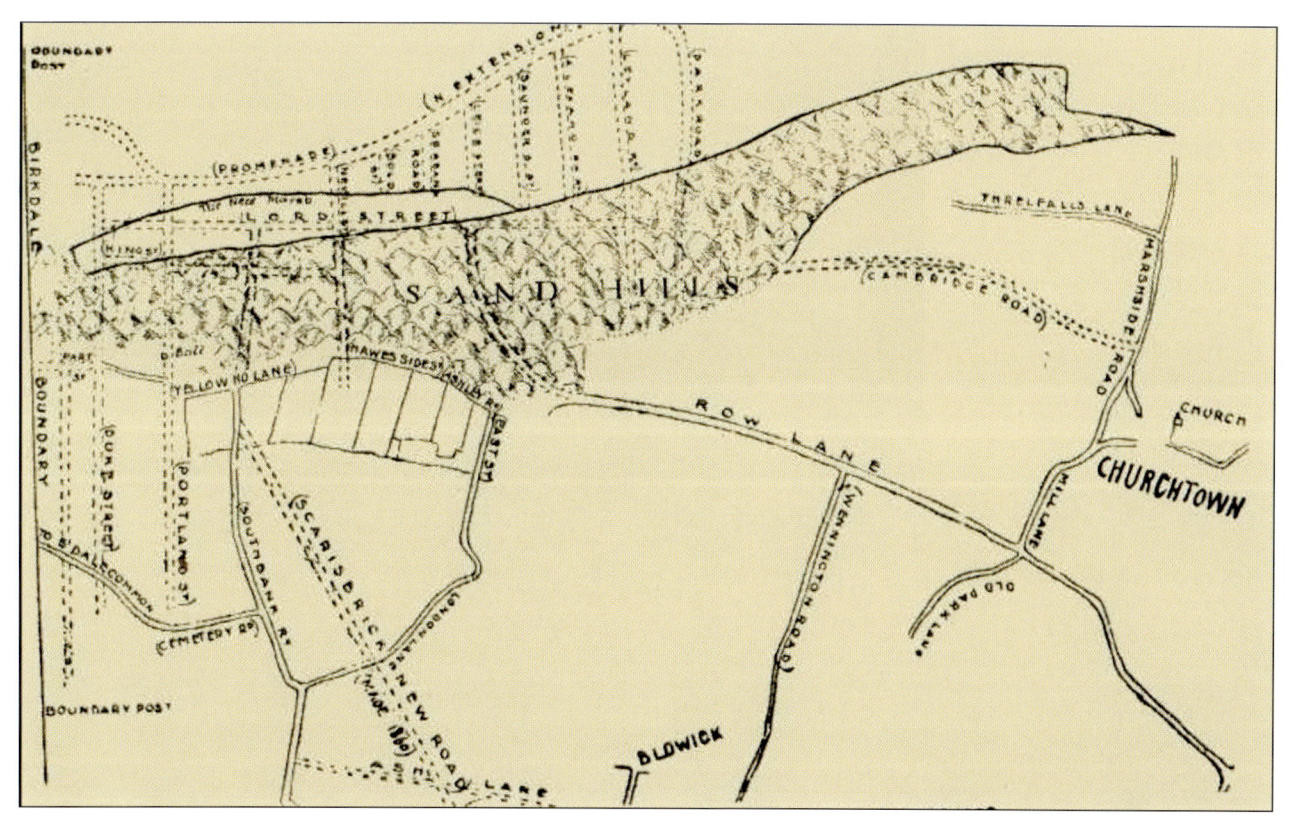

Figure 1: North Meols (Southport) coastline in 1736-1920 (Source: Ashton 1920, p100).

applied to the later embankments which extended the Coastal Road north to Crossens (1968 to 1974) and south to Weld Road (1974 to 1976). The level of the central section of Marine Drive remained at its original low level of 5.2m AOD (Plate 2). The embankments were built on the shore, close to the high water mark. The earliest embankments used sand from the adjoining shore. During the 1960's, a series of embankments were built with a sand bund on the seaward side and a core of refuse tipped directly from the town's collection vehicles (Plates 3 and 4). In the 1970's embankments were built from tipped demolition rubble faced with beach sand. All these structures were lightly armoured on their seaward side with concrete no more than 150mm thick or with reclaimed highway stone setts. Clearly, cost saving was a major consideration.

Plate 2. Marine Drive, Southport, in 1966 looking north from the Pier, with Marine Lake to the right behind the embankment. (Image held in SMBC archive).

Plate 3. Construction of Marine Drive, looking south east, estimated mid 1960s. The black dots above the construction site are large numbers of seagulls in the picture due to the road being backfilled with household waste. (Image held in SMBC archive).

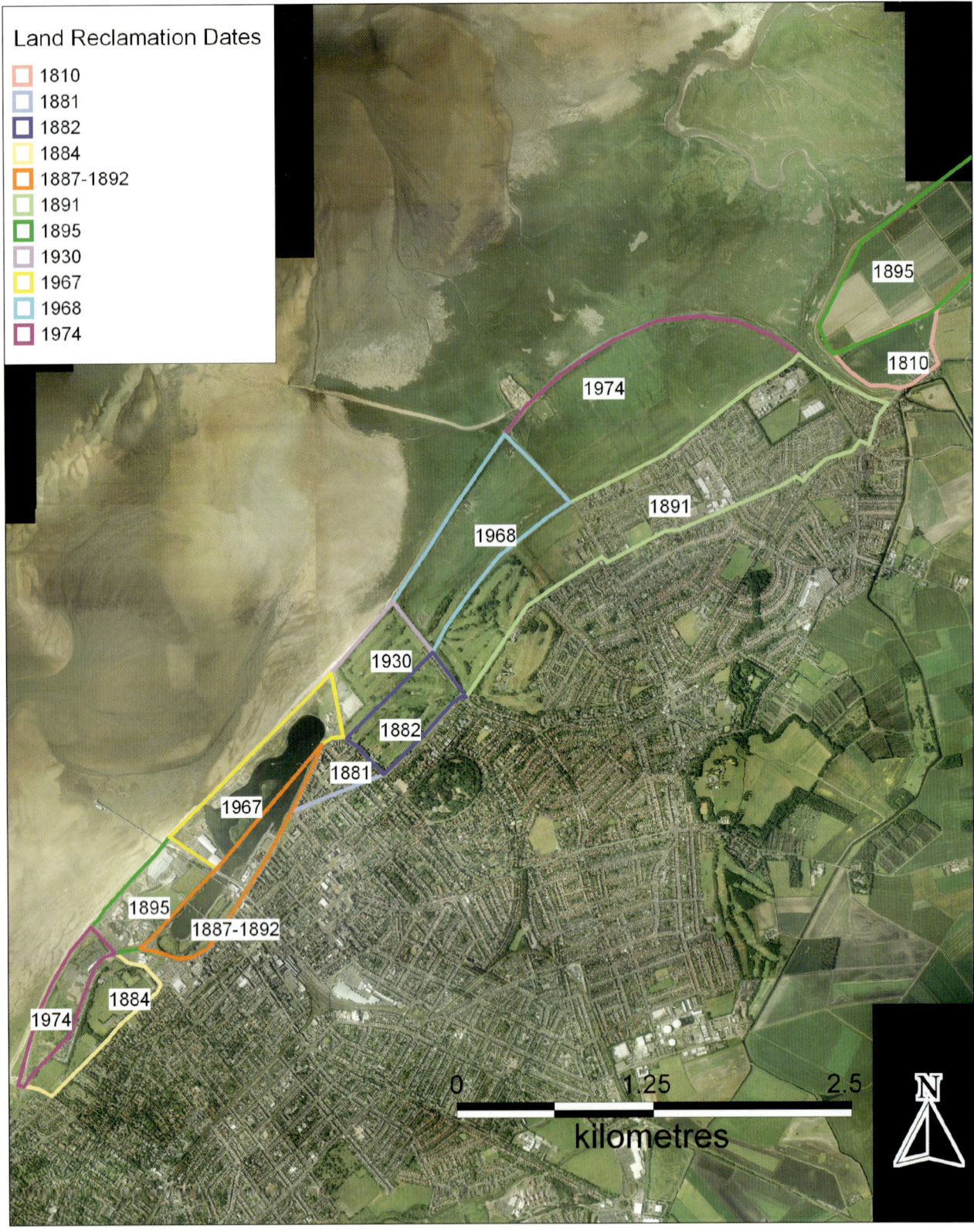

Figure 2: Southport land reclamation from 1810 to 1974. © Crown copyright. All rights reserved. Sefton Council Licence number 100018192 2010.

The Problem

The need to secure Marine Drive against tidal flooding between Esplanade and Marine Parade had been evident since the completion of the Coastal Road/Marine Drive highway route in 1976 (Plate 5). This route acts as both a town centre by-pass and an important feeder for the central sea front leisure zone. Because of its low level, just 0.8 metres above high water mean spring tides and 0.2 metres below highest astronomic tide, it was subject to frequent precautionary

Plate 4. Photograph of construction of Marine Lake extension, with ground being dewatered during building of perimeter wall, looking west, with the Promenade in the background, dated 13th January 1962. (Image held in SMBC archive).

closure and was often flooded by the sea. Moreover, the leisure related property adjoining the road was also subject to flooding during storm surge events (Plate 6). Proposals for sea defence works were first brought to the attention of central Government in 1982 during Sefton Council's response to a request for preliminary information in connection with the DoE (Department

Plate 5. The problem of flooding on Marine Drive (estimated year 1970). (Image held in SMBC archive).

of Environment), Herlihy, coastal defence survey. This survey sort to collate information about the countries coastal defences - their form, location and condition. In 1988 consulting engineers were commissioned to investigate the condition of Southport's coastal embankments because of concern over their continued settlement. The consultants were also instructed to prepare preliminary proposals for the Marine Drive

Plate 6. Flooding of Marine Drive posed a threat to leisure facilities 1991 (Image held in SMBC archive).

sea defence works. Their report led to outline proposals which were used to initiate discussions both locally and with the Regional Engineer of MAFF (Ministry of Agriculture, Food and Fisheries).

Planning Process

The initial proposal made by Sefton Council in 1993 was for a stand alone project, entailing the construction of a flood wall to protect the low lying 1,260m central sea section of Marine Drive. The project included a promenade which was designed as an integral part of the hydraulic structure, replacing the public footway on the seaward side of Marine Drive. Discussion with the Borough Town Planning Department highlighted the importance of high standards of visual design and the need to mitigate the possible 'canyon' effect of an existing sea front highway in close proximity to a raised promenade with proposed development to the landward side of the highway; it was perceived that driving along the road with high structures to either side would resemble driving through a canyon. This 'canyon' effect would have been particularly noticeable when compared to the existing state where the road was at beach level with wide open vistas. The resulting design entailed a small incursion into the Ribble and Alt Estuary Special Protection Area.

An Environmental Statement was prepared in support of the Planning Application which was submitted in May 1993 and approved by the Planning Authority after due consideration of all submissions and objections.

Concerns were expressed by English Nature (now Natural England) and the RSPB about the necessity of incursion into the Special Protection Area (SPA). The Council explained its design strategy and environmental management policies and activities to both agencies. It also agreed (outside of the Planning mechanism) to undertake a range of environmental management activities on the beaches and in the seafront zone including Marine Lake to compensate for this incursion, and after further extensive discussions, both agencies withdrew their objections.

A few local pressure groups advocated an alternative design based on a sand dune defence, stating that the natural form of sea defence in this area is sand dunes. However, this was not a tenable argument in this situation. The Sefton sand dunes are a feature of the open coastline between the rivers Mersey and Ribble, whilst the central Southport seafront is transitional between the dune coastline south of Esplanade and the Ribble Estuary marsh coastline. The morphology of the coastline fronting Southport has been highly influenced by humans, with sand dunes no longer being a feature since development began in the 1800's. The artificial construction of a range of sand dunes on Southport beach would, presuming it could be sustained, lead to greatly increased amounts of sand moving inland, exacerbating existing problems of wind blown sand that would require regular removal from around the Marine Lake, highways and drains. For similar reasons, beach recharge is not a viable option. Adding further material to the existing shore could potentially change the character and ecological interest of the beach as it presently exists. The habitat interest of this area is not founded solely on dunes but is based on the total near-estuary environment, which includes emerging sand dunes, salt marshes, intertidal sand flats, enclosed wetlands and the wide variety of plant and animal life which exists in and visits the area.

The Planning Consent for Phase 1 called for a high standard of surface finishes and street furniture in order to provide a high quality, co-ordinated appearance to the sea front area. The inclusion of a wide promenade, with decorative and other highway works was considered essential in order to provide an impetus for the attraction of appropriate redevelopment of a declining sea front amenity zone.

MAFF was not approached to fund works other than those needed for the purpose of sea defence. Southport was included within the Merseyside ERDF (European Regional Development Fund) Objective 1 area which was a potential source of grant aid to fund the highway and decorative elements of the scheme. Subsequently, an application for £1.1 million Objective 1 grant under Driver 1 (Action for Industry), Measure 1 (Inward investment and key corporate business development) was submitted via the Government Office for Merseyside and after appropriate evaluation the application was approved in principle. Matching funding was sourced from the Supplementary Credit Approval which accompanies MAFF flood defence grant aid.

A local pressure group concerned with increases to traffic raised objections during all stages of the consultative and planning process. The group found a powerful blocking mechanism with the enactment of the Habitats Regulations in 1994. The Ribble and Alt Estuary is a European Conservation Site under the regulations, by virtue of its status as a Special Protection Area (SPA) for birds. In addition the Sefton coast was also now a candidate Special Area of Conservation (SAC) on the basis of the international importance of both the sand dune and intertidal areas. The Phase 1 incursion into the SPA enabled the group to lodge an objection with the European Community which required lengthy investigation. MAFF would not grant any approval until the objection was resolved. This proved to be an extremely difficult process, even though the environmental assessment submitted with the Planning application demonstrated negligible adverse environmental effects. Moreover the declaration of the site's candidate SAC status during 1995 was a trigger for a further objection. The main thrust of the objections was that alternative "soft" forms of defence should be used with there being no justification for the need to protect the Marine Drive from flooding by extreme high tides. The objections were resolved by

Sefton Council to the satisfaction of the European Community through the provision of additional evidence. MAFF approved all three phases in principle and gave detailed grant approval for Phase 1 in June 1996, before the lodging of the pSAC (proposed Special Area of Conservation) based objection. Final confirmation the objection had been resolved, and approval of the Objective 1 funding essential to commence the scheme, were both received late 1996. The construction Contractor began work on site on 10 February 1997 during the highest storm tide since 1990.

Discussion of Design Issues

Concerns were expressed about the visual impact of the structure both in terms of loss of sea view by car passengers and the creation of a 'canyon' effect as the road ran through walls to either side, with a potentially great expanse of grey concrete. In order to appreciate the first concern it is important to remember that prior to construction of the new seawall, the beach ran up to the pavement on the seaward side and was only marginally lower than it. The introduction of a structure that was to be 2.3m higher than the road was going to have a considerable impact on the landscape that the public had become used to. Part of the mitigation for this within the design was to switch the drainage for the road to the landward side, allowing the carriageway and pavement to be raised on the seaward side thus reducing the apparent height of the seawall. The maintenance of pedestrian views was achieved through the introduction of a raised promenade which not only provided good views but also separated the pedestrians from the road traffic making for a more pleasant experience. The second issue was addressed through the appointment of a firm of landscape architects, Partnership Art (now Eaton Waygood Associates), to work with the design team. Key aspects designed or influenced by Partnership Art were the street furniture (seats, bins, hand railing, street lighting), the artistic elements (such as the shoal of fish, the obelisks, the Eel and the patterned concrete) and the plinths upon which the street lighting is mounted and seating fitted to (Plate 7). These features helped to address some of the concerns and turn a potentially dull engineering structure into a feature of interest.

Plate 7. Street furniture along the raised promenade, Marine Drive, 2000. Designed by Partnership Art. (Image held in SMBC archive).

Concerns were also expressed in relation to the need for the seawall to be so high. To some extent this was linked to the previous concern over the landscape impact, but also reflected a lack of understanding of the technical issues related to designing the height of the seawall. The standard of defence that was chosen for the wall, was for it to be able to protect against the highest tide we would expect to occur once every twenty years; this equates to a still water level of 6.09m AOD. The design of the structure is not only required to prevent flooding by this level of water but also to prevent significant overtopping by the predicted wave activity that would be expected with such an event. Since construction, the largest storm that the central section of the seawall has had to cope with had a return period of 10 years. It coped with this event with no unacceptable overtopping, but it was apparent that it was approaching its design standard. It is worth noting that where the design of the defence changes at the sluice gates for the Marine Lake more overtopping was evident (Plate 8). This is despite the crest level in this section of wall being higher than the surrounding seawall, reflecting the impact of having a vertical structure here as opposed to the sloping revetment used for the rest of the structure.

Plate 8. Wave overtopping at the sluice gates, Marine Drive, 2005. (Image courtesy of Terry Hedges).

Issues were raised regarding access by the public to the beach. The location of access points was discussed in detail with the Planning Department to ensure that there were sufficient access points, and that they were located so as to be appropriate for potential future developments on the landward side of the seawall. Whilst it was not possible to coordinate funding with other budgets for inclusion of pedestrian crossings at the time, the ducting for connections was included in the design so that unnecessary excavation was avoided at a later date. The area under the pier was raised as a concern because the height of the wall here would be more apparent, as the promenade had to drop down to the level of the footway to allow adequate headroom for pedestrians. The street lighting in this area was designed to minimise any dark spot and the 'Eel' was included to bring some interest to a potentially large blank spot and deter graffiti (Plate 9).

Discussion of Benefits

It was predicted that the construction of the seawall would lead to £30 million of investment in the seafront area, and was critical to the regeneration of what had become an area that was at best suffering from underinvestment and at worst was derelict. There were other benefits predicted if the seawall was constructed, the obvious ones being the reduction in the number of flooding events and the associated damages and road closures; the not so obvious being the amenity value of the seawall itself.

Prior to construction of the seawall, the seafront area, all the area to seaward of the Marine Lake and surrounding the Marine Lake was to varying degrees at risk of flooding from the sea; the construction of the seawall has significantly reduced this risk. Whilst some of the investment in this area might have been possible without the construction of the seawall it is unlikely that it would have been as viable without the whole of the seafront being regenerated. It is unlikely that the Pier (cost circa £7 million in 1999) would have been restored if it had been crossing a derelict site; the Ocean Plaza development (a retail and leisure development, cost c. £30 Million) would not have been feasible, and without this there would have been no requirement for the now landmark Marine Way Bridge (cost circa £8 Million). All this development has supported the case for the expansion of the Floral Hall to include a new conference centre and four star hotel (cost circa £30 Million) and general improvements around the Marine Lake (cost £1.5m) (Plate 10). The leisure facility of Pleasureland did invest money on their site as a result of the seawall being constructed and although they have now chosen to vacate the site, the location may now be considered prime and as such will attract a significant amount of investment.

Following completion of construction of the seawall in (2002), there has been no incidence of road closure

Plate 9. Artistic element - Eel designed by Partnership Art. (Image held in SMBC archive).

Plate 10. Development behind the sea wall showing the Pier, Marine Way Bridge, the Marine Lake and the retail development. (Oblique aerial photography 2008; Copyright North West and North Wales Coastal Group).

due to tidal flooding. The biggest storm event has been of a scale expected once every ten years with the design standard of the wall being for a once in every twenty year event, therefore it has yet to be tested by its design event. One aspect that was not anticipated was the amount of sand that would accumulate on the seawall, which requires an ongoing effort on the part of the Coast and Countryside Service within Sefton Council to remove from the seawall, often on a daily basis (Plate 11).

The amenity value of the new Promenade has not been studied analytically and any views here are only subjective views from the author. The Promenade appears to be a highly utilised facility, but that could be because of the increased attractions in the area. There has been little negative press about the seawall and promenade following on from completion. Any negativity has generally related to details such as residual rubble on the beach or insufficient access points. Generally, it appears to be considered a positive feature.

SUMMARY

It is past human actions in terms of reclamation and development that exacerbated the flooding risks that were being experienced in the seafront area. Having recognised not only that the problem existed, but that the seafront area could not be economically regenerated with this scale of risk, it was necessary to take positive actions to reduce the risk. A design phase allowed an appropriate solution to be identified to address the principal issue of reducing flood risk, along with other concerns in relation to the setting for the structure and public concerns; the public identifying a number of issues.

Plate 11. Accumulating wind blown sand on the sea wall. (Image held in SMBC archive).

Part of the development process for the design was to justify the scheme not only to the funding bodies but also to the public and elected Members in Southport as key stakeholders. The justification included not only the benefits in terms of reduced flood risk but also regeneration. Whilst the arguments for the scheme were robust there were a small number of people who were strongly opposed to the scheme and used the system of checks and balances that exist to try and halt the scheme and substantially delay it; their arguments included challenging the need for the scheme.

With the scheme now complete and having been in place for some years it is possible to reflect on the development process and whether or not the issues were addressed satisfactorily and the benefits realised. The development process took a significant amount of time and effort, which was further extended as a result of the need to co-ordinate funding and address formal complaints. The need for the scheme being informally identified as early as 1976 and formally identified in 1982; with the third and final phase being completed in 2002. The identification and resolution of issues was an essential element of the design process, although the perception (by the public) of some of the issues, such as the dark area under the pier, may have been overstated when considered in hindsight. Of the issues identified they all appear to have been addressed satisfactorily based on the way the scheme has been received and subsequently reported in the press. The economic benefits to the area that were identified have been realised, with the reduction of the flood risk and significant investment in the seafront area, in the region of £85 million.

It is reasonable that people have concerns or fears over developments such as the seawall and that they have an opportunity to express them. Such consultation forms a key part of the design and planning process. The benefits clearly have accrued and illustrate the key role that the seawall has played in the regeneration of the Southport seafront.

REFERENCES

Ashton, W. (1920) *The evolution of the coastline, Barrow to Aberystwyth & the Isle of Man.* Southport: WM. Ashton & Sons Ltd., Grosvenor Works. 100.

Barron, J. (1938) *A history of the Ribble navigation.* Fishergate: Guardian Press. 172p.

Newton, M., Wisse, P. and **Lymbery, G.** (2007) *Report on the evolution of the Bog Hole Channel located on the intertidal foreshore at Southport, North West England.* Sefton Coast Database.

Smith, A.J. (1982) *A Guide to the Sefton Coast Database.* Sefton Council: Bootle.

The development of Integrated Coastal Zone Management (ICZM) in the UK: the experience of the Sefton Coast

John Houston

Department of Geography, Faculty of Sciences and Social Sciences, Liverpool Hope University, Hope Park, Liverpool, L16 9JD dunes@hope.ac.uk

ABSTRACT

The Sefton Coast Management Scheme was one of the first coastal management initiatives in North West England. Since its establishment in 1978 the partnership approach to management on the Sefton Coast has reflected the development of Integrated Coastal Zone Management (ICZM) regionally, nationally and internationally. In the early 1990s Sefton Council was one of several local government bodies calling for better national coordination of coastal planning and management. The flexible ICZM approach in Sefton has allowed the continuous partnership to develop local responses to subsequent national planning guidance, international nature conservation obligations and government policy for shoreline management. From 2001, with the change of emphasis to the Sefton Coast Partnership, the principles of IZCM have become central to the development of plans and strategies. The Sefton Coast Partnership is a good example of how the management of a distinctive area of coast has been able to adapt to changing circumstances whilst maintaining the fundamental aspects and benefits of partnership working.

The outline of this paper, covering the period from 1988 to 2000, was presented at the 'Waves of Change' research seminar held at Liverpool Hope University in 2001. It has been updated to cover some of the key initiatives from 2001-2008 in relation to Integrated Coastal Zone Management including the establishment of the Sefton Coast Partnership.

INTRODUCTION

The modern approach to coastal management in Sefton was established in 1978 with the launch of the Sefton Coast Management Scheme (Houston and Jones, 1987). Over the following thirty years the Sefton Coast has provided a case study in the development of approaches to Integrated Coastal Zone Management (ICZM) in the UK. The development of coastal planning and management in England in the 20th Century, and how this relates to the Sefton Coast, has been described by Houston and Jones (1987). The authors describe how the early years of the Sefton Coast Management Scheme were based on the Heritage Coast style of management (pioneered in the mid 1970s by the Countryside Commission), and, importantly, the Sefton experience demonstrated that the approach was transferable to other coastal areas (Houston and Jones, 1987).

The Sefton Coast is characterised by a 14 km arc of dunes lying between Crosby and Southport. Although the current ICZM Plan for the Sefton Coast covers a comprehensive range of issues, the broad picture, when viewed over decades, is still driven by a requirement to find a balance in the functions of the dunes, between coastal defence, being a popular recreation location, and its biodiversity value for wildlife (as already identified by Ranwell 1972, Chapter 13).

The countryside management approach to resolving local issues on the Sefton Coast, as advocated by the Countryside Commission, was effective in terms of dune management, landscape conservation and the provision of access. However, by the late 1980s coastal management initiatives, such as the Sefton Coast Management Scheme, and local government associations, identified a lack of national co-ordination on coastal issues in the UK. Whilst several projects were demonstrating good practice there was an absence of national policy, especially on those issues which required a long-term perspective. Sefton Council, with the support of the Royal Town Planning Institute, took an initiative to host the 1989 symposium Planning and

Management of the Coastal Heritage to identify the most urgent coastal planning issues at that time and to agree a programme for working towards better integrated coastal planning (Houston and Jones, 1990).

The 1989 coastal planning symposium and its five-point Agenda for Action (see box this page) came at a time when the conservation and sustainable use of the coast was becoming the subject of intense national and European Union interest (e.g. European Coastal Conservation Conference 1991: Proceedings 1992). As a consequence, in the 1990s, the knowledge of coastal systems, the understanding of the process of integrated coastal zone management and the number of coastal initiatives grew considerably. From a situation in the 1980s where there was an absence of opportunities for debate on coastal issues, in the 1990s it became difficult to keep pace with the many policies, strategies and initiatives generated by the European Commission, the UK Government and its agencies, interest groups, coastal partnerships and associations.

**Planning and Management of the Coastal Heritage
Southport Symposium 1989**

Agenda for Action

1. Update 'The Planning of the Coastline' (HMSO, 1970). Using the 1970 report of the Countryside Commission as a model, survey the current issues for coastal planning and management and recommend action.

2. Produce national guidance on coastal planning. In parallel with (1) above issue planning guidance for structure and other development plans (there is no guidance available at all at present).

3. Coordinate coastal information and action. Link existing data-bases and ensure access to information; identify gaps in knowledge and promote integrated studies. Improve both central and local government coordination.

4. Promote local involvement and action. Use 'key-worker', multi-agency models to help cross-boundary planning and management; improve coordination between coastal authorities, professional, academic, voluntary and business organisations.

5. Raise profile of coastal issues. Stop the coast being 'taken for granted' and reinforce the drive towards better integrated planning and management.

The 1990s was an influential decade for assembling knowledge and for developing new thinking on habitat management, in attitudes to coastal change and in mechanisms for delivering ICZM. Ideas, declarations and legal instruments at the global, European and national level all had implications for the conservation and management of the Sefton Coast and, in return, the experience of the Sefton Coast made an active contribution to national debates (e.g. House of Commons Environment Committee 1992).

This paper summarises some of the key milestones in the development of ICZM from the international to the local level as seen from the perspective of the Sefton Coast. It is intended partly to be a historical record of significant events and partly to show how the Sefton Coast Partnership continues to evolve as it adapts to changing priorities.

International and European initiatives in Integrated Coastal Zone Management

The Rio Earth Summit in 1992 set the scene for a global approach to sustainable development and, through Agenda 21, supported the need to develop new approaches to marine and coastal area management, and developments which "are integrated in content" (UNCED, 1992, Chapter 17.1). Guidance for Local Agenda 21 action on the coast in the UK was published by the Local Government Management Board (LGMB, 1995). The guidance was designed to help local authorities understand some of the issues affecting the coast and to promote opportunities for coastal protection and enhancement.

In response to these global concerns in 1996 the European Union funded the EU Demonstration Programme on Integrated Management of Coastal Zones (http://ec.europa.eu/environment/iczm/). The demonstration programme was based on 35 projects testing innovative solutions and developing best practice in coastal zone management (European Commission, 1999a; 1999b). The experience of these projects was used to prepare a communication on Integrated Coastal Zone Management: a Strategy for Europe (Commission Communication COM/2000/547) published in 2000. The EU strategy "aims to promote a collaborative approach to planning and management of the coastal zone, within a philosophy of governance by partnership with civil society" (European Commission, 2000, Page 2). The strategy defines the EU's role as one of providing leadership and guidance to support the implementation of ICZM by the Member States, at local, regional and national levels.

The Sefton Coast Partnership has adopted the principles of ICZM as set out in Annex I of the Commission Communication COM/2000/547 (European Commission, 2000). The European Commission proposes that successful coastal zone management is based on the following eight principles;

1. A broad 'holistic' perspective (thematic and geographic)
2. A long term perspective
3. Adaptive management during a gradual process
4. Reflect local specificity
5. Work with natural processes
6. Participatory planning
7. Support and involvement of all relevant administrative bodies
8. Use of a combination of instruments

A Recommendation on ICZM was adopted by the European Parliament and Council in 2002 (European Commission, 2002) committing Member States to the preparation of national strategies for coastal management. In 2003, as part of this process, the UK Government commissioned an initial stocktake of the framework for management of the coastal zone in the UK (Atkins, 2004). This UK response is set against a background of continuing pressure from various interest groups arguing for improved governance and decision making in the coastal zone within Europe. For example, the trans-national coastal partnership Coastlink published its own best practice guidelines (Bridge and Gilbert, 2001; Gallagher, 2001) and non-governmental organisations such as the Coastal and Marine Union (EUCC), World Wildlife Fund (WWF) and the RSPB (also with Birdlife International) continue to push for greater recognition of coastal issues at the European level.

In 2007 the European Commission reviewed the experience of the demonstration programme and published a communication on the evaluation of ICZM in Europe setting out the main policy directions for the development of ICZM in Europe (European Commission, 2007).

Whilst the EU communications on the development of ICZM have offered guidelines and promoted good practice they are not backed up by the force of law. The EU 'Birds Directive' of 1979 and the 'Habitats Directive' of 1992 on the other hand have required the identification, selection and protection of large areas of the coastal zone, including foreshores, sand dunes, salt marshes and marine areas as part of the European Natura 2000 network. The EU Habitats Directive is translated into UK law through the Habitat Regulations of 1994 (UK Government, 1994). The regulations interpret the obligations of the EU Habitats Directive and set out the process for establishing Special Protection Areas (SPAs) and Special Areas of Conservation (SACs) and for the setting up of 'schemes of management' for European 'marine sites'.

The protection of nature in the marine environment had lagged behind that for terrestrial areas but is being addressed at EU level through guidelines for the establishment of the Natura 2000 network in the marine zone and through the Marine Strategy Framework

Directive (European Commission, 2008). European nature conservation designations are particularly relevant to an area of high biodiversity value such as the Sefton Coast and its adjacent estuaries and inshore zone. Strict conservation protection, whilst being challenging, is not intended to stop economic activity or people's use of the land, with initiatives such as the Sefton Coast Partnership being well-placed to develop plans and strategies that aim to accommodate a wide range of interests.

Integrated Coastal Zone Management in the UK

Coastal Zone Protection and Planning

The Government's definition of Integrated Coastal Zone Management was that it is the process which brings together all those involved in the development, management, and use of the coast within a framework which facilitates the integration of their interests and responsibilities to achieve common objectives (Department of the Environment, 1996).

A review of coastal zone management in the UK was carried out by the House of Commons Environment Committee in 1991-92, with a subsequent report Coastal Zone Protection and Planning (House of Commons Environment Committee, 1992). Many coastal interest groups, including Sefton Council, contributed to the study and evidence was given to the Committee during a visit to the Sefton Coast. The report made recommendations under 27 headings, including ideas for a national coastal unit, a national coastal strategy, the establishment of regional coastal groups particularly for coastal defence, that controlled retreat of the coastline be considered and that there should be a hierarchy of Coastal Zone Management Plans supported by Coastal Zone Management Groups. However, the Government's response also titled Coastal Zone Protection and Planning (Department of the Environment, 1992a) rejected many of the recommendations to the frustrations of campaigning organisations, such as the RSPB who responded by publishing their own vision for the coast A Shore Future in 1993 (RSPB, 1993).

The Government, whilst not wholeheartedly endorsing the report of the Environment Committee did act on a number of issues in support of planning policy guidance, shoreline management, byelaw-making powers and communication. It sponsored a number of reviews of coastal management activity (Department of the Environment; 1993, Department of the Environment and Welsh Office, 1993a, 1993b) and also published the document Policy Guidelines for the Coast in 1995 (Department of the Environment, 1995) which highlighted Government policy in England on key issues affecting the coastal zone.

Coastal planning

The Government introduced Planning Policy Guidance PPG20 Coastal Planning in 1992 (Department of the Environment and Welsh Office, 1992) which is adopted by local authorities in the preparation of development plans and advising planning decisions in the coastal zone.

At the end of the 1990s the Local Government Association Special Interest Group on Coastal Issues published On the edge - the coastal strategy (Local Government Association, 2000). The LGA strategy was timely in respect of the discussions taking place at the European level and recognised and followed the landmark report of the House of Commons Select Committee in 1992. The principal aim of the strategy was to "establish improved governance, management and community well-being to ensure that the UK has the best managed coast in Europe" (Local Government Association, 2000, page 5). The strategy advocates a strategic approach to governance with recommendations directed especially at the European Union, national government, the regions and local government. It was the LGA's view that "local authorities are in the position to sustain the vital partnerships and develop the expertise and commitment to meet the duty of care and be successful champions of the coast" (page 24) and that "local authorities play a vital role in building partnerships, resolving conflicts and supporting the community on the coast" (page 24). The document reflected many of the issues and concerns raised in the Planning and Management of the

Coastal Heritage conference held in Southport a decade earlier.

Nature conservation

A particular focus of coastal zone initiatives in the late-1980s and early-1990s was on estuaries, given their international importance for nature and continued threats from, amongst others, recreational pressures, marinas, pollution, land claim and barrages. RSPB launched its Turning the Tide - a future for estuaries in 1990 (Rothwell and Housden, 1990). This was complemented by the Nature Conservancy Council's Estuaries Review (Davidson, 1991a; 1991b) and the promotion of estuary management plans which led to management strategies for the Dee, Mersey and Ribble estuaries. By 2000 estuary management plans covered most of the important sites in the UK for internationally important assemblages of birds.

In 1992 English Nature re-packaged a number of its marine, estuaries and coastal initiatives under the banner Campaign for a Living Coast (English Nature, 1992a, 1992b, 1992c). The thrusts of the campaign were to advocate the adoption of flexible coast defence policies through, for example, 'managed retreat', to seek management plans for four out of five estuaries by 2000 and to create a national network of protected marine sites. Much of this work was updated in 2002 through the publication of a 'state of nature' report for the maritime issues (Covey and Laffoley, 2002). During the 2000s English Nature and Natural England continued to develop a range of coastal policies including promotion of the 'ecosystem approach' for holistic coastal zone management (English Nature, 2004). Natural England's maritime strategy Our coasts and seas - making space for people, industry and wildlife (English Nature, 2005) sets out priorities for action at different scales including developing local solutions to local issues.

Heritage coasts

In 1970 the Countryside Commission proposed that nearly 1300 km of the undeveloped coastline of England and Wales should be defined as heritage coast. In 1972 the Government endorsed the concept and asked local authorities to define heritage coasts in their development plans and to prepare management plans for them in consultation with the Countryside Commission. It was this approach that the Countryside Commission promoted when it first encouraged the establishment of the Sefton Coast Management Scheme in 1977.

The support of the Countryside Commission, in part-funding a Project Officer for the Sefton Coast, was critical to the success of the project in its first five years and throughout the 1980s the management of the Sefton Coast followed the heritage coast model.

In the 1990s the Countryside Commission continued to support the management of Heritage Coasts. However in 1992, in the publication of Heritage Coasts: Policies and Priorities, the impetus of the initial designation of 45 Heritage Coasts had been somewhat lost. The Commission's view "that it had no plans to extend the present coverage of heritage coasts" (Countryside Commission, 1992, page 4) was in contrast to the more campaigning tone of organisations such as the National Trust which has continued to purchase coastal properties. The Countryside Commission was prepared to drop its view that Heritage Coasts should be a statutory designation and would "seek reaffirmation of government recognition of the national importance of Heritage Coasts through inclusion of appropriate references to Heritage Coasts in Planning Policy Guidance Notes" (Countryside Commission, 1992, page 4).

The Government, in its response to the Heritage Coast policy paper (Department of the Environment, 1992b) supported the experience of the Sefton Coast when it proposed that the planning and management principles developed in the context of Heritage Coasts could be applied to other stretches of the undeveloped coast.

Heritage Coasts in a sense lost some of their special status and in 1993 a UK Coastal Heritage Forum was established to reflect a new coast wide approach and, in time, the word 'heritage' was lost with the formation of CoastNet (www.coastnet.org.uk). With little new impetus in relation to Heritage Coasts the most recent guidance is found in Heritage Coasts: a Guide for Councillors and Officers (Countryside Commission, 1995).

Coastal defence and flood-risk management

An issue in the early 1990s was the future role of the Ministry of Agriculture Fisheries and Food (MAFF) in coastal defence. The House of Commons Environment Committee recommended that "the Department of Environment would provide a more suitable lead on coastal defence issues, operating through the National Rivers Authority" (House of Commons Environment Committee, 1992, page xiii). Responsibility for coastal defence had, in fact, been transferred to MAFF from the Department of the Environment in 1985 to bring together the supervision of coast protection (against erosion) and sea defence (against flooding) under a single ministry. The Government, however, did not accept any reason for change. Perhaps stirred by an impending review, in late 1991 MAFF announced its intention to develop a national flood and coastal defence strategy.

In the early 1990s came a growing awareness that the processes of change that affect the coastline are both complex and inter-related (Clayton, 1993). For example, it began to be recognised that action to protect one part of the coast from erosion may starve an adjacent part of vital sediment thus accelerating erosion 'downstream', and that immovable sea defences could cause loss of intertidal habitat through 'coastal squeeze'. At the time of the House of Commons Environment Committee's report a number of organisations including English Nature and the Countryside Commission were taking advice from coastal geomorphologists. In 1995, for example, English Nature published Managed Retreat: a practical guide based on trials and research in partnership with the National Trust and the National Rivers Authority (Burd, 1995).

The modern approach to coastal defence was established in the early 1990s with the identification of natural coastal cells within which would be developed sustainable coastal defence policies. As part of its Strategy for Flood and Coastal Defence in England and Wales, published in 1993, MAFF and the Welsh Office encouraged the formation of groupings of relevant authorities to prepare 'shoreline management plans'. The concept was supported by guidance (e.g. MAFF, 1993a, 1993b; MAFF and Welsh Office, 1995). The value of Shoreline Management Plans (SMPs) continues to be supported by Government (http://www.defra.gov.uk/environment/flooding/policy/guidance/smp.htm) Following a review of the strengths and weaknesses of the first generation SMPs, updated guidance was published in 2001(Defra and Welsh Office, 2001) and again in 2006 (Defra and Welsh Office, 2006). The updated guidance recommends that options should be appraised over a 100-year horizon, rather than 50 years as previously advised. A number of national research studies (such as Foresight www.foresight.gov.uk and Futurecoast www.defra.gov.uk/environ/fcd/Futurecoast.htm) have advised the revised guidance.

Coastal defence planning in England and Wales is based on the Shoreline Management Plans which identify general policies and implementation requirements, Strategies which identify the nature and timing of works and Schemes which identify the design and construction of major works. The overall approach to flood management in England is given in the strategy document Making Space for Water launched in 2005 (see http://www.defra.gov.uk/ environment/flooding/policy/strategy/index.htm for more information).

Responsibility for coastal defence in England is divided between the Environment Agency with respect to the prevention of flooding, and maritime local authorities (e.g. Sefton Council) with respect to the prevention of coastal erosion. The Environment Agency has prepared the Alt Crossens Catchment Flood Management Plan which includes the Sefton Coast (Environment Agency, 2009).

Integrated Coastal Zone Management in the North West of England

The Sefton Coast Management Scheme, formally established in 1978, was one of the earliest coastal management projects in the North West. Partners on the Sefton Coast supported and welcomed the development of the adjacent Mersey Estuary Strategy and Ribble Estuary Strategy in the 1990s. From the start, however, the estuary strategies relied on partnership and year-on-year funding leading to short term contracts, turnover of staff and problems with continuity and funding. The establishment of the Partnership of Irish Sea Coast and Estuary Strategies (PISCES) in the mid 1990s came partly out of the need to form a lobby to promote the work of the projects, to seek additional funding and to put forward a common view on regional coastal matters.

In 2000 the Government Office for the North West, the North West Development Agency and the North West Regional Assembly supported a North West Coastal Conference in Blackpool. The conference was welcomed by organisations already active in the North West such as the Irish Sea Forum, English Nature and PISCES. As a result of the meeting, the North West Coastal Forum was established to address the strategic issues in the region (www.nwcoastalforum.co.uk). The report of the conference (Irish Sea Forum, 2000) includes a summary of a study on coastal planning in the North West, commissioned by The Department of the Environment, Transport and the Regions (DETR) and carried out by Liverpool University. The main report (DETR, 2000) informed the review of Regional Planning Guidance (RPG) for the northwest. The University of Liverpool study looked at a wide range of plans and strategies at the regional, sub-regional and local level and reported that "coast-related plans and strategies, including Regional Planning Guidance itself, are products of the last decade and for the most part they have developed independently of each other. As a result, the current pattern of coastal planning and management in the North West is fragmented, with connections between the various exercises ill-defined and ill-developed both at the local and sub-regional levels and at the regional scale" (DETR, 2000, page 15).

Subsequent to this report and the Blackpool conference, the coordination of coastal planning and management in the north-west region has been well-supported through the work of the North West Coastal Forum. The region is one of the strongest in the UK in terms of links between regional economic strategies and coastal zone management (North West Coastal Forum, 2008).

Several issues concerning the management of the Sefton Coast also have a regional dimension including, for example, the monitoring of coastal change and the development of Shoreline Management Plans, the delivery of the England Biodiversity Strategy, the development of a North West Coastal Trail and the North West Bathing Water Forum.

Integrated Coastal Zone Management in Sefton

Since its inception in 2001, the Sefton Coast Partnership's purpose has been to bring landowners together in joint working arrangements whilst also increasingly setting the coast in the context of Integrated Coastal Zone Management. The strength of coastal management in Sefton has always been that the development of policy and strategy is matched by partnership activity on the coast through the delivery of projects. The development of strategies, a priority in the early years of the Sefton Coast Partnership, has subsequently helped to attract project funding from regional, national and European sources.

Coastal management in Sefton will always be influenced by external initiatives such as EU Directives, national policies, regional issues, and by local priorities set out in development plans, community strategies and biodiversity action plans. Responding to these in relation to the interests of the whole coast is central to the work of the Sefton Coast Partnership.

Partly as a result of experience with practical coastal management Sefton Council was one of the first local authorities in the UK to develop policies for a Coastal Planning Zone in its Unitary Development Plan (Sefton Council, 1995; 2006). The key planning issues include;

- The control of development to protect the coastal environment, to reduce the risk of flooding or the need for expensive coast defences and to protect groundwater reserves
- The implications of European, national and local nature conservation designations
- Supporting economic regeneration projects in the coastal zone

The statutory development plan process supports the non-statutory approach to integrated coastal zone management. In 2001 the Sefton Coast Partnership inherited the agreed delivery mechanism of the Sefton Coast Management Plan, Second Review, 1997-2006 (Sefton Council, 1998) and following this developed a rolling ICZM 'business plan' supported by strategy documents. These documents, which include strategies for nature conservation, access and the promotion of tourism, are important tools for securing external funding for the work of the Sefton Coast Partnership.

In Sefton the development of approaches to coastal management continue to be influenced by policy changes and trends in nature conservation, recreation and tourism, and shoreline management. The next section of the paper summarises some of the key milestones in relation to the Sefton Coast.

Nature conservation and land management

Much of the work on the Sefton Coast in the 1980s was linked to landscape restoration and access provision through the support of bodies such as the Countryside Commission. In the 1990s, however, attention turned to the need to develop a more strategic approach to nature conservation. The catalyst for activity came from the identification of the Sefton Coast as a possible Special Area of Conservation (SAC) under the EU Habitats Directive of 1992. The proposed designation required new approaches to management and in 1995 Sefton Council, English Nature and the National Trust were successful in a bid for funding from the EU LIFE-Nature programme to develop a conservation strategy for the sand dunes of the Sefton Coast. The three-year project included:

- The purchase of dune land at Ravenmeols
- Emergency habitat restoration for fixed dunes, dune slacks and dune heath
- The continuation of habitat management work and promotion of livestock grazing
- Actions for European protected species
- The preparation of management plans for golf courses and military sites
- The development of a Geographical Information System (GIS)
- The dissemination of best practice (e.g. Houston,1994)
- Publications including The Sands of Time (Smith, 1999)

The project's final recommendations helped to extend the boundary of the candidate SAC to include the dune heath habitat which forms the easternmost part of the dune system.

Following the LIFE project, English Nature completed a review of site designations in 2001 which resulted in the amalgamation of five former SSSIs (Sites of Special Scientific Interest) and their enlargement to form the single Sefton Coast SSSI covering almost the same area as the European designation. The amalgamation of sites helps to eliminate the 'nature reserve' image of nature to plans which embrace nature conservation (biodiversity) at a landscape scale. Additional sites for nature conservation identified in the 1990s were extensions to the Ribble Estuary National Nature Reserve and the Ribble Estuary SSSI, designation of the Mersey Narrows SSSI and proposals for a Liverpool Bay SPA.

The 1990s were a period of consolidation of nature conservation as a major land use and strategy driver on the Sefton Coast. Land managed for nature conservation, however, also has high amenity value and the land managers of the Sefton Coast help to support the attractiveness of the coast as a destination for visitors. The most significant changes have been:

- The establishment of the Marshside Nature Reserve in 1994 when the RSPB leased two

areas of freshwater grazing marsh from Sefton Council.
- The enlargement of Sefton Council's Local Nature Reserves when a substantial area of private land at Ravenmeols was bought in 1995.
- The consolidation of the National Trust's landholding at Formby with the purchase of former agricultural land.
- The establishment of the Lancashire Wildlife Trust's Freshfield Dune Heath reserve in the early 2000s.

This level of activity has ensured that sustainable techniques for conservation land management are now well established on the Sefton Coast and Ribble Estuary. On many sites grazing with domestic stock, or natural grazing by rabbits and wildfowl, helps to maintain a good mix of habitats.

In 2001 the North Merseyside Biodiversity Action Plan was launched as part of a nationwide approach to the delivery of the UK Biodiversity Action Plan, itself published in 1994 in response to the Earth Summit of 1992. Within the local plan the Sefton Coast is a key site and conservation targets have been set out in a series of habitat and species action plans.

Until the 2000s woodland management had been seen as a rather separate land management activity from 'nature conservation'. Although there is still a need for a prescriptive woodland management plan there is a growing appreciation that the woodland and open habitat components of the Sefton Coast should be seen as complementary components of the landscape. Successive management plans from the late 1970s have been written to conserve and enhance the woodlands. A woodland and scrub management strategy (Joint Countryside Advisory Service, 1999) identified the need to develop a long-term approach to woodland regeneration through thinning and replanting. This was taken forward through the Sefton Coast Woodlands Forest Plan coordinated through The Mersey Forest team following a national approach developed by the Forestry Commission. The Sefton Coast Woodlands Forest Plan is supported by the main woodland owners on the coast and owners of smaller woodland areas. It sets out a vision for the next 20-50 years in an integrated manner whilst taking account of the interests of access, nature conservation, landscape and local concerns.

The development of a revised nature conservation strategy by the Sefton Coast Partnership in the mid-2000s was an opportunity to bring together the interests arising from the statutory obligations to protect nature (SSSIs etc), the non-statutory and broader aspects of the North Merseyside Biodiversity Action Plan and the silvicultural and nature conservation aspects of the Forest Plan (Sefton Coast Partnership, 2007). The result was a more open and comprehensive view of the nature conservation value of the Sefton Coast and an understanding that no areas are fixed for all time on this dynamic coast.

Nature conservation itself has become more integrated in its approach and, in respect to some of the challenges of climate change, there is a move nationally towards larger, landscape-scale projects which take an ecosystems approach to giving nature room to adapt to change. There is also more interest in the ecosystem services which nature provides such as the provision of a sea defence function or the protection of water resources. These are clearly relevant issues for the Sefton Coast.

Recreation, Leisure and Tourism

A second key strand of management on the Sefton Coast is providing for the needs of local people and visitors in a way which does not damage the landscape value of the coast. The establishment of the Sefton Coast Management Scheme in the 1970s was partly in response to the very severe damage being done to the landscape through uncontrolled visitor use. The initial focus of work in the 1980s was in constructing access infrastructure, repairing damage and establishing ranger services. The improved control on access allowed the development of a more positive partnership between the interests of tourism and conservation in the 2000s. Through the development of a Sefton Coast Tourism Development Plan in 2005 the brand 'Sefton's Natural Coast' has been developed as a marketing tool promoting outdoor activities and the attractions of open spaces.

The Sefton Coast Management Scheme helped to raise awareness of the importance of the coast as a recreational resource and protecting this access has been supported by planning policies. 'Strategic' routes (which are given additional protection) include the Sefton Coastal Footpath (opened in 1990) and the Trans-Pennine Trail (inaugurated in 1989). With a general right of access to the foreshore and much of the coastal area open to the public, the Sefton Coast is well served in terms of access provision.

The diversity of landownership and land-uses on the Sefton Coast is one of its interests and creates a pattern of access and zoning so that some areas are intensively used and others are only lightly used. In some instances new access opportunities are sought and, for example, when land is purchased for conservation use, a degree of access is usually provided. The public's use of the Ravenmeols dunes was defended against the proposed development of a golf course by the Formby Land Company at a public enquiry in 1995 and open access was secured following purchase of the land. Similarly the National Trust has opened up new access routes on former asparagus fields. But there are concerns that if visitor numbers increase the coast could come under renewed threat from recreation pressure.

The 1998 Coastal Management Plan stressed that provision for access must be integrated with other interests in a sustainable manner and that "the long-term sustainability of the landscape, wildlife and recreation value of the coast could be impaired if demand led to provision. New ways must be sought to manage vehicles on the coast through adequate provision at urban centres, car-free zones, and support for public transport and car parking provision set back from sensitive zones." (Sefton Council 1998, pages 20-21). In 1999 Sefton Council, and the Sefton Coast Partnership, joined a trans-national study with Dutch, Belgian and other UK partners to address such issues as sustainable tourism, visitor management and transportation. The project entitled Quality of Coastal Towns was funded through the EU Interreg IIc programme and allowed a comprehensive programme of visitor research and a study of visitor facilities along the Merseyside coast to be carried out (Sefton Council, 2001).

Visitor pressures on the coast are predicted to increase, but not greatly and generally to the economic benefit of the region. The fastest growth sectors are likely to be in niche markets such as 'green tourism' (e.g. bird watching and walking holidays) and visits to friends and relatives (Atkins, 2001). Visitor research carried out in 2000 found general satisfaction amongst visitors and a low level of conflict between different user-groups suggesting that there is still some capacity to absorb a slight increase in visitor numbers. Climate change, however, may affect visitor use of the coast as people are drawn by the "more pleasant climate on the coast" (McEvoy et al., 2006, page 5).

The Sefton Coast Partnership identifies the need for a coordinated approach to access provision, visitor facilities, including visitor centres and toilets, information and interpretation. The Partnership, with the support of Mersey Waterfront, developed an Access Strategy in 2007. This added value to the Quality of Coastal Towns studies by confirming the concept of 'Gateway Sites' which would be managed to accommodate the bulk of visitor numbers. The 'Gateways Study' has prepared detailed plans for access and investment at Lifeboat Road, the National Trust property at Victoria Road and at Ainsdale-on-Sea.

A particular aspect of coastal management is related to beach management. In the early 1990s Sefton Council reviewed the management issues affecting the beaches and prepared a Beach Management Plan. The Plan was introduced in 1993 with the zoning of the beaches between Ainsdale and Southport (see box page 299). The strategy has been valuable in helping to accommodate an increase in beach-based sports whilst reducing the overall impact of recreation pressure.

Beach Management Plan

In the 1980s concerns were expressed about the damage to dune habitats by public pressure, car parking and mechanical beach cleansing on the amenity beaches. In 1987 an experimental beach barrier was erected to the south of Shore Road, Ainsdale to reduce pressure on the Ainsdale and Birkdale Sandhills Local Nature Reserve. Elsewhere sand-trapping fencing was used to help build out the foredunes and mechanical cleansing was restricted to seaward of a line of posts which marked the extent of the dune habitat.

An inter-departmental working group was set up in 1990 to prepare a management plan for the beaches of the Sefton Coast. The Beach Management Plan was introduced in 1993 with the zoning of the beaches between Ainsdale and Southport using vehicle barriers. At the same time an integrated beach management service for Sefton Council was established with the amalgamation of the Ranger Service, Lifeguard Unit and Beach Patrol under the Leisure Services Director.

Since 1993 the beach management operation has been a considerable success in terms of recreation management through zoning, investment in staff and resources, improved management of beach cleansing and the establishment of a Beach Management Consultative Group to liaise with interest groups. Bathing water quality has generally improved from investment in wastewater treatment by United Utilities (formerly North West Water).

Measures introduced since 1993 include:
- A reduction in the area used for car parking between Ainsdale and Southport
- Directing events and sports to those areas of the beach set aside for such purposes
- Promoting 'codes of conduct' for beach staff and beach users
- Introducing seasonal restrictions on certain activities
- Management of beach cleansing to minimise disruption or damage to habitats
- Developing an integrated and corporate approach to beach management

Sefton Council's Leisure Services Department launched a Beach Management Strategy in 2001. The introduction of recreation management policies has allowed the shoreline to develop in a more natural manner. To the north of Ainsdale there has been rapid sediment accretion with the formation of a series of new dune ridges and embryonic slacks. The dune formation continues northwards to the southern part of Southport beach. As well as habitat gains the zoning has allowed birds, such as the ringed plover, to breed on the upper beach area. The beach management strategy is thus a good example of integrated management in practice.

An important aspect of integrated management is developing coordinated approaches to the provision of information, interpretation and education for the benefit of local communities, schools and visitors. In the 1960s the Nature Conservancy developed a close link with Liverpool Museum to promote its nature trail at Freshfield. In the 1980s the nature trail continued to operate but more and more educational visits were being attracted to the National Trust and to Sefton Council's dune areas. In 1988 Sefton Council commissioned a review of educational provision on the coast and this work (Huddart, 1988) supported the establishment of the Sefton Coast Education Project. The coastwide education project ran from 1991 to 1994 and, through its project officer, helped with the development of curriculum-based resources for the coast. Following the work of the project the educational role of both the National Trust and the Sefton Coast and Countryside Service with primary and secondary level students has increased with dedicated staff assisted by site staff and volunteers. Educational visits to the National Nature Reserves are now more targeted to tertiary level groups studying conservation management.

There has been less progress with the development of a coastwide interpretive strategy partly due to the difficulty in securing funding for this work. In the early 1990s an Interpretive Strategy was prepared for the coast but from 1995 was not updated. The 2000 study of coastal visitor facilities (Atkins, 2001) proposed the preparation of an interpretive strategy to guide the development of visitor centres and a coordinated and thematic interpretive programme.

Ideas for a Merseyside interpretive programme were brought together in 1997 under the banner of Sands of Time. The overall approach received the support of the National Museums and Galleries on Merseyside but the hoped for funding through the English Tourist Board's Celebration of the Coast Millennium Lottery proposal did not materialise when the umbrella bid was withdrawn. Nevertheless the brand-name Sands of Time has been identified with ideas for interpretation on the Sefton Coast and has been used to raise awareness about the importance of the dune coast (Smith, 1999).

Mersey Waterfront supported the development of a tourism-led approach to visitor services through the Communication, Interpretation and Visitor Product Development strategy (referred to as the CIPD strategy). This strategy helped to identify and confirm economic developemnt and tourism as a key interest in the current approcah to ICZM in Sefton. The tourism sector is keen to support walking, cycling and special-interest breaks and has established a web-site at www.seftonsnaturalcoast.org.uk .

The CIPD strategy formed a key element of a bid for Heritage Lottery Funding in 2008 and it helps to support a range of educational and interpretive initiatives, including those which address people's perceptions of the coast in relation to future adaptation to climate change

Shoreline Management

The third key area which forms part of the approach to Integrated Coastal Zone Management in practice is the need to maintain the sea defence function of the dune system. The Sefton Coast is only protected by hard sea defences in the southern section at Crosby and in the northern area around Southport. Along most of its length it is the dune system and its integral dune aquifer which protects the low-lying hinterland from flooding. The importance of maintaining the natural flood protection service of the dunes is made all the more relevant when predicted sea level rise is considered.

As part of the national approach to coastal defence planning in England and Wales a Shoreline Management Plan will be prepared to identify general policies for the coast. The Sefton Coast falls within the identified national Coastal Cell 11. The regional cell is sub-divided into five sub-cells and these in turn are divided into 'coastal process units'. In the second generation SMP being prepared for North West England and North Wales Sub-cell 11a runs from Great Orme's Head to Southport Pier and sub-cell 11b runs from Southport Pier to Rossall Point.

Work on developing the first generation Shoreline Management Plans started in the mid-1990s with the Liverpool Bay and Ribble Estuary Plans completed in 1999. In 2008 work commenced to review the Shoreline Management Plan in North West England and North Wales. The plan is led by local authorities (the North West and North Wales Coastal Group) and will ensure that future defence works along the entire stretch of coast are managed in relation to their impact on one another. The revised Shoreline Management Plans take account of the most up-to-date knowledge on flood and coast erosion risks. They also give consideration to the impact of any works on protected European habitats and species.

Shoreline Management Plans are a key component of the approach to Integrated Coastal Zone Management and the recommendations of the plans and the results of annual monitoring will have implications for the way the coast is managed for nature, amenity and economic activity.

Other aspects of Integrated Coastal Zone Management

This paper has focused on the main activity of the Sefton Coast Partnership, which is to encourage a coordinated approach to management (involving landowners, statutory bodies, users and other interests) and an integrated approach to management (to address sometimes conflicting issues between nature conservation, access and shoreline management). In Integrated Coastal Zone Management at the sub-regional or regional level the Sefton Coast Partnership may be one of several bodies with an interest in, for example, sea fisheries, marine safety, pollution prevention, bathing water quality, planning policies etc. The Sefton Coast Partnership maintains its interest in these wider issues through communication within the partnership and through links to regional and national groups and fora.

The issue of marine spatial planning has been debated for many years. In the 1990s the Government view was for no change and that there was not a strong case for marine management arrangements (Department of the Environment, 1992). This will change, however, with the introduction of the Marine and Coastal Access Act 2009 setting out new arrangements for the control of activity in the marine and coastal zone (see http://www.defra.gov.uk/marine/legislation/index.htm for updates).

CONCLUSIONS

The Sefton Coast Management Scheme provided a mechanism for a coordinated approach to coastal zone management from 1978 to 2001. Now, with continuity provided through the Sefton Coast Partnership the Sefton Coast benefits from over 30 years of sustained management. In the first decade (the 1980s) the focus of activity was on landscape restoration and recreation management on the dune coast between Formby and Southport. In 1983 the management scheme was extended to include the dune area to the south of Hightown and further extended in 1991 to cover the whole of the Sefton coast including the beach area. It has remained as this area for several years, a zone recognised in the Unitary Development Plan as the Coastal Planning Zone. In practice, however, the Sefton Coast Partnership now operates within whichever boundaries are appropriate to the issue.

Since the mid 1980s management has focused on sustainability, especially on achieving a balance between the needs of visitors, the requirements of nature conservation and the need to maintain coastal defences. The coastal partnership approach in Sefton allows the development of coast wide projects and the development of strategies, such as those for access, nature conservation and tourism. The partnership has been successful because it has achieved tangible results and has brought in additional resources such as the successful bid to the Heritage Lottery Fund in 2009 for a Landscape Partnership project.

In 2000 the Board of the Sefton Coast Management Scheme agreed to restructure its operation to bring it more up-to-date with similar coastal initiatives in Britain. Since 1984 the Sefton Coast Management Scheme had been led by Sefton Council and its operation had been incorporated into the structure of its Committees. With the restructuring of the Council itself and the introduction of a Cabinet and Area Committee system there was a good opportunity to look at the running of the partnership. The key partners met in June 2000 and unanimously agreed to move forward with a new name, Sefton Coast Partnership, and a new structure where Sefton Council would be more of an equal partner with others. The Council agreed with the proposals put forward by the partners and the new partnership began in 2001. The Partnership Board was immediately strengthened with representation from new sectors including tourism, higher education, golf courses, archaeology and civic societies.

The Sefton Coast Partnership is a good case study of how the management of a distinctive area of coast has been able to adapt to changing circumstances whilst maintaining the fundamental aspects and benefits of partnership working. In the 1990s the Sefton Coast Management Scheme was able to keep abreast of changes and, in some instances, led on the development of best practice. The scheme was able to make the transfer from the Heritage Coast management approach to a more holistic ICZM style with comparative ease. The development of the Sefton Coast Partnership in the 2000s has provided a further opportunity to follow the principles of ICZM and to develop a long-term strategic approach to coastal management.

ACKNOWLEDGEMENTS

The author would like to thank Graham Lymbery for helpful comments on the first draft and for information on Shoreline Management Plans in Sefton. The author would also like to thank the reviewers for useful suggestions for additional information.

REFERENCES

Atkins (2001) *Sustainable Tourism on Merseyside: assessment of coastal visitor facilities.* Report to Sefton Council, Quality of Coastal Towns Project. (see www.seftoncoast.org.uk).

Atkins (2004) *ICZM in the UK: A Stocktake.* http://www.defra.gov.uk/environment/marine/documents/iczm/st-full-report.pdf

Bridge, L. and **Gilbert, C.** (2001) *Taking Action on the Coast: an introductory guide for local authorities.* Coastlink, Kent.

Burd, F. (1995) *Managed Retreat: a practical guide.* English Nature, Peterborough.

Clayton, K. (1993) *Coastal Processes and Coastal Management.* Countryside Commission, Cheltenham.

Countryside Commission (1992) *CCP 397 Heritage Coasts: Policies and priorities.* Countryside Commission, Cheltenham ISBN 0 86170 353 7.

Countryside Commission (1995) *CCP 475 Heritage Coasts: A Guide for Councillors and Officers.* Countryside Commission, Cheltenham ISBN 0 86170 456 8.

Covey, R. and **Laffoley, D. D'A.** (2002) *Maritime State of Nature Report for England: getting on to an even keel.* Peterborough, English Nature.

Davidson, N.C. (1991a) *Nature conservation and estuaries in Great Britain.* Nature Conservancy Council, Peterborough.

Davidson, N.C. (1991b) *Estuaries, wildlife and man.* Nature Conservancy Council, Peterborough.

Defra and Welsh Office (2001) *Shoreline management plans: a guide for coastal defence authorities.* Defra (replaced by 2006 guide).

Defra and Welsh Office (2006) *Shoreline management plan guidance: Volumes 1 and 2.* http://www.defra.gov.uk/environment/flooding/policy/guidance/smp.htm

Department of the Environment (1992) *Coastal Zone Protection and Planning.* HMSO, London.

Department of the Environment (1992) *Heritage Coasts: the Government's response to the Countryside Commission's revised Policy statement.* HMSO, London.

Department of the Environment and Welsh Office (1992) *Planning Policy Guidance Note 20: Coastal Planning.* HMSO, London.

Department of the Environment (1993) *Coastal Planning and Management: A Review.* Rendel Geotechnics. HMSO.

Department of the Environment and Welsh Office (1993a) *Development below low water mark: a review of Regulation in England and Wales.* HMSO.

Department of the Environment and Welsh Office (1993b) *Managing the coast: a review of Coastal Management Plans in England and Wales and the powers supporting them.* HMSO.

Department of the Environment (1995) *Policy Guidelines for the Coast.* DOE, Bristol.

Department of the Environment (1996) *Coastal Zone Management - towards best practice.* DOE, London.

Department of the Environment, Transport and the Regions (2000) *Research into Integrated Coastal Planning in the North West Region.* Government Office for the North West and University of Liverpool, Liverpool.

English Nature (1992a) *Caring for England's estuaries.* An agenda for Action. English Nature, Peterborough.

English Nature (1992b) *Coastal Zone Conservation.* English Nature's rationale, objectives and practical recommendations. English Nature, Peterborough.

English Nature (1992c) *The Seas of England. An agenda for Action.* English Nature, Peterborough.

English Nature (2004) *The Ecosystems Approach: coherent actions for marine and coastal environments.* English Nature, Peterborough.

English Nature (2005) *Our coasts and seas - making space for people, industry and wildlife.* Peterborough, English Nature.

Environment Agency (2009) *Alt Crossens Catchment Flood Management Plan.* Environment Agency, Warrington.

European Coastal Conservation Conference 1991 (1992) Proceedings. Ministry for Agriculture, Nature Management and Fisheries, the Netherlands & European Union for Coastal Conservation (EUCC). The Hague/Leiden.

European Commission (1999a) Lessons from the European Commission's Demonstration Programme on Integrated Coastal Zone Management (ICZM). ISBN 92-828-6471-5.

European Commission (1999b) Towards a European Integrated Coastal Zone Management (ICZM) Strategy: General Principles and Policy Options. ISBN 92-828-6463-4.

European Commission (2000) Communication from the Commission to the Council and the European Parliament on Integrated Coastal Zone Management: A Strategy for Europe. COM(2000) 547 http://ec.europa.eu/environment/iczm/

European Commission (2002) Recommendation of the European Parliament and of the Council of 30 May 2002 concerning the implementation of Integrated Coastal Zone Management in Europe (2002/413/EC). *Official Journal L 148 , 06/06/2002 0024 – 0027* http://ec.europa.eu/environment/iczm/

European Commission (2007) Commission Communication on the evaluation of Integrated Coastal Zone Management (ICZM) in Europe. COM(2007)308 http://ec.europa.eu/environment/iczm/

European Commission (2008) Directive 2008/56/EC of the European Parliament and of the Council of 17 June 2008 establishing a framework for community action in the field of marine environmental policy (Marine Strategy Framework Directive). http://ec.europa.eu/environment/water/marine/index_en.htm

Gallagher, A. (2001) *Putting Sustainability into Practice in the Coastal Environment.* Coastlink, Kent.

HMSO (1970) *The Planning of the Coastline.* HMSO, London.

House of Commons Environment Committee (1992) Coastal Zone Protection and Planning. HMSO, London.

Houston, J.A. and **Jones, C.R.** (1987) The Sefton Coast Management Scheme: Project and Process. *Coastal Management*, 15(4), pp.267-298.

Houston, J.A. and **Jones, C.J.** (1990) *Planning and Management of the Coastal Heritage.* Sefton Council, Southport.

Houston, J.A. (1994) Conservation management practice on British dune systems. *British Wildlife* 8, 297-307.

Huddart, D. (1988) *Environmental education on the Sefton Coast: its past and current development and possible future management strategies.* Unpublished report to Sefton Council by Liverpool Polytechnic Science and Outdoor Education Section.

Irish Sea Forum (2000) *Report on the North West Coastal Conference - Challenges and Opportunities.* Liverpool University Press, Liverpool.

Joint Countryside Advisory Service (1999) *Sefton Coast Woodland and Scrub Management Strategy.* Sefton Council.

Local Government Association (2000) *On the edge - the coastal strategy.* LGA, London.

Local Government Management Board (1995) *Local Agenda 21 Roundtable Guidance - Action on the Coast.* LGMB, Luton.

MAFF (1993a) *Coastal defence and the environment - a guide to good practice.* MAFF, London.

MAFF (1993b) *Shoreline management plans - interim guidance on contents and procedures.* MAFF, London.

MAFF and Welsh Office (1993) *Strategy for flood and coastal defence in England and Wales.* MAFF, London.

MAFF and Welsh Office (1995) *Shoreline management plans - a guide for coastal defence authorities.* MAFF, London. (Replaced by 2001 SMP guidance).

McEvoy, D., Handley, J.F., Cavan, G., Aylen, J., Lindley, S., McMorrow, J. and **Glynn, S.** (2006) Climate change and the visitor economy: the challenge and opportunities for England's Northwest. Sustainability Northwest (Manchester) and UKIP (Oxford).

North West Coastal Forum (2008) *Making the most of the North West Coast: Coastal Management challenges and success.*

Ranwell, D.S. (1972) *Ecology of Salt Marshes and Sand Dunes.* Chapman and Hall, London.

Rothwell, P. and **Housden, S.** (1990) *Turning the Tide: A future for estuaries.* RSPB, Sandy.

RSPB (1993) A *Shore Future - RSPB vision for the coast.* RSPB, Sandy.

Sefton Coast Partnership (2007) *Nature Conservation Strategy and Biodiversity Delivery Plan*, Draft 2007.

Sefton Council (1995) *Unitary Development Plan.* Sefton Council, Southport.

Sefton Council (1998) *Sefton Coast Management Plan*, second review, 1997-2006. Sefton Council, Bootle.

Sefton Council (2001) *Merseyside Coast Visitor Research 2000: summary document.* Prepared as part of the Interreg IIc Project-Quality of Coastal Towns. Sefton Council.

Sefton Council (2006) *Sefton Unitary Development Plan*, Sefton Council.

Smith, P.H. (1999) *The Sands of Time: An introduction to the Sand Dunes of the Sefton Coast.* NMGM, Liverpool.

UK Government (1994) *The Conservation (Natural Habitats, &c.) Regulations 1994*. HMSO, London.

UNCED (1992) United Nations Conference on Environment and Development: Agenda 21:The United Nations Programme of Action from Rio. http://www.un.org/esa/dsd/agenda21/index.shtml

The Pinewoods Project – an innovative approach to improving consultation with younger members of the community

David Bill

North Sefton City Learning Centre, Sandringham Road, Ainsdale, Southport, PR8 2PJ.
E-mail:david.bill@nsclc.co.uk

ABSTRACT

The Forest Plan (2003) for the management of the Sefton Coast Woodlands, including the Formby Pinewoods, called for an interim five year review on progress against 13 identified criteria. The North Sefton City Learning Centre (NSCLC) was asked by The Mersey Forest to assist in the public consultation aspect of the review by working with primary and secondary schools in Formby. The NSCLC and the Mersey Forest worked with pupils to develop an innovative website (www.pinewoods-project.org.uk) featuring radio shows, posters, videos and even pop songs based around the children's experiences in the pinewoods. This increased the numbers involved in the consultation, engaged more young people and widened the ways in which they could express their opinions. This paper describes the methodology used and reflects upon the benefits and problems of the approach adopted and suggests how the approach could be developed.

INTRODUCTION

The NSCLC was established in 2003 under the Government's Excellence in Cities initiative (www.nsclc.co.uk). Its aim is to provide schools with opportunities to use the latest learning technologies in a high quality and supportive environment. The main partners are the secondary and special schools in Southport, Crosby, Formby and Maghull, whilst also working with the secondary schools in South Sefton and the borough's primary schools. The NSCLC broadly works towards two interlinked goals: it aims to inspire schools to use the latest in learning technologies and techniques by putting on special events where teachers can see for themselves the benefit of the new developments; and secondly, it responds to needs identified by individual schools by running bespoke projects.

The Mersey Forest (www.merseyforest.org.uk) is an environmental regeneration initiative covering 465 square miles of Merseyside and North Cheshire, working to increase tree cover and involve local communities to bring environmental, social and economic benefits. The regeneration of the Forest is driven by The Mersey Forest Partnership, made up of nine local authorities (including Sefton), the Forestry Commission, Natural England and a range of other partner organisations. The Partnership is coordinated by The Mersey Forest Team, who act as facilitators and as a 'hub' resource for the partners. In Sefton, The Mersey Forest Team coordinate the Sefton Coast Woodlands Forest Plan, a 20-year Plan bringing together all of the woodland landowners in the Sefton Coast Partnership to ensure a healthy future for the Coast Woodlands (Nick Roche, 2003).

In 2003, The Mersey Forest developed a Forest Plan for the management of the Sefton Coast Woodlands, including the Formby Pinewoods, which created a vision of how the woods should look in 2023. In order to achieve this, five long term objectives were identified as shown in Table 1. Detailed accounts of the writing and implementation of the Plan are available from the Mersey Forest website (Nick Roche, 2003).

The Project

The NSCLC was approached to assist in this task following its work in other multi-agency environmental projects (North Sefton City Learning Centre, 2005). The NSCLC subsequently collaborated with the Forthright Group, a group consisting of primary and secondary schools in Formby and Thornton with a remit

Table 1. Summary of the long term objectives of the Sefton Coast Woodlands Plan.

The Sefton Coast Woodlands Plan: Long Term Objectives.

The Formby Forest Plan was created in 2003 with the following five long-term objectives which were then broken down to thirteen short-term targets:

1. To maintain the same area of woodland as we have at present, bearing in mind natural processes.
    a. To involve additional landowners in the planting of trees as appropriate.
    b. To ensure an income from the woodland where possible.
    c. To maintain fire risk reduction strategies.
2. To maintain pine trees as a habitat for the red squirrel.
    a. To maintain the predominantly pine woodlands with trees of all ages, through thinning and select felling.
    b. To keep about 10% of the total area as broadleaf woodland.
3. To maintain the existing character of the Sefton coast woodlands.
    a. To maintain a patchwork within the woodland that includes everything from grass to mature trees.
    b. To provide places within the woodland for plants and animals to survive – especially those named as 'Priority Species'.
4. To ensure that people work together while managing the woodland to help check progress and quality.
    a. To provide a structure that allows owners and managers to work together.
    b. To keep the public informed of progress and developments.
    c. To monitor the progress of the Plan.
    d. To involve the public and local community in the management of the woodland mosaic.
5. To utilise the woodlands as an educational and amenity resource
    a. To utilise the woodlands as a resource for local educational establishments.
    b. To maintain the woodlands as a safe area for use as an amenity resource.

---

to promote projects involving all or some of the schools in the area. Initial discussions led to two secondary schools in Formby and five primary schools becoming involved. The schools are shown in Table 2.

As part of the project management it was essential to:
- provide a general framework within which schools could use a range of investigative skills,
- offer specific targets that could be analysed for monitoring,
- link to a specific area of Pinewoods for field visits, and
- maintain the overall integrity of the Forest Plan.

In completing this work, the schools went through the following five stage process:
1. Allocation of a theme to examine (Table 2).
2. Fieldwork visit to the Pinewoods to find out if the objectives of the forest plan were being met.
3. Analysis of the information.
4. Visit to the NSCLC to record thoughts using video, radio and printed media.
5. Preparation time.

From the outset, it was felt that the results of the project should be presented to the Sefton Coast Woodlands Landowners' Group to provide focus and a "real world" feel to the initiative. The results would also be displayed on a dedicated website, using one of the key strengths of the NSCLC, the Centre's video and radio studios as a way of communicating information. It was also planned for an overview of the project to be presented at the 2008 Sefton Coast Forum, an annual meeting of stakeholders on the coast.

The schools were supported throughout their fieldwork by Rangers from the Sefton Coast and Countryside Service, the National Trust and the Ainsdale National Nature Reserve, who developed bespoke fieldtrips to help the pupils gain a better understanding of the issues

Table 2. Participating schools in the Pinewoods project.

Participating schools and their allocated strands	
St Jerome's Catholic Primary	Involving public and local community.
St Luke's CE Primary	Generating an income from the woodland.
Woodlands Primary	Using the woodlands to help people learn and providing wildlife habitats
Freshfield Primary	Making the woodlands safe.
Redgate Primary	Maintaining the 'patchwork' of pinewoods and broadleaf woodland.
Formby High School Range High School	Both supporting the work of the primary schools.

surrounding each strand. Where possible, the Ranger visited the school beforehand to discuss the programme, ensuring clear lines of communication and responsibility over health and safety. The fieldtrip routes were carefully chosen, allowing the children to make field sketches, take pictures and notes, with demonstrations where appropriate. For example one of the Rangers showed the safety procedures associated with tree felling. Plate 1 shows a fieldtrip in action. Further examples of photographs of fieldwork activities can be found at www.pinewoods-project.org.uk/patchwork-pictures.html

The primary schools were also assisted by student mentors from Range and Formby High Schools (including sixth form students), who helped with the fieldwork and devised and built the website www.pinewoods-project.org.uk. Some mentors used their specialist A Level knowledge, such as geography, to support the primary schools and others assisted in the creation of the website.

Plate 1. The fieldwork in action. (Photograph by D. Bill).

Most schools integrated the project across the curriculum, to include creative writing, poster design, drama, history, geography and citizenship. Some schools added numeracy by developing questionnaires to be completed by friends and family in the Formby area and then analysing the results. A typical visit to the NSCLC comprised three elements: (i) the children worked independently on the computers on a Pinewoods related task, which included designing posters or writing poems; (ii) groups of pupils then visited the radio studio to record adverts or discussions; (iii) the drama studio was used to record short plays. Each of these elements were judged to be very successful based on feedback from those involved. The success of all the visits was in no small part due to the high levels of planning and preparation (Figure 1).

The landowners' meeting was an undoubted highlight as it reflected the culmination of the considerable effort from all concerned. Each school was allotted seven minutes to present their findings to those with the responsibility to implement the Forest Plan. They used audio and video clips taken from the website, posters, poems, and in one case, a rap. It was viewed as highly successful, was very well attended and it clearly reflected the amount of effort and creativity that had gone into the project.

Discussion

Due to the complexity of factors feeding in to the Forest Plan five year review, it is difficult to isolate and quantify the impact the project has had in terms of

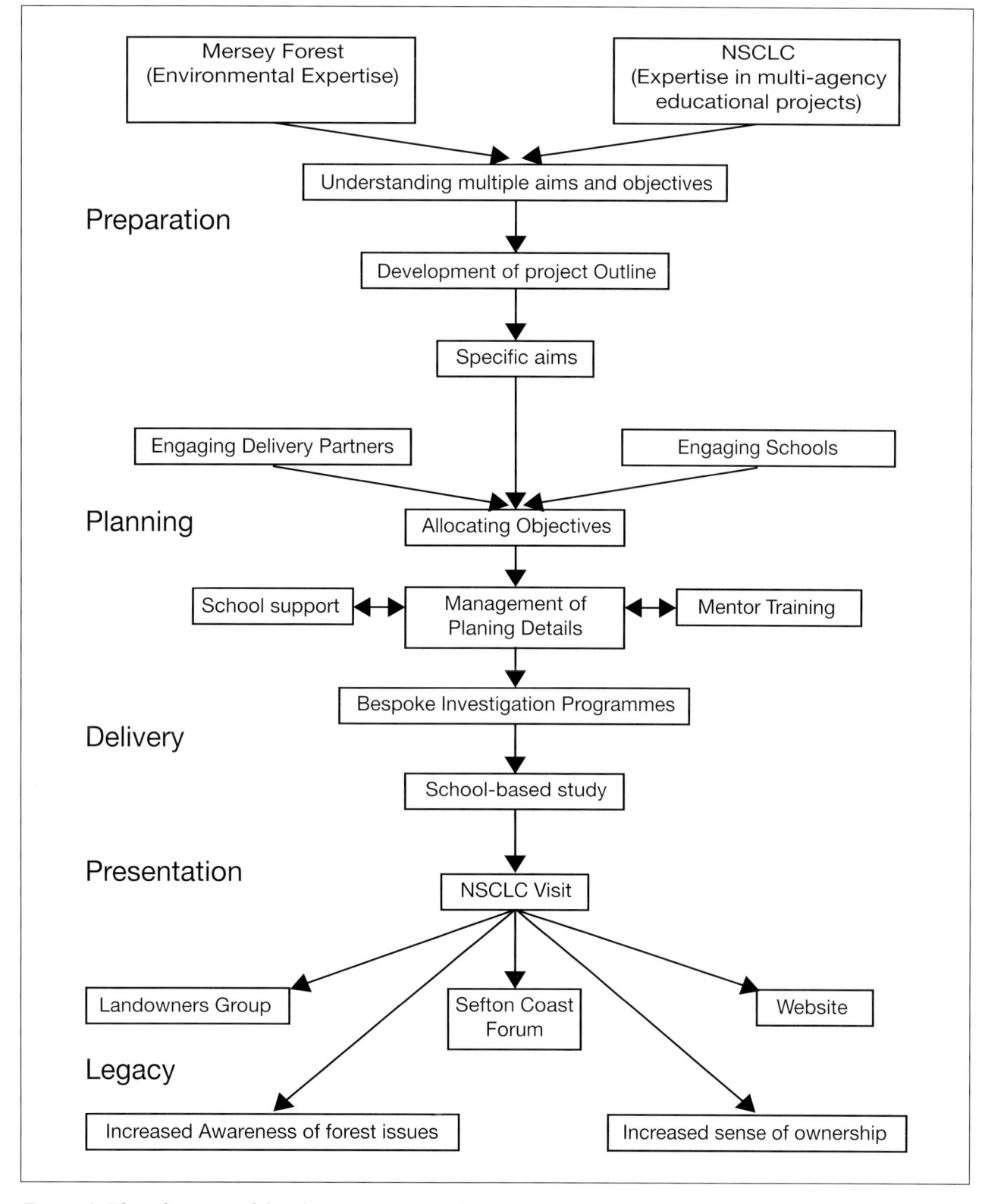

Figure 1. Flow diagram of the planning process of the Pinewoods project.

specific policy details at this stage; a follow-up review will assess this. However, it is clear that the project gave the five year review's public consultation the opinions and ideas of an almost entirely new demographic in terms of the audiences previously reached. Young people were engaged on a significant scale, not only increasing the richness of the five year review, but also providing an exciting platform for long-term interest and involvement in the future of the woodlands from a new generation of the local community. The involvement of the young people had an additional benefit of cascading information and interest to their families.

The website provides the project legacy; it is not intended that it is updated annually as it was specifically designed to support the five year review. However, it provides participating schools with the resources offered by the Pinewoods project and provides a case study for similar projects.

As with any new initiative there are areas that could be improved for similar work in the future. For example, more time should be allocated to the dedicated planning stage, with interim meetings between teachers and organisers being undertaken earlier. We also felt that there was a potential future opportunity to involve the parents of the participating children. However, there is no doubt that the participants (from the youngest of the pupils to the oldest of the organisers) gained hugely from the project. An awareness of the Forest Plan, and the land use pressures that lie behind it, have now been brought to the attention of a wider, younger, and very open-minded generation. It would be hoped that the appreciation of place and the need to manage what is of value will be beneficial to the pupils as they get older. It is hoped they use this understanding to influence future decision making both here in Formby, and in a wider regional or national setting.

CONCLUSIONS
The success of the project can be illustrated in a number of ways:
- A lively debate about the development of the Pinewoods was developed, especially amongst the town's younger residents who will reap the greatest benefits from the improvements.
- A "real world" problem was addressed which allowed the pupils to make realistic suggestions to the people who own and manage the woodlands.
- A participatory framework was developed which could be adapted to meet the needs of the individual schools and focus on selected areas of the curriculum.
- Primary to secondary school links were enhanced within Formby.
- Sixth form students were able to apply their ideas, knowledge and understanding developed in A levels to a real project. This included both environmental and IT based topics.

As a method of both improving community engagement and providing benefits to local schools this project has proved to be successful. Whilst there are improvements that can be made it is clear that there is scope for repeating this approach both on the Sefton Coast and with other environmental projects.

REFERENCES
Nick Roche (2003) 'Sefton Coast woodlands plan' http://databases.euccd.de/files/000203_Case_Study_for_Upload_-_SCWP.doc

North Sefton City Learning Centre (2005) 'Innovations' http://www.nsclc.co.uk/?_id=282

ACKNOWLEDGEMENTS
This project was planned and managed in partnership with Nick Roche and Michael Bray at The Mersey Forest.

This project relied on the goodwill, enthusiasm and commitment from a large number of supporters. These include colleagues at the NSCLC, the landowners and coordinators outlined above, but special thanks must go to all the pupils and staff from all the participating schools which made this project such a success.

All participating organisations have been named in this article. Full contact details can be found at: www.pinewoods-project.org.uk/thanks.html

Thanks must also go to colleagues at both Sefton Council and The Mersey Forest for checking this work and making valuable suggestions and contributions.

Marketing Sefton's natural coast – Evaluating the impact

Peter Sandman

Sefton Council, Magdalen House, 30 Trinity Road, Bootle L20 3NJ
Peter.sandman@tourism.sefton.gov.uk

ABSTRACT

Tourism is of significant economic importance to the Sefton Coast and through the Sefton Coast Partnership there has been a continuing effort to coordinate the promotion of the coast and in particular its value as a natural coast. This paper sets out recent efforts relating to the promotion of Sefton's Natural Coast highlighting successes and setting this within recent information relating to visitor activity.

INTRODUCTION

As part of the Integrated Coastal Zone Management (ICZM) initiative coordinated through the Sefton Coast Partnership, the Tourism Department of Sefton Metropolitan Borough Council (MBC) developed the Sefton Coast Tourism Development Plan in 2005 (Sefton Council, 2005a). Its aim was to improve the structure and coordination of tourism activity on the Sefton Coast. Part of the plans output was the creation of a marketing plan for the coast which complimented the branding of the coast as 'Sefton's Natural Coast' (SNC) (Sefton Council, 2006a). The broad aim of the marketing plan was the promotion of 'responsible' tourism growth within the coastal area. In addition the plan aimed to provide clear guidance for coastal land managers and stakeholders in order to promote what the coast has to offer to new and existing customer target audiences.

Plate 1. Anthony Gormley's 'Another Place', Crosby, Merseyside. Copyright Sefton Council.

Since that time, there have been a number of developments in regional and sub regional policy that have recognised the importance of developing opportunities for tourism within the Northwest's natural environment, to enhance the performance of the region's economy. In particular, the adoption of the final report 'Marketing the Natural Environment of the Northwest' (TEAM, 2006), produced for the North West Development Agency, has placed much greater onus on The Merseyside Partnership (Merseyside's Tourist Board) to channel funding and support into sub regional marketing activity. This is aimed at raising the profile of the natural environment within the Liverpool City Region. Similarly the work being conducted by the Northwest Regional Assembly has seen a number of potential tourism related projects being established through the European Union's INTERREG initiative. These have the potential to secure further funding for sustainable tourism initiatives ranging from quality assurance, to marketing and product development programmes.

Within the context of the above, it is encouraging to see that the many assets located within Sefton's Natural Coast, for example, the Gormley Statues (Plate 1), red squirrels, sand dunes and wide-open spaces are at the forefront of these programmes with their images used as a key element of the marketing. Furthermore, there is also genuine recognition at all levels, of the progress made by the Sefton Coast Partnership in structuring what the coast has to offer to visitors. This has been through the creation of the Sefton's Natural Coast brand (Sefton Council, 2006b), developing research initiatives (Sefton Council, 2005b) to better understand the behaviour of current and future audiences and the importance of structuring tourism activity so that it does not damage the natural environment.

The intention of this review is to assess the impact of activity designed to promote Sefton's Natural Coast as a destination for tourism between 2005 and 2006.

Visitor Audiences

There are many aspects of Sefton's Natural Coast that will appeal to and attract visitors. These range from general recreation associated with the coastline and its beaches, to red squirrels and the art installation "Another Place" at Crosby and Waterloo. To provide a degree of structure to this review, the key audiences targeted during the marketing campaign were as follows:

- Groups & North West Leisure Day Trip Market – these visitors are defined as those taking round trips made from home for leisure purposes, to locations anywhere in the UK. People start and return to home within the same day. Visits to overseas destinations are excluded. In Sefton's case, leisure day-tripper's will be predominantly people living in Wirral, Liverpool, Warrington and the Northwest in general.
- Tourism Day Trip Market – these visitors are a subset of all home-based leisure day trips. They refer to those visits undertaken by people travelling to and staying in places outside their usual environment. Visitors might be on holiday in the area – e.g. staying in Southport, Liverpool or Chester accommodation - but take a day trip excursion to the Sefton Coast. Trips last three or more hours and were not taken on a regular basis.
- Specialist/Niche Markets
    o Bird watching.
    o Cycling & walking.
    o General recreation - wind sports, bathing etc.

Performance against objectives

Contained within the 2006/2007 Sefton's Natural Coast marketing plan (Sefton Council, 2006a) were 3 main objectives on which progress was to be monitored. Outcomes were as follows:

Objective 1: To clearly position Sefton's Natural Coast within the sub region (Manchester & Greater Merseyside) as a unique destination for leisure, recreation and niche/natural tourism by 2007.

Outcome: The launch of the Sefton's Natural Coast brand provided a much needed identity that allowed the Sefton Coast to be positioned in consumers' minds as a single destination rather than a collection of individual smaller parts.

Whilst brand recognition has been difficult to measure given the limited resources available for this kind of research, the marketing campaign included the production of a series of destination guides (generic, events, cycling & walking) and the creation of the coastal visitor web site www.seftonsnaturalcoast.com using the new Sefton's Natural Coast brand guidelines (Sefton Council, 2006b). The key attributes of the campaign lay in the identification of key access points to the coast and the associated offer of popular nature, wide-open spaces and iconic structures. Targeting audiences within a 1/1.5 hour drive time, the potential impact in terms of brand awareness has seen a marginal increase in visitors to the coast from outside the immediate Sefton area (see below, objective 2).

In terms of marketing value, the Sefton's Natural Coast brand was subject to a significant amount of regional and national press attention during the last campaign. The Daily Express, Daily Telegraph, The Observer, The Sunday Tribune and the Financial Times all undertook familiarisation visits to the coast. Subsequent editorial pieces focussed on the key attributes of the destination reaching audiences across the UK (and beyond). In addition Sefton's Natural Coast featured in the 100 days campaign run as part of the North West Development Agency's Marketing the Natural Environment of the Northwest initiative and featured on national TV including Coast and Nature Watch. In addition the regular 'nature' features on local radio have further helped establish the brand.

Whilst there was significant national TV and press exposure associated with the Gormley Statues during the review period, it is not thought the brand was heavily associated in terms of establishing the attraction's geographical location. A key priority for future promotional and marketing activity will be to ensure the statues play a significant role in anchoring the brand as a means of capitalising on the national and regional recognition they now command.

Objective 2: Increase the number of 'non' Sefton visitors (tourist day and leisure day trip visitors) using Sefton's Natural Coast for leisure, recreational and niche/natural tourism by 20% by March 2007.

Outcome: Destination (quantitative and qualitative) research undertaken during the campaign has not proved effective in measuring this objective given the associated costs of the research and in gathering meaningful data that can be translated into visitor figures. That said; research conducted in 2006 showed a considerable increase in spend per capita (44%) from 2005, supporting the overall view that a higher proportion of visitors came from outside Sefton; visitors from outside of Sefton will typically spend more. With regard to visitor numbers, monitoring indicates that day visitors to Sefton (of which the coast is a key component) and excluding visits to Southport increased by 4% in 2006 to 1,909,600.

Objective 3: Attract short break niche (walking & bird watching) visitor markets through integration with Southport tourism product marketing initiatives, via the work of Sefton Council Tourism Department and the Southport Tourism Business Network Partnership (STBN).

Outcome: The Sefton's Natural Coast brand featured heavily in the resort marketing for Southport and was included within the main resort market guide. In addition, links from www.visitsouthport.com to www.seftonsnaturalcoast.com were established, with significant promotion at key exhibitions such as Great Days Out, British Travel Trade Fair. The generic value of what the coast can offer visitors was included within the resort's group marketing initiatives.

In terms of performance, visitors using serviced or non serviced accommodation in Southport increased by approximately 6% in 2006. However, it is not possible to determine what influence coastal marketing activity had on these figures given that other factors such as the resort's events and conference programme are also included.

In terms of destination research conducted in 2006 by The Mersey Partnership Research Services, there are notable increases in customers visiting the coast to walk (5% increase) and for nature (10% increase) when compared with the 2005 report.

RSPB Visitor Information indicated that during the period 2006/2007, visitors to Marshside increased to 25000 from 12000 during the period 2005/2006. Of those visiting, 77% were RSPB members, 23% were non-members, with 58% originating from the wider North West (excluding Sefton).

Summary of findings of local research activity

As part of the coastal research action plan adopted by the Tourism Department in 2005, a number of annual 'coast wide' surveys have been undertaken during mid August/early September (2005 & 2006). The purpose of the research being to gain quantitative and qualitative information about the use and attitudes of those who are currently visiting the Sefton Coast. Some of the key points of the 2006 survey (The Mersey Partnership, 2006) were:-

Visitor Profile

- Respondents represented a wide range of age bands, with 20% being aged 16-24 and 43% being aged 55+, (Figure 1).
- 58% of respondents were female, 42% male.
- The sample used was particularly representative of the higher social classes, 46% being A/B.
- 70% of respondents were visiting with their family (40% including children), and 19% were visiting on their own, (Figure 2).
- 5% of respondents considered themselves to have a disability.

Visitor Origin

Almost half of the sample comprised Sefton residents, with 17% from other areas of Merseyside.

Profile of Visit

- Reflecting active use of the coast, 20% of all respondents were visiting the coast to walk, 9% for exercise or fresh air, 4% to walk a dog and 1% to ride a bike or a horse. Nature was an attraction for 14% who came to see the red squirrels, and 5% who were bird watching.
- "Knowledge of the area" was the main influence of the visit (66% of respondents); and 17% being influenced by talking to friends and relatives. 3% had seen an advert for the area and 2 respondents out of 368 questioned in the 2006 survey were influenced by the "Coast" TV series.

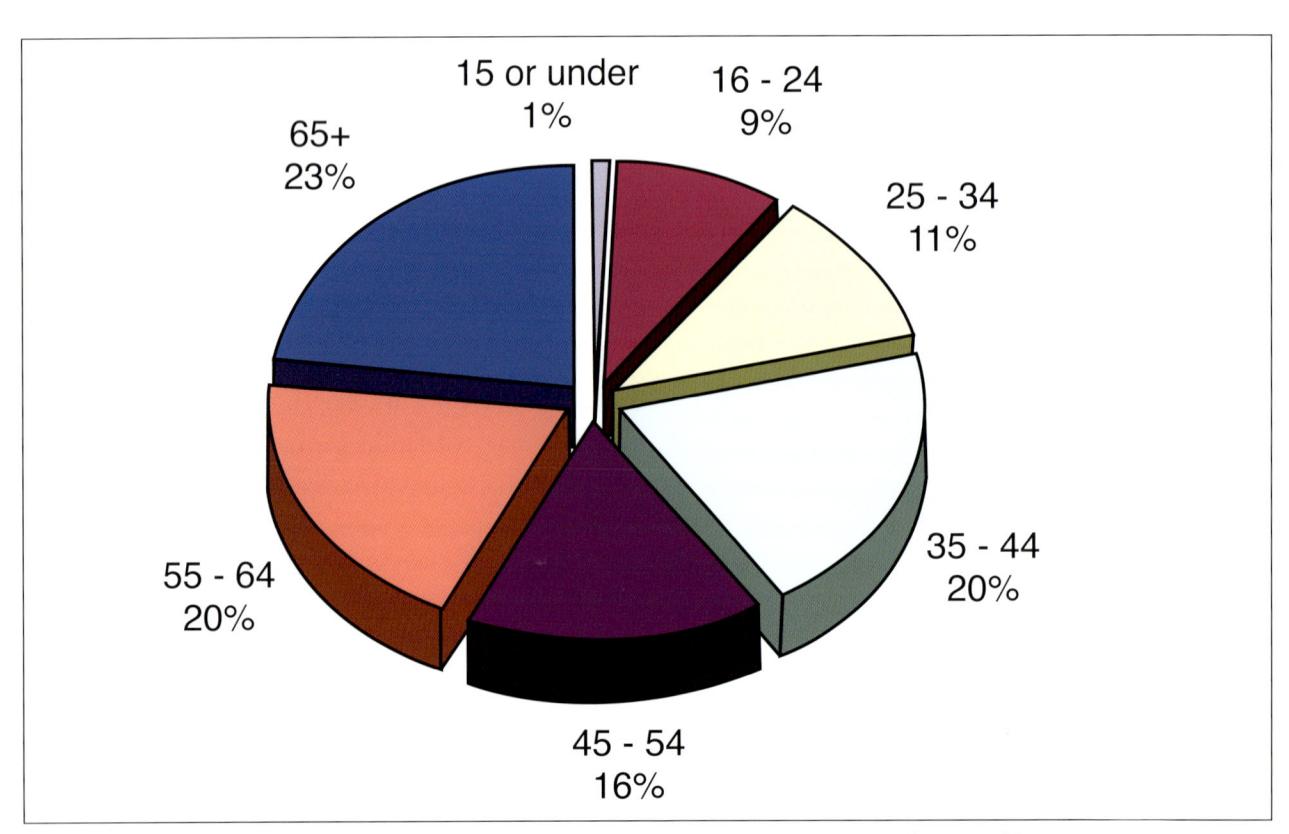

Figure 1. Pie-chart of 2006 coastal visitor survey to the Sefton Coast, respondent profile age.

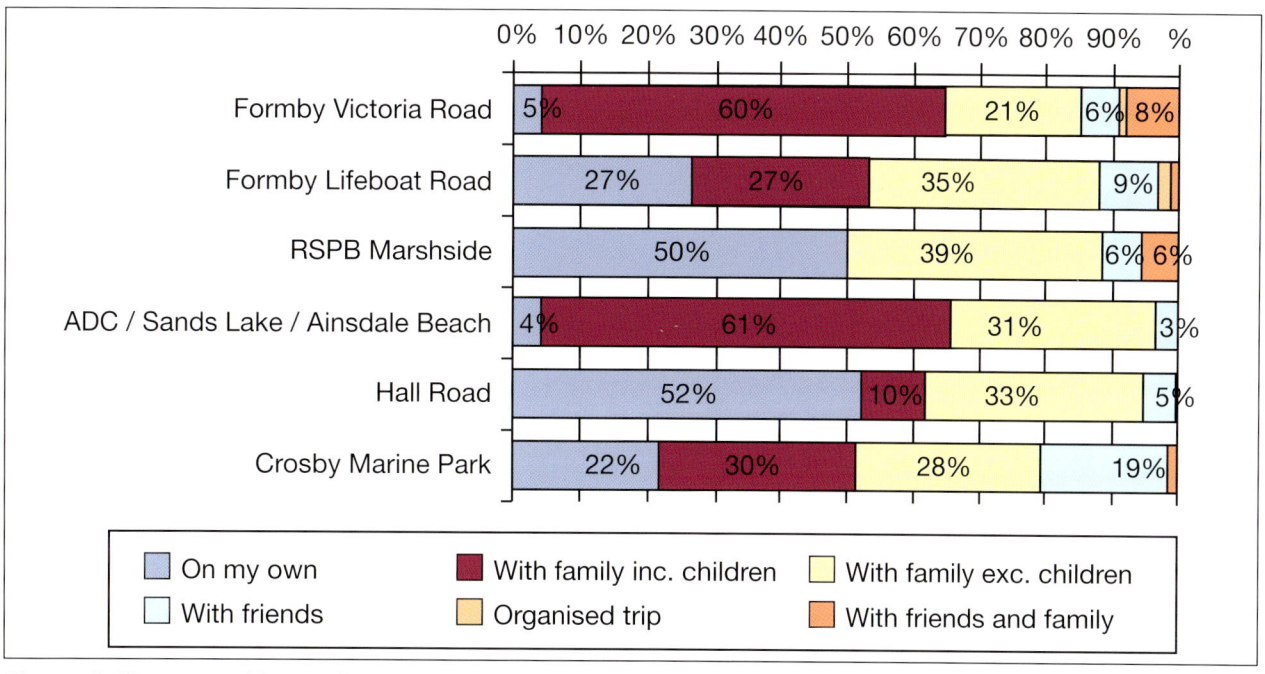

Figure 2. Visitor profile at selected locations along the Sefton coast, 2006.

Visitor Opinions

- Very high levels of satisfaction were recorded by respondents with the overall enjoyment of their visit (4.67 on a scale of 1 to 5). However, toilet facilities and signposting were less favourably viewed (2.96 out of 5). This was mainly the case at Formby Lifeboat Road, Crosby Marine Park and Ainsdale locations. There was some dissatisfaction with maps and information boards at locations in Ainsdale, (Figure 3).

- Changes to current facilities that respondents suggested were: better or more toilet facilities (19%); improved refreshment facilities/shop (10%); with 6% wanting better visitor centres.

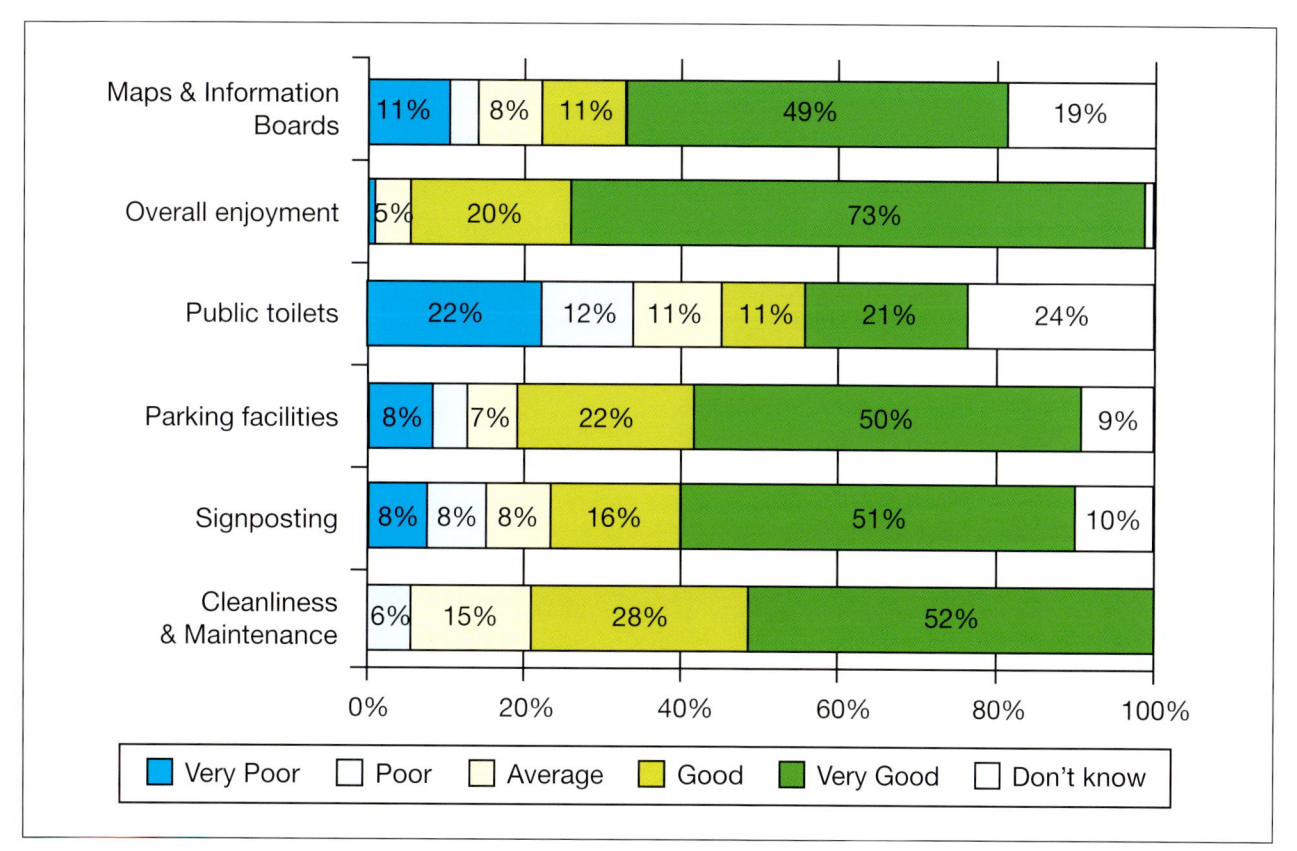

Figure 3. Visitor opinions of facilities for visitors to the Sefton coast, 2006.

CONCLUSIONS

Leisure Day & Tourism Day Visitors

- Although the situation is improving, the predominant coastal visitors are Sefton residents rather than visitors from the wider North West region. Whilst local and regional research, to a certain degree conflict, the general consensus is that the current profile of day/tourism day visitors to the coast are prominently ABC1 social groups over the age of 55. However, there is still a sizeable element of our current audience visiting with children. Therefore it would appear the current offer has a significant degree of appeal among both these core segments of the day visitor market.

- The combination of local knowledge and those visiting friends and relatives is confirmed as a major influence in attracting non-Sefton residents to the Sefton's Natural Coast. As a consequence, emphasis must continue to be placed on informing local markets of the coastal offer to sustain demand and attract new expenditures.

- Following 'non user' (people not visiting the coast) research undertaken in 2005 and the limited reach of marketing activity undertaken to date, there still appears to be a significant proportion of potential leisure day visitors within the North West who are unaware of the coastal attractions and/or do not have enough information to make a choice. This fact, coupled with the satisfaction levels and the modest increase in regional/sub regional visits to the Sefton Coast shows an opportunity still exists for marketing to further raise the profile of the destination and increase demand from previously untapped markets. However, it is imperative that this strategy must be balanced against issues such as carrying capacity at more sensitive destinations and those already at capacity in peak season.

- The concept of visiting Sefton's Natural Coast (among those familiar with the destination) is a popular one, indicating that Sefton's Natural Coast has a unique offering in the form of its most important visitor attractions: the Gormley art installation "Another Place", Formby's iconic Red Squirrels, the extensive system of beaches and dunes, and the Sefton Coastal Path.

- The high level of Northwest residents willing to drive to the Sefton Coast for a day trip demonstrates that there is a misconception that the destination is too far away for a day trip other than for Sefton residents. This indicates that the Sefton Coast is not sufficiently well recognised as a regional asset – lack of knowledge often encourages people to assume that it is not a local or accessible destination.

- High satisfaction levels among those visiting the coast demonstrate that the core coastal offer of wide-open spaces, nature, leisure/recreation and scenery are central to the coastal visitor experience. However there is a need for improvements in basic visitor infrastructure such as better signage, cycle paths, access, catering and toilet provision, in order for the marketed offer to match visitor expectations. These improvements will also help encourage return visits and spread word of mouth recommendations.

- The high satisfaction levels among those surveyed in 2006 indicates that the branding, content, look and feel of promotional literature is correct, strong and has the potential to act as a major influence on the decision to visit the coast.

- The marketing and growth of Southport as England's Classic Resort and the development of natural environment promotional initiatives for the Mersey Waterfront being undertaken by The Mersey Partnership and North West Development Agency should have 'spin off' benefits for Sefton's Natural Coast in reaching the leisure and tourist day visitor market. Close association should be made with these initiatives in all marketing literature as a means of raising Southport visitor awareness of the coastal offer and to develop links between the Mersey Waterfront destinations. Similarly, high profile attractions including the Gormley Statues should be used as key image assets within these wider marketing initiatives as a means of attracting these audiences to the coast.

Specialist Niche Visitor Markets

- Marshside and Formby (Victoria Road) proved to be the most popular destinations among bird/nature watchers. The most likely explanation is the quality of product at these sites combined with the strength of the RSPB and National Trust brands – this factor should not be overlooked in future coastal marketing initiatives.

- Sefton's Natural Coast has advantages of being relatively flat and therefore accessible by all age groups, having specialist wildlife interests (e.g. red squirrels, birdlife) and good public transport links to many sites, making it attractive for specific segments of the nature, walking and cycling markets; in particular, family and elderly visitors (e.g. mid-week walking groups, enthusiastic day visitor bird watching segments).

- Research indicates the Sefton Coastal Path is increasingly becoming a key destination among those walking Sefton's Natural Coast and with further development (in terms of signage and access), could play a much greater role in attracting these audiences.

- Food and beverage is increasingly becoming an important factor in the decisions being taken by consumers in order to have the 'complete' visitor experience which includes relaxation and stress release. There would be benefit in packaging the borough's facilities in this area with marketing of the coast.

- The walking, cycling and birdwatching products may also be suited to marketing within/through individual organisations specialising in such markets e.g. rambling associations, RSPB member groups, travel and cycle groups.

It can be seen that tourism is an important economic activity for the Borough of Sefton with the coast being a key element of the attraction. However there are issues with marketing such a large area hence the need to develop a single brand that can represent the whole coast and its attractions. Research such as that discussed above helps to guide marketing campaigns and investment in improving infrastructure but will always be limited in its usefulness due to the complexity of the range of influences on visitor's decisions. Further to this, whilst it is clear that there are opportunities to increase visitor numbers, this needs to be balanced with the management requirements of the coastal areas which are being visited. Key to maintaining this balanced approach will be the continuation of working in partnership with all the agencies involved with managing the coast.

REFERENCES

Sefton Council (2005a) *Sefton Coast Tourism Development Plan*. Sefton Council, Southport.

Sefton Council (2005b) *Sefton Coast Research Plan*. Sefton Council, Southport.

Sefton Council (2006a) *Sefton Coast Tourism Marketing Plan*. Sefton Council, Southport.

Sefton Council (2006b) *Sefton's Natural Coast Brand Guidelines & Architecture*. Sefton Council, Southport.

TEAM (Tourism Enterprise and Management) (2006) *Marketing the Natural Environment of the Northwest – Final Report*. Edinburgh.

The Merseyside Partnership (2006) *Sefton's Natural Coast Research*. Liverpool.

Sefton Beach Management - Twenty Years of Progress

Dave McAleavy

Head of Coast and Countryside Service, Sefton Council
E-mail: Dave.McAleavy@leisure.sefton.gov.uk

ABSTRACT

Sefton's beaches form part of the much larger beach complex within the major estuaries of the Ribble, Alt and Mersey. The Sefton beach complex is an important regional and local tourism resource, and includes the coastal resort town of Southport, Ainsdale beach, the National Trust and Lifeboat Road beaches at Formby, and the urban coastal strip at Crosby.

In the late 1980's it became obvious that to manage recreational activities seen to be both socially undesirable and environmentally damaging operations on the northern coast from the River Alt to Marshside, the foreshore from Ainsdale to Marshside would require positive management and a radical review of services.

This resulted in the implementation of a Beach Management Plan in May 1993. Coinciding with the commencement of the summer season, it included actions such as the installation of physical barriers and the transfer of cleansing contract supervision. Most importantly, it resulted in the formal integration of coastal operational staff and transfer of the responsibility for managing the beaches to Leisure Services, the Sefton Council department already responsible for managing the dune areas. During the first six months of implementing the plan, it became obvious that issues associated with many decades of custom and practise, such as un-restricted access for vehicles and a cleansing operation that had been in place for many years, were not going to be resolved easily.

This paper does not attempt to explain the success of the Beach Management Plan in respect of nature conservation, this is well documented elsewhere and in this volume, but it describes how the mechanisms for positive management were put into place and have since surpassed expectations across all areas of management. It is written from the personal perspective of the author as someone involved in implementing the changes and managing the resistance that these changes met with from some sectors.

INTRODUCTION

In 1991, the then Coast Management Scheme (Since 2000 the Sefton Coast Partnership) was extended to include the coast between Hall Road and the Freeport and all the Sefton beach areas. Up until the late 1980's there was little interaction between coastal landowners and the Council Departments who were responsible for the management of the foreshore. In the case of Sefton Council, there was, in addition, no effective working between the individual Departments that were responsible for the management of Council owned dune and beach areas.

This paper reports on twenty years of developing and implementing a beach management plan across a number of areas including; beach zoning, car parking and access, coordinated management, bathing and lifeguard services, provision of toilets, beach cleansing, enforcement of byelaws and an integrated beach management plan.

It sets out some of the problems leading to the development of the Beach Management Plan, some of the issues with its implementation in 1992 and discusses three specific areas of management: beach zoning, beach cleansing and coordinated management. Whilst it describes some of the detail of the plan, it includes opinions based on the personal experience of the manager overseeing implementation to provide a context that is otherwise often lost.

Rationale for the Beach Management Plan

The Sefton Coast (Figure 1) is designated as a Site of

Special Scientific Interest (SSSI) and is adjoined by two further SSSI's the Ribble Estuary SSSI and the Mersey Narrows SSSI. Each SSSI has an associated list of potentially damaging operations (PDOs) that need to be considered in the day to day management of the site. These PDOs are considered during the process of agreeing management planning objectives. The outcome is usually a Consent from Natural England for identified actions to be carried out under the provisions of the Wildlife and Countryside Act 1981 (As amended), and in Sefton can include actions such as mowing dune slacks or allowing beach car parking. In 1986 provisions for management and the basis for a five-year Consent were agreed with the then Nature Conservancy Council (now Natural England). These were most successfully implemented in relation to the protection and management of the dunes. However, the objectives agreed in respect of the management of the foreshore area proved to be unsatisfactory, and the recommendations for management that were implemented, quickly deteriorated. This included the initial beach barrier 1 kilometre south of Shore Road that delineated the first vehicle free zone in 1987, prior to which, vehicles could drive on the beach south of this point. Also beach cleansing guidelines, which although well intended, seemingly allowed the cleansing contractor to continue depositing strandline debris along the top of the beach regardless of its content.

In the first ten years of the Sefton Coast Management Scheme, considerable success was achieved and effective dialogue resulted in agreement with the then Nature Conservancy Council (NCC) on appropriate management measures to reduce activities damaging to the dune element of the SSSI's along the coast. In particular, this included the dune areas between the Ainsdale National Nature Reserve and Southport Pier where public recreation, including use of the beach for car parking, was very high.

It was in 1989 that the increasingly uncontrolled recreational pressures on the Sefton Council owned beaches between Ainsdale and Marshside were causing increasing concern, with attempts to encourage joint working between a number of Council departments

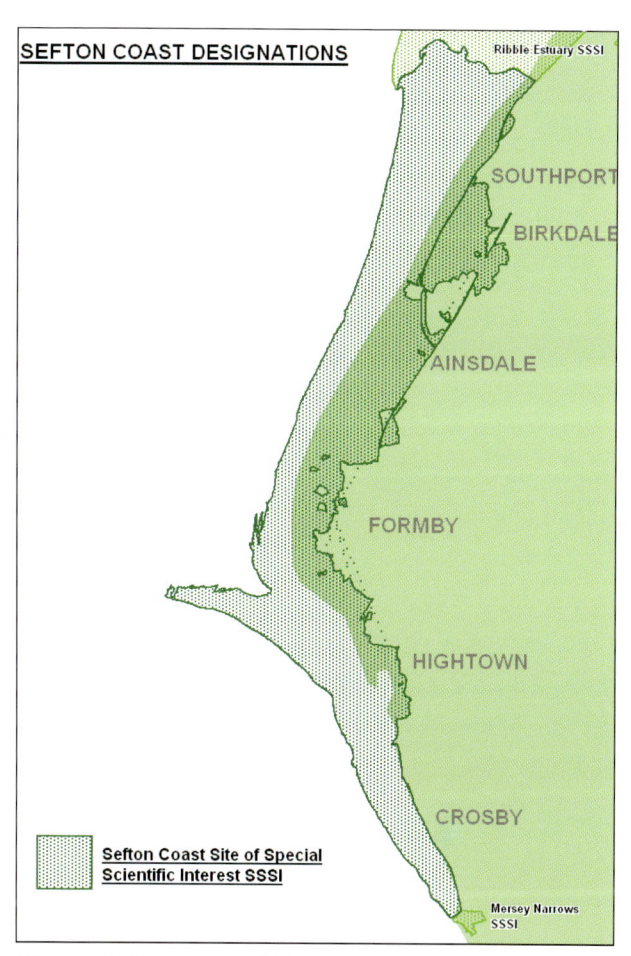

Figure 1. Location of SSSI dessignations. © Crown copyright. All rights reserved. Sefton Council Licence number 100018192 2010.

with responsibilities for beach management proving difficult to implement – mostly due to limited resources, differing priorities and more importantly entrenched custom and practise. It was not unusual to see in excess of thirty scrambler motorcycles and many off-road vehicles on the beaches between Southport and Ainsdale on a Sunday afternoon, not only causing considerable disturbance to wildlife and adding to the workload of landowners, but also (based on comments from visitors in recent times) deterring a considerable number of people who wanted to visit the coast.

The need to deal with short-term management issues, including the need to balance tourism and nature conservation, resulted in an internal Environmental Issues Working Party establishing a Beach Management Sub-Group to consider the development of a Beach Management Plan. It provided the impetus for the Sub-Group to consider longer-term trends, external issues and management structure (Houston and

Jones, 1987). A 'Beach Management in Sefton' Issues Report prepared by the Coast Management Officer on behalf of the group was wide ranging and covered all aspects of beach management coast-wide - from toilets to sand winning; this led to the development of the beach management plan.

Beach Zoning

The real and potential conflicts between nature conservation interests and recreational use of the beaches can only be solved through the introduction of spatial and temporal zoning (Sefton Council, 1992). Land ownership boundaries defined the main beach zones, but it was on the beaches managed by Sefton Council where significant changes were introduced through the installation of physical barriers that changed decades of unmanaged use by motor vehicles (Plate 1). Not only were vehicles (except fishermen and others with permission), prevented from driving outside of the new car parking areas, for the first time vehicles were prohibited from driving more than fifty metres from the front of the sand dunes or the sea wall (defined in Seashore Byelaws, in place for some years).

Plate 1. Beach management zoning sign north of Shore Road, Ainsdale, 1995. (Copyright Sefton Council).

Vehicle barriers were erected in May 1993 as the first stage of implementing the beach management plan. A metal barrier was erected a kilometre to the north and to the south of the main beach access at Shore Road, Ainsdale and a third erected to the south of Weld Road, Birkdale. Interestingly, the zoning map of the beach areas was starting to show similarities to the fishing stalls figure described in Old Birkdale and Ainsdale (Harrop, 1985). By 1993 the beach at Birkdale was beginning to show signs of natural change and a line of Common Saltmarsh Grass (*Puccinellia maritima*) was becoming established on the beach seaward of Taggs Island. This change quickly accelerated, developing into a major new habitat, causing considerable interest due to the speed of its development. The Birkdale 'Green Beach' (as the area is commonly known) is described as a 'biodiversity hotspot' by Smith (2007) and Smith (This volume).

Two later amendments were made to the location of the beach barriers. In 1995, for health and safety reasons, car parking ceased south of Weld Road between the frontal dunes and the *Puccinellia maritima* ridge forming on the beach. Cars were allowed to continue parking seaward of this ridge for a number of years, but this was discontinued as part of a later review of winter beach parking. In 2004 areas of beach vegetation were treated and removed at Southport to maintain it as an important amenity beach, and the resulting habitat compensation required resulted in the beach barrier north of Shore Road being relocated 400m further south in late summer 2004. Again, the rate at which this zone was colonised by plants and the Natterjack Toad, (*Epidalea calamita*) was rapid and is described by Smith in this volume.

In March 1998, the Leisure Services Committee considered a report on the first five years of beach management (Sefton Council, 1998). It was requested that the Committee agree to reduce the area of beach used by vehicles during the winter months both due to the reduced demand and to reduce impact on the site. Subsequently, there was agreement for a pilot scheme between January and March 1999. In November 1999, the Leisure Services Ratification Committee considered a report that recommended adopting the pilot scheme as part of the annual management arrangements. The report proposed that during the winter, Southport Beach should be closed to cars, Weld Road should be defined by a small car park at the beach entrance and at Ainsdale Beach the winter car parking area should be reduced to considerably less than the 2 kilometres available during the summer months (Sefton council, 1999) (Plate 2). This proved to be the most contentious decision yet associated with the Beach Management Plan and coincided with the Council adopting a 'Cabinet' system approach to local government. This resulted in the decision being 'called in' (when a decision is challenged by other Councillors) and referred to a Scrutiny and Review Committee, which referred it back to the Leisure Services Ratification Committee who confirmed their original decision. In accordance with the decision-making process the matter was referred to the Sefton Cabinet at its meeting in December 1999 (Sefton Council, 1999) for consideration and approved. To many the appearance of beach barriers and zones in 1999 was welcome but there were a minority who were unhappy with this despite it being an obvious improvement with respect to managing the site and having been approved by the public's elected representatives. A well organised campaign by a small group of local residents ensured continuing opposition to every decision taken including beach zoning, with the debate continuing today; but many of the opponents of beach management policy, both residents and elected representatives, have seen the improvements to the beaches – notably safety, cleansing and protection of wildlife. Although many accept the growth of beach vegetation is a natural process, there is disappointment in some sections of the public at the perceived loss of the upper beach along part of the Southport seafront and there will be discussion for years to come on the issues relating to the growth of sand dunes and associated problems of beach levels and sand-blow at Crosby. However, there is now a mellowed view and general understanding of the management issues facing coastal landowners and to a large extent many who were suspicious of the Beach Management Plan now support the policies. Twenty years on and it is noticeable that the plan is now gaining the support of important user group's such as the Beach User's Group and has local support through volunteers.

Beach Cleansing

The cleaning of beaches on Sites of Special Scientific Interest (SSSI), especially those backed by dune systems has always been contentious amongst practitioners as there is a perceived conflict between what is best for nature and what is best for tourism and this was no different on the Sefton Coast.

Sefton also had its own peculiarities when it came to the mechanical removal of material from the beach as none of it actually left the coastal area, being deposited either within the frontal dunes or amongst the embryo dunes at Birkdale (Plate 3). The tipping in the dunes is well documented as it developed an interesting flora.

Before 1992, two methods of mechanical cleansing were used. The first, a method that continues today, utilises a mechanical surf rake towed behind a tractor using a sprung-tine method that flicks debris into a hopper, when necessary, penetrating up to 5cm into the

Plate 2. Ainsdale Beach revised car parking 2006. (Copyright Sefton Council).

surface of the beach. The second method involved a tractor with a large bucket scraping along the strandline and scooping debris into large heaps before being deposited in the area of the dunes. The tipping of strandline material within the dunes ceased in the late 1980's and the mechanical scraping or scooping method of beach cleansing ceased during a review of the cleansing contract in the early 1990's. However, the tipping of strandline material at the top of the beach or in the embryo dunes continued unabated until 1993.

Most Local Authority beach-cleansing operations around the country are usually part of a larger cleansing contract carried out in-house by the department responsible for refuse collection and street cleansing. It was difficult to influence specifications mid-contract to take account of nature conservation interests or low-level pollution; the latter being at such significant levels on some beach areas that hand-picking before being deposited in the dunes was not economical due to it being labour intensive. There was speculation that Local Authorities in pursuit of beach awards was also having a detrimental effect on natural coastal habitats (Llewellyn and Shackley, 1996), but beach managers had little option as the requirements of the Environmental Protection Act 1990 dictated the level of cleansing on amenity beaches which included the removal of all organic material including the strandline.

An attempt to improve the operation was the development of Beach Cleansing Guidelines (Sefton Council, 1987). These guidelines included advice to the contractor on methods to protect embryo dunes and promote their growth by the tipping and spreading of strandline material along the top of the beach. These guidelines were never managed or monitored and therefore had very little influence on beach cleansing where it affected the dunes.

The Sefton beach cleansing operation was under-funded in comparison to other local authorities and could be up to 75% less than some of the major resorts such as Blackpool. Manual cleansing was at a low level and seen as a lesser priority in the overall cleansing contract for the Borough: it was usual for operatives to be moved to other areas of work within the resort. Generally, the beaches at Ainsdale and Southport received both manual and mechanical cleansing in the morning during the summer months. The strandline of higher tides at Crosby were mechanically cleaned utilising an external contractor with a Bobcat excavator to remove material during the summer with manual cleansing also taking place; the rangers and volunteers cleansed Lifeboat Road beach at Formby.

One of the major successes in recent years, and since the implementation of the Beach Management Plan, is associated with beach cleansing. The plan identified the need to consider beach cleansing again and review the guidelines, but the Council went further and transferred the responsibility for the budget and monitoring of the operation to the Leisure Services Department which includes the Coast and Countryside Service who were responsible for other aspects of management of the beach and sand dunes.

The contract to clean the beaches was originally developed as part of a Voluntary Competitive Tendering approach within the main Sefton Cleansing Contract and, although the operation was transferred in April 1993, it was only to be subject to review as part of Compulsory Competitive Tendering (CCT) process when it was successfully retained in-house by the Council Cleansing Section.

Plate 3. Beach cleansing mounds of litter, plastics, organic material and sand tipped in the embryo dunes at Birkdale 1993. (Photograph by Sefton Council).

On transfer in 1993, the main objective was to make the most effective use of the limited resources available whilst also working with the contractor to bring in methods that also took into account the sensitive nature of the coast and a requirement to reduce disturbance to wildlife and protect habitats. In 1997 the beach zoning designations in the Beach Management Plan (1992) were adopted and sub-zones created as part of a revised specification for the Beach Cleansing Contract. This revision also resulted in a small increase in funding and an obvious improvement in delivery on the main amenity beaches. The revised 1997 specification consolidated some of the voluntary arrangements from 1993 in respect of mechanical cleansing, including the cessation of this activity on Birkdale Beach, although it is debatable as to whether or not mechanical cleansing had had any real impact on beach or dune habitats prior to this. The tipping of strandline material removed from Ainsdale and Southport beaches, based on the authors experience, did not have a major impact on dunes habitats; the impact was mainly aesthetic due to the content such as toiletry plastics and such like, which was unpleasant for visitors picnicking nearby during hot summer months.

The ability of the cleansing specification to deliver was hampered by there only being limited provision for taking material off-site, with every short-cut aiming to retain as much cleansing material on the beaches as possible. Once the contractor was instructed to concentrate resources on the main amenity beaches, rather than other parts of the Borough, the problem reduced, and when resources for manual cleansing were fully concentrated on the beaches maintaining a full daily operation, the levels of beach litter and general appearance of the main beaches improved.

A new Government process for Local Authorities, known as 'Best Value', replaced CCT although this meant that services could still be out-sourced in a process designed to improve Council services. Fortunately, the beach cleansing element of the Borough–wide specification was removed from the contract and fully integrated into the Coast and Countryside Service, allowing the coast-wide beach cleansing operation to be far more flexible and able to respond to variations in daily need and specific requirements depending on location such as embryo dunes.

Later improvements to the service included the operation to remove litter and sand from the promenades at Crosby and Southport. The additional staff employed as part of the Coast and Countryside beach management cleansing operation contributed considerably across the service as they became trained to carry out byelaw enforcement, habitat work and woodland management when required.

Since the late 1990's the Mersey Estuary Pollution Alleviation Scheme (MEPAS) and a major sewer improvement scheme on the Southport coast (implemented by United Utilities, the regional water company) contributed to other major investment to improve bathing waters in the North West and the impact was an improvement in bathing water quality and a reduction in sewage derived litter on the shore, the latter influencing a change in the specification for mechanical cleansing that allowed for organic material to be left on some beach areas. An ongoing problem for beach manager's in respect of marine and riverine litter is that it can occur in high volumes after periods of sustained rainfall, coinciding with high tides and onshore winds, especially during the summer months when high levels of organic material mixed with marine and riverine flotsam and jetsam is deposited on the beach and embryo dunes.

Coordinated Management

The Beach Management Plan discussed the possibility of better coordination based on the Coast Management Scheme and it sought better integration of coastal staff to provide a coordinated Borough-wide service.

The success of the plan is due solely to actions taken to amalgamate all the services involved in the day to day management of Sefton Councils coast and countryside land and this included the Ranger Service, Beach Patrol, Lifeguards, cleansing, education and first aid staff. It is likely that without this radical approach the key aspects of the plan would have failed as they had done in the past.

All aspects of the integrated service and land management actions were reviewed after 1993 and this resulted in a greatly improved service. Notable improvements were the cleansing operation, a permit system for vehicles and kiting sports, and guidelines for horse owners. The cessation of a wildfowling permit on Council land at Marshside resulted in a challenge by British Association for Shooting and Conservation (in the High Court, they were unsuccessful in this challenge).

The review of the Lifeguard Service and the review of the use of antiquated World War 2 DUKW amphibious craft (used as vehicles for lifeguards) saw improvements in beach safety with the Coast and Countryside Service adopting a new Beach Lifeguard qualification, for a time one of the Sefton Lifeguards was the oldest person to take and pass this qualification at the age of sixty. The work to review safety on Sefton beaches was supported by the publication of guidelines for safety on British beaches (ROSPA, 1993) and at the time of writing Sefton Council is discussing out-sourcing the Lifeguard operation so that an improved service, operated by the RNLI, may be in place for summer 2010.

There is too much to include in detail to describe the improvements overall, but in addition to that described above the following has also taken place:

- A Beach Consultation Group was established and held its first meeting in October 1999. It includes all coast communities, representatives of all users from fishermen to naturalists, Merseyside Police, land managers and regional/national agencies. The group meets three times a year and is chaired by the Sefton Council Cabinet Member for Leisure and Tourism.

- Merseyside Police has raised its profile on the coast with the use of quad motorcycles, offering considerable support for land managers.

- The RSPB are planning to expand the Marshside Nature Reserve to include most of the outer marsh and include improved visitor facilities.

- The first major artwork on the coast was installed on Crosby Beach in 2005 with Antony Gormleys' 'Another Place'.

- Southport Pier underwent a major restoration and a new seawall was constructed between Weld Road and Fairways at Southport.

- The Coast and Countryside Service played a considerable role in ensuring that new coastal awards being developed by ENCAMS (now Keep Britain Tidy) included stronger emphasis on nature conservation (Plate 4). The author is the current Chair of the United Kingdom Beach Management Forum (UKBMF).

The Coast and Countryside Service is recognised as the only fully integrated service of its type in the UK and beach management in Sefton is seen as best practice, but the progress described in this paper has not come easily.

SUMMARY

Considering the many decades of custom and practice and unmanaged access to the coast it was always going to be radical when the Beach Management Plan was implemented in May 1993 and there was going to be considerable apprehension and distrust regardless of how beneficial the actions may be. There was a minority of people who thought and always will feel that they lost a right in not being able to drive in a motor vehicle the whole length of the shore from Southport to Formby and that permits and zoning to manage activities is a restriction. Natural accretion although welcome as a habitat and coastal defence in most areas where it is occurring, has affected parts of the tourist beach at Southport and although it is well documented that this is a process that has been ongoing for centuries, people will still resist change, highlighting the need to consider how we manage not only the changing coast but how people understand and appreciate living with a changing coast.

Twenty years on and the majority of people with an interest in, or who visit the coast in Sefton realise that the implementation of new beach management policies is a positive, successful means of making the Sefton Coast a safer, cleaner and more pleasant place to visit because of this.

REFERENCES

Atkinson, D. and **Houston, J.** (eds) (1993) *The Sand Dunes of the Sefton Coast.* National Museums and Galleries on Merseyside.

Harrop, S. (1985) *Old Birkdale and Ainsdale: Life on the South West Lancashire Coast 1600 – 1851.* The Birkdale and Ainsdale Historical Research Society, Southport.

Houston, J.A. and **Jones, C.R.** (1987) The Sefton Coast Management Scheme: project and process. *Coastal Management* 15 267 - 297.

Llewellyn, P. J. and **Shackley, S.E.** (1996) Effects of Mechanical Beach-Cleaning on Invertebrate Populations. *British Wildlife* 7 pp.147-155.

ROSPA (1993) *Safety on British Beaches – Operational Guidelines.*

Sefton Council (1987) *Beach Cleansing Guidelines.* Unpublished.

Sefton Council (1988) *Beach Management Issues Report.* Unpublished.

Sefton Council (1992) *Beach Management Plan.* Sefton Council.

Sefton Council (1998) *Beach Management the First Five Years.* Sefton Council.

Sefton Council (1999) *Review of Winter Beach Car Parking.* Sefton Council.

Smith, P.H. (2009) *The Sands of Time Revisited – an introduction to the sand dunes of the Sefton Coast.* Amberley Publishing, Gloucester.

Plate 4. Ainsdale Beach Quality Coast Award Sign 2008. (Copyright Sefton Council).

Smith, P.H. (This volume) Birkdale Green Beach In: *Sefton's Dynamic Coast: proceedings of the conference on coastal geomorphology, biogeography and management 2008.* Worsley, A.T., Lymbery, G., Holden, V.J.C. and Newton, M. (eds) Coastal Defence: Sefton MBC Technical Services.

Smith, P.H. (2007) The Birkdale Green Beach – a sand dune biodiversity hotspot. *British Wildlife* 19 (1) 11-16.

Sefton's Dynamic Coast – our response

Graham Lymbery, Michelle Newton and Paul Wisse

The Ainsdale Discovery Centre, The Promenade, Shore Road,
Ainsdale-on-Sea, Shore Road, Southport, PR8 2QB
Coastaldefence@technical.sefton.gov.uk

ABSTRACT

A number of the other papers in this volume have considered past events and current knowledge regarding the Sefton Coast. This paper seeks to expand upon some of the work that is currently being undertaken through the coastal defence team to consolidate our current knowledge and to inform our approaches to living with a dynamic coast.

INTRODUCTION

The physical form of the Sefton Coast has a history of change which ranges over a series of timescales. For the purpose of this paper we will limit our consideration to human history, as longer timescales are covered in the introduction to these proceedings (Lewis, This volume). Recent human history is covered by both Holden (This volume) and Plater *et al.* (This volume). Together they illustrate the physical changes at the coast but also the significantly greater influence that human actions have had in recent history through activities such as land reclamation and dredging.

The coastal environment changes on a daily or even hourly basis. These changes are occurring on the coast whether it is as a result of a high tide or a strong wind redistributing sediment. As we go forward the rate of change on the Sefton Coast will increase as we start to see the impacts of climate change, with changes in sea level, storminess, temperature and precipitation patterns expected (Wisse and Lymbery, 2007).

A question often raised is: "Are changes brought forward by climate change considered important?" A typical response may only consider it to be of importance if it impacts upon our way of life or upon something that we have deemed to be important. Changes occurring two or three thousand years ago might have been considered important in the context of changing patterns of food sources, but the population at that time would have had little choice but to try and adapt to this change. More recently (20 to 40 years ago), the coast was not only valued as an economic resource but humans also had a greater capacity to alter the coast themselves. This meant that as the coast changed, humans were presented with a choice to either adapt to the change, or to try and adapt the coast in some form. By adapting the coast, this could either resist further change or, where perceived as beneficial, to exploit and encourage further change. Examples might include encouraging accretion around Formby Point with sand trapping and controlling wind blown sand through the planting of trees. It is only within the last 40 years that as a society we appear to have recognised the value of the coast as a natural feature. It is only within the last 10 to 20 years that as a society we have recognised that we cannot exert our control on the entire length of the coast in the U.K. because it is unsustainable both economically and in terms of the damage it does to this important environment.

We are now at the stage as a society where we recognise that our environment is changing and that we need to adapt to that change (Stern, 2006; Department of the Environment, Heritage and Local Government, 2007; DEFRA, 2008a; Pitt, 2008; Commonwealth of Australia, 2009). In some areas this adaptation is difficult and may lead to people losing their homes in the near future; we're fortunate not to be in that position in Sefton. However, that does not mean that we do not need to adapt, only that it will be easier for us as we have more time in which to adapt.

The rest of this paper will set out the approach that we will be taking to develop our adaptation strategy, outline work that we have or are planning to undertake to support the development of the strategy and make use of examples from across the coast to illustrate some of the issues that we face.

Adaptation

As stated in the introduction, there is recognition at a national level that within the coastal zone we need to adapt to climate change. There is a range of guidance available at the national and international level to support this process (IPCC, 2001; DEFRA, 2008b) but broadly speaking there are three common elements:

(i) developing the evidence base to support decisions - which includes developing future scenarios;
(ii) building capacity - in order to engage effectively with stakeholders there is a need for all concerned to understand and broadly agree how and why the coast is changing and the implications of this change;
(iii) development of an adaptation strategy in consultation with stakeholders - including a risk based prioritised action plan.

Dealing with these elements in turn:

(i) Developing the evidence base

The Sefton Coast is fortunate in being relatively well monitored and analysed when compared to other areas in the UK; there have been a range of technical reports commissioned by the relevant authorities to consider concerns such as the location of the navigable channel in the Ribble to erosion at Formby Point (e.g. Barron, 1938; Gresswell, 1953; Neal, 1993; Jay 1998; Blott *et al* 2006; van der Wal *et al* 2002). There are academics who have taken a personal interest in the coast, such as those contributing to this conference, with prominent researchers such as Gresswell (1953) and Parker (1969). There are still, however, a number of areas where we need to improve our understanding in order to be able to make better informed decisions, this requires us to undertake research. A complimentary aspect of the need to undertake research is to collate all our present understanding and data in a secure and accessible manner. There is a major limitation to undertaking research and data management which is the availability of financial resource, this means that we have to be opportunistic in our approach and this does not always reflect our priorities. Key research areas and data collection and storage projects are shown in Tables 1 and 2.

Whilst we will never have as much data and understanding as we would like, we have far more than ever before. We need to manage this knowledge and data as a valuable resource which introduces issues such as secure storage, accessible formats, metadata and presenting the knowledge in such a way that it informs decisions rather than confuses.

(ii) Capacity building

Capacity building refers to the process of engaging with stakeholders and communicating the evidence to them. Stakeholders are all those who can influence or implement a decision, those who are impacted upon by the decision and those who have an interest in the coast. Clearly it makes sense for the decision makers to understand the evidence. When considering those who are impacted upon by decisions or just have an interest in the coast the need for them to understand the evidence relates to the democratic process and the need for us to be able to debate decisions in an informed manner. If those debating the issue all understand and can agree on the evidence it makes the debate far more productive. It is also important to be able to debate issues with a wide range of stakeholders as there will often be conflicting views as to what is valuable and it will be necessary to attempt to identify acceptable compromises that reconcile these conflicting views.

From a coastal defence perspective Sefton Council have done less than other Partners on the coast to communicate with and engage stakeholders; many of the Partners have dedicated officers working with stakeholders on extensive programmes of activities, on-site interpretation, newsletters and leaflets. Given the context of developing an adaptation strategy and the degree to which this relies on an understanding of coastal processes, it is appropriate for us develop our communication and engagement with stakeholders. As a starting point for this we have developed some key messages to communicate:

Table 1. Key research areas.

Title / Area of Study	Activity
CoFEE (Coastal Flooding from Extreme Events) 2006-10	Research into coastal flooding from extreme events working with Proudman Oceanographic Laboratory, Plymouth University, Liverpool University and Edge Hill University through funding from NERC (Natural Environment Research Council). This will improve our understanding of onshore responses to offshore conditions.
IMCORE (Innovative Management of Coastal Resources) 2008-11	Working with 17 other Partners in North West Europe funded through Interreg IVB. Looking at the development of adaptation strategies at a range of sites with a view to informing European approaches. This will facilitate the development of an adaptation strategy within Sefton along with further research on Hydrology within the dune system, improving our understanding of past change and improving our understanding of the impacts of future change on the areas of habitat.
Saltmarsh 2004-09	Through sponsorship of a PhD at Edge Hill University we have improved our understanding of the evolution of the saltmarsh at Marshside.
Sand Dune Soils 2006-10	Through sponsorship of a PhD at Wolverhampton University we have improved our understanding of the soils within the dune system.
Sand Dune Management 2009-11	Through Environment Agency funding research is to be undertaken at Lifeboat Road looking at the net impact of sand dune management techniques on the rate of coastal erosion and the ecological and geomorphological resilience of the sand dune system.
CETASS (Cell Eleven Tidal and Sediment Study) 2010-11	This is a study funded through the Environment Agency and promoted through the North West and North Wales Coastal Group that will improve our understanding of sediment movement and tidal currents within the region. All of our defences rely on sediment so it is important that we understand its movement.
JPS (Joint Probability Study) 2010-11	This is a study funded through the Environment Agency and promoted through the North West and North Wales Coastal Group that will improve our understanding of the probability of storm events occurring at the same time as high tides within the region. Important for coastal defence as a storm at low tide will have significantly less direct impact on defences.
CERMS (Cell Eleven Regional Monitoring Strategy) 2007-11	This monitoring is funded through the Environment Agency and promoted through the North West and North Wales Coastal Group. It provides baseline data for all other studies and as the dataset improves it will improve our understanding of the systems we are trying to work with.

Table 2. Key data storage and collection projects

Title / Area of Study	Activity
Sefton Coast Database	Originally developed in the early 1980s and updated over the last five years through an Interreg funded project (Corepoint, 2004-07) this database collates a wide range of reference materials relating to coastal change and stores them in a secure and accessible manner. A number of reports having been compiled to summarise the information relating to key areas. Examples of output are given in figures 1 to 5.
CERMS (Cell Eleven Regional Monitoring Strategy)	This monitoring is funded through the Environment Agency and promoted through the North West and North Wales Coastal Group. It provides baseline data for all other studies and as the dataset improves it will improve our understanding of the systems we are trying to work with.
Sefton coastal monitoring	In addition to the monitoring undertaken through CERMS we also collect weather data, additional local topographic surveys and hydrographic data.
SANDS (Shoreline and Nearshore Data System)	A bespoke system used within the CERMS project for the storage and analysis of the data collected.

The coast is changing: always has and always will – This includes not only the physical change in the coast as a landform but also changes in human use.

There will be implications arising from these changes – Changes not only in the extent and quality of the physical landform but also in human usage.

We will need to consider and evaluate how we adapt to change – The decision to take no action or defer action might be appropriate but should be a conscious decision.

Adaption strategies should be developed in accordance with some guiding principles –
- Inter-generational equality
- a balance between environmental, economic and social needs.

Delivery of these messages will include material specifically developed for use in the educational system and a range of material developed to address the varying needs within the stakeholder group, ranging from technical reports through to more accessible summaries and interpretive material.

We are fortunate in Sefton to have a good selection of aerial photography that illustrates our changing coast especially around Formby Point (Figure 1). At this location the imagery is supported by ground based surveys dating back to 1958. We also have a good record of human usage of the coast as illustrated in the papers by Holden (this volume) and Plater et al. (this volume). These allow us to present an understanding of current processes based on past change (Figures 2, 3 and 4) and human influences (Figure 5).

Having clearly illustrated that the coast has changed and is still changing it is then reasonable to start extrapolating these changes and attempting to build in additional adjustments as a result of climate change in the form of sea-level rise and increased storminess. An example of this predictive work for Formby is illustrated below (Figure 5) and whilst it has limitations

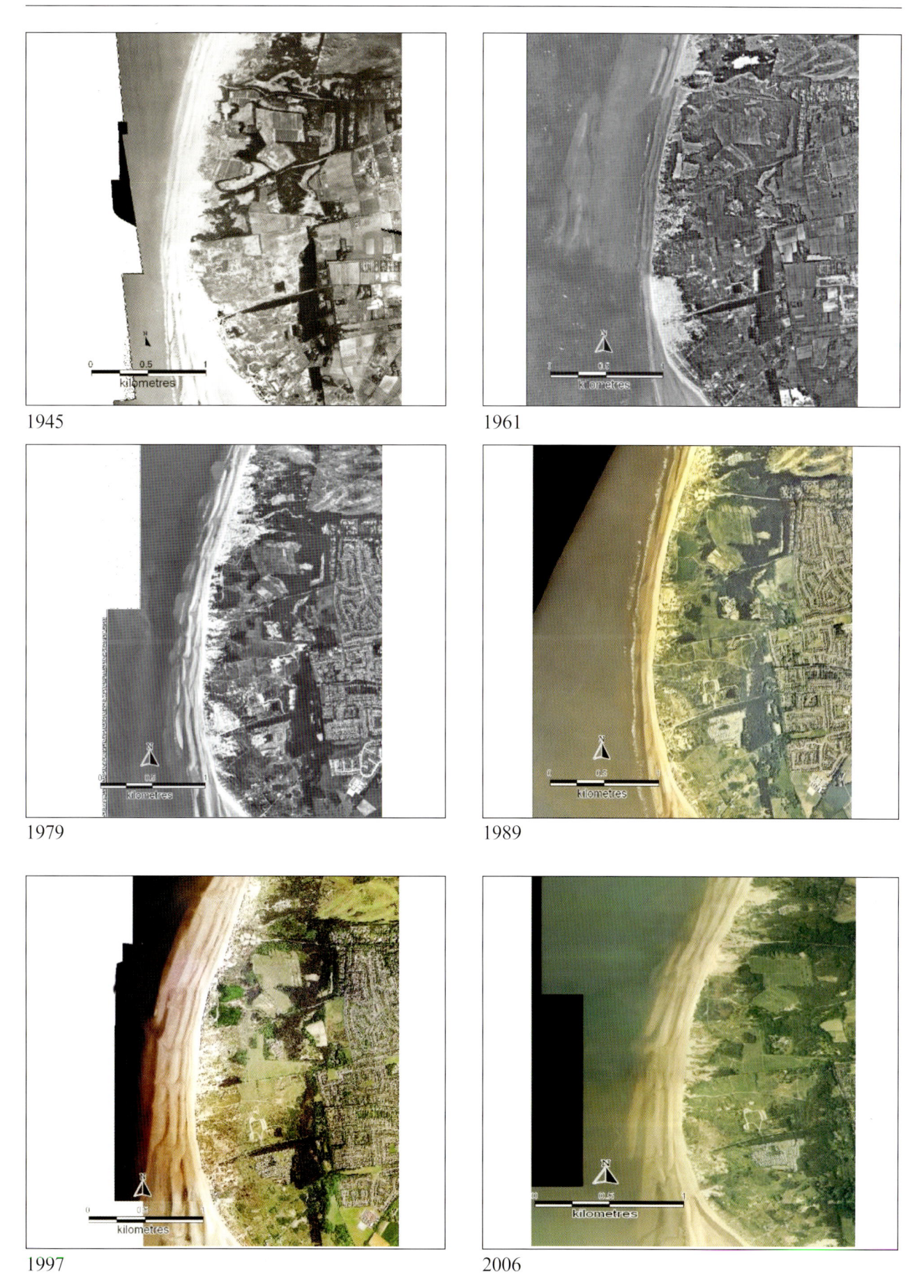

Figure 1. Sefton Coast Database aerial photography archive of Formby Point illustrating coastal change. Copyright Sefton Council. © Crown copyright. All rights reserved. Sefton Council Licence number 100018192 2010.

it provides an initial basis for considering potential implications. This would provide the basic evidence for the first two messages but there would be more detailed issues as discussed below.

For those who wish to consider the changing coast and the implications arising from this in more detail there is a range of supplementary evidence. Some of this has already been described in the section on developing the evidence base, such as an improved understanding of hydrology, better modelling of off-shore wave conditions related to the on-shore response, and a quantitative understanding of the impact of erosion on the extent of habitats and impact on infrastructure. The key issue is that it is not sufficient to undertake the research alone; there is a need to be able to make it relevant at a practical level and to then communicate it to the relevant stakeholders.

It can be seen from the previous paragraphs that the first two messages are relatively simple to communicate. The latter two can be communicated as principles but the detail of them will need to form part of the development of the Adaptation Strategy, as they are reliant upon discussion between a wide range of stakeholders based upon the evidence communicated to them.

(iii) Development of an Adaptation Strategy

This section will focus on three key elements of the Strategy and a brief outline of what might be expected of the finished document. The three key elements are: (i) the identification of impacts, (ii) their prioritisation, and (iii) a consideration of timing of actions or decision points.

Identification of impacts is dependent upon both the evidence base and upon the perspective that you take; as an archaeologist you might be excited by the exposure of footprints as a result of coastal erosion; as someone playing on the beach you may see no difference; as someone trying to access the beach you might struggle due to cliffed dunes and car-parks swamped in wind blown sand; as someone responsible for nature conservation you might be concerned at the loss and fragmentation of habitat; as a site manager you might be worried about the rubble (used as a foundation for the car park) being exposed from the eroding dunes and what you are going to do about your car-park; as someone responsible for coastal defence you might be worried about a diminishing resource; as someone promoting tourism you might be excited by the opportunities that warmer summers present. The list goes on but this is adequate to illustrate the complexity in terms of range of issues along one short frontage.

Having identified and documented the impacts it will then be necessary to prioritise them both in terms of opportunities and threats. Opportunities are likely to be prioritised on a cost benefit basis but will need reconciling against any potential impacts arising from exploiting them. Threats will need to be prioritised on a risk basis; in this context risk equals probability times consequence, with the probability being derived from the evidence base and consequence being derived from the value that stakeholders place on the loss. But risks from climate change or the more general coastal change are not the only risks that we have to manage so it will be necessary to reconcile the risks we identify here against other risks we face and allocate our limited resources accordingly.

All of the above is further complicated by the timescale over which the changes are being considered and the potentially long time periods over which actions might be implemented. This means that the timing of both decisions to act and actions themselves need to be considered in relation to the time that we expect the risk to occur. For example we may consider that there is a probability of one (i.e. it is a certainty) that an area will be lost to erosion and there may be a significant consequence associated with this loss but if it is anticipated to occur eighty years hence we may not need to do anything about it at this moment in time. When considering the options for managing this risk they may require different timescales to implement; for example if a house were at risk it may be possible to buy it and rent back to the occupant over a long time period that covers the costs (requiring early intervention) or it may be decided that it just needs

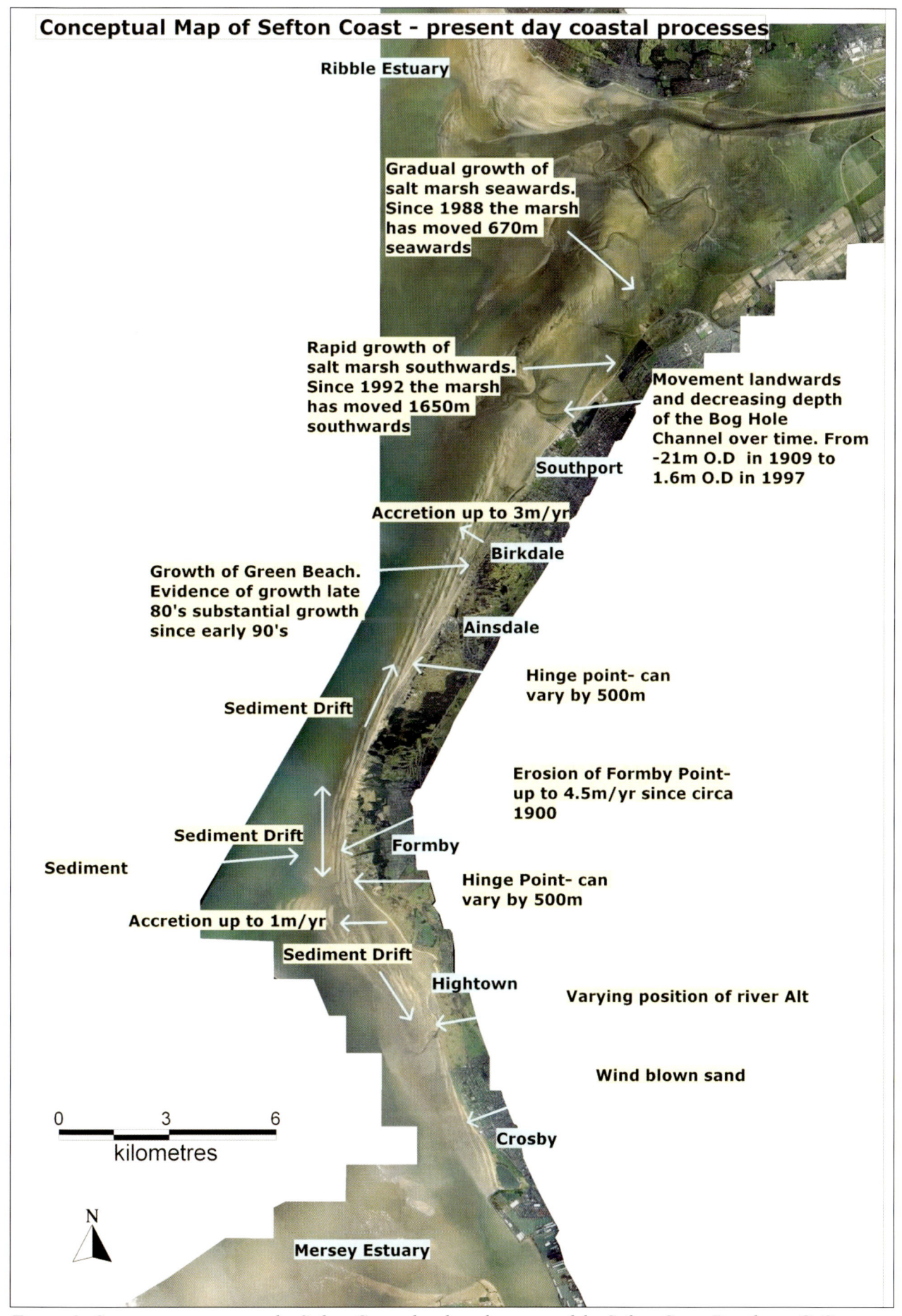

Figure 2. Current processes on the Sefton Coast developed as part of the Sefton Coast Database. © Crown copyright. All rights reserved. Sefton Council Licence number 100018192 2010.

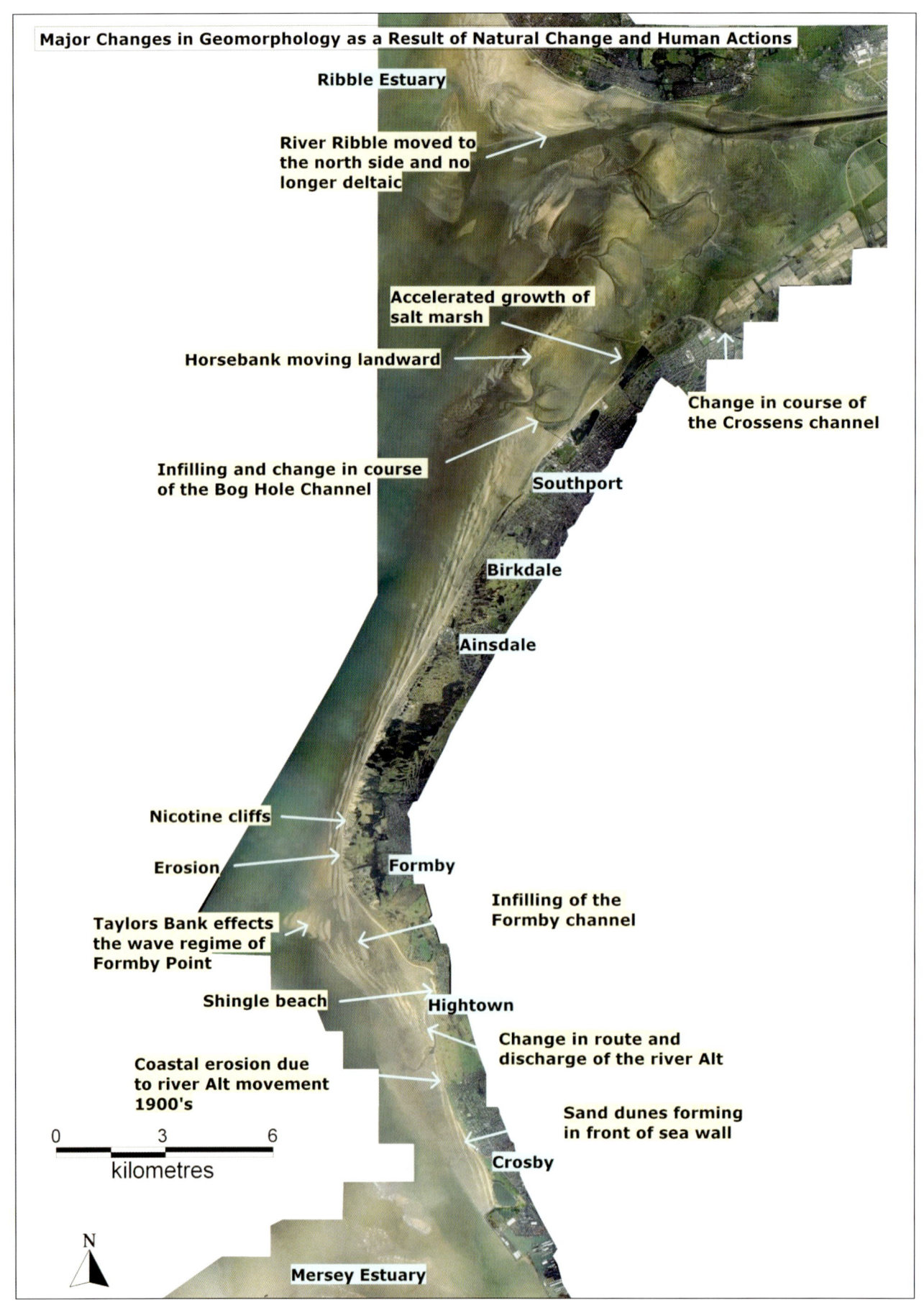

Figure 3. Past changes in geomorphology as a result of natural change and human actions, developed as part of the Sefton Coast database. © Crown copyright. All rights reserved. Sefton Council Licence number 100018192 2010.

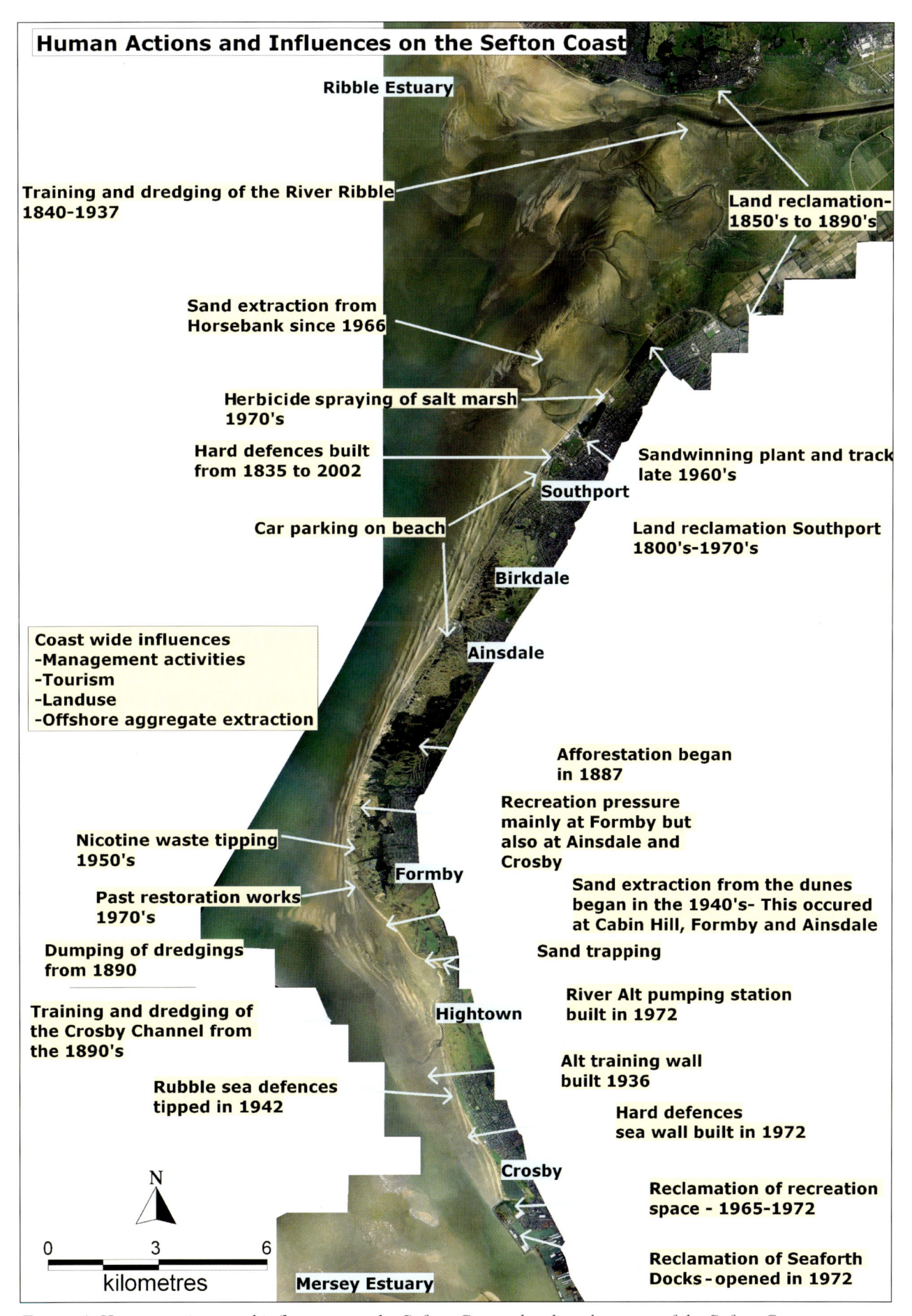

Figure 4. Human actions and influences on the Sefton Coast, developed as part of the Sefton Coast Database. © Crown copyright. All rights reserved. Sefton Council Licence number 100018192 2010.

Figure 5. Extrapolated coastal change as a result of sea-level rise and increased storminess. (Taken from Lymbery et al, 2007). Aerial Photography 2006 Formby Point. © Crown copyright. All rights reserved. Sefton Council Licence number 100018192 2010.

demolition closer to the time (late intervention). And whilst we discuss this in terms of timing of intervention this is subject to uncertainty and it might be better to consider 'trigger' points such as when the coast erodes back to a particular point it 'triggers' an action or the need for a decision.

In practice the Strategy itself has a straightforward structure in that it will set out the potential change, it will identify the impacts, prioritise them and set them over a time period of a hundred years so that future actions or decisions are identified. Given its nature as a forward looking document it should inform but not define current actions, as such it will compliment current plans such as site management plans.

SUMMARY

As a society, we have always adapted to change on the coast, whether it be to exploit change through reclamation or to respond to change in terms of relocation of our activities. We also know that the coast will continue to change and that this rate of change will be accelerated through the impacts of climate change. Given this, it is sensible to plan to adapt to this change in order to minimise the negative impacts and maximise the beneficial impacts. We have a long history of basing our decisions on scientific evidence and understanding and this will be continued as we develop our adaptation strategy. The identification of the issues arising from coastal change would appear to be achievable, what remains to be seen is how prepared people and organisations are to take action now that benefits future generations rather than themselves.

REFERENCES

Barron, J. (1938) *A History of the Ribble Navigation.* Guardian Press, Fishergate.

Blott, S.J., Pye, K., van der Wal, D. and **Neal, A.** (2006) Long-term morphological change and its causes in the Mersey Estuary, NW England. *Geomorphology.* 81, 185-206.

Commonwealth of Australia (2009) Department of the Environment, Heritage and Local Commonwealth of Australia Change Strategy 2007-2012. *Climate Change Adaptation Actions for Local Government.* Department of Climate Change. Canberra, Australia 71p.

DEFRA (2008a) *Adapting to Climate Change in England: A Framework for Action.* Department for Environment, Food and Rural Affairs, London. 52p.

DEFRA (2008b) *NI188 planning to adapt to climate change: Indicator guidance.* DEFRA, Bristol. http://www.defra.gov.uk/environment/local-govindicators/documents/ni188-guidance-2008.pdf

Department of the Environment, Heritage and Local Government (2007) National Climate Government, Dublin, Ireland 62.

Gresswell, R.K. (1953) *Sandy shores of South Lancashire.* University of Liverpool Press, UK.

Holden, V.J.C. (This volume) The historical development of the North Sefton Coast. In: Worsley, A.T., Lymbery, G., Holden, V.J.C. and Newton, M. (eds) *Sefton's Dynamic Coast: proceedings of the conference on coastal geomorphology, biogeography and management 2008.* Coastal Defence: Sefton MBC Technical Services.

IPCC (2001) *Climate Change 2001: Impacts, Adaptation, and Vulnerability.* Cambridge University Press, Cambridge.

Jay, H. (1998) *Beach dune sediment exchange and morphodynamic responses: Implications for shoreline management, the Sefton Coast, NW England.* Unpublished PhD Thesis, University of Reading, UK.

Lewis, J. (This volume) Archaeology and history of a changing coastline. In: Worsley, A.T., Lymbery, G., Holden, V.J.C. and Newton, M. (eds) *Sefton's Dynamic Coast: proceedings of the conference on coastal geomorphology, biogeography and management 2008.* Coastal Defence: Sefton MBC Technical Services.

Lymbery, G., Wisse, P. and **Newton, M.** (2007) *Report on Coastal Erosion Predictions for Formby Point, Formby, Merseyside.* Sefton Council, Bootle, UK.

Neal, A. (1993) *Sedimentology and morphodynamics of a Holocene coastal dune barrier complex, North West England.* Unpublished PhD Thesis, University of Reading, UK.

Parker, W.R. (1969) *A report on research conducted into aspects of the marine environment affecting coast erosion between Ainsdale and Hightown, Lancashire.* Unpublished PhD Thesis, University College Swansea, UK.

Pitt, M. (2008) *Learning lessons from the 2007 floods.* Cabinet Office, London.

Plater, A.J., Hodgson, D., Newton, M. and **Lymbery, G.** (This volume) Sefton South Shore: Understanding Coastal Evolution from Past Changes and Present Dynamics In: Worsley, A.T., Lymbery, G., Holden, V.J.C. and Newton, M. (eds) *Sefton's Dynamic Coast: proceedings of the conference on coastal geomorphology, biogeography and management 2008.* Coastal Defence: Sefton MBC Technical Services.

Stern, N. (2006). Short Executive Summary. *Stern Review Report on the Economics of Climate Change*, pre-publication edition. HM Treasury, London.

van der Wal, D., Pye, K. and **Neal, A.** (2002) Long term morphological change in the Ribble Estuary, North West England. *Marine Geology* 189, 249-266.

Wisse, P. and **Lymbery, G.** (2007) *Climate change and the Sefton Coast: Implications for coastal geomorphology.* Sefton Council, Merseyside, UK.